Basic Concepts of

Vegetable Science

A Compendium for JRF, SRF, NET, ARS, Ph.D. and other Competitive Exams

Syllabus

National Eligibility Test Conducted by ASRB, New Delhi

Subject : Vegetable Science

Unit 1. Production technology of Cool season Vegetable Crops

Introduction, botany and taxonomy, climatic and soil requirements, commercial varieties/ hybrids, sowing/planting times and methods, seed rate and seed treatment, nutritional and irrigation requirements, intercultural operations, weed control, mulching, physiological disorders, harvesting, post harvest management, plant protection measures and seed production of potato, Cole crops, cabbage, cauliflower, knoll kohl, sprouting broccoli, Brussels sprout, root crops: carrot, radish, turnip and beetroot, bulb crops: onion, garlic, peas and broad bean, green leafy cool season vegetables.

Unit 2. Production technology of Warm season Vegetable Crops

Introduction, botany and taxonomy, climatic and soil requirements, commercial varieties/ hybrids, sowing/planting times and methods, seed rate and seed treatment, nutritional and irrigation requirements, intercultural operations, weed control, mulching, physiological disorders, harvesting, post harvest management, plant protection measures, economics of crop production and seed production of: tomato, eggplant, hot and sweet peppers, okra, beans, cowpea, and cluster bean, cucurbitaceous crops, tapioca and sweet potato, green leafy warm season vegetables

Unit 3. Breeding of Vegetable Crops

Origin, botany, taxonomy, cytogenetics, genetics, breeding objectives, breeding methods (introduction, selection, hybridization, mutation), varieties and varietal characterization, resistance breeding for biotic and abiotic stress, quality improvement, molecular marker, genomics, marker assisted breeding and QTLs, biotechnology and their use in breeding in vegetable crops, issue of patenting, PPVFR act. Potato and tomato, eggplant, hot pepper, sweet pepper, and okra, peas, beans, amaranth, chenopods and lettuce, gourds, melons, pumpkins and squashes, cabbage, cauliflower, carrot, beetroot, radish, sweet potato and tapioca.

Unit 4. Growth and Development

Cellular structures and their functions; definition of growth and development, growth analysis and its importance in vegetable production; physiology of dormancy and germination of vegetable seeds, tubers and bulb; role of auxin, gibberellins, cytokinins, and abscissic acid; application of synthetic hormones, plant growth retardants and inhibitors for various purposes in vegetable crops; role and mode of action of morphactins, antitranspirants, anti-auxin, ripening retardant and plant stimulants in vegetable crop production; role of light, temperature and photoperiod on growth, development of underground parts, flowering and sex expression in vegetable crops; apical dominance, physiology of fruit set, fruit development, fruit growth, flower and fruit drop; parthenocarpy in vegetable crops; phototropism, ethylene inhibitors, senescence and abscission; fruit ripening and physiological changes associated with ripening; plant growth regulators in relation to vegetable crops. Sex expression in cucurbits and checking flower and fruit drops and improving fruit set in solanaceous vegetables; growth analysis techniques in vegetable crops.

Unit 5. Seed Production

Introduction; modes of propagation in vegetables; seed morphology and development in vegetable seeds; floral biology of these plant species; classification of vegetable crops based on pollination and reproduction behaviour; steps in quality seed production; identification of suitable areas/locations for seed production of these crops; classification based on growth cycle and pollination behaviour; methods of seed production; comparison between different methods e.g. seed to seed vs. root to seed method in radish; seed multiplication ratios in vegetables; pollination mechanisms; sex types, ratios and expression and modification of flowering pattern in cucurbits; nursery raising and transplanting stage; seed production technology of vegetables viz. solanaceous, cucubitaceous, leguminous, malvaceous, cole crops, leafy vegetables, root, tuber and bulb crops and spices; harvesting/picking stage and seed extraction in fruit vegetable; clonal propagation and multiplication in tuber crops e.g. potato, sweet potato, colocasia, tapioca; seed- plot technique in potato, tuber seed production; hybrid seed production technology of vegetable crops, TPS (true potato seed) and its production technique; hybrids in vegetables; maintenance of parental lines; use of male sterility and self incompatibility in hybrid seed production, environmental factors related to flowering/bolting in vegetable crops; Share of vegetable seeds in seed industry; importance and present status of vegetable industry; intellectual property rights and its implications, impact of PVP on growth of seed industry.

Unit 6. Systematics of Vegetable Crops

Principles of classification; different methods of classification; salient features of international code of nomenclature of vegetable crops; Origin, history, evolution and distribution of vegetable crops, botanical description of families, genera and species covering various tropical, subtropical and temperate vegetables; Cytological level of various vegetable crops; descriptive keys for important vegetables; Importance of molecular markers in evolution of vegetable crops; molecular markers as an aid in characterization and taxonomy of vegetable crops.

Unit 7. Production Technology of Underexploited Vegetable Crops

Introduction, botany and taxonomy, climatic and soil requirements, commercial varieties/ hybrids, owing/planting times and methods, seed rate and seed treatment, nutritional and irrigation requirements, intercultural operations, weed control, mulching, physiological disorders, harvesting, post harvest management, plant protection measures and seed production of: Asparagus, artichoke and leek; Brussels sprout, Chinese cabbage, broccoli, kale and artichoke; Amaranth, celery, parsley, parsnip, lettuce, rhubarb, spinach, basella, bathua (chenopods) and chekurmanis; Elephant foot yam, lima bean, winged bean, vegetable pigeon pea, jack bean and sword bean; Sweet gourd, spine gourd, pointed gourd, Oriental pickling melon and little gourd (kundru).

Unit 8. Post Harvest Technology of Vegetable Crops

Importance and scope of post harvest management of vegetables; Maturity indices and standards for different vegetables; methods of maturity determinations; biochemistry of maturity and ripening, enzymatic and textural changes, ethylene evolution and ethylene management, respiration, transpiration, regulation methods; Harvesting tools, harvesting practices for specific market requirements; post-harvest physiological and biochemical changes, disorders-chilling injury in vegetables, influence of pre-harvest practices and other factors affecting post harvest losses, packaging house operations, commodity pre-treatments- chemicals, wax coating, pre-packaging and irradiation; packaging of vegetables, post harvest, diseases and prevention from infestation, principles of transport; Methods and practices of storage- ventilated, refrigerated, MA, CA storage, hypobaric storage, pre-cooling and cold storage, zero energy cool chamber; storage disorders.

Contents

CHAPTER 1

Importance of Vegetable Crops

S.K. Tyagi
Scientist (Horticulture)
RVSKVV, Krishi Vigyan Kendra, Khargone (M.P.)

- Vegetable growing is one of the major branch of horticulture and from the point of view of value of the products, it is the most important branch.
- The science of growing vegetable crops is called **Olericulture**. The term Olericulture is derived from the **Latin word *'oleris'*** meaning pot herbs (vegetables) and the **English word *'culture'*** meaning to cultivate. Thus, Olericulture is the science of growing vegetable crops.
- "Vegetables are defined as edible herbaceous plants/plant parts consumed either raw or after cooking and rich in vitamins and minerals and low in calorific value.
- In the technical sense all parts of vegetables are used for consumption. The term, however, is usually applied to the edible plants which store up reserve food in their roots, tubers, bulbs, stems, petioles, leaves, buds, flowers, fruits, and seeds and which are eaten either cooked or raw.
- Growing vegetables is not only important us providing the protective food but it also serves as an important subsidiary food playing a more significant role in food consumption. They are rich in nutrients and are essential components of a balanced diet. Vegetables contribute vitally to our general well being due to the following reasons:
  - By supplying us the most important natural elements, viz. vitamins, minerals, fibres, carbohydrates and supplementary amount of protein and colour.
  - Vegetables are a rich source of vitamins, antioxidants, minerals, salts, carbohydrates, roughages and other useful chemicals which play an important role in reducing the risk of many diseases.
- The low availability of vegetables in India is due to low productivity and heavy post-harvest losses.
- Vegetables are the most sustainable and affordable source of micronutrients including vitamins and minerals.
- In the new millennium vegetable production and consumption have to meet the national demand of the increasing population.

- Our demand of vegetables will be 225 million tonnes by 2020 and 350 million tonnes by 2030. India has been gifted with diverse climatic conditions which make it possible to grow a disverse array of vegetables.

Current Scenario of Vegetable Production in India (2014-15)

- Now, India is the second largest producer of vegetables in the world, next only to China, with cultivation of vegetables in about 9.5 million hectares (2014-15).
  - Current area = 9.54 million ha.
  - Production = 169.47 million tonnes.
  - Percent area under vegetables = 6.77% of the total net sown area (140.71 million ha).
  - Only 12 commercial vegetables occupy 65% of our production.
  - Post-harvest losses=20-25%.
  - Less than 2% vegetables are commercially processed as compared to 70% in Brazil and 65% in the USA.
  - India exports onion, potato, tomato, brinjal, green beans, carrot, pea, gherkin, chillies and capsicum as processed vegetables.
  - Major export markets are the Gulf countries, the UK, Srilanka, Malaysia and Singapore.

Importance of Vegetable Crops

- A large number of vegetable crops are being cultivated in India in temperate, tropical and sub-tropical regions.
- Most of the vegetables are quick growing, high yielding, and give very high remuns.
- India is the **world's second largest producer of vegetables** next only to China.
- India shares 15% of the world's production of vegetables with a productivity of about 17.76 t/ha which is quite low as compared to many countries.
- The present production is not sufficient to meet the requirement of 300 g of vegetables on an average per capita per day.
- According to the recommendations given by ICMR, an average person with vegetarian or non vegetarian food habit should consume 125 g leafy vegetables, 100 g of roots and tubers and 75 g other vegetables.
- Since the availability of vegetables per day per capita is very low, it is necessary that the vegetable production and consumption in India will have to be increased three to four-fold.
- Area, production and productivity of leading vegetable producing states are given in Table 1.1.

Table 1.1: *Area, production and productivity of vegetables in leading states of India* (2014-15)

S. No.	States/UTs	Area in '000 ha	Production in '000 MT	Productivity in MT/ha
1	West Bengal	1387.2	26354.61	19.00
2	Uttar Pradesh	1163.55	26120.18	22.45
3	Bihar	842	14467.15	17.18
4	Madhya Pradesh	672.3	14199	21.12
5	Gujarat	603.11	11861.23	19.67
6	Odisha	668.53	9413.54	14.08
7	Karnataka	485.88	8828.37	18.17
8	Maharashtra	595.21	8783.01	14.76
9	Tamil Nadu	284.81	7521.02	26.41
10	Chhattisgarh	425.07	5812.32	13.67
Total including others		**9542**	**169478.23**	**17.76**

Source: http://nhb.gov.in/Default.aspx

Global Scenario of Vegetable Production

- India is the second largest producer of vegetables (15%) in the world, however the average production low. (Table 1.2)

Table 1.2: *Major fruits and vegetable producing countries in the world* (2012-13)

Country	Area (in ha)	Production (in Tonnes)	Productivity (in tonnes/ha)
China	24560900	573935000	23.4
India *	9205186	162186567	17.6
USA	1104640	35947720	32.5
Turkey	1111702	27818918	25.0
Iran	876830	23485675	26.8
Italy	450186	12297645	27.3
Russian Fed.	790500	16084372	20.3
Egypt	772487	19825388	25.7
Spain	318971	12531000	39.3
Mexico	683294	13599497	19.9
Others	19096425	261467661	13.7
World+	**58971121**	**1159179443**	**19.7**

Source: FAO Website except for Indias data (for Indias data source: India Horticulture Database 2012, Ministry of Agriculture)*In India during 2013-14, in case of vegetables the figure for Area, Production & Productivity are 9396057, 162896911 & 17.3 respectively.

- The vegetable industry contributes heavily to the national economy. Millions of farmers, business people (marketing) and industries (seeds, fertilizer, pesticides, herbicides, machinery manufactures etc.) are directly or indirectly dependent on vegetable cultivation.
- Vegetables have great potential to contribute directly to national income, employment and nutritional needs. Due to varied geographical situation and favourable climatic conditions, India can develop into one of the leading countries exporter of fresh vegetables and their seeds.
- In addition to traditional vegetables, non-traditional vegetables which have export potential are asparagus, celery, sweet pepper, sweet corn, green pea, french bean and tomato.
- Cole crops and root vegetables are much in demand in temperate countries during winter. In India, these vegetables are grown in the open during winter and thus the cost of production is less as compared to those grown under protective cover. There is a lot of opportunity for exporting these vegetables to the European and North American countries from India.
- Indian vegetable industry is developing at a fast rate. Many aspects on of vegetable improvement and production practices are developed still the following limitations are there of vegetable farming in India.
  - **India is the world's second largest producer of vegetables,** but hardly **2% of the produce is processed.**
  - About **25-30%** of the production is wasted due to lack of adequate processing and infrastructure facilities. Besides meeting domestic requirement, the country is meeting the adhoc demands of various foreign markets from the vegetables being produced every year and exports a sizeable quantity of vegetables and processed products after completing domestic requirement of vegetables.
  - The non -traditional items meet the requirement of Indian expatriates in South East Asia and Gulf countries and to some extent in the UK and also go to European countries where growing conditions are not ideal during November. There is a great demand for vegetable products for export because vegetables have shown to earn 20-30 times more foreign exchange per unit area than cereals.
  - In global vegetable markets, Thailand, Jordan, Syria, Lebanon, Kenya, Zimbabwe, Guatemala, China, Argentina, Indonesia, Egypt. Turkey, Iran, Cyprus, Australia, New Zealand and Holland are the main competitive countries, which export vegetables to different countries.
  - In India, vegetables are grown in the open and thus their cost of production is much less as compared to those grown under protected condition. There is a lot of opportunity to export these vegetables to the European and North American countries from India. Presently our share in world export market is negligible. If we are to make our presence felt in the export market, considerable development is required in infrastructure and export from India.

Importance of Vegetable Production

- Vegetables are the most sustainable and affordable source of micronutrients including vitamins and minerals.
- In the new millennium vegetable production and consumption have to meet the national demand.
- Our demand of vegetables will be 225 million tonnes by 2020 and 350 million tonnes by 2030. We have been gifted with diverse climatic conditions which make it possible to grow an array of vegetables.
- Vegetables are known to be the cheapest source of nourishment and antioxidants.
- As per nutritionists, per capita requirements of vegetables should be **300g,** in which **115 g leafy vegetables, 70g root vegetables** and **115 g others** are to be inlcluded.
- Deficiency in macro and micro nutrients deads to various diseases.
  - Deficiency of vitamin A causes night blindness.
  - Deficiency of vitamin B1 causes beriberi.
  - Scurvy disease is caused due to deficiency of vitamin C.
  - Vitamin E is also known as anti-sterility vitamin.
  - Agathi (Sesbania) leaves are the rich source of protein and calcium.
  - Rajgira (Amaranth) leaves are the richest source of vitamin A followed by colocasia.
  - Radish leaves are a rich source of riboflavin followed by fenugreek leaves.
  - Vitamin D is essential for the prevention of rickets, osteomalacia and dental diseases.
  - The synthesis of prothrombin and normal blood clotting are regulated by vitamin C.
  - The deficiency of calcium in the body causes rickets and osteomalacia.
  - Phosphorus is essential for cell multiplication of bones and soft tissue.
  - Goitre is caused due to deficiency of iodine.
  - For good health, the requirement of vitamin A per day is more than 2000 IU.
  - The requirement of vitamin 13 per capita/day for good health is above 0.17 mg.
  - For good health per capita/day requirement of vitamin C is above 20 mg.
  - Iron requirement per capita/day for good health is above 3.0 mg.
  - Calcium requirement per capita/day is above 20 mg.
  - Magnesium deficiency plays a key role in cardiovascular diseases.
  - Zinc deficiency leads to growth retardation
  - Chromium deficiency leads to impaired glucose tolerance.
  - Manganese deficiency leads to abnormality in skeletal bone mineralization.
  - Selenium deficiency is also implicated as a risk factor in cancer.
  - Megaloblastic anaemia in living organism is due to deficiency of vitamin B_{12}.

- Inadequate intake of vitamins like riboflavin result in sore tongue, cracking at the corners of mouth (angular stomatitis), redness of the eye and burning sensation in eyes, scaling of the skin in the region between the nose and the angles of the lips (seborroic dermatitis).
- Psychomotor development in children may be impaired due to riboflavin deficiency.
- Vitamin A is incorporated in rhodapsin (eye pigment)
- Tapioca is one of the richest source of carbohydrate (38.1g/100g) and calories followed by sweet potato.
- Chilli one of is the richest source of fiber (6.8 g / 100 g).
- Giant chillies are the richest source of thiamine followed by peas.
- **China** is the largest producer of vegetables in world followed by India.
- India occupies first position in cauliflower, second in onion and third in cabbage in the world.
- India shares 15% of world production of vegetables.
- Vegetable crops occupy only 6.77% of the total net sown area (140.71 million ha.).
- The production-wise largest vegetable producing states are West Bengal, Uttar Pradesh, Bihar, Madhya Pradesh, and Gujarat.
- Parsley is one of the richest source of vitamin C (281mg/ 100 g edible part).
- Sweet potato is one of the richest source of vitamin A.
- Protein-energy malnutrition (PEM), vitamin A deficiency, iodine deficiency disorders (IDD) and nutritional anaemia- mainly resulting from iron deficiency or iron losses - are the most common serious nutritional problems in almost all countries of Asia, Africa, Latin America and the Near East.

Export and Processing of Vegetable Crops

- India's diverse climate ensures availability of all varieties of fresh fruits and vegetables. It ranks second in fruits and vegetables production in the world after China.
- As per National Horticulture Board, during 2014-15 India produced 86.602 million metric tonnes of fruits and 169.478 million metric tonnes of vegetables. The area under cultivation of fruits stood at 6.110 million hectares while vegetables were cultivated at 9.542 million hectares.
- India is the largest producer of ginger and okra amongst vegetables and ranks second in production of potato, onion, cauliflower, brinjal, cabbage, etc. Amongst fruits, the country ranks first in production of bananas (22.94%), papayas (44.03%) and mangoes (including mangosteens and guavas) (37.57%).
- The vast production base offers India tremendous opportunities for export. During 2015-16, India exported fruits and vegetables worth Rs. 8,391.41 crores which comprised fruits worth Rs. 3,524.50 crores and vegetables worth Rs. 4,866.91 crores.

- Mango, walnuts, grapes, bananas, pomegranates account for a larger percentage of fruits exported from the country while onions, okra, bitter gourd, green chillies, mushrooms and potatoes contribute largely to the vegetable export basket.
- The major destinations for Indian fruits and vegetables are a UAE, Bangladesh, Malaysia, Netherland, Sri Lanka, Nepal, UK, Saudi Arabia, Pakistan and Qatar.
- Though India's share in the global market is still nearly 1% only, there is increasing acceptance of horticultural produce from the country. This has occurred due to concurrent developments in the areas of state-of-the-art cold chain infrastructure and quality assurance measures.
- Apart from the large investments pumped in by the private sector, the public sector has also taken initiatives and with APEDA's assistance several Centers for Perishable Cargoes and integrated post-harvest handling facilities have been set up in the country. Capacity building initiatives at the farmers, processors and exporters' levels have also contributed towards this effort.
- The export of horticultural products and foreign exchange earnings during 2015-16 and 2016-17 are given in Table 1.3.

Table 1.3: *Export of fresh and processed vegetables*

Products	2015-16		2016-17	
	Quantity in MT	Value in Rs. Crore	Quantity in MT	Value in Rs. Crore
Fruits vegetables & seeds	10925.75	493.54	11638.46	527.42
Fresh onion	1201245.29	2747.41	2415757.11	3106.50
Other fresh vegetables	699600.34	2119.50	1002396.86	2815.38
Cucumber and gherkins(Prepd. & Presvd)	202926.91	999.17	180820.87	942.72
Dried & preserved vegetables	66189.62	914.21	87279.99	1088.55
Other processed fruits & vegetables	320732.58	2900.33	351834.58	3116.08

Source: http://apeda.gov.in/apedawebsite/index.html

Vegetables Having Export Potential

- The Agricultural and Processed Food Products Export Development Authority (APEDA) has identified traditional vegetables like onion, potato, okra, bitter gourd, chilli and non-traditional vegetables like asparagus, celery, sweet pepper, sweet corn, baby corn, green peas, French bean, cucumber and gherkin and cherry tomato as having good export potential.

- Among other vegetables, potato, okra, tomato, baby corn, cucumber and gherkins, chillies, French beans, capsicum, bitter gourd & bottlegourd are being exported.
- Vegetables are also being exported in processed form. The total export of vegetables in processed form is over Rs. 1088 crores annually.
- **Importing Countries:** Onions and traditional vegetables are being exported to Malaysia, Singapore, Gulf countries, Sri Lanka, Bangladesh, Pakistan and Nepal. Non-traditional vegetables are exported to European countries and Australia in addition to Gulf countries and South East Asian countries.
- Onions and garlic are transported by ships and other vegetables by air. Kuwait Airways, Saudi Airways, Air India and TWA are the important airlines which transport vegetables.
- **Variety and Quality Requirement:** The requirement of foreign markets in terms of varieties and qualities differ from domestic market.
- The requirement also differs country to country. European markets, Japan, Australia, USA, etc., require different qualities and varieties than Gulf countries and South East Asian countries. For example, in case of onions, European and Japanese markets require yellow/brown coloured mild pungent onions of big size, whereas Gulf countries and South East Asian countries require light red to dark red coloured strong pungent onions of varying sizes.
- Demand small roses and multiplier onions are there in South East Asian and Sri Lanka.
- Parbhani Kranti, Arka Anamika, Varsha hybrid, Varsha Uphar varieties of okra; Awsari, Panvel, G 4 and Pusa Jwala varieties of chilli; Pusa Navbahar variety of cluster beans; Contender, Premier, Tweet Wonder, Astrel and Monel varieties of French beans; Varad bottlegourd, MBTH-1 bittergourd, Agrifound Dark Red, Agrifound Light Red, Arka Niketan, Baswant 780 and N-2-4-1 varieties of common big onion in red types, Granex 33, Spanish Brown, Tana F-1 in yellow types, Agrifound Rose, Arka Bindu in small rose type, Agrifound Red, CO, & CO4 in multiplier type of onion; Sugarbaby in watermelon; Great Lake in Lettuce; Nantes and Zino Carrot; Agrifound Parvati & G-282 in garlic are some of varieties of different crops which are in good demand. Some specific requirements of size, colour, shape etc. are given in Table 1.4.

Table 1.4: *Some specific requirements for the export of vegetables*

Crops	Specific requirement
Okra	Green, tender and 6-9 cm long
Chillies	Green and 6-7cm long
Cluster beans	Green, tender and 7-10 cm long
Bittergourd	Green, 20-25 cm long having a short neck
Bottlegourd	Green and 20-25 cm long having a short neck
Bottlegourd	Light green, straight cylindrical in shape and 25-30 cm long

Crops	Specific requirement
Gherkin	Green, small sized having 160-300 fruits/kg in premium grade
Tomato	Round, medium-sized red coloured in Middle East; cherry tomatoes in Europe
French beans	Straight, 10-12 cm long, round green pods in bush beans, flat beans, flat beans having 12-13 cm length and pods are also in demand in the European markets.
Big onion	4-6 cm in diameter, light to dark red in colour, round shaped, strong pungency in Gulf markets and South East Asian markets; 3-4 cm in diameter, light red and round shaped in Bangladesh. Yellow/brown colour, 7-8 cm in diameter, round or spindle shape in European and Japanese markets.
Small onion	Dark red, 2-3 cm in diameter and round shaped
Multiple onions	2.5-3.5 cm in diameter and round shaped
Garlic	White, round, 5 cm or above in diameter, big cloves of 10-12 mm and more than 10-15 in. For Bangladesh and Srilanka 4-5 cm size bulbs are in demand.
Potato	White, oval, 4.5-60 cm in size. In Bangladesh there is demand red types in that Iran & Iraq demand for potatoes with yellow flesh.

Source: Pandey and Singh (2000).

- Fresh vegetables export other than onion is allowed freely. Under this category any person may export vegetables to any country except to a country where export is prohibited by law is enforced from time to time.
- Government has also started the policy of giving cash incentive of 10-20% depending upon the kind of vegetable being exported.
- Export of onion up to 20 kg per consignment is allowed by air as a part of assorted vegetables. Onion export has been canalized through National Agricultural Co-operative Marketing Federation of India (NAFED) under the present policy of export.
- In this system, the exporters are required to register with NAFED and get an NOC for export Other vegetables are exported on consignment basis under OGL (Open General License).

RULES AND REGULATIONS FOR VEGETABLE

Firm Name Registration

- If anyone is doing any business like clinic, pharmacy, agri-business under a registered name with specific then no one can object. The party has right to go to the court of law. Firm registration must be done under the Indian Registration Act, 1958.

Trade Mark/Trade Name

- A trademark means identification of your goods. It is a symbol which a person uses in the course of trade in order to that the purchasing public from similar goods of other traders may distinguish his goods. Registration of trademark done is under Trade Mark and Merchandise Act, 1958.

Patent

- A patent is granted under the Patent Act, 1970. The patent can be required is for an article or a substance and also for his agents or licensees to make, use, exercise, sell or distribute such articles.

Agmark

- Vegetable produces are covered: under Agmark registration, Vegetable oil cakes, dried edible mushroom, table potato, Kanchan (Bathua), seed potatoes, table potatoes (export variety), water chestnut, curry powder, chilliy powder, celery seeds, etc.

FPO

- Under FPO registration, vegetable processed products are **covered** including dehydrated vegetables, tomato products, ketchup and sauces, etc.

Quality Labelling of Produce

- Vegetable growers and exporters in India have not equipped themselves with the qualitative aspects of production and pre- and post-harvest core, which play a very vital role in the export of vegetable produce. This requires government intervention and also sizeable investment in research and development. It is also necessary to familiarize the growers and exporters with the technologies involved to ensure quality.
- While promoting vegetable exports in India, quality should be the watchword. We must ensure such quality as would be the consumer' s delight. It is no longer sufficient if we ensure just consumer satisfaction. In the export world, the sum is fast transforming from customer satisfaction to consumer delight. And to achieve to this, the entire outlook of the agricultural industry and business should be radically transformed to adopt the proven three-pronged commitment, consistency and competence; commitment to lift agricultural from its present state of stagnated growth to achieve a buoyant agrarian economy; consistency in terms of supply of quality products and competence to sell India's products at competitive rates by increased productivity and an adoption of innovative and appropriate technology and expertise.

About APEDA

- The Agricultural and Processed Food Products Export Development Authority (APEDA) was established by the Government of India under the Agricultural and Processed Food Products Export Development Authority Act passed by the Parliament in December, 1985.
- The Act (2 of 1986) came into effect from 13th February, 1986 by a notification issued in the Gazette of India: Extraordinary: Part-II [Sec. 3(ii): 13.2.1986). The Authority replaced the Processed Food Export Promotion Council (PFEPC).

World Trade Organization

- The World Trade Organization (WTO) is the only global international organization dealing with the rules of trade between nations.
- At its heart are the WTO agreements, negotiated and signed by the bulk of the world's trading nations and ratified in their parliaments. The goal is to help producers of goods and services, exporters, and importers conduct their business

Future thrust

- Enhancement of good quality seed availability of temperate vegetables.
- Introduction of high value exotic vegetables.
- Standardization of agro-techniques for organic production of temperate vegetables.
- Need for introduction and development of new varieties of temperate vegetable crops for specific climatic conditions.
- Exploration of new potential belts for off-season production of temperate vegetables.
- Credit system to be in place.
- Profitable and sustained crop management packages need to be developed and disseminated to farmers.
- Simple affordable and durable post-harvest technologies to small-scale farmers needed.
- Need for development of mechanism for marketing intelligence to regulate gluts and optimum profit to farmers.
- Promotion of small-scale collectives and pre-cooling, grading and packaging centers in rural areas.

❍❍❍

CHAPTER 2

Classification of Vegetable Crops

S.S. Singh
Assistant Professor (Horticulture)
MGCGVV, Chitrakoot, Satna (M.P.)

- There are more than 240 plants in the world which are consumed as vegetables.
- Hence, their classification is essential to understand the nature of the vegetable crops and their commonness or distinctness as pre-requisite for commercial production.
- The classification is mainly to show the relationship between individual vegetables and to avoid repetition while describing their cultural operations.
- Any method of classification systematizes to some extent the preparation and presentation of the material and eliminates unnecessary repetition of some of the principles of culture.
- Vegetable crops have been classified into various groups, but the classification based on the cultural practices has been proved to be the best adoptable in the field conditions, since it is very convenient to growers.
- No single method of classification would serve the purpose for different groups like agronomists, breeders, taxonomists and so on. For convenience the following methods of classification are suggested.

1. Botanical classification

Class: **Spermatophyta**

Division-I: **Gymnospermae:** Ovules naked, not enclosed in ovary

Division- II: **Angiospermae:** Ovules in an ovary

- All vegetables belong to the group Spermatophyta and sub-division II Angiospermae. Botanical classification is most useful from breeder's point of view and helps in identification of vegetables.
- It provides information on morphological and cytological similarities and dissimilarities, floral biology and other details of crop improvement.

Table 2.1 provides a list of botanical names along with the famely of vegetable crops.

Table 2.1: *Botanical names along with the family of common vegetable crops*

S. No	Family	Common name	Botanical name
A.	**Monocotyledonae**		
1.	Amaryllidaceae	Onion	*Allium cepa* L.
2.	Amaryllidaceae	Garden leek	*Allium porrum*
3.	Araceae	Colocasia/ Taro	*Colocasia esculenta* (L.) Schott.
4.	Araceae	Elephant foot yam	*Amorphophallus paeoniifolius* (Dennst.) Nicolson
5.	Araceae	Tannia	*Xanthosoma sagittifolium* (L.) Schott
6.	Asparagaceae	Asparagus	*Asparagus officinalis* L.
7.	Dioscoreaceae	Yam	*Dioscorea* spp.
B.	**Dicotyledonae**		
8.	Asteraceae (Compositae)	Globe artichoke	*Cynara scolymus* L.
9.	Apiaceae (Umbelliferae)	Coriander	*Coriandrum sativum* L.
10.	Apiaceae (Umbelliferae)	Celery	*Apium graveolens* L.
11.	Apiaceae (Umbelliferae)	Carrot	*Daucus carota* L.
12.	Amaranthaceae	Spinach/Vilayati palak	*Spinacia oleracea* L.
13.	Amaranthaceae	Beet leaf/ Palak	*Beta vulgaris* var. *bengalensis* L.
14.	Asteraceae (Compositae)	Lettuce	*Lactuca sativa* L.
15.	Basellaceae	Ceylon spinach Malabar spinach Indian spinach	*Basella alba* L.
16.	Brassicaceae (Cruciferae)	Sprouting broccoli	*Brassica oleracea* var. *italica* Plenck
17.	Brassicaceae (Cruciferae)	Brussels sprouts	*Brassica oleracea* var. *gemmifera* DC.
18.	Brassicaceae (Cruciferae)	Cabbage	*Brassica oleracea* var. *capitata* L.
19.	Brassicaceae (Cruciferae)	Cauliflower	*Brassica oleracea* var. *botrytis* L.
20.	Brassicaceae (Cruciferae)	Chinese cabbage	*Brassica rapa, subspecies pekinensis and chinensis*
21.	Brassicaceae (Cruciferae)	Knol khol	*Brassica oleracea* var. *gongylodes* L.
22.	Brassicaceae (Cruciferae)	Radish	*Raphanus sativus* L.
23.	Brassicaceae (Cruciferae)	Turnip	*Brassica rapa* L.
24.	Convolvulaceae	Sweet potato	*Ipomoea batatas* (L.) Lam.
25.	Cucurbitaceae	Ash gourd	*Benincasa pruriens* f. *hispida* (Thunb.) de Wilde & Duyfjes
26.	Cucurbitaceae	Bitter gourd	*Momordica charantia* L.
27.	Cucurbitaceae	Bottle gourd	*Lagenaria siceraria* (Molina) Standl.
28.	Cucurbitaceae	Chow Chow	*Sechium edule* Swaetz.
29.	Cucurbitaceae	Cucumber	*Cucumis sativus* L.

S. No	Family	Common name	Botanical name
30.	Cucurbitaceae	Ivy gourd	*Coccinia grandis* (L.) Voigt
31.	Cucurbitaceae	Longmelon	*Cucumis melo* var.*utillisimus*
32.	Cucurbitaceae	Muskmelon	*Cucumis melo* L.
33.	Cucurbitaceae	Pointed gourd	*Trichosanthes dioica* Roxb.
34.	Cucurbitaceae	Pumpkin	*Cucurbita moschata* (Duchesne) Duchesne ex Poir.
35.	Cucurbitaceae	Ridge gourd	*Luffa acutangula* (L.) Roxb.
36.	Cucurbitaceae	Round gourd/ Tinda	*Praecitrullus fistulosus* (Stocks) Pangalo
37.	Cucurbitaceae	Snake gourd	*Trichosanthes cucumerina*
38.	Cucurbitaceae	Snapmelon	*Cucumis melo* var.*momordica*
39.	Cucurbitaceae	Spine gourd	*Momordica dioica* Roxb. ex Willd.
40.	Cucurbitaceae	Sponge gourd	*Luffa cylindrica* (L.) Roem.
41.	Cucurbitaceae	Summer squash	*Cucurbita pepo* L.
42.	Cucurbitaceae	Watermelon	*Citrullus lanatus* (Thunb.) Matsumura & Nakai
43.	Cucurbitaceae	Winter squash	*Cucurbita maxima*
44.	Euphorbiaceae	Cassava (Tapioca)	*Manihot esculenta* Crantz
45.	Fabaceae (Leguminoceae)	Broad bean	*Vicia faba* L.
46.	Fabaceae (Leguminoceae)	Cluster bean	*Cyamopsis tetragonoloba* (L.)Taub.
47.	Fabaceae (Leguminoceae)	Cowpea	*Vigna unguiculata* (L.)Walp.
48.	Fabaceae (Leguminoceae)	French bean	*Phaseolus vulgaris* L.
49.	Fabaceae (Leguminoceae)	Lablab bean/ Indian bean/	*Lablab purpureus* (L.)Sweet, *Dolichos lablab* L. (ambiguous synonym)
50.	Fabaceae (Leguminoceae)	Garden pea	*Pisum sativum* L.
51.	Fabaceae (Leguminoceae)	Winged bean	*Psophocarpus tetragonolobus* (L.)DC.
52.	Fabaceae (Leguminoceae)	Fenugreek	*Trigonella foenum-graecum* L
53.	Lamiaceae	Coleus (Chinese potato)	*Plectranthus rotundifolius* (Poir.) Spreng.
54.	Malvaceae	Okra	*Abelmoschus esculentus* (L.) Moench
55.	Marantaceae	Arrowroot	*Maranta arundinacea* L.
56.	Moringaceae	Drum stick	*Moringa oleifera* Lam.
57.	Phyllanthaceae	Chekkurmanis	*Sauropus androgynus* (L.) Merr.
58.	Polygonaceae	Rhubarb	*Rheum rhabarbarum* L.
59.	Rutaceae	Curry leaf	*Murraya koenigii* (L.) Spreng.
60.	Solanaceae	Brinjal	*Solanum melongena* L.

S. No	Family	Common name	Botanical name
61.	Solanaceae	Capsicum	*Capsicum annuum* L.
62.	Solanaceae	Chilli	*Capsicum annuum* L.
63.	Solanaceae	Potato	*Solanum tuberosum* L.
64.	Solanaceae	Tomato	*Solanum lycopersicum* L. (Syn. *Lycopersicon esculentum* Mill.)

II. According to the part consumed

1. **Stem:** Asparagus, Kohlrabi
2. **Leaves:** Methi, palak, coriander, amaranthus, spinach, cabbage
3. **Fruits:** Tomato, brinjal, okra, peas, beans, watermelon, pumpkin, chillies
4. **Flower:** Globe artichoke
5. **Roots:** Beetroot, radish, carrot
6. **Tuber:** Potato
7. **Bulb:** Onion, garlic
8. **Corm:** Yam
9. **Immature seed:** Peas, beans
10. **Immature inflorescences** [broccoli (*B. oleracea* var. *italica*), cauliflower (*B. oleracea* var. *botrytis*)].
11. **Buds:** Brussels sprout
12. **Underground stem:** Potato

III. Life-cycle

1. **Annuals:** Vegetables which complete their life-cycle in one season. E.g. Broad bean, Lima bean, water melon, musk melon, Indian spinach, Chinese cabbage, Cress etc.
2. **Biennials:** Those vegetables which take two seasons to complete their life cycle. E.g. Cabbage, cauliflower, turnip, carrot, etc.
3. **Perennials:** Those vegetables which take more than two seasons complete their life cycle. E.g. Artichoke, chicory, asparagus, chow chow etc.

IV. According to Season/Climatic area/ area of cultivation

1. **Cool season vegetables:** such as cabbage, garlic, onion, radish, spinach, lettuce, potato, carrot, etc.
2. **Warm season vegetables:** such as tomato, pepper, cucumber, okra, eggplant, melon, pumpkin, sweet potato, etc.

V. According to photoperiod

1. **Short-day vegetables:** Sweet potato
2. **Long-day vegetables:** Flowering is induced by period longer than critical units. E.g.: Spinach, beet, Chinese cabbage, lettuce, radish etc.

3. **Day neutral vegetables:** Flowerings induced by a period of 10-18 hours of even continuous illumination. e.g.: Tomato, squash, pumpkin, asparagus and pepper, etc.

VI. According to growing season

1. **Kharif:** Cucurbitaceous vegetables which complete their life-cycle during rainy season.
2. **Rabi:** Root crops, cole crops, potato, lettuce, etc. which grow from October to February.
3. **Jayad:** Melons etc., which grow from February to May.

VII. According to methods of raising

1. **Direct sown crops:** Okra, carrot, radish, beans, peas, garlic.
2. **Transplanted crops:** Tomato, brinjal, chillies, cabbage, cauliflower.
3. **Vines and cuttings:** Sweet potato, cassava, pointed gourd, coccinia.
4. **Bits of tubers and corms:** Potato, yams.

VIII. According to hardiness

- Vegetables which can tolerate frost are known as hardy and vice versa are known as tender.

Table 2.2: *The degree of frost tolerance makes the crop hardy, semi hardy and tender*

Hardy vegetable crops	Semi-hardy vegetable crops	Tender vegetable crops
Asparagus	Beetroot	Amaranthus
Broccoli	Carrot	Okra
Brussels sprout	Cauliflower	Chilli
Cabbage	Celery	Tomato
Garlic	Globe artichoke	Cluster bean
Knol khol	Lettuce	Cowpea
Leek	Palak	Cucurbits
Onion	Potato	Snap bean
Peas		Sweet potato
Radish		Tapioca
Spinach		Yams
Turnip		

IX. According to rate of respiration

- After harvest, the rate of respiration of the parts harvested enhanced depending on the climatic conditions.

Table 2.3: *The vegetable having the lowest rate of respiration possess the longest storage ability*

Very high	High	Moderate	Low	Very low
Asparagus	Bean	Beet	Cabbage	Onion
Broccoli	Lettuce	Carrot	Sweet potato	Potato
Pea	Lima bean	Celery	Turnip	Pumpkin
Spinach		Cucumber		Ash gourd

X. According to soil reaction (pH)

Slightly tolerant (6-6.8 pH)	Moderately tolerant (5.5 to 6.0pH)	Highly tolerant (5.0-5.5 pH)
Asparagus	Carrot	Potato
Onion, cauliflower	Cucumber	Sweet potato
Broccoli, leek	Brinjal, garlic	Water melon
Cabbage,	Pea, chilli, radish	Chicory
Lettuce, muskmelon	Pumpkin, tomato	Rhubarb
Celery	Turnip, parsley	

XI. According to Rooting depth

Shallow rooted (30-40 cm)	Moderately deep rooted (50-60 cm)	Deep rooted (80-100 cm and above)
Broccoli	Beet	Artichoke
Brussels sprout	Carrot	Asparagus
Cabbage	Cucumber	Cluster bean
Cauliflower	Brinjal	Cow pea
Garlic, leek, lettuce	Muskmelon	French bean
Onion, potato, radish	Pea, pepper	Lima bean
Spinach	Summer squash, turnip	Pumpkin, sweet potato, tomato, water melon

XII. Classification on the basis of production and utilization of the products

Vegetable gardening can broadly be classified into the following types on the basis of production and utilization of the products:

- The popular and commonly used classification of types of vegetable garden was suggested by Thompson and Kelly as follows:

  1. **Home or Kitchen garden/Nutrition garden**
     - Vegetables are grown in areas surrounding the house for fresh supply to kitchen for family consumption.

- The main purpose of kitchen or home garden is to provide the family daily requirement with fresh vegetables rich in nutrients and energy.
- There is to be more or less continuous supply of a variety of vegetables throughout the year according to the season. It also provides recreation and exercise especially for ladies.

2. **Commercial vegetable gardening**
   - The vegetables are grown on large scale for sale in the market. This is further divided into the following types:

   (i) Market gardening

   (ii) Truck gardening

   (iii) Vegetable forcing

   (iv) Vegetable growing for processing

   Crops suitable for various methods of preservation

   (*a*) ***Canning:*** Tomatoes, peas, beans, okra, pointed gourd, sweet corn, asparagus.

   (*b*) ***Freezing:*** Peas, sweet corn, lima beans, asparagus, cauliflower, and spinach.

   (*c*) ***Dehydration:*** Onion, potato, cauliflower, peas, cauliflower, etc.

   (*d*) ***Pickling and fermentation:*** Turnip, cucumbers, cabbage, cauliflower, carrot, chilli, and radish.

   (v) Vegetable seed production

❍❍❍

CHAPTER 3

Vegetable Plant Nutrients

R.C. Jain
Associate Professor (Soil Science)
R.A.K. College of Agriculture, Sehore (M.P.)

Plant nutrition is the study of the essential and beneficial inorganic and organic chemical and compounds necessary for plant growth, plant metabolism and their external supply.

Plants must obtain the following mineral nutrients from their growing medium:

- **Macronutrients:** Nitrogen (N), phosphorus (P), potassium (K), calcium (Ca), magnesium (M) and sulfur (S).
- **Micronutrients (or trace minerals):** Zinc (Zn), iron (Fe), copper (Cu), manganese (Mn), boron (B), molybdenum (Mo), chlorine (Cl) and nickel (Ni).
- **Beneficial elements but not essential:** Cobalt (Co), vanadium (Va), sodium (Na), selenium (Se) and iodine (I) etc.
- The macronutrients are consumed in larger quantities for Hydrogen, Oxygen, Nitrogen, and Carbon contributes to over 95% of a plants' entire biomass on a dry matter weight basis.
- Micronutrients are present in plant tissue in quantities measured in parts per million, ranging from 0.1 to 200 ppm, or less than 0.02% dry weight.

Plant Nutrients

Nutrients which are required by plants for their normal growth and development and which are not replaceable in their function by any other nutrients are referred to as **essential** (Mengel, 1982). Seventeen plant elements have been recognized as being essential to all plants for their normal growth and development. These include Carbon (C), hydrogen (H) and oxygen (O) and are derived from the air and water. Nitrogen (N), phosphorus (P) and potassium (K) are also needed in large quantities by the plants and are called as **major** or **"primary nutrients"**. Calcium (Ca), magnesium (Mg) and sulphur (S) are required in relatively smaller but in appreciable quantities and hence named as **"secondary nutrients"**. Iron (Fe), zinc, (Zn), manganese (Mn), copper (Cu), boron (B), molybdenum (Mo), chlorine (Cl) and nickel (Ni) are needed in small quantities and are called as **"micronutrients"** or trace elements. In addition to the above, some elements are required by a particular group or species of plants. They are silicon (Si), cobalt (Co), sodium (Na) and vanadium (Va).

Criteria of essentiality

In order to identify a type of nutrient as essential nutrient three criteria must be met **(Arnon and Stout 1939)**:

1. A deficiency of an essential nutrient makes it impossible for the plant to complete the vegetative or reproductive stage of its life cycle.
2. Such deficiency is specific to the element in question and can be prevented or corrected only by supplying this element.
3. The element is involved directly in the nutrition of the plant quite apart from its possible effects in correcting some unfavourable microbiological or chemical condition of the soil or other culture medium.

Many individuals have contributed to the discovery of the essential elements in plant nutrition. Much of the early research focused on the beneficial effects or sometimes on the toxic effects of the elements. Generally, an element was accepted as a plant nutrient after the body of evidence suggested that the element was essential for plant growth and reproduction.

Table 3.1: *Listing of essential elements, their date of acceptance as essential, and discoverers*

Element	Date of essentiality	Researcher
Nitrogen	1804	De Saussure
Phosphorus	1839	Sprengel
Potassium	1839	Sprengel
Calcium	1839	Sprengel
Magnesium	1839	Sprengel
Sulfur	1851	Salm-Horstmann
Iron	1943	Gris
Manganese	1922	McHargue
Copper	1931	Sommer, Lipman
Boron	1923	Warington
Zinc	1931	Sommer, Lipman
Molybdenum	1939	Arnon & Stout
Chlorine	1954	Broyer, Carlton, Johnson, & Stout
Nickel	1987	Brown, Welch, & Cary

Classification of plant nutrients

There are numerous ways in which plant nutrients can be categorized.

A. Classification on the basis of quantity of nutrient required

1. **Primary nutrients:** Among macronutrients, Nitrogen (N), Phosphorus (P) and Potassium (K) are known as primary nutrients which are required in a proper ratio for a successful crop.

2. **Secondary nutrients:** Next to primary nutrients, there are three elements such as Calcium (Ca), Magnesium (Mg) and Sulphur (S) which are known as secondary nutrients.
3. **Micronutrients:** These nutrients are required by plants in small quantities and are also known as minor or trace elements. These are seven in number and include Fe, Mn, Zn, Cu, B, Mo, Cl, Ni

B. Classification on the basis of mobility of nutrient in the soil

1. **Mobile nutrients:** These nutrients are highly soluble and are not adsorbed on clay complexes.
   Examples: NO_3^-, $SO4^{2-}$, BO_3^{2-}, Cl^- and Mn^{+2}
2. **Less mobile nutrients:** They are soluble, but they are adsorbed on clay complex, so their mobility is reduced.
   Examples: NH_4^+, K^+, Ca^{2+}, Mg^{2+}, Cu^{2+}
3. **Immobile nutrients:** Nutrient ions are highly reactive and get fixed in the soil.
   Examples: $H_2PO_4^-$, HPO_4^{2-}, Zn^{2+}

C. Classification on the basis of mobility within the plant

1. Highly mobile: N, P and K
2. Moderately mobile: Zn
3. Less mobile: S, Fe, Mn, Cl, Mo and Cu
4. Immobile: Ca and B

D. Classification on the basis of functions in the plant

1. Elements that provide basic structure to plant. Example: Ca, H, O.
2. Elements useful in energy storage, transfer and bonding: These are accessory structural elements which are more active and vital for living tissues. Example: N, S and P.
3. Elements necessary for charge balance: K, Ca and Mg.
4. Elements involved in enzyme activation and electron transfer: Fe, Mn, Zn, Cu, B, Mo and Cl.

Table 3.2: *Essential plant nutrients*

S. No.	Nutrient	Source	Chemical symbol	Form taken up by plant	Typical concentration in plant dry matter	
					%	ppm
A.	Primary nutrients					
1.	Carbon	Carbonate and air	C	CO_3^{-2}, H_2CO_3, CO_2	45	
2.	Hydrogen	Air and water	H	H^+, OH^-, H_2O	6	

S. No.	Nutrient	Source	Chemical symbol	Form taken up by plant	Typical concentration in plant dry matter	
					%	ppm
3.	Oxygen	Air and water	O	O_2, OH^-	45	
4.	Nitrogen	Organic matter	N	NO_3^-, NH_4^+	1.5	
5.	Phosphorus	Apatite, Fe or Al Phosphate, organic matter	P	HPO_4^{-2}, $H_2PO_4^-$	0.1-0.4	
6.	Potassium	Micas, Feldspars	K	K^+	1.0-5.0	
B.	**Secondary nutrients**					
7.	Calcium	Dolomite, Calcite, Apatite, Gypsum	Ca	Ca^{+2}	0.2-1.0	
8.	Magnesium	Dolomite, Biotite, Muscovite, Augite	Mg	Mg^{+2}	0.1-0.4	
9.	Sulphur	Pyrite, Gypsum	S	SO_4^{-2}, SO_2	0.1-0.4	
C.	**Micronutrients**					
10.	Iron	Pyrite, Magnesite	Fe	Fe^{+2}		50-250
11.	Manganese	Manganite, Augite, Pyrolusite, Olivine	Mn	Mn^{+2} , $MnSO_4^-$		20-500
12.	Zinc	Sphalerite, Augite, Biotite, Hornblend	Zn	Zn^{+2}		21-150
13.	Copper	Chalcopyrite, Olivine, Augite, Biotite	Cu	Cu^{+2}		5 to 20
14.	Boron	Tourmaline	B	H_3BO_3, H_2BO_3		6 to 60
15.	Molybdenum	Olivine	Mo	MoO_4^{-2}		Below 1
16.	Chlorine	Apatite	Cl	Cl^-	0.2 -2	
17.	Nickel		Ni	Ni^{+2}		

Roles of the 17 Essential Nutrients in Crop Development

- Seventeen plant food nutrients are essential for proper crop development.
- Each is equally important to the plant, yet each is required in vastly different amounts.
- These differences have led to the grouping of these essential elements into three categories; primary (macro) nutrients, secondary nutrients, and micronutrients.

Primary (Macro) Nutrients

- Primary (macro) nutrients are nitrogen, phosphorus, and potassium.
- They are the most frequently required in a crop fertilization program. Also, they are need in the greatest total quantity by plants as fertilizer.

Nitrogen

- Necessary for formation of amino acids, the building blocks of protein
- Essential for plant cell division, vital for plant growth
- Directly involved in photosynthesis
- Necessary component of vitamins
- Aids in production and use of carbohydrates
- Affects energy reactions in the plant

Phosphorus

- Involved in photosynthesis, respiration, energy storage and transfer, cell division, and enlargement
- Promotes early root formation and growth
- Improves quality of fruits, vegetables, and grains
- Vital to seed formation
- Helps plants survive harsh winter conditions
- Increases water-use efficiency
- Hastens maturity

Potassium

- Carbohydrate metabolism and the break down and translocation of starches
- Increases photosynthesis
- Increases water-use efficiency
- Essential for protein synthesis
- Important in fruit formation
- Activates enzymes and controls their reaction rates
- Improves quality of seeds and fruit
- Improves winter hardiness
- Increases disease resistance

Secondary Nutrients

- The secondary nutrients are calcium, magnesium, and sulphur.
- For most crops, these three are needed in lesser amounts that the primary nutrients.
- They are growing in importance in crop fertilization programs due to more stringent clean air standards and efforts to improve the environment.

Calcium

- Utilized for continuous cell division and formation
- Involved in nitrogen metabolism
- Reduces plant respiration
- Aids translocation of photosynthesis from leaves to fruiting organs

- Increases fruit set
- Essential for nut development in peanuts
- Stimulates microbial activity

Magnesium

- Key element of chlorophyll production
- Improves utilization and mobility of phosphorus
- Activator and component of many plant enzymes
- Directly related to grass tetany
- Increases iron utilization in plants
- Influences earliness and uniformity of maturity

Sulphur

- Integral part of amino acids
- Helps develop enzymes and vitamins
- Promotes nodule formation on legumes
- Aids in seed production
- Necessary in chlorophyll formation (though it isn't one of the constituents)

Micronutrients

- The micronutrients are boron, chlorine, copper, iron, manganese, molybdenum, zinc and nickel.
- These plant food elements are used in very small amounts, but they are just as important to plant development and profitable crop production as the major nutrients.
- Especially, they work "behind the scene" as activators of many plant functions.

Boron

- Essential of germination of pollen grains and growth of pollen tubes
- Essential for seed and cell wall formation
- Promotes maturity
- Necessary for sugar translocation
- Affects nitrogen and carbohydrate uptake

Chlorine

- Not much information about its functions
- Interferes with P uptake
- Enhances maturity of small grains on some soils

Copper

- Catalyzes several plant processes
- Major function in photosynthesis

- Major function in reproductive stages
- Indirect role in chlorophyll production
- Increases sugar content
- Intensifies colour
- Improves flavour of fruits and vegetables

Iron

- Promotes formation of chlorophyll
- Acts as an oxygen carrier
- Reactions involving cell division and growth

Manganese

- Functions as a part of certain enzyme systems
- Aids in chlorophyll synthesis
- Increases the availability of P and CA

Molybdenum

- Required to form the enzyme "nitrate reductas" which reduces nitrates to ammonium in plant
- Aids in the formation of legume nodules
- Needed to convert inorganic phosphates to organic forms in the plant

Zinc

- Aids plant growth hormones and enzyme system
- Necessary for chlorophyll production
- Necessary for carbohydrate formation
- Necessary for starch formation
- Aids in seed formation

Nickle

It is needed for grain filling and seed vitality.

Table 3.3: *Essential plant nutrients, function in plants and their deficiency symptoms*

S. No.	Name	Function	Deficiency symptoms
Primary (Macro) Nutrients			
1.	Nitrogen	• Necessary for formation of amino acids, the building blocks of protein • Essential for plant cell division, vital for plant growth • Directly involved in photosynthesis • Necessary component of vitamins • Aids in production and use of carbohydrates • Affects energy reactions in the plant	Pale green to light yellow colour (chlorosis) appearing first on older leaves

S. No.	Name	Function	Deficiency symptoms
2.	Phosphorus	• Involved in photosynthesis, respiration, energy storage and transfer, cell division, and enlargement • Promotes early root formation and growth • Improves quality of fruits, vegetables, and grains • Vital to seed formation • Helps plants survive harsh winter conditions • Increases water-use efficiency • Hastens maturity	Dark to blue-green coloration to appear on older leaves of some plants
3.	Potassium	• Carbohydrate metabolism and the break down and translocation of starches • Increases photosynthesis • Increases water-use efficiency • Essential to protein synthesis • Important in fruit formation • Activates enzymes and controls their reaction rates • Improves quality of seeds and fruit • Improves winter hardiness • Increases disease resistance	The most common symptom is chlorosis along the edges of leaves (leaf margin scorching). This occurs first in older leaves, because K is very mobile in the plant
Secondary Nutrients			
4.	Calcium	• Utilized for continuous cell division and formation • Involved in nitrogen metabolism • Reduces plant respiration • Aids translocation of photosynthesis from leaves to fruiting organs • Increases fruit set • Essential for nut development in peanuts • Stimulates microbial activity	Symptoms first appear on the younger leaves and leaf tips. The growing tips of roots and leaves turn brown and die
5.	Magnesium	• Key element of chlorophyll production • Improves utilization and mobility of phosphorus • Activator and component of many plant enzymes • Directly related to grass tetany • Increases iron utilization in plants • Influences earliness and uniformity of maturity	Symptom of interveinal chlorosis first appears in older leaves. Leaf tissue between the veins may be yellowish, bronze, or reddish, while the leaf veins remain green

S. No.	Name	Function	Deficiency symptoms
6.	Sulphur	• Integral part of amino acids • Helps develop enzymes and vitamins • Promotes nodule formation on legumes • Aids in seed production • Necessary in chlorophyll formation (though it isn't one of the constituents)	Younger leaves are chlorotic with evenly, lightly coloured veins
Micronutrients			
7.	Iron	• Promotes formation of chlorophyll • Acts as an oxygen carrier • Reactions involving cell division and growth	Interveinal chlorosis in younger leaves
8.	Manganese	• Functions as a part of certain enzyme systems • Aids in chlorophyll synthesis • Increases the availability of P and CA	Symptoms first appear as chlorosis in young tissues
9.	Zinc	• Aids plant growth hormones and enzyme system • Necessary for chlorophyll production • Necessary for carbohydrate formation • Necessary for starch formation • Aids in seed formation	Interveinal chlorosis occurs on younger leaves, similar to Fe deficiency.
10.	Copper	• Catalyzes several plant processes • Major function in photosynthesis • Major function in reproductive stages • Indirect role in chlorophyll production • Increases sugar content • Intensifies color • Improves flavour of fruits and vegetables	Reduced growth, distortion of the younger leaves
11.	Boron	• Essential of germination of pollon grains and growth of pollen tubes • Essential for seed and cell wall formation • Promotes maturity • Necessary for sugar translocation • Affects nitrogen and carbohydrate	Generally, B deficiency causes stunted growth, first showing symptoms on the growing point and younger leaves
12.	Molybdenum	• Required to form the enzyme "nitrate reductas" which reduces nitrates to ammonium in plants • Aids in the formation of legume nodules • Needed to convert inorganic phosphates to organic forms in the plant	Older and middle leaves become chlorotic, and the leaf margins roll inwards

S. No.	Name	Function	Deficiency symptoms
13.	Chlorine	• Not much information about its functions • Interferes with P uptake • Enhances maturity of small grains on some soils	Chlorosis of younger leaves and wilting of the plant
14.	Nickel	• It is needed for grain filling and seed vitality	Visual deficiency symptoms of Ni have not yet been defined adequately

In addition to the 14 nutrients listed above, plants require carbon, hydrogen, and oxygen, which are extracted from air and water to make up the bulk of plant weight.

Table 3.4: *Deficiency symptoms*

S. No.	Deficiency symptoms appearing first on	Nutrients
1.	Old leaves	N,P,K, Mg, Mo
2.	New leaves	S, Fe, Mn, Cu
3.	Old and new leaves	Zn
4.	Terminal bud	Ca, B

Points to be Remembered

1. Total essential nutrients-**17**. Recently added is **Ni**.
2. Ultra micro nutrient- **Molybdenum.**
3. The smell of soils after fresh showers is due to **actinomycetes.**
4. Element involved in nitrogen fixation is **molybdenum.**
5. Co- factor needed for nitrogen fixation is **cobalt.**
6. **Gypsum** is used for reclamation of alkaline soils.
7. Phosphate solubilising fungi is – ***Aspergillus.***
8. Solubility of rock phosphate can be improved by- ***Bacillus polymexa***.
9. ***Mycorrhiza*** is a symbiotic association between- Fungi and roots of higher plants.
10. Which pulse crop doesn't fix nitrogen from atmosphere –Rajma (***Phaselous vulgaris***).
11. Nitrogen fixing enzyme is **Nitrogenase.**
12. Nitrogen fixing gene is **Nif gene.**
13. Red pigment in the root nodules is known as **Leg haemoglobin.**
14. A bacterium not responsible for N fixation is **E. coli.**

❍❍❍

CHAPTER 4

Manures and Fertilizers

Vikas Chaurasiya
Field Officer
Indian Farmers Fertiliser Cooperative Limited

Fertilizer is any material of natural or synthetic origin added to the soil to supply one or more plant nutrients.

Classification of Fertilizers

1. **Straight fertilizers:** Straight fertilizers are those which supply only one primary plant nutrient, namely nitrogen or phosphorus or potassium. For example: Urea, ammonium sulphate, potassium chloride and potassium sulphate.
2. **Complex fertilizers:** Complex fertilizers contain two or three primary plant nutrients of which two primary nutrients are in chem cal combination. These fertilizers are usually produced in granular form. For example: Diammonium phosphate, nitrophosphates and ammonium phosphate.
3. **Mixed fertilizers:** These are mixtures of straight fertilizers. They contain two or three primary plant nutrients. Mixed fertilizers are made by thoroughly mixing the ingredients either mechanically or manually.

Table 4.1: *Composition of fertilizers*

S. No.	Fertilizer	N	P_2O_5	K_2O	Miscellaneous
(A)	**Nitrogenous fertilizers**				
I.	**Ammonium fertilizers**				
1.	Ammonium sulphate $(NH_4)2\ SO_4$	20.6			24 % S
2.	Ammonium chloride NH_4Cl	25			
3.	Anhydrous ammonia NH_3	82			
II.	**Nitrate fertilizers**				
1.	Calcium nitrate $Ca\ (NO_3)_2$	15.5			20 % Ca
2.	Sodium nitrate $NaNO_3$	11.6			

S. No.	Fertilizer	N	P_2O_5	K_2O	Miscellaneous
III.	**Ammonium + nitrate fertilizers**				
1.	Ammonium nitrate NH_4NO_3	35			
2.	Calcium ammonium nitrate $NH_4NO_3 + CaCO_3$	27			10.2 % Ca
3.	Ammonium nitrate sulphate $NH_4NO_3 + (NH_4)_2\ SO_4$	26			15% S
IV.	**Amide fertilizer**				
1.	Calcium cyanamide $CaCN_2$	22			
2.	Urea $CO(NH_2)_2$	46			
(B)	**Phosphatic fertilizers**				
1.	Single superphosphate $Ca(H_2PO_4)_2 + CaSO_4\ .2H_2O$		16		12% S, 18-21% Ca
2.	Double superphosphate		32		
3.	Triple superphosphate $Ca(H_2PO_4)2 + CaHPO_4$		46		
4.	Basic slag (citric-acid-soluble concentration)		14-18		
5.	Phosphate rocks, finely ground (< 0.16 mm): evaluated according to solubility in citrate or formic acid		20-38		
(C)	**Potassic fertilizers**				
1.	Potassium chloride (KCL)			60	
2.	Potassium sulphate (K_2SO_4)			50	17.5 % S
3.	Potassium magnesium sulphate ($K_2SO_4.2MgSO_4$)			22	11% Mg and 18%S
(D)	**Complex fertilizers**				
1.	Di-ammonium phosphate	18	46	0	
2.	Ammonium phosphate sulphate	16	20	0	
		20	20	0	
3.	Ammonium nitrate phosphate	20	20	0	
		23	23	0	
4.	Urea ammonium phosphate	14	35	14	
		24	24	0	
		28	28	0	
5.	Mono-ammonium phosphate	11	52	0	

S. No.	Fertilizer	N	P_2O_5	K_2O	Miscellaneous
6.	Ammonium polyphosphate	10	34	0	
7.	Nitro phosphate	20	20	0	
		23	23	0	
8.	Nitro phosphate with potash	15	15	15	
(E)	**NPK fertilizers**				
9.	10-26-26	10	26	26	
10.	12-32-16	12	32	16	
11.	14-35-14	14	35	14	

Table 4.2: *Materials containing calcium, sulphur and magnesium*

Nutrients	Materials	Content (%)
Calcium	CAN	10 to 20
	Single superphosphate	18-21
	Calcium nitrate [$Ca(NO_3)_2$]	20
	Gypsum ($CaSO_4.2H_2O$)	23
	Dolomite ($CaCO_3.MgCO_3$)	24-32
	Agricultural limestone ($CaCO_3$)	30-38
Magnesium	Potassium magnesium sulphate ($K_2SO_4.2MgSO_4$)	11
	Magnesium sulphate monohydrate ($MgSO_4.H_2O$)	16
	Magnesium sulphate anhydrous ($MgSO_4$)	20
	Magnesite ($MgCO_3$)	24-27
	Magnesium oxide (MgO)	42
Sulphur	Single superphosphate	12
	Gypsum/phosphogypsum ($CaSO_4.2H_2O$)	13-18
	Pyrites (FeS_2)	18-22
	Potassium sulphate (K_2SO_4)	18
	Potassium magnesium sulphate ($K_2SO_4.2MgSO_4$)	22
	Magnesium sulphate monohydrate ($MgSO_4.H_2O$)	22
	Ammonium sulphate ($(NH_4)2\ SO_4$)	24

Table 4.3: *Nutrient contents of micronutrient fertilizers*

S. No	Name	Formula	Element/Forms	Contents (%)
1	Chelated Zn	Zn-EDTA	Zn	12
2	Zinc sulphate*	$ZnSO_4.7H_2O$	Zn	21
3	Zinc sulphate monohydrate	$ZnSO_4.H_2O$	Zn	33
4	Chelated Mn	Mn-EDTA	Mn	12
5	Manganese sulphate*	$MnSO_4.3H_2O$	Mn	30.5
6	Sodium molybdate	($NaMoO_4.2H_2O$)	Mo	40

S. No	Name	Formula	Element/Forms	Contents (%)
7	Ammonium molybdate	$[(NH4)_6 Mo_7O_{24}.4H_2O]$	Mo	52
8	Borax	$Na_2B_4O_7.10H_2O$	B	10.5
9	Boric acid	(H_3BO_3)	B	17
10	Solubor (foliar spray)	$Na_2B_4O_7.5H2O$ + $Na_2B_4O_7.10H2O$	B	19
11	Chelated Fe	Fe-EDTA	Fe	12
12	Copper sulphate*	$CuSO_4.5H_2O$	Cu	24
13	Copper sulphate*	$CuSO_4.H_2O$	Cu	35
14	Ferrous ammonium sulphate	$(NH_4)_2SO_4.FeSO_4.6H_2O$	Fe	14
15	Ferrous sulphate*	$FeSO_4.7H_2O$	Fe	19
16	Potassium chloride	KCL	Cl	47
17	Ammonium chloride	NH_4Cl	Cl	66
18	Cobalt sulphate	$CoSO_4$	Co	21

* Average sulphur content
(i) $ZnSO_4$ = 15% (ii) $MnSO_4$ = 17%
(iii) $CuSO_4$= 13% (iv) $FeSO_4$ = 19%

4. **Manures:** The excreta of animals, with or without an admixture of bedding or litter, fresh or at various stages of further decomposition or composting. In some countries, may denote any fertilizer material.
5. **Compost**:
   - A mixture of organic residues and soil that is allowed to decompose biologically.
   - Organic residues, or a mixture of organic residues and soil which have been piled, moistened, and allowed to undergo biological decomposition.
6. **Green manure crop:** A crop grown mainly as a source of organic matter. It is usually incorporated into the soil and contributes nutrients and soil binding qualities.

Table 4.4: *Average nutrient content of bulky organic manures*

Type of manure	N (%)	P_2O_5 (%)	K_2O (%)
Poultry manure (fresh)	1.0-1.8	1.4-1.8	0.8-0.9
Urban compost	1.0-2.0	0.9-3.0	1.0-2.0
Rural compost(dry)	0.5-1.0	0.4-0.8	0.8-1.2
Farmyard manure (dry)	0.5-1.5	0.4-0.8	0.5-1.9
Sheep dung	0.95	0.35	1.00
Cattle dung and urine mixed	0.60	0.15	0.45

Table 4.5: *Percent nutrient content of manure of animal origin*

S. No.	Manure	N	P_2O_5	K_2O
1	Dried blood	10.0-12.0	1.0-1.5	0.6-0.8
2	Fish manure	4.0-10.0	3.0-9.0	0.3-1.5
3	Raw bone-meal	3.0-4.0	20.0-25.0	–
4	Steamed bone-meal	1.0-1.2	25.0-30.0	–

Table 4.6: *Percent nutrient contents of oil cakes*

S. No.	Oil cake sources	N	P_2O_5	K_2O
(A)	**Non-edible oil cake**			
2	Castor cake	5.5-5.8	1.8 1.9	1.0-1.1
3	Neem cake	5.2-5.3	1.0-1.1	1.4-1.5
4	Safflower cake (undecorticated)	4.8-4.9	1.4-1.5	1.2-1.3
5	Karanj cake	3.9-4.0	0.9-1.0	1.3-1.4
6	Mahua cake	2.5-2.6	1.8-1.9	1.8-1.9
(B)	**Edible oil cake**			
8	Groundnut cake	7.0-7.2	1.5-1.6	1.3-1.4
9	Cotton seed (decorticated)	6.4-6.5	2.8-2.9	2.1-2.2
10	Cotton seed (undecorticated)	3.9-4.0	1.8-1.9	1.6-1.7
11	Sesame cake	6.2-6.3	2.0-2.1	1.2-1.3
12	Linseed cake	5.5-5.6	1.1-1.5	1.2-1.3
13	Rapeseed cake	5.1-5.2	1.8-1.9	1.1-1.3

Table 4.7: *Some green manure crops and their N contribution*

S. No.	Crop	Botanical name	Duration in days	N added (Kg/ha)
1	Sunnhemp	*Crotalaria juncea* L.	45-50	120
2	Sesbania	*Sesbania rostrata* L.	45-50	100
3	Cowpea	*Vigna unguiculata* L.	45-60	60
4	Black gram	*Vigna mungo* L.	70	60
5	Mung bean	*Vigna radiata* L.	60-65	55

Source: *Fertilizer Statistics*. **2003-04**; and *Handbook of Fertilizer,* Fertilizer Association of India, New Delhi.

❍❍❍

CHAPTER 5

Water Management in Vegetable Crops

V.K. Jain
Scientist (Agriculture Engineering)
KVK, Ashok Nagar (M.P.)

Water Requirement of Crops

The estimation of the water requirement (WR) of crops is one of the basic needs for crop planning on a farm and for the planning of any irrigation project. Water is mainly needed to meet the demands of evapotranspiration (ET) and the metabolic activities of plant, both together known as consumptive use (C or U). Since the water used in the metabolic activities of the plant are negligible, ET is practically considered equal to Cu.

Water Requirement Definition

Water requirement may be defined as the quantity of water, regardless of its source, required by a crop or diversified pattern of crops in a given period of time for its normal growth under field conditions at a place.

Water requirement includes the losses due to ET (or CU) plus the losses during the application of irrigation water and the quantity of water required for special operations such as land preparation. It may be formulated as follows:

WR= ET or CU + application losses + special needs.

Based on the sources of water supply to meet the water requirement, numerically it is represented as, WR=IR+ER+S, i.e., Irrigation water (IR), effective rainfall (ER) and soil profile contribution (S).

Suitability of Water for Irrigation

The suitability of irrigation water mainly depends on the amounts and type of salts present in water. The main soluble constituents are calcium, magnesium, sodium as cations and chloride, sulphate, bicarbonate as anions. The other ions present in minute quantities are boron, selenium, molybdenum and fluorine which are harmful to animals fed on plants grown with excess concentration of these ions.

Quality of Irrigation is Judged with Three Parameters

1. Total salt concentration 2. Sodium adsorption ratio 3. Boron content.

Total Salt Concentration

Salt concentration of irrigation water is measured as electrical conductivity (EC). Conventionally, water containing total dissolved salts to the extent of more than 1.5 m mhos/cm has been classified as saline. Saline waters are those which have sodium chloride as the predominant salt.

Table 5.1: *Classification of irrigation water based on total salt content*

Class	EC (ds/m)	Quality characterization	Soils for which suitable
C1	<1.5	Normal waters	All soils
C2	1.5 – 3	Low salinity waters	Light and medium textured soils
C3	3 – 5	Medium salinity waters	Light and medium textured soils for semi – tolerant crops
C4	5 – 10	Saline waters	Light and medium textured soils for tolerant crops
C5	> 10	High salinity waters	Not suitable

Sodium Adsorption Ratio

Sodium Adsorption ratio (SAR) and residual sodium carbonate (RSC) are also the main criterion to determine the quality of irrigation water.

Boron Content

Irrigation water which contains more than 3 ppm boron is harmful to crops, especially on light soils.

Table 5.2: *Classification of irrigation water based on boron content*

Class	Boron (ppm)	Characterization	Soils suitable
B1	3	Normal waters	All soils
B2	3 – 4	Low boron waters	Clay soils and medium textured soils
B3	4 – 5	Medium boron waters	Heavy textured soils
B4	5 – 10	Boron waters	Heavy textured soils
B5	>10	High boron waters	Not suitable

Water Requirement of Different Crops

Amount of water required by a crop in its whole production period is called water requirement. The amount of water taken by crops varies considerably.

Table 5.3: *Water requirement of vegetable crops*

Crop	Water requirement (mm)
Tomato	600-800
Potato	350-550
Onion	350-550
Chillies	500

Crop	Water Requirement (mm)
Bean	300-500
Cabbage	380-500
Pea	350-500
Ginger	350-400

Critical Stages of Irrigation Requirement

- The water balance in ID crops is referred to the soil water storage in the root zone and not to the level of standing water in the field, as in case of paddy.
- The critical stages or otherwise known as sensitive stages of different crops for irrigation water requirement are as follows.

Table 5.4: *Critical stages of irrigation of different crops*

Crop	Critical stages of irrigation
Chillies	Flowering.
Tomato	From the commencement of fruit set.
Potato	Tuber initiation to tuber maturity.
Cabbage	Head formation until becomes firm.
Carrot	Root enlargement.

Water Use Efficiency

- Water use efficiency is the yield of marketable crop produced per unit of water used in evapotranspiration.

  WUE = Y/ET

- Water use efficiency is also known as crop water use efficiency or consumptive water use efficiency (Ecm) if the water used for metabolic purpose of the crop (G) and is included with ET.

  ECU = Y/G+ET

- If yield is proportional to ET, water-use efficiency has to be a constant but it is not so. Actually, Y and ET are influenced independently or differently by crop management practices, while ET is mainly dependent on climate and soil moisture. Fertilization and other cultural practices for high crop yields usually increases WUE. The factors affecting WUE are nature of the plant, agronomic practices, climate, ET, irrigation, fertilization and plant population.
- There are considerable differences between plant species to produce a unit dry matter per unit amount of water used resulting in widely varying values of water use efficiency. The water use efficiency for few crops is listed below.

Drip Irrigation

Drip irrigation, also known as "trickle" irrigation, is one of the methods of water management. Under this system, water is carried to the plant under low pressure, through small diameter plastic pipes and delivered at the root zone, drop by drop, through drippers. Drip irrigation is a widely practised and established method of irrigation in developed countries and is slowly gaining popularity in India. It is most suited for horticulture crops, vegetables etc. and finds applicability in hard rocky areas where groundwater is scarce and helps in optimization of the limited water resources. The system has its advantages and limitations. Its advantages are in terms of saving of water (50-60%) of that required for flow irrigation, effective use of fertilizers and less labour and energy cost. The limitation for adopting this method is its high initial cost which is beyond the purchasing capacity of small and marginal farmers and is thus mainly adopted by large farmers.

Drip Irrigation System

Precise and regulated application of irrigation water and plant nutrients at low pressure and frequent intervals through drippers/emitters directly into the root zone of plant with the help of close network of pipes is known as drip irrigation system.

Advantages

- Increase in production and productivity.
- Improves quality and ensure early maturity of the crops.
- Water saving up to 40% - 70%.
- Controls weed growth, saving of fertilizer (30%) and labour cost (10%).
- Fertigation / Chemigation can be made efficiently.
- Control diseases.
- Use of saline water is possible.
- Soil erosion is eliminated.
- Suitable for uneven / undulating land.
- High water use efficiency.

Table 5.5: *Major Components of Drip Irrigation System*

1.	Pump station	2.	By-pass assembly
3.	Control valves	4.	Filtration system
5.	Fertilizer tank /Venturi	6.	Pressure gauge
7.	Mains / Sub-mains	8.	Laterals
9.	Emitting devices	10.	Micro tubes

Table 5.6: *Horticultural Crops Suitable for Drip Irrigation System*

1.	Orchard crops	Grapes, Banana, Pomegranate, Orange, Citrus, Mango, Lemon, Custard Apple, Sapota, Guava, Pineapple, Coconut, Cashewnut, Papaya, Amnla, Litchi, etc.
2.	Vegetables	Tomato, Chilly, Capsicum, Cabbage, Cauliflower, Onion, Okra, Brinjal, Bitter gourd, Ridge gourd, Cucumber, Peas, Spinach, Pumpkin, Watermelon, Muskmelon etc.

4.	Flowers	Rose, Carnation, Gerbera, Anthurium, Orchids, Jasmine, Dahlia, Marigold etc.
5.	Plantation	Tea, Rubber, Coffee, Coconut etc.
6.	Spices	Turmeric, Cloves, Mint etc,

Table 5.7: *Response of Different Crops to Drip Irrigation System*

Crops	Water saving (%)	Increase in yield (%)
Banana	45	52
Cauliflower	68	70
Chilly	68	28
Cucumber	56	48
Grapes	48	23
Pomegranate	45	45
Sweet lime	61	50
Tomato	42	60
Watermelon	66	19

Source: https://www.ncpahindia.com/drip-irrigation

Fertigation

Fertigation is a method of fertilizer application in which fertilizer is incorporated within the irrigation water by the drip system. In this system fertilizer solution is distributed evenly in irrigation. The availability of nutrients is very high therefore the efficiency is more. In this method liquid fertilizer as well as water soluble fertilizers are used. By this method, fertilizer use efficiency is increased from 80 to 90%.

Table 5.8: *Fertilizer use efficiencies of various application methods*

Nutrient	Fertilizer use efficiency (%)	
	Soil application	Fertigation
Nitrogen	30-50	95
Phosphorous	20	45
Potassium	50	80

Advantages of fertigation

- Nutrients and water are supplied near the active root zone through fertigation which results in greater absorption by the crops.
- As water and fertilizer are supplied evenly to all the crops through fertigation there is possibility for getting 25-50% higher yield.
- Fertilizer use efficiency through fertigation ranges between 80-90%, which helps to save a minimum of 25% of nutrients.
- By this way, along with less amount of water and saving of fertilizer, time, labour and energy use is also reduced substantially.

Fertilizers used in fertigation

- Urea, potash and highly water soluble fertilizers are available for applying through fertigation.
- Application of super phosphorus through fertigation must be avoided as it makes precipitation of phosphate salts. Thus phosphoric acid is more suitable for fertigation as it is available in liquid form.
- Special fertilizers like mono ammonium phosphate (Nitrogen and Phosphorus), poly feed (Nitrogen, Phosphorus and Potassium), Multi K (Nitrogen and Potassium), Potassium sulphate (Potassium and Sulphur) are highly suitable for fertigation as they are highly soluble in water. Fe, Mn, Zn, Cu, B, Mo are also supplied along with special fertilizers.

Table 5.9: *Fertilizers commonly used in fertigation*

Name	N – P_2O_5 – K_2O content	Solubility (g/l) at 20°C
Ammonium nitrate	34-0-0	1830
Ammonium sulphate	21-0-0	760
Urea	46-0-0	1100
Monoammonium phosphate	12-61-0	282
Diammonium phosphate	18-46-0	575
Potassium chloride	0-0-60	347
Potassium nitrate	13-0-45	316
Potassium sulphate	0-0-50	110
Monopotassium phosphate	0-52-34	230
Phosphoric acid	0-85-0	457

Table 5.10: *Specialty water soluble fertilizers*

Name	N %	P_2O_5 %	K_2O %
Polyfeed	19	19	19
Polyfeed	20	20	20
Polyfeed	11	42	11
Polyfeed	16	8	24
Polyfeed	15	15	30
MAP	12	61	0
Multi-K	13	0	46
MKP	0	52	34
SOP	0	0	50

N fertigation

Urea is well suited for injection in micro irrigation systems. It is highly soluble and dissolves in non-ionic form, so that it does not react with other substances in the water. Also urea does not cause precipitation problems. Urea, ammonium nitrate, ammonium sulphate, calcium ammonium sulphate, calcium ammonium nitrate are used as nitrogenous fertilizers in drip fertigation.

P fertigation

Application of phosphorus to irrigation water may cause precipitation of phosphate salts. Phosphoric acid and mono ammonium phosphate appears to be more suitable for fertigation.

K fertigation

Application of K fertilizer does not cause any precipitation of salts. Potassium nitrate, potassium chloride, potassium sulphate and mono potassium phosphate are used in drip fertigation.

Micronutrients

Fe, Mn, Zn, Cu, B, Mo could be used as micronutrients in drip fertigation.

Fertigation equipments

Three main groups of equipments used in drip system are:

- Ventury
- Fertilizer tank
- Fertilizer pump

Ventury

Construction in the main water flow pipe causes a pressure difference (Vacuum) which is sufficient to suck fertilizer solution from an open container into the water flow. It is very easy to handle and it is affordable even by small farmers. This equipment is most suitable for smaller area.

Fertilizer tank

A tank containing fertilizer solution is connected to the irrigation pipe at the supply point. Part of the irrigation water is diverted through the tank diluting the nutrient solution and returning to the main supply pipe. The concentration of fertilizer in the tank thus becomes gradually reduced.

Fertilizer pump

The fertilizer pump is a standard component of the control head. The fertilizer solution is held in a non-pressurised tank and it can be injected into the irrigation water at any desired ratio. Therefore the fertilizer availability to each plant is maintained properly.

Economics of drip irrigation system

The initial investment in drip irrigation system mainly depends upon the spacing of the crops. The initial cost will be almost Rs 75000-80000 per hectare for wider spacing crops such as mango, sapota, guava, citrus fruits and pomegranate. The initial cost is approximately Rs 1.0 lakh per hectare for close spacing crops such as banana, papaya, mulberry, turmeric, tapioca, vegetables and flower crops.

Sprinkler Irrigation

- In the sprinkler method of irrigation, water is sprayed into the air and allowed to fall on the ground surface somewhat resembling rainfall.
- The spray is developed by the flow of water under pressure through small orifices or nozzles. The pressure is usually obtained by pumping.
- With careful selection of nozzle sizes, operating pressure and sprinkler spacing, the amount of irrigation water required to refill the crop root zone can be applied nearly uniformly at the rate to suit the infiltration rate of soil.

Table 5.11: *Response of different vegetable crops to sprinkler irrigation*

Crops	Water saving (%)	Yield increase (%)
Okra	28	23
Cabbage	40	3
Cauliflower	35	12
Chillies	33	24
Cowpea	19	3
Fenugreek	29	35
Garlic	28	6
Gram	69	57
Onion	33	23
Potato	46	4

Table 5.12: *Irrigation Efficiencies under Different Methods of Irrigation (%)*

Irrigation efficiencies	Method of irrigation		
	Surface	Sprinkler	Drip
Conveyance efficiency	40-50 (canal), 60-70 (well)	100	100
Application efficiency	60-70	70-80	90
Surface water moisture evaporation	30-40	30-40	20-25
Overall efficiency	30-35	50-60	80-90

Source: Sivanappan, R. K. (1998)

❍❍❍

CHAPTER 6

Plant Growth and Development

I.S. Tomar
Senior Scientist and Head
Krishi Vigyan Kendra, Jhabua (M.P.)

The development of a mature plant from a single fertilized egg follows a precise and highly ordered succession of events. The fertilized egg cell, or zygote, divides, grows, and differentiates into increasingly complex tissues and organs. In the end, these events give rise to the complex organization of a mature plant that flowers, bears fruit, senescencs, and eventually dies. These events, along with their underlying genetic programs and biochemistry, and the many other factors that contribute to an orderly progression through the life-cycle, constitute development. Three terms routinely used to describe various changes that a plant undergoes during its life-cycle are *growth, differentiation*, and *development*.

Growth

- Growth is an irreversible increase in volume or size.
- Growth is a quantitative term, related only to changes in size and mass.
- For cells, growth is simply an irreversible increase in volume. For tissues and organs, growth normally reflects an increase in both cell number and cell size.
- Growth can be assessed by a variety of *quantitative measures*.
- Growth of cells such as bacteria or algae in culture, for example, is commonly measured as the fresh weight, cell number or packed cell volume in a centrifuge tube.
- For higher plants, however, fresh weight is not always a reliable measure.

Differentiation

- Differentiation refers to qualitative changes that normally accompany growth.
- Differentiation occurs when cells assume different anatomical characteristics and functions, or form patterns.
- Differentiation begins in the earliest stages of development, such as, when division of the zygote gives rise to cells that are destined to become either root or shoot.
- Later, unspecialized parenchyma cells may differentiate into more specialized cells such as xylem vessels or phloem sieve tubes, each with a distinct morphology and unique function.

- Differentiation is a two-way street and is not determined so much by cell lineage as by cell position with respect to neighbouring cells.
- Thus, even though some plant cells may appear to be highly differentiated or specialized, they may often be stimulated to revert to a more embryonic form.
- It is as though the cells have been genetically reprogrammed, allowing them to reverse the differentiation process and to differentiate along a new and different path.
- This ability of differentiated cells to revert to the embryonic state and form new patterns without an intervening reproductive stage is called ***totipotency.***
- Most living plant cells are totipotent – something akin to mammalian stem cells – and retain a complete genetic program even though not all of the information is used by the cell at any given time.

Development

- Development is the sum of growth and differentiation.
- Development is an umbrella term, referring to the sum of all of the changes that a cell, tissue, organ, or organism goes through in its life-cycle.
- Development is most visibly manifested as changes in the form of an organ or organism, such as the transition from embryo to seedling, from a leaf primordium to a fully expanded leaf, or from the production of vegetative organs to the production of floral structures.
- *Embryogenesis, vegetative, and reproductive development* are the stages of sporophytic development of higher plants.

Cell Cycle and Cell Division

- According to the cell theory, cells arise from pre-existing cells. The process by which this occurs is called cell division.
- Any sexually reproducing organism starts its life-cycle from a single-celled zygote. Cell division does not stop with the formation of the mature organism but continues throughout its life-cycle.
- The stages through which a cell passes from one division to the next is called the cell cycle.
- Cell cycle is divided into two phases called (i) Interphase – a period of preparation for cell division, and (ii) Mitosis (M phase) – the actual period of cell division.
- Interphase is further subdivided into G1, S and G2. G1 phase is the period when the cell grows and carries out normal metabolism.
- Most of the organelle duplication also occurs during this phase. S phase marks the phase of DNA replication and chromosome duplication.
- G2 phase is the period of cytoplasmic growth.
- Mitosis is also divided into four stages, namely, prophase, metaphase, anaphase and telophase.
- Chromosome condensation occurs during prophase. Simultaneously, the centrioles move to the opposite poles.

- The nuclear envelope and the nucleolus disappear and the spindle fibres start appearing.
- Metaphase is marked by the alignment of chromosomes at the equatorial plate.
- During anaphase the centromeres divide and the chromatids start moving towards the two opposite poles.
- Once the chromatids reach the two poles, the chromosomal elongation starts, the nucleolus and the nuclear membrane reappear. This stage is called the telophase.
- Nuclear division is then followed by the cytoplasmic division and is called cytokinesis.
- Mitosis thus, is the equational division in which the chromosome number of the parent is conserved in the daughter cell.
- In contrast to mitosis, meiosis occurs in the diploid cells, which are destined to form gametes.
- It is called the reduction division since it reduces the chromosome number by half while making the gametes.
- In sexual reproduction when the two gametes fuse, the chromosome number is restored to the value in the parent.
- Meiosis is divided into two phases – meiosis I and meiosis II.
- In the first meiotic division the homologous chromosomes pair to form bivalents, and undergo crossing over.
- Meiosis I has a long prophase, which is further divided into five phases. These are leptotene, zygotene, pachytene, diplotene and diakinesis.
- During metaphase I the bivalents arrange on the equatorial plate. This is followed by anaphase I in which homologous chromosomes move to the opposite poles with both their chromatids.
- Each pole receives half the chromosome number of the parent cell.
- In telophase I, the nuclear membrane and nucleolus reappear.
- Meiosis II is similar to mitosis. During anaphase II the sister chromatids separate.
- Thus at the end of meiosis four haploid cells are formed.

Plant Growth Regulators for Vegetable Crops

Introduction

- Plant hormones are the organic substances which are produced naturally in the plants, controlling growth and other physiological functions at a site remote from its place of production and active in minute amounts. They are termed as phytohormones.
- These plant hormones control cellular processes in targeted cells. Plant hormones regulate the upward and downward growth of tissues, leaf formation, stem growth, fruit development and ripening, plant longevity and even plant death.
- Therefore, hormones are vital to plant growth and are also called as growth factors or growth hormones.

- These hormones usually move within the plant from the site of production to the site of action.
- Plant hormones may be defined as organic compounds other than nutrients, which in small amounts promote/inhibit or otherwise modify any physiological response in plants.
- Plant hormones are naturally produced within plants. Similar chemicals are also produced by fungi and bacteria which can affect plant growth.
- A large number of related chemical compounds are artificially synthesized to regulate the growth of different plants both under open field conditions and artificial or laboratory conditions. These manmade compounds are called Plant Growth Regulators or PGRs.
- The external application of these substances on plants can bring about modification by improved seed germination, rooting, better plant growth, better fruit set, increased rate of ripening and increased yield.
- The plant hormones are identified as promoters (auxins, gibberellins and cytokinins), inhibitors (abscissic acid and ethylene) and other hypothetical growth substance (florigen, flowering hormone, etc.)

Classes of plant hormones

In general, there are five main classes of plant hormones. Each class has stimulatory as well as inhibitory action. These five major classes of plant hormones are given in Table 6.1.

Table 6.1: *Plant growth regulator class and associated function(s)*

Class	Discovered by	Year	Function(s)	Examples
Auxins	Charles Darwin	1881	Auxins positively influence root initiation, stem elongation, bud formation, fruit development and apical dominance	Indolacetic acid (IAA), Indol butyric acid (IBA), Napthalene acetic acid (NAA), 2,4-dichloro phenoxy acetic acid (2,4-D)

Class	Discovered by	Year	Function(s)	Examples
Gibberellins	Gibberellin was first recognized in 1926 by a Japanese scientist, Eiichi Kurosawa, studying *bakanae*, the "foolish seedling" disease in rice. It was first isolated in 1935 by Teijiro Yabuta and Sumuki, from fungal strain (*Gibberella fujikuroi*) provided by Kurosawa.	1926	They are active in regulating dormancy, flowering, fruit setting and stimulating seed germination and extending growth of shoots after seed germination. They stimulate cell division or cell elongation or both, especially in genetically dwarf species, beans, peas, etc. Gibberellins also reverse the inhibition of shoot growth and dormancy induced by ABA.	Gibberellic acid (GA)
Cytokinins	F. Skoog, C. Miller and co-workers during the 1950s	1950s	They have similar effects as those of gibberlins in breaking the dormancy of a wide range of seeds and in increased fruit set. They mainly stimulate cell division, and prevent chlorophyll degradation. They also help in delaying senescence or the aging of tissues. They are responsible for mediating auxin transport throughout the plant and affect internodal length and leaf growth. Cytokinins counter the apical dominance induced by auxins. They in conjunction with ethylene promote abscission of leaves, flower parts, and fruits.	Kinetin, Zeatin

Class	Discovered by	Year	Function(s)	Examples
Ethylene	Doubt discovered that ethylene stimulated abscission in 1917 (Doubt, 1917). It wasn't until 1934 that Gane reported that plants synthesize ethylene (Gane, 1934). In 1935, Crocker proposed that ethylene was the plant hormone responsible for fruit ripening as well as inhibition of vegetative tissues (Crocker, 1935).		It plays an important role in the ripening of fruits, inhibition of root growth, and abscission or the shedding of plant parts etc. Ethylene is also known to play a role in seed and bud dormancy, induction of roots, flowering, and stem elongation.	Ethylene
Abscisic acid (ABA)	F.T. Addicott and his associates discovered abscisic acid	early 1960s	It is involved in the abscission of plant organs, retardation of vegetative buds, regulation of fruit ripening and generally in reduction of growth. It accumulates within seeds during fruit maturation and prevents seed germination within the fruit. In plants, ABA plays important role in closing the stomata under water stress.	Abscisic acid

Non-traditional plant hormones

These are present in some plant species, not universally available in all plant species. These are

1. **Aromatic compounds:** They include phenols and benzoic acid to the polymeric tannins, the cinnamic acid, the commarins and the flavanoids. They affect flowering and such other physiological characteristics. The commarins are strong inhibitors of seed germination.
2. **Nitrogen containing compounds** *e.g.* colchicines
3. **Terpenoid compounds** *e.g.* monoterpene essential oil (growth inhibitor), -methylene lactones (both growth inhibitor and root stimulator)
4. **Aliphatic alcohols** *e.g.* Tricontanol

5. **Triazoles:** It is a new type of plant growth retardant and has a great broad spectrum retarding ability e.g. **Paclobutrazol, uniconazol, triapentanol** etc.
6. **Polyamines:** They are essential for cell division, normal morphology and hence plant development.

Growth retardants: They are new type of organic chemicals which in general check the rate of growth without any adverse effect. They may retard cell division and cell elongation, and thus plant height is affected without causing mal formation of leaves and stems e.g. CCC (Chlorocholine chloride).

Table 6.2. *List of growth regulators along with their, concentration, time of application and purpose in vegetable crops*

S. No.	Crops	Active ingredient	Concentration	Time of application	Purpose
1.	Brinjal	2,4-D	30 ppm	At intervals of one week over a period of 60-70 days from commencement of flowering	To increase fruit set percentage
2.	Brinjal	PCPA	20 ppm	At intervals of one week over a period of 60-70 days from commencement of flowering	To increase fruit set percentage
3.	Brinjal	NAA	60 ppm	At intervals of one week over a period of 60-70 days from commencement of flowering	To increase fruit set percentage
4.	Bottle gourd	Ethephon	250 ppm	At 2- true leaf and secondly 4- true leaf stage	To enhance number of female flowers
5.	Capsicum	NAA	10 ppm	At flower initiation followed by another spray after 15 days	To check flower drop and improved fruit set and fruit size.
6.	Chilli	NAA	10 ppm	At flower initiation followed by another spray after 15 days	To check flower drop and improved fruit set and fruit size.
7.	Coriander	CCC	250 ppm	At one month after sowing	Inducing drought tolerance in rainfed crops
8.	Cucumber	Ethephon	150-200 ppm	At 2- true leaf and secondly 4- true leaf stage	To enhance number of female flowers
9.	Leafy vegetables	Cytokinin[#]			Maintaining freshness in cut leafy vegetables

S. No.	Crops	Active ingredient	Concentration	Time of application	Purpose
10.	Muskmelon	Ethephon	250 ppm	At 2- true leaf and secondly 4- true leaf stage	To enhance number of female flowers
11.	Potato	Mepiquat chloride		one spray at 45 DAP	To restrict the excessive vegetative growth of potato and increasing its yield
12.	Potato	GA_3	5 ppm	dipping tubers in 5ppm solution of GA_3 for 10 seconds	To break dormancy
13.	Potato	Thiourea	1%	Dip of potato tubers in 1% solution for one hour	To break dormancy
14.	Pumpkin	Ethephon	250 ppm	At 2- true leaf and secondly 4- true leaf stage	To enhance number of female flowers
15.	Ridge gourd	Ethephon	250 ppm	At 2- true leaf and secondly 4- true leaf stage	To enhance number of female flowers
16.	Sponge gourd	Ethephon	250 ppm	At 2- true leaf and secondly 4- true leaf stage	To enhance number of female flowers
17.	Summer squash	Ethephon	250 ppm	At 2- true leaf and secondly 4- true leaf stage	To enhance number of female flowers
18.	Tomato	Ethephon	2500 ppm	Fruit dip at fully green maturity stage	Uniform ripening
19.	Tomato	4-PCPA	30 ppm	In winter and early spring under poor natural light conditions	To improve fruit formation, ripening and fruit yield
20.	Watermelon	Tri-iodo benzoic acid (TIBA)	25-250 ppm	At 2- true leaf and secondly 4- true leaf stage	To enhance number of female flowers

Note: 1% - 1 gram dissolved in 100ml (made up to 100 ml), 1ppm – 1 mg dissolved in 1 litre (made up to 1 lit.)

CHAPTER 7

Breeding of Vegetable Crops

A.R. Khire
Senior Scientist (Plant Breeding & Genetics)
Zonal Agriculture Research Station, Khargone (M.P.)

Definition

- Plant breeding can be defined as "an art and science" and technology of improving the genetic makeup of plants in relation to their economic use for the mankind.
- Plant breeding is the art and science of improving the heredity of plants for the benefit of mankind.
- Plant breeding deals with the genetic improvement of crop plants also known as science of crop improvement.
- Plant breeding is the science of changing and improving the heredity of plant.
- Plant breeding is the art and science of changing the traits of plants in order to produce desired characteristics.

History and development of plant breeding

- About 10,000 years ago man is believed to have started agriculture.
- Plant breeding began when man first chose certain plants for cultivation.
- The process of bringing a wild species under human management is referred to as domestication.
- Domestication may be the most basic method of plant breeding.
- All other breeding method become applicable to a plant species only after it has been successfully domesticated.
- Domestication continuous till today and is likely to continue for some time in future- Ex: In case of timber trees, medicinal plants, microbes, etc.
- During the long period of cultivation natural selection has definitely acted on the domesticated species.
- Movement of man from one place to another brought about the movement of his cultivated plant species.
- 700 BC - Babylonians and Assyrians pollinated date palm artificially.

- 17th century - several varieties of **heading lettuce** were developed in France.
- **1717:** Thomas Fairchild - produced the first artificial hybrid, popularly known as Fairchild's mule, by taking pollen from a carnation and inserted it into a Ssweet William.
- **1727:** The first Co. plant breeding company was established in France was the Vilmorins.
- **1760-1766:** Joseph Koelreuter, a German, made extensive crosses in tobacco.
- **1759-1835:** Knight was perhaps the first man to use artificial hybridization to develop several new fruit varieties.
- Le couteur and Shireff used individual plant selections and progeny tests to develop some useful cereal varieties.
- **1873:** The work of Patrick Shireff was first published.
- He concluded that only the variation heritable nature responded to selection, and that their variation arose through 'natural sports' (= mutation) and by 'natural hybridization' (= recombination during meiosis in the hybrids so produced).
- **1856:** Vilmorin developed the progeny test and used this method successfully in the improvement of sugarbeets.
- **1900:** Nilson-Ehle, his associates, developed the individual plant selection method in Sweden.
- **1903:** Johannsen proposed the pureline theory that provided the genetic basis for individual plant selection.
- The science of genetics began with the rediscovery of Gregor Johan Mendel's paper in 1900 by Hugo de veris, Tshermark and Correns, which was originally published in 1866.
- The modern plant breeding methods have their bases in the genetic and cytogenetic principles.
- Numerous workers who determined the various modes of inheritance have contributed to the development and understanding of plant breeding.
- The discovery of chromosomes as carriers of genes has led to the development of specialized plant breeding methods for chromosome engineering.
- The totipotency of plant somatic and gametic cells allows regeneration of complete plants from single cells. This, coupled with the development of recombinant DNA technology, has enabled the transfer of desirable genes from any organism into plants. Crop varieties developed in this manner are already in cultivation in several countries.

Table 7.1: *Scientific contributions of eminent scientists*

Contributions	Name of the Scientist
G x E interaction	Allard and Bradshaw
Grater initial capital hypothesis	Ashby
Modified bulk method	Atkins
Chemical mutagen	Aurbach
Colchicine	Blackslee and Nebel

Contributions	Name of the Scientist
Dominance hypothesis	Davenport
Soyabean breeder	E.E. Hartwig
Components of genetic variance classification	Fischer
Gene for gene hypothesis	Flor
Forage breeder	G.W. Burton
Single Seed Descent Method	Goulden
Microcentre	Harlan
Backcross method	Harlan and Pope
Ear to row method	Hopkins
Sprophytic system of self-incompatibility	Hughes and Babcock
Mutation	Hugo de Vries
Recurrent selection for SCA	Hull
Recurrent selection for GCA	Jenkins
Multiline concept	Jenson
Pureline concept	Johannsen
Double cross scheme	Jones
Cytoplasmic genetic male sterility	Jones and Davis
Male sterility	Jones and Davis
Raphano brassica	Karpenchenko
Self-incompatibility	Lewis
Self-incompatibility classification	Lewis
Green revolution in India	M.S. Swaminathan
Bulk method	Nilsson & Ehle
Mechanism of insect resistance	Painter
Male sterility in maize	Rhoades
Triticale	Rimpau
Heterosis	Shull
Forage breeder	T.J. Jenkin
Semidwarf rice varieties at IRRI	T.T. Chang
First artificial hybrid	Thomas Fairchild
Vertifolia effect	Van Der Plank
Centres of diversity, Law of homologus series	Vavilov
Progeny test	Vilmorin
Father of hybrid rice	Yuan Long Ping

Table 7.2: *Some Indian plant breeders and their contributions*

Name of the Scientist	Contributions
B.P. Pal	An eminent Wheat breeder, developed superior disease resistant N.P. varieties of wheat
Bosisen	An eminent maize breeder
C.T. Patel	Famous cotton breeder who developed world's first cotton hybrid in 1970
D.S. Athwal	Famous pearlmillet breeder
Dharampal Singh	An eminent oil-seed breeder
K. Ramaiah	A renowned rice breeder
M.S. Swaminathan	Responsible for green revolution in India, developed high yielding varieties of Wheat and Rice
N.G.P. Rao	An eminent sorghum breeder
Pushkarnath	Famous potato breeder
Ram Dhan Singh	Famous wheat breeder
T.S. Venkatraman	An eminent sugarcane breeder, he transferred thick stem and high sugar contents from tropical noble cane to North Indian Canes. This process is known as noblization of sugarcane
V. Santhanam	Famous cotton breeder

Concepts of breeding self-pollinated, cross-pollinated and asexually propagated crops

The mode of pollination and reproduction play an important role in plant breeding. Based on this, crop plants are divided into three groups:

1. Self-pollinated
2. Cross-pollinated
3. Vegetatively propagated

Self-pollinated Species

In these species development of seed take place by self-pollination (autogamy). Hence self-pollinated species are also known as autogamous species or inbreeders. Various plant characters such as homogamy, cleistogamy, chasmogamy, bisexuality, etc. favour self-fertilization. Some important features of autogamous species are

1. They have regular self-pollination
2. They are homozygous and have advantage of homozygosity, means they are true breeding.
3. Inbreeders do not have recessive deleterious genes, because deleterious genes are eliminated due to inbreeding by way of gene fixation.

4. Inbreeding does not have any adverse effects on inbreeders.
5. In autogamous species, new gene combinations are not possible due to regular self pollination.
6. Inbreeders are composed of several component (homozygous) lines. Hence variability is mostly among component lines.
7. Inbreeders have generally narrow adaptation and are less flexible.

Methods of Breeding in Autogamous Species

1. Plant introduction
2. Pureline selection
3. Mass selection
4. Pedigree method
5. Bulk method
6. Single seed descent method
7. Backcross method
8. Heterosis breeding
9. Mutation breeding
10. Polyploidy breeding
11. Distant hybridization
12. Transgenic breeding

Four breeding approaches viz., recurrent selection, disruptive selection, diallel selective mating and biparental mating are used for population improvement.

Cross-pollinated species

This group refers to cross-fertilizing species. These species produce seed by cross-pollination (allogamy) hence, referred to as allogamous species or out breeders. Various plant characters which promote cross pollination include dichogamy, monoecy, dioecy, heterostylely, herkogamy, self-incompatibility and male sterility. Some important features of out breeders are

1. They have random mating. In such population, each genotype has equal chance of mating with all other genotypes
2. Individuals are heterozygous and have advantage of heterozygosity
3. Individuals have deleterious recessive genes which are concealed by masking the effect of dominant genes.
4. Out breeders are intolerant to inbreeding. They exhibit a high degree of inbreeding depression on selfing.
5. Cross-pollination permits new gene combinations from different sources.
6. In these species, variability is distributed over entire population.
7. They have wide adaptability and more flexibility to environmental changes due to heterozygosity and heterogenety.

Methods of breeding allogamous species

- Plant introduction
- Mass and progeny selection
- Back cross method.
- Heterosis breeding
- Synthetic breeding
- Composite breeding
- Polyploidy breeding
- Distant hybridization
- Transgenic breeding
- Mutation breeding (rarcly)

Three breeding approaches, viz., *recurrent selection, disruptive mating* and *biparental mating* are used for population improvement.

Asexually Propagated Species

Some crop plants propagate by asexual means, i.e. by stem or root cuttings or by other means. Such species are known as asexually propagated species or vegetatively propagated species. Such species are found in both self, and cross pollinated groups. Generally asexually propagated species are highly heterozygous and have broad genetic base, wide adaptability and more flexibility.

Methods of breeding asexually propagated species

- Plant introduction
- Clonal selection
- Mass selection (rarely used)
- Heterosis breeding
- Mutation breeding
- Polyploidy breeding
- Distant hybridization
- Transgenic breeding

Breeding Populations

The genetic constitution of plants is determined by its mode of pollination. Self-pollination leads to homozysity and cross-pollination results in heterozygosity to exploit homozygosity in self pollinated crops and heterozygosity in cross pollinated species, because inbreeders have advantage of homozygosity and outbreeders have advantage of heterozygosity. Based on genetic constitution, breeding populations are of four types-*viz.*

- Homogenous
- Heterogenous
- Homozygous
- Heterozygous

Homogenous population

Genetically similar plants constitute homogenous populations. Examples of homogeneous populations are pure lines, inbred lines, F1 hybrid between two pure lines or inbred lines and progeny of a clone. Pure lines and inbred lines generally have narrow adoption.

Heterogenous populations

Genetically dissimilar plants constitute heterogenous populations. Examples are land races, mass selected populations, composites, synthetics and multilines. Heterogenous populations have wide adaptability and stable performance under different environments.

Homozygous populations

Individuals with like alleles at the corresponding loci are known as homozygous. Such individuals do not segregate on selfing. Thus non-segregating genotypes constitute homozygous populations. Examples are pure lines, inbred lines and mass selected populations in self-pollinated plants. Thus pure lines and inbred lines are homozygous and homogeneous and mass selected varieties of self-pollinated crops and multi lines are homozygous and heterogenous, because they are mixtures of several pure lines.

Heterozygous populations

Individuals with unlike alleles at the corresponding loci are referred to as heterozygous. Such individuals segregate into various types on selfing. This includes F1 hybrids, composites and synthetics. Thus F1 hybrids are heterozygous but homogeneous and composites and synthetics are heterozygous and heterogenous population. Such populations have greater buffering capacity to environmental fluctuations.

Table 7.3: *Different types of genetic populations in plant breeding*

Population	Brief description / definition	Examples
Homogeneous	Genetically similar population	Purelines, inbred line, F1 hybrids progeny of a clone
Heterogeneous	Genetically dissimilar population	Land races, composites synthetics and multilines
Homozygous	Non-segregating populations	Purelines, inbred lines mass selected autogamous varieties and multilines
Heterozygous	Populations segregate on selfing	F1 hybrids, composites, synthetics and a clone
Combinations		
Homogeneous and Homozygous	Genetically similar and non segregating populations	Purelines and inbred lines
Homogeneous and heterozygous	Genetically similar but segregating on selfing	F1 hybrids between inbred lines and progeny of a clone
Heterogeneous and homozygous	Genetically dissimilar but non segregating populations	Multilines and mass selected varieties in autogamous species
Heterogeneous and Heterozyous	Genetically dissimilar and segregating populations	Composites and synthetics

Strategies for Breeding and Improvement of Vegetable Crops

- That vegetables an important article of daily human diet is well known since time immemorial as they supply all the main components of human diet.
- They are commonly called **"protected food"** because of their protective effects against degenerative diseases.
- Although the main breeding objective will continue to be increasing yield to meet the food requirement of an ever increasing population, but in order to ensure health security, it is imperative that nutrition-rich varieties are bred.
- Conventional breeding in conjunction with molecular biology has bright prospects of developing vegetable varieties with high nutraceuticals and bio-active compounds suitable for the fresh market.
- India is the second largest producer of vegetables in the world but current production is not sufficient to meet the demand of an ever increasing population.
- The improvement of vegetable crops has until recently, been largely confined to conventional breeding approaches and such programmes rely on interspecific sexual hybridization of plants which have desirable heritable characteristics and on naturally or artificially induced random mutations.
- The introduction of new genetic information can result in increased resistance to insect pests, disease tolerance to environmental condition, improved quality, etc.
- India is bestowed with varied agro-climatic conditions which allow growing of all types of vegetables in one part of the country or other.
- Export potential of vegetables to generate valuable foreign exchange requires to be fully tapped.

Need For Improvement of Vegetable Crops

Crop improvement in vegetables is needed to develop varieties/ cultivars/ hybrids having:

- Higher yield and better quality.
- Extended harvest period and areas of adaptability.
- Photo insensitivity, salt tolerance and moisture stress.
- Synchronous maturity and non-shattering characteristics.
- Better nutritional, processing quality and seed production.
- Resistance against biotic and abiotic stresses.
- Long shelf-life and export quality produce.

History of Vegetable Breeding

- **1940:** Successful attempt of seed production of temperate vegetables at Quetta (now in Pakistan).
- **1949:** Establishment of a vegetable breeding station at Katrain in Kullu Valley.
- **1955:** Vegetable breeding station, Katrain came under IARI, New Delhi.
- **1956:** Creation of Division of Horticulture at the IARI, New Delhi.
- **1960:** Establishment of first agriculture university at Pantnagar, UP.
- **1968:** Establishment of IIHR at Bangalore.

- **1970:** Initiation of All-India Coordinated Vegetable Improvement Project
- **1987:** Start of Project Directorate of Vegetable Research (PDVR) at IARI.
- **1992:** Shifting of the headquarters of PDVR from New Delhi to Varanasi.

Breeding Methods

A. Conventional Breeding methods

- In conventional breeding, progenies inherit genes for both desirable and undesirable traits from both parents.
- Breeders conserve the desired characteristics and suppress the undesirable ones by repeatedly selecting meritorious individuals from each generation to be the parents of the next.

B. Non-conventional Breeding methods

- Non-conventional breeding encompasses all essential cell and tissue culture techniques that assist in propagating, studying and manipulating the plant gene without use of sexual cycle.

Table 7.4: *Differences between conventional breeding and non-conventional breeding*

Particulars	Conventional breeding	Non-conventional breeding
Type of markers used	Morphological markers	Molecular or DNA markers
Laboratory required	Simple laboratory	Sophisticated laboratory
Effect of environment	High effect of environment on conventional markers	No effect of environment on identification of DNA markers
Accuracy of method	Medium to high	Very high
Time required to develop new variety	Very long (10-15 years)	Very short (3-5 years)
Cost involved	Low to medium	Very high
Health hazards	Only in mutation breeding	With technique involving radioactive labelling
Effect of gene interaction	Very high	No effect
Mapping of QTL	Not possible	Possible
Screening at seedling stage for economic traits	Not possible	Possible

Breeding Methods for Self-pollinated Crops

Pure line selection

- Pure line is the progeny of single self-fertilized homozygous individuals.
- In this method, a large number of plants are selected on the basis of phenotypic performance and their individual progenies are evaluated till homozygosity is attained.

- The progeny is developed as a pure line variety. This procedure may be applied to improve any local variety or exotic collections.
- In case there is variation within plants of a progeny, single plant selection is followed till complete homozygosity is attained.

Single-plant selection

- It is used to select chance variants, off types or superior types from mixed homozygous population in self-pollinated vegetable crops.

Hybridization

- It is one of the techniques used to generate variability in the segregating progenies, advancing population and selection of desired segregants by using procedures such as pedigree, bulk, single seed descent, back cross and their combinations.
- In a hybridization programme, parents are selected on the basis of phenotypic performance, morphological characteristics and general combining ability of the varieties.
- Generally, parents exhibiting high gca, where there is involvement of additive genetic variance, are selected in self-pollinated crops.
- After selecting the parents, single cross, three-way cross or double crosses are attempted. Sometimes, F1s are backcrossed to their respective parents for traits of interest.

Pedigree method

- This method has been adopted to develop material as well as a variety for different nutritional quality.
- In several vegetable crops, improvement has been accomplished through pedigree method and as a result a large number of varieties have been developed.
- In this method, single plant selection is followed upto F5 or F6 generations and in advance generations, families are selected on the basis of their phenotypic performance.
- A pedigree record of the selected plants and their families is maintained.
- The F2 population, which is highly heterozygous, is grown in rows and single plant selections are made. Larger the population, more chances there are of getting superior segregants.

Bulk method

- This is one of the most economical methods of handling segregating population based on natural selection.
- In natural selection, plants having poor survival value are eliminated.
- This method consists of growing a large population in each generation and harvesting the seeds in bulk and planting a sample of the seeds in the next generation.
- No selection is made upto F5 or F6 generations. In this case only natural selection operates. Generally there is no selection of plants on the basis of phenotypic performance, but sometimes superior plants are selected and their seeds are composited.

- This type of mass selection will help in the chances of getting superior plants and their composition of seeds upto F3 and after that pedigree method is very effective.
- In F6, single plants are selected and their progenies are grown in the next generations. Superior families are selected on the basis of phenotypic performance, which are further evaluated against standard checks.

Single seed descent method

- Single seed descent method is based on the assumption of additive genetic variance, a rule in self-fertilizing crops. In this method all plants in F2 generation are advanced to the next generations by harvesting and growing on seed from each plant.
- This process is continued till F5 or F6 in which superior individual plants are selected.

Backcross method

- This method is applied to transfer the characters controlled by a single gene.
- It is more convenient to execute this method in self-fertilizing crops.
- The backcross method is based on the principle of quick and efficient transfer of a gene through repeated backcrossing and then selection in F2 and subsequent generations.
- With continuous backcrossing with the recurrent parent, the backcross population becomes more like a recurrent parent with additional trait of the donor parent.

Breeding methods for Cross-pollinated vegetable crops

Introduction

- The introduced materials may be valuable in selecting desirable plants.
- In case of non-uniform introductions the desirable plants can be selected, their progeny increased, purified and later tested against the standard or local varieties for selecting the most promising lines.
- The chances of an introduction depends to a great extent upon the relationship between the agro climates, particularly temperature and day length of the donor and receptor areas.

Mass selection

- In this method the best individual plants are selected in the population and their seeds are composited for the following generation.
- Since the mass selection is made exclusively based on the phenotype of the plants without any progeny testing, the success of selection depends upon the heritability of the characters under selection.

Family breeding

- There is elaborate testing of the progenies not only in F1 generation but also in later 2 or 3 generations and usually more than one cycle of selection.

- This method is practised in beet and can also be adopted in radish, carrot and cauliflower.

Recurrent selection

- The source population of recurrent selection is heterozygous.
- It may be an open pollinated variety, single cross or double cross F1 hybrid, intercrossed progenies of selected inbred lines, a synthetic variety or a composite.
- There are four types of recurrent selection viz. simple recurrent selection, recurrent selection for general combining ability and recurrent selection for specific combining ability, recurrent selection for general and specific combining ability and reciprocal recurrent selection.

Heterosis breeding

- The hybrid varieties are developed by exploiting the dominance variance in heterosis breeding.
- Hybrids are available for commercial cultivation in all the cross-pollinated crops like cole crops and cucurbits etc.
- For the last two decades, vegetable breeders in India hence diverted their attention toward the development of hybrids and disease-resistant varieties/hybrids.
- Vegetables have great potential for export and after value addition these can become an important commodity of agricultural export of India.

Production of high-yielding varieties/hybrids

- Many open pollinated varieties in different vegetable crops have been identified for the cultivation under different agro climatic zones of the country.
- For the past one and half decade, vegetable breeders of India have paid more attention towards the development of F1 hybrids in important vegetables particularly tomato, brinjal, bell pepper, watermelon, muskmelon etc.
- The hybrids have been found more suitable, stable and tolerant to major diseases under varied agroclimatic conditions.
- The credit for popularization of hybrid vegetables in India goes to the seed companies and first tomato and bell pepper hybrids were marketed by Indo-American Hybrid Seed Company in 1973.
- A major breakthrough in the past few years have been realized in tomato production from F1 hybrids.

Biotic stresses

- The major problem in vegetable production is the heavy incidence of diseases and pests.
- Amongst the many alternatives of controlling insect pests and diseases, chemical control is being widely followed all over the world.
- But due to the carcinogenic nature of these chemicals many countries including India are interested in vegetables produced without using chemicals.

- Therefore, the breeding work has been directed toward development of insect pests and disease-resistant varieties in okra and bacterial wilt-resistant varieties in brinjal have been developed and recommended for cultivation in disease-prone areas. These varieties have made sustainable impact in the production of tomato, brinjal and okra in many parts of the country.
- On the contrary, the progress in the development of insect resistance is very limited. Varieties released till date are having one or more desirable undesirable characters associated with them and hence have not gained popularity.
- In some varieties of brinjal, biochemical analyses have revealed high glycoalkaloid content, peroxidase and polyphenol oxidase activity than the susceptible cultivars.
- This requires strategies to generate technology for economic control of major pests of vegetable crops.
- Host plant resistance is an economical method but such resistance against insect pests is stable due to population pressure of insects on the host. As a result there has been evolution of new biotypes and breakdown of resistance.
- Under this situation there is need to search tolerant lines showing low degree of incidence and application of chemicals at critical stages.

Abiotic stresses

- In India, approximately 70% cultivable land is under cultivation.
- Varieties which can withstand drought, cold, heat, salinity, alkalinity, environmental pollution or other climatic stress of growth, will have wider adaptability as well as stability.
- Very limited efforts have been made to develop vegetable varieties resistant/tolerant to these abiotic stresses. Since area of such stresses is increasing day by day due to a number of ecological factors, comprehensive breeding strategies will have to be developed to produce varieties/hybrids resistant /tolerant to these stresses.
- Furthermore, there is a need to standardize technology and package of practices for growing of vegetables under abiotic stress conditions.

Breeding for export

- India is the **second largest producer of vegetables next to China**, yet less than 1% of the produce is exported.
- The reason can be found in lack of quality, poor post-harvest management and lack of market intelligence. So, to remain competitive, quality has to be maintained which is of prime importance in horticultural crops like vegetables.
- The quality being most important for export, the variety chosen has to be acceptable internationally/nationally and the improvement in which it is produced should be conducive for the development of inherent characters of the variety to the best.
- However in India no systematic breeding work has been done to produce varieties/ hybrids acceptable in the international market.
- Among vegetables, onion is the major item of export. Exports of onion from India is primarily directed to Malaysia, Singapore, Srilanka and Gulf countries.

- Light red and dark red varieties are in demand in Gulf countries, Srilanka and the far East while small onions are in demand in Malaysia and Srilanka.
- Arka Bindu and yellow onion evolved by IIHR, Bangalore are suitable for export. Arka Bindu has good demand in Middle East countries.
- Similarly Co-4 variety of multiplier onion developed by TNAU, Coimbatore, is suitable for export to South East Asian countries.
- In respect of fresh vegetable export, nearly 60% of trade centres around okra. Varieties Arka Abhay, Varsha and Uphar, which are resistant to YVMV, are suitable for export.
- The export demand centres on fruits having 7-9 cm length. In tomato fresh fruits can be exported to Middle East. Varieties Arka Alok and Arka Shakti from IIHR have been bred to be resistant to wilt and can be grown in sick soil in Kerala, parts of Karnatka, Maharashtra and Orissa for export.

Application of Biotechnological Techniques

- Although the main emphasis has been till now on conventional breeding methodology to develop high yield, good quality and resistant varieties, work in several laboratories has been initiated on biotechnological aspect also.
- Techniques in regeneration are available and these are useful to regenerate or multiply male sterile, gynoecious and self-incompatible lines.

Future Needs

- There is a need to undertake intensive research programme to make the best use of available germplasm in the country and above all to conserve this valuable genetic resource.
- Due emphasis has not been given to breeding for integrated management of diseases, insect pests such as leaf curl, TMV in tomato, fruit and shoot borer in brinjal, virus complex in sweet pepper as well as control of insect pests in cole crops etc.
- An effective collaboration of breeders, plant pathologists and entomologists is of high priority to tackle this problem. Future of vegetable breeding hinges on developing varieties resistant to biotic and abiotic stresses, particularly in India.
- There is a need for heterosis breeding of onion, cauliflower, cabbage and root crops, Inspite of the fact that mechanisms like cytoplasmic genetic male sterility, self-incompatibility and suitable sex expression are available in these crops.
- Indian breeders have failed to exploit them for raising yield levels and quality of these crops. Similarly, work on breeding varieties of okra, pea tomato and onion for processing has not received due attention of the breeders.
- Presently the processing industry survives mainly on the availability of surplus vegetables during the glut period in the market. Such vegetables are of poor quality from the processing point of view, since the requirement of these industries is very specific for a quality produce.

- However, work on breeding for quality traits is lacking. Vegetable varieties with prolonged shelf life are needed for distant transportation and export to eastern countries and Australia.
- For instance, in tomato, efforts should be made to improve shelf life by using slow ripening genes (Nor and Rin). Due emphasis has also not been given to the development of fresh vegetables for export.
- The Agricultural and Processed Food Products Export Development Authority (APEDA) has identified traditional vegetables like onion, okra, bitter gourd, chilli and non-traditional vegetables like asparagus, celery, bell pepper, baby corn, greens and lima beans having good potential.
- There is a great demand for export of snow peas. The future need should, therefore, be more on developing disease and insect pests resistant varieties and increasing the productivity per unit area per unit time.
- Developing vegetable varieties and seeds for export is yet another area where breeding is required to be done.

Vegetable Breeding

- Each of the world's basic vegetable crops originated in a relatively confined geographic region.
- The regions overlap for a number of crops.
- Vavilov (1926) suggested that centres of origin of crop plants coincide with the areas of greater genetic diversity.
- Vavilov identified 8 centres, some of which were subdivided.
- However, Zhukovsky (1965), a close associate of Vavilov, proposed 12 mega gene centres of crop plant diversity.
- The new areas added to Vavilov's eight centres were Australia, whole of Africa and Siberia followed by revision of boundaries to make 12 centres.
- The global genetic wealth comprising about 3000 cultivated taxa, which mainly include both primitive/native cultivars and their wild and weedy species is distributed to these 12 primary centres of diversity (Zeven and Zhukovsky, 1975).
- This includes centres of origin and distribution of diversity in 218 vegetable crop species in different regions of diversity.
- During the process of domestication of vegetable crops other than the centres of origin/diversity, the primitive species moved to other areas during ancient past and due to continuous mutations, crossing with the related species, evolution, and selection by man and nature would have resulted in development of the secondary centre of diversity.
- The criteria for assigning a particular geographic region to a specific vegetable are the availability of considerable variability, presence of wild forms and existence of source of resistance to many of the diseases.
- Historical records, fossil records and archaeological findings add to this information.

- Knowledge regarding gene centres of cultivated plants and their possible wild ancestors (progenitors) serve as guidelines to collect the main sources of genetic resistance to parasitic and non-parasitic diseases, insect pests and nematodes.
- In these centres of origin, the vegetable crops have long been exposed to the selective pressure of local pathogens and insect pests, and have developed inherent basic resistance.
- Leppik (1970) reviewed the role of gene centres of plants as sources of disease and pest resistance. He opined that systematic exploration of primary and secondary gene centres of a particular vegetable may provide additional gene pools for resistance breeding.
- Exploration and exploitation of gene centres therefore must for resistance proceed with systematic cooperation of taxonomists, geneticists and plant pathologists.

Centre of origin of crop plants

The **center of origin** is a geographical area where a group of organisms, either domesticated or wild, first developed its distinctive properties. Centers of origin are also considered centers of diversity.

Vavilov Center of Diversity is a region of the world first indicated by **Dr. Nikolai Ivanovich Vavilov** to be an original center for the domestication of plants. Vavilov developed a theory on the **centers of origin of cultivated plants**. He stated that plants were not domesticated somewhere in the world at random but there are regions where the domestication started.

Table 7.5: Centres of diversity of the major vegetable crop species

Mega gene centre	Primary centres of origin and domestication	Secondary centres
Chinese-Japanese	Eggplant, wax gourd, Chinese cabbage, kangkong,welsh onion	Water melon, amaranth
Indo-Chinese	Wax gourd, sponge gourd, ridge gourd, bitter gourd, sword bean, winged bean, Taro, cucumber, bottle gourd, chayote, yam	Cucumber, bottle gourd, chayote, shallot, yam bean, yard long bean, Chinese cabbage, amaranth, kangkong
Hindustani	Eggplant, wax gourd, cucumber, ridge gourd, sponge gourd, bittergourd, hyacinth bean, kangkong, okra, drumstick	Watermelon, melon, bottle gourd, amaranth, roselle
Central Asia	Onion, garlic, mustard, spinach , carrot	Eggplant, watermelon, melon, cauliflower
Near East	Onion, garlic, leek, mustard beet	Okra
Mediterranean	White cabbage, cauliflower, broccoli, bean, radish	Sweet pepper, garlic, okra
African	Eggplant, watermelon, melon, bottle gourd, locust bean, cowpea, okra, roselle	Onion, shallot, lima bean, mustard, amaranth, corchorus

Contd...

Mega gene centre	Primary centres of origin and domestication	Secondary centres
European-Siberian	Lettuce	Onion, common bean, white cabbage, cauliflower, spinach, carrot
South American	Tomato, hot pepper, pumpkins and squashes, lima bean, common bean, cassava, xanthosoma, chayote	Common bean
Central American and Mexican Region	Tomato, hot pepper, pumpkins and squashes, yam bean, common bean, amaranth, sweet potato	-
North America	-	Tomato, eggplant, pepper, melon, watermelon, pumpkin and squashes, onion, lettuce, common bean, lima bean, okra

Indian spices has been one of the prime sources of fabulous wealth and prosperity and there is glowing accounts by Pliny, Ptolemy, Macropolo, Vasco da Gama and other travellers.

- The Malabar Coast of India with its abundance of the world's best pepper, cinnamon and ginger was a vast emporium of spices, which attracted foreign traders.
- Fortune seekers and invaders by overcoming huge nature's barriers travelled to India with the main object of controlling the spice trade.
- Indian spices travelled far and wide to traditional destinations to flavour foods, beverages and to beautify women of many countries.

Biodiversity and Conservation

Biological diversity or **biodiversity** refers to the variety and variability of life on earth. Biodiversity is expressed at three levels on earth: genetic diversity, species diversity and ecosystem diversity. Its direct and indirect services are crucial for the sustenance of life on this planet.

Biodiversity ensures food, fuel, shelter, medicines and other resources which are vital for our survival. Most of the crop pests are controlled by a variety of other organisms, including insects, birds and fungi, which are certainly superior natural pesticides than their chemical equivalents. The pesticides which are extensively used are extremely harmful to human beings and the environment.

Many flowering plants rely on animal species such as insects, bats and birds for pollination and seed dispersal. Biodiversity, more importantly, offers a wide range of resources and services for any location at any specific time and thus providing manwith a variety of options to make it possible to switch over from one resource to another and prefer another if something is scarce, or if market demands changes.

Access to diverse species enables diversification of livelihood through, planting multiple crops and thus enabling food production throughout the year, or engaging in alternative income-generating activities. The availability of diverse resources also allows different genders, cultural or age groups to engage in and benefit from the resources.

Plant Biodiversity

- Biodiversity or biological diversity is a neologism and a portmanteau word, from *bio* and *diversity*. It is the diversity of living nature. Diversity, at its heart, implies the number of different kinds of objects, such as species. However, defining biodiversity or measures of biodiversity, is not so simple.
- The term 'biological diversity', was coined by Thomas Lovejoy in 1980, while the word 'biodiversity' itself, was coined by the entomologist E.O. Wilson in 1986, in a report for the first American Forum on biological diversity organized by the National Research Council (NRC). The word biodiversity was suggested to him by the staff of NRC, to replace 'biological diversity' a word which was considered to be less effective in terms of communication.
- Since 1986, the terms and the concept have achieved widespread use among biologists, environmentalists, political leaders, and concerned citizens world-wide. This use has coincided with the expansion of concern over extinction observed in the last decades of the 20th century.

Biodiversity Definitions

- Biological diversity has no single standard definition. One definition holds that biological diversity is a measure of the relative diversity among organisms present in different ecosystems. "Diversity" in this definition includes diversity among species, within species and comparative diversity among the types in different ecosystems.
- Another definition, simpler and clearer, but more challenging, is "the totality of genes, species, and ecosystems of a region". An advantage of this definition is that it seems to describe most instances of its use, and one possibly unified view of the traditional three levels at which biodiversity has been identified.
- Genetic Diversity - diversity of genes within a species. There is a genetic variability among the populations and the individuals of the same species

Species Diversity—diversity among species

- *Ecosystem Diversity* - diversity at a higher level of organization, the ecosystem (richness in the different processes to which the genes ultimately contribute).
- The latter definition, which conforms to the traditional five organization layers in biology, provides additional justification for multilevel approaches.
- The 1992 Earth Summit in Rio de Janeiro defined biodiversity as

  > "The variability among living organisms from all sources, including, inter alia, terrestrial, marine and other aquatic ecosystems and the ecological complexes of which they are part: this includes diversity within species, between species and of ecosystems."

- This is in fact the closest we come to a single legally accepted definition of biodiversity, since it is the definition adopted by the United Nations Convention on Biological Diversity. The parties to this convention include all the countries on Earth with the exception of Andorra, Darussalam Brunei, the Holy See, Iraq, Somalia, Timor-Leste, and the USA.

- If the gene is the fundamental unit of natural selection, thus of evolution, some scientists like E.O. Wilson, say that the real biodiversity is the genetic diversity. However, species diversity is the easiest one to study.
- For geneticists, biodiversity is the diversity of genes and organisms. They study processes such as mutations, gene exchanges, and genome dynamics that occur at the DNA level and generate evolution.
- For biologists, biodiversity is the diversity of populations of organisms and species, but also the way these organisms function. Organisms appear and disappear; sites are colonized by organisms of the same species or by another. Some species develop social organisations to improve their reproduction goals or use neighbour species that live in communities. Depending on their environment, organisms do not invariably use the same strategies of reproduction,
- For ecologists, biodiversity is also the diversity of durable interactions among species. It not only applies to species, but also to their immediate environment (biotope) and the ecoregions the organisms live in. In each ecosystem, living organisms are part of a whole; they interact with one another, but also with the air, water, and soil that surround them.

Origin of life and biodiversity evolution

- Biodiversity found on Earth today is the result of 3.5 billion years of evolution. Until the emergence of humans, the Earth supported more biodiversity than in any other period in geological history. Since the advent of humans, however, biodiversity has begun a rapid decline, with one species after another suffering extinction.
- Estimates of global species diversity vary from 2 to 100 million species, with a best estimate of somewhere near 10 million.
- New species are regularly discovered (on an average about three new species of birds each year) and many, though discovered, are not yet classified (an estimate gives that about 40% of freshwater fishes from South America are not classified yet). Most of the diversity is found in tropical forests.

Benefits of biodiversity

- Biodiversity has contributed in many ways to the development of human culture, and, in turn, human communities have played a major role in shaping the diversity of nature at the genetic, species, and ecological levels.
- There are three main reasons commonly cited in the literature for the benefits of biodiversity.
  - Ecological role of biodiversity
  - Economic role of biodiversity
  - Ethical role of biodiversity
  - Scientific role of biodiversity

Ecological role of biodiversity

- All species provide some kind of function to an ecosystem. They can capture and store energy, produce organic material, decompose organic material, help to cycle water and nutrients throughout the ecosystem, control erosion or pests, fix atmospheric gases, or help regulate climate.
- Ecosystems also provide us various supports of production (soil fertility, pollinators of plants, predators, decomposition of wastes) and services such as purification of the air and water, stabilization and moderation of the climate, decrease of flooding, drought and other environmental disasters. These functions are important to an ecosystem, and to human survival.
- Research shows that the more diverse an ecosystem, the better it can withstand environmental stress and eventually the more productive it is. The loss of a species thus decreases the ability of the system to maintain itself or to recover in case of damage. The mechanisms underlying these effects are complex and hotly contested. In recent years, however, it has become clear that these ecological effects of biodiversity do indeed exist.

Economic role of biodiversity

- For all humans, biodiversity is first a resource for daily life. Such a diversity in crops is also called agro biodiversity. Most people see biodiversity as a reservoir of resources to be drawn upon for the manufacture of food, pharmaceutical, and cosmetic products.
- This concept of biological resources management probably explains most fears of resources disappearance related to the erosion of the biodiversity. However, it is also the origin of new conflicts dealing with rules of division and appropriation of natural resources.
- Some of the important economic commodities that biodiversity supplies to humankind are

Food Biodiversity

- Food crops, livestock, forestry, and fish contribute to the development of high yielding and disease resistant varieties.
- **Medication:** Wild plant species have been used for medicinal purposes from time immemorial. For example, quinine comes from cinchona tree (used to treat malaria), digitoxin from the foxglove plant (to treat chronic heart trouble), and morphine from the poppy plant (pain relief). According to the National Cancer Institute, over 70 % of the promising anti-cancer drugs come from plants in the tropical rainforests. Animals may also play a role, in particular in research. It is estimated that of the 250,000 known plant species, only 5,000 have been researched for possible medical applications.
- **Industry:** Biodiversity may be a source of energy (such as biomass) for example fibres for clothing and wood for shelter and warmth. Other industrial products are oils, lubricants, perfumes, fragrances, dyes, paper, waxes, rubber, latex, resins, poisons and cork can all be derived from various plant species. Supplies from

animal origin are wool, silk, fur, leather, lubricants and waxes. Animals may also be used as a mode of transportation.

- **Tourism and recreation:** Biodiversity is a source of economic wealth for many areas, such as many parks and forests, where wild plants and animals are a source of beauty and joy for many people. Ecotourism in particular, is a growing outdoor recreational activity.
- Ecologists and environmentalists were the first to insist on the economic aspect of biological diversity protection. Thus, Edward O. Wilson wrote in 1992, that biodiversity is the one of the biggest wealths of the planet, but the least recognized.
- Estimation of the value of biodiversity is a necessary precondition to any discussion on the distribution of biodiversity richnesses. This value can be divided into use value (direct such as tourism or indirect such as pollination) or non-use or intrinsic value.
- If biological resources represent an ecological interest for the community, their economic value is also increasing. For society, biodiversity also is a field of activity and profit. It requires a proper management set up to determine how these resources are to be used. The majority of species are yet to be evaluated for their current or future economic importance.

Ethical role of biodiversity

- Finally, the role of biodiversity is to be a mirror of our relationships with the other living species, an ethical view with rights, duties, and education. If humans consider species have a right to exist, they cannot cause voluntarily their extinction. Besides, biodiversity is also part of many cultures of spiritual heritage.

Scientific role of biodiversity

- Biodiversity is important because each species can give scientists some clue as to how life evolved and will continue to evolve on Earth. In addition, biodiversity helps scientists understand how life functions and the role of each species in sustaining ecosystems.

Plant biodiversity Merits

- Gene sanctuaries offer the following two advantages:
  - It not only conserves the existing genetic diversity present in the population, but also allows evolution to continue. As a result, new allele and gene combinations would appear with time.
  - The risks associated with ex situ conservation are not operative.

Plant biodiversity Demerits

- Gene sanctuaries suffer from certain limitations as listed below;
  - They are easiest to demarcate, difficult to establish and very difficult to maintain. This is particularly so in a country like India, which has an ever increasing population pressure.

- These cannot conserve the variability found in crop plants, for which ex situ conservation is the only answer.

Germplasm Conservation

- Conservation refers to protection of genetic diversity of crop plants from genetic erosion.
- There are two important methods of germpalsm conservation or preservation viz., *in situ* conservation and *ex situ* conservation.

In situ conservation

- Conservation of germplasm under natural conditions is referred to as *in situ* conservation. This is achieved by protecting the area from human interference, such an area is often called natural park, biosphere reserve or gene sanctuary. The NBPGR, New Delhi have established gene sanctuaries in Meghalaya for citrus and other North Eastern regions for musa, citrus, oryza and saccharum.

Merits

- In this method of conservation, the wild species and the complete natural or seminatural ecosystems are preserved together.

Demerits

- Each protected area will cover only a very small portion of total diversity of a crop species, hence several areas will have to be conserved for a single species.
- The management of such areas also poses several problems.
- This is a costly method of germplasm conservation.

Ex Situ Germplasm Conservation

- Conservation of germplasm away from its natural habitat is called *ex situ* germplasm conservation. It can be achieved in the following five ways:
    1. Seed gene banks.
    2. Plant or field gene banks.
    3. Shoot-tip gene banks.
    4. DNA gene banks.
    5. Germplasm is stored as seeds of various accessions in seed gene banks.
- Virtually all gene banks are essentially seed gene banks. Seed conservation is quite easy, relatively safe and ordinarily needs minimum space. Under suitable conditions, seeds of many species can be stored for up to 50-100 years.
- Containers of glass, tin, plastic or a combination of these may be used for seed storage. Seeds are classified, mainly on the basis of their storability, into two major groups:
    1. Orthodox and
    2. Recalcitrant

- This grouping of seeds was proposed by Roberts in 1973.

Orthodox Seeds

- Seeds of this type can be dried to a moisture content of 5% or lower without lowering their viability.
- Most crop seeds belong to this category. Such seeds can be easily stored for long periods; their longevity increases in response to lower humidity and storage temperature.

Recalcitrant Seeds

- The viability of this group of seeds drops drastically if their moisture content is reduced below 12-30%. Seeds of many forest and fruit trees, and of several tropical crops like citrus, cocoa, coffee, rubber, oil palm, mango, jackfruit, etc. belong to this group.
- Such seeds present considerable difficulties in storage. Therefore, germplasm of such plants are conserved by alternative approaches.
- The conditions for seed storage depend mainly on the duration of storage. Generally, seed bank collections are classified into three groups:
  1. Base collections,
  2. Active collections and
  3. Working collections.
- This grouping increases the efficiency of use and the level of management of the collections.

Base collections

- These consist of all the accessions present in the germplasm of a crop, which are stored at about - 20°C with 5% moisture content; they are disturbed only for regeneration.
- Germination tests are done every 5-10 years. When the germination of an accession falls below, usually, 95% of its germination at the start of storage, the accession is regenerated. For reasons of safety, duplicates of base collections should be conserved in other germplasm banks as well.
- High quality orthodox seeds can maintain good viability upto 100 years.

Active collections

- The accessions in an active collection are stored at temperatures below 15°C (often near 0°C) and the seed moisture is kept at 5%.
- The storage is for medium duration, i.e. 10-15 years.
- These collections are used for evaluation, multiplication and distribution of the accessions.
- Active collection are usually maintained by multiplying the seeds of their own accessions. But from time to time, base collection material should be used for regeneration of these collections.

- This is essential to prevent any appreciable shift in the genetic makeup of the collections.

Working collections

- The accessions being actively used in crop improvement programmes constitute working collections.
- Their seeds are stored for 3-5 years at less than 15°C and they usually contain about 10% moisture.
- These collections are maintained by the breeders using them.

Field Gene Banks

Essentially, a field or plant gene bank is an orchard or a field, in which accessions of fruit trees or vegetatively propagated crops are grown and maintained. Field banks suffer from the following serious limitations:

1. They require large areas,
2. They are expensive to establish and maintain, and are prone to damage from
3. disease and insect attacks,
4. They are man-made
5. There are natural disasters and human errors in handling.
6. Few good plant banks exist in India.

Shoot Tip Gene Banks

- In such gene banks, germplasm is conserved as slow growth cultures of shoot-tips and nodal segments. Their regeneration consists of sub-culturing the cultures, which may be done every 6 months to 3 years. This approach offers the following advantages for the conservation of germplasm of vegetatively propagated crops and tree species.
- Genotypes of the accessions can be conserved indefinitely, free from diseases and pests.
- They can be used for such crops, which either do not produce seeds or produce recalcitrant seeds.
- Subculture becomes necessary only after relatively long periods (every 6-36 months).
- Regeneration, i.e., sub-culturing, requires a comparatively very short time.
- In addition, cuttings, bulbs and tubers can be maintained under controlled humidity and temperature conditions; however, this approach is practical the short- and medium-term storage, and it should be used in conjunction with a field gene bank.

Cryopreservation

Cryopreservation is a process where cells, whole tissues, or any other substances susceptible to damage caused by chemical reactivity or time are preserved by cooling to sub-zero temperatures. At low enough temperatures, any enzymatic or chemical activity which might cause damage to the material in question is effectively stopped. Cryopreservation methods seek to reach low temperatures without causing additional damage caused by the formation of ice during freezing. Traditional cryopreservation has relied on coating the material to be frozen with a class of molecules termed cryoprotectants. New methods are constantly being investigated due to the inherent toxicity of many cryoprotectants. By default it should be considered that cryopreservation alters/compromises the structure and function of cells until and unless it is proven otherwise for a particular cell population.

Temperature

Cryogenic storage at very cold temperatures is presumed to provide an indefinite longevity to cells, although the actual effective life is rather difficult to prove. Researchers experimenting with dried seeds found that there was noticeable variability of deterioration when samples were kept at different temperatures – even ultra-cold temperatures. Temperatures less than the glass transition point (*Tg*) of polyol's water solutions, around –136 °C (137 K; –213 °F), seem to be accepted as the range where biological activity very substantially slows, and –196 °C (77 K; –321 °F), the boiling point of liquid nitrogen, is the preferred temperature for storing important specimens. While refrigerators, freezers and extra-cold freezers are used for many items, generally the ultra-cold of liquid nitrogen is required for successful preservation of the more complex biological structures to virtually stop all biological activity

Plant quarantine

Plant quarantine is a technique for ensuring diseases- and pest-free plants by isolating them for a period while performing tests for the presence of any pathogens.

Directorate of Plant Protection Quarantine & Storage, Faridabad (Haryana)

The mandate of Plant Quarantine Service within the Directorate of Plant Protection, Quarantine and Storage is to prevents the entry, establishment and spread of exotic pests in India as per the provisions of the Destructive Insects & Pests Act, 1914 and the notifications issued thereunder.

- The first legal restrictions to hinder the spread of disease were enacted against human disease.
- It was the notorious outbreak of bubonic plague which swept through Europe during the 14th century that led the Venetian Republic to appoint three guardians of public health, to exclude infected and suspected ships and to make the first quarantine of infected areas in 1403.
- The term "quarantine" has been derived from the Latin words *'quaranta giorni'* meaning fourty.
- Travellers from the Levant and Egypt, where plague was endemic, were isolated in a detection hospital for 40 days.
- In its strict sense, plant quarantine refers to the holding of plants in isolation until they are believed to be healthy.
- The meaning of this term has been broadened, and plant quarantine is now taken to mean all aspects of the regulation of the movement of living plants, living

plant parts or plant products between politically defined territories or ecologically distinct parts of them.

- Plant quarantine may be defined as the restriction imposed by duly constituted authorities on the production, movement and existence of plants or plant materials, or animals or animal products or any other article or material or normal activity of persons and is brought under regulation in order that the introduction or spread of a pest may be prevented or limited or in order that the pest already introduced may be controlled or to avoid losses that would otherwise occur through the damage done by the pest or through the continuing cost of their control.
- When plant pathogens are introduced into an area in which they did not previously exist, they are likely to cause much more catastrophic epidemics than do the existing pathogens.
- Examples of some of the worst plant disease epidemics that have occurred throughout the world are the downy mildew of grapes epidenic in Europe and the bacterial citrus canker epidenic in Florida.
- It is extremely difficult to predict accurately whether an exotic organism will become established, and once established, become economically important.

Factors which affect entry and establishment of an organism are

- Hitchhiking potential compared with natural dispersal,
- Ecological range of the pest as compared with ecological range of its host,
- Weather,
- Ease of colonization including reproductive potential, and
- Agricultural practices including pest management.

History of plant quarantine regulations

- In 1660, a law was passed in Rouen, France, ordering the eradication of burberry plants from the vicinity of grain fields. The first British legislation against a disease in animals or plants was an Act of 1866 granting emergency powers for the destruction of all cattle affected by rinderpest which had been introduced into Britain by cattle imported from Russia the previous year.
- An embargo was passed in Germany to prevent importation of plant and plant products from the US to prevent the introduction of the Colorado potato beetle (*Leptinotarsa decemlineata*) in 1873. In 1877, the United Kingdom Destructive Insects Act was passed to prevent the importation of the Colorado beetle.
- In North America, the first legislative measures against plant disease were promulgated by states in 1875. Such measures included a series of laws against peach yellows. In 1891, the first plant quarantine measure was initiated in US by setting up a seaport inspection station at San Padro, California. In 1912, the US Congress enacted the Federal Plant Quarantine Act, which among other things, prohibits the entry of plants into the United States.
- It was phylloxera gall louse (*Phylloxera vastatrix*) that provided the initial stimulus for the establishment of effective international cooperation on plant disease

legislation. The aphid was introduced into Europe from America in 1865 and caused major losses in the vineyards of France.

- The first international plant protection convention, the Phylloxera Convention, was signed at Berne on 3 November 1881 by five countries. This convention remained in force till 1951, when International Plant Protection Convention under FAO was established at Rome. This agreement was constituted with the purpose of securing common and effective action to prevent the introduction and spread of pests and diseases of plants and plant products.
- Within the framework of the cooperation, there are six regional groups.
  - European Plant Protection Organization.
  - Inter-African Phytosanitary Commission for South Africa, South of Sahara.
  - Plant Protection Committee for South East Asia and the Pacific Region.
  - Organizmo International Regional de Sanidad Agroperciane for Central America and Mexico.
- The PFS (Plants Fruits & Seeds) Order (1989) was revised in order to fulfil India›s legal obligations under the WTO Agreement and the Plant Quarantine (Regulation of Import into India), 2003 (referred to as PQ Order) came into force with effect from 1 January, 2004.
- Under the DIP (Destouctive Insects and Pess) Act the Directorate of Plant Quarantine and Storage (DPPQS), Faridabad, headed by Plant Protection Advisor to the Govt. of India (under the Department of Agriculture and Cooperation, Ministry of Agriculture) is responsible for enforcing quarantine regulations and also for making rules for quarantine inspection and disinfestations of any article or class of articles in respect of which a notification has been issued by the Central Government.
- There are 29 plant quarantine stations under DPPQS.
- The National Plant Quarantine Station at Rangpuri, New Delhi and four Regional Plant Quarantine Stations at Amritsar, Chennai, Kolkata and Mumbai are the major stations and are located at places having international airport/seaport/ land frontier with neighbouring countries.
- National Bureau of Plant Genetic Resources, New Delhi, which is the nodal institution for exchange of plant genetic resources, has been empowered under PQ (Plant Quarantine) order to handle quarantine processing of germplasms and transgenic planting material being imported for research purposes in the country.
- As per the recent amendments made under the PQ order, the Advanced Centre for Plant Virology at IARI, New Delhi, Indian Institute of Horticultural Research, Bangalore and Institute of Himalayan Bioresource Technology, Palampur have been identified for ensuring virus-free status in imported *in vitro* material.
- Under the DIP Act, there is a provision of Domestic Quarantine to restrict the inter-state movement of nine invasive pests *viz* flute scale, sanjose scale, coffee berry borer, codling moth, banana bunchy top and mosaic viruses, potato cyst nematode, potato wart and apple scab. Keeping in view the biosafety concerns associated with

the growing of imported transgenic material, a containment Facility of Level -4 has been established at NBPGR, New Delhi for their quarantine processing.

Modes of Reproduction

- Knowledge of the modes of reproduction and pollination is essential for a plant breeder, because these aspects help in deciding the breeding procedures to be used for the genetic improvement of a crop species.
- Choice of breeding procedure depends on the mode of reproduction and pollination of a crop species. Reproduction refers to the process by which living organisms give rise to offspring of similar kind (species).
- In crop plants, the mode of reproduction is of two types *viz*:
    1. Sexual reproduction and
    2. Asexual reproduction
- In the sexually reproducing crop plants, definite male and female sexual organs namely androecium and gynoecium produce male gametes and female gametes respectively. The male gametes are formed from a haploid cell called microspore and the female gametes are formed from a haploid celled megaspore.

1. Sexual Reproduction

- Multiplication of plants through embryos which have been developed by fusion of male and female gametes is known as sexual reproduction. All the seed propagating species belong to this group.

Sporogenesis

- Production of microspores and megaspores is known as sporogenesis. In anthers, microspores are formed through *microsporogensis* and in ovules, the megaspores are formed through *megasporogenesis*.

Microsporogenesis

- The sporophytic cells in the pollen sacs of anther which undergo meiotic division to form haploid microspores are called microspore mother cells (MMC) or pollen mother cells (PMC) and the process is called microsporogenesis. Each PMC produces four microspores and each microspore after thickening of the wall transforms into pollen grain.

Megasporogenesis

- A single sporophytic cell inside the ovule, which undergoes meiotic division to form haploid megaspore, is called megaspore mother cell (MMC) and the process is called megasporogenesis. Each MMC produces four megaspores out of which three degenerate resulting in a single functional megaspore.

Gametogenesis

- The production of male and female gametes in the microspores and megaspores is known as gametogenesis.

Microgametogenesis

- Microgametogenesis is the production of male gametes or sperm. On maturation of the pollen, the microspore nucleus divides mitotically to produce a generative and a vegetative or tube nucleus. The pollen is generally released in this binucleate stage. The reach of pollen over the stigma is called pollination.
- After pollination, the pollen germinates. The pollen tube enters the stigma and travels down the style. The generative nucleus at this phase undergoes another mitotic division to produce two male gametes or sperm nuclei. The pollen along with the pollen tube possessing a pair of sperm nuclei is called the *microgametophyte*. The pollen tube enters the embryo sac through the micropyle and discharges the two sperm nuclei.

Megagametogenesis

- The nucleus of the functional megaspore undergoes three mitotic divisions to produce eight or more nuclei. The exact number of nuclei and their arrangement varies from one species to another. The megaspore nucleus divides thrice to produce eight nuclei. Three of these nuclei move to one pole and produce a central egg cell and two synergid cells on either side. Another three nuclei migrate to the opposite pole to develop into three antipodal cells.
- The two nuclei remaining in the centre, the polar nuclei, fuse to form the secondary nucleus. The megaspore thus develops into a mature female gametophyte called *megagametophyte* or embryo sac. The development of the embryo sac from a megaspore is known as *megagametogenesis*. The embryo sac generally contains one egg cell, two synergids with the apparent function of guiding the sperm nucleus towards the egg cell and three antipodals, which form the prothalamus cells and one diploid secondary nucleus.

Fertilization

- The fusion of one of the two sperms with the egg cell producing a diploid zygote is known as fertilization.
- The fusion of the remaining sperm with the secondary nucleus leading to the formation of a triploid primary endosperm nucleus is termed as *triple fusion*.
- The primary endosperm nucleus after several mitotic divisions develops into a mature endosperm (consistsing of triploid cells) and nourishes the developing embryo which is a diploid.

2. Asexual Reproduction

- Multiplication of plants without the fusion of male and female gametes is known as asexual reproduction. Asexual reproduction can occur either by vegetative plant parts or by vegetative embryos (diploids) which develop without sexual fusion (apomixis). Thus asexual reproduction is of two types: *viz*. (a) vegetative reproduction and (b) apomixis. Vegetative reproduction refers to multiplication of plants by means of various vegetative plant parts. Vegetative reproduction is again of two types: viz.

1. Natural vegetative reproduction and
2. Artificial vegetative reproduction.

Natural vegetative reproduction

- In nature, multiplication of certain plants occurs by underground stems, sub-aerial stems, roots and bulbils. In some crop species, underground stems (a modified group of stems) give rise to new plants. Underground stems are of four types: viz. rhizome, tuber, corm and bulb.
- The examples of plants which reproduce by means of underground stems are given below:
  - Rhizome: Turmcric (*Curcuma longa*), Ginger (*Zingiber officinale*)
  - Tuber: Potato(*Solanum tuberosum*)
  - Corm: Arvi (*Colocasia esculenta*), Bunda (*C. antiquorum*)
  - Bulb: Garlic (*Allium sativum*), onion (*A. cepa*)
  - Sub-aerial stems include runner, sucker, stolon, etc.
  - These stems lead to vegetative reproduction in mint (*Mentha* sp), strawberry, banana, etc.
  - Bulbils are modified forms of flowers. They develop into plants when they fall on the ground. Bulbils are found in garlic.

Artificial Vegetative Reproduction

- Multiplication of plants by vegetative parts through artificial mcthods is known as artificial vegetative reproduction.
- Such reproduction is done by cuttings of stem and roots, and by layering, grafting and budding.

Examples of such reproduction are given below: Stem cuttings: Tapioca (*Manihot esculenta*) grapes (*Vitis vinifera*), roses, etc.

- **Root cuttings:** Bread fruit, Sweetpotato, etc.
- Layering, grafting and budding are used in fruit and ornamental crops.

Apomixis

- Apomixis refers to the development of seeds without sexual fusion (fertilization). In apomixes, the embryo develops without fertilization. Thus apomixis is also an asexual means of reproduction. Apomixis is found in many crop species. Reproduction in some species occurs only by apomixis. This apomixis is termed as *obligate apomixes,* e.g. mangosteen. But in some species sexual reproduction also occurs in addition to apomixis. Such apomixis is known as *facultative apomixis*.

There are four types of apomixis, viz.

1. Parthenogenesis
2. Apogamy
3. Apospory and
4. Adventitious embryony

Parthenogenesis

- Parthenogenesis refers to development of an embryo from the egg cell without fertilization. e.g. mangosteen

Apogamy

- The development of an embryo from either synergids or antipodal cells of the embryo sac is called apogamy.

Apospory

- In apospory, the first diploid cell of ovule lying outside the embryo sac develops into another embryo sac without reduction. The embryo then develops directly from the diploid egg cell without fertilization.

Adventitious embryony

- The development of an embryo directly from the diploid cells of ovule lying outside the embryo sac belonging to either nucellus or integuments is referred to as adventitious embryony. e.g. citrus.

Pollination

Introduction

- Pollination is the transfer of pollen from the stamen (male reproductive part of a flower) to the pistil (female reproductive part of a flower).
- Since both stamen and pistil are necessary for seed formation, they are called essential organs. Most flowers contain two other parts, sepals and petals to attract insect pollinators.
- A complete flower contains all four parts. If only one of the essential organs is present in a flower, it is called an incomplete flower.
- Plants develop seeds through the process of pollination and fertilization.
- Pollination can be accomplished by cross-pollination or by self-pollination. Pollen grains from an anther may fall on to the stigma of the same flower leading to self-pollination or autogamy.
- When pollen grains from flowers of one plant are transmitted to the stigmas of flowers of another plant, it is known as cross-pollination or allogamy.

Self-pollination

- Many cultivated plant species reproduce by self-pollination. Plant species showing self-pollination, i.e., inbreeding, are generally annuals or, in any case, short-lived.
- These species, as a rule, must have hermaphrodite flowers. But in most of these species, self-pollination is not complete and cross-pollination may occur up to 5%. The degree of cross-pollination in self-pollinated species is affected by several factors, e.g., variety and environmental conditions like temperature, humidity and location.

Mechanisms promoting self-pollination

- The various mechanisms that promote self-pollination are:
  1. **Cleistogamy:** In this case, flowers do not open at all. This ensures complete self-pollination since foreign pollen cannot reach the stigma of a closed flower.
  2. **Chasmogamy:** In some species, the flowers open, but only after pollination has ocurred.
  3. In crops like tomato (*S.lycopersicum*) and brinjal (*S. melongena*), the stigma is closely surrounded by anthers and pollination generally occurs after the flowers open. But the position of anthers in relation to stigma ensures self-pollination.
  4. In some species, the flowers open but the stamens and the stigma are hidden by other floral organs.
  5. In peas (*P. sativum*), the stamens and the stigma are enclosed by two petals forming a keel.
  6. In a few species, the stigma become receptive and elongates through the staminal columns.This ensures predominant self-pollination.

Cross-pollination

- In cross-pollinating species, the transfer of pollen from a flower to the stigmas of another flower may be brought about by wind (anaemophily), water (hydrophily) or insects (entomophily). Many of the crop plants are naturally cross-pollinated. In many species, a small amount (upto 5-10%) of selfing may also occur. Mechanism promoting cross pollination: There are several mechanisms that facilitate cross-pollination. These are:

1. Dicliny

- Dicliny or unisexuality is a condition in which the flowers are either staminate (male) or pistillate (female). Dicliny is of two types:

  (*i*) Monoecy and

  (*ii*) Dioecy

  (i) **Monoecy:** Staminate and pistillate flowers occur in the same plant separately, e.g. cucurbits

  (ii) **Dioecy:** The male and the female flowers are present on different plants, e.g. spinach

Dichogamy

- In the same hermophrodite flower, stamens and pistils may mature at different times, thereby hindering self-pollination eventually facilitating cross- pollination.

Dichogamy is of two types

- **Protogyny:** The pistils mature before the stamens

- **Protandry:** Stamens mature before the pistils

Self-incompatibility

- It refers to failure of pollen from a flower to fertilize the same flower or other flowers on the same plant. Self-incompatibility is of two types; sporophytic and gametophytic. In both the cases, flowers do not set seed on selfing.
- Self-incompatibility is common in cauliflower and radish. It is highly effective in preventing self-pollination.

Male sterility

- It refers to the absence of functional pollen grains in otherwise hermaphrodite flowers

Often cross-pollinated species

- In some crop plants, cross-pollination often exceeds 5% and may reach 30%.
- Such species are generally known as '*often cross-pollinated species*'.
- Cross-pollination, also called allogamy, occurs when pollen is delivered to a flower from a different plant.
- Plants adapted to outcross or cross-pollinate often have taller stamens than carpels or use other mechanisms to better ensure the spread of pollens to other plant flowers.

Methods of pollination

Vegetables may be grouped into four categories according to the way they pollinate:

Group 1

- Vegetables in Group 1 have flowers that can receive only their own pollens. They are called self-pollinizers.

Examples	Bush, pole and lima beans
	Chicory
	Endive
	English and Southern peas
	Tomatoes

Group 2

- Vegetables in Group 2 form seed only with pollens from an unrelated plant.

Examples	Cabbage
	Radish

Group 3

- The cross-pollinated vegetables in Group 3 may either set seed from their own pollens (self-pollinated) or from pollens received from another plant (cross-pollinated). They can be divided into two sub-groups: (A) vegetables pollinated by air-borne pollens and (B) vegetables pollinated by insect-borne pollens.

(A) pollinated by air-borne pollen:

Examples	
Beets	Corn
Carrots	Onions
Celery	Spinach
Swiss chard	

(B) pollinated by insect-borne pollen:

Examples		
Broccoli	Kale	Parsnip
Brussels sprouts	Kholrabi	Hot pepper
Cauliflower	Lettuce	Sweet pepper
Citron	Muskmelon	Pumpkins
Collards	Mustard	Rutabaga
Cucumbers	Okra	Squashes
Eggplant	Parsley	Watermelons
Gourds		

Group 4

- Vegetables in Group 4 have both male and female plants. Seeds are formed only when male plants furnish pollens.

Examples	Asparagus
	Spinach
	Some Hybrid Cucumbers

- Pollination may occur between vegetable cultivars, creating new cultivar varieties. For example, plants in the Cucurbitaceae or gourd family belong to four species among which crosses may occur. The success of such crossing depends on the species to which a variety belongs.
- Plants belonging to the Cucurbitaceae family produce separate male and female blooms on the same plant. Insects are usually required to cross-pollinate blooms. Pumpkins and squashes do not cross-pollinate with cucumbers, watermelons or citron. Watermelons and citron both belong to the genus *Citullus* and, therefore, will cross-pollinate each other. Muskmelons and Casaba melons will cross since both belong to the genus *Cucumis* and the same species, *melo*.

The pollination process

- Monoecious plants have both male and female flowers on the same plant. Dioecious plants have only one sex of flower per plant.
- Both monoecious and dioecious plants generally require cross-pollination. When a pollen grain lands on a receptive stigma, it forms a pollen tube down the style to the ovary.
- Male genetic material passes down the pollen tube and fertilizes an ovule. Ovules become seeds and the surrounding ovary develops into the fruit.
- Cucurbit plants, for example, have separate male (staminate) and female (pistillate) flowers. Male flowers generally appear on the plants several days before the female flowers.
- The female flower is easily recognized by the presence of a miniature fruit below the flower petals. Pollen from the male flower must be transferred to the female flower for pollination and subsequent fruit development to occur.
- Many insects visit flowers to collect pollen and nectar as food.
- These insects spread pollen grains among flowers, accomplishing pollination.
- Many flowers offer a sugary liquid nectar as an added enticement for these pollinating insects.
- Among insect pollinators, bees are especially efficient because they eat pollen and nectar exclusively, visit many flowers of the same species during a single trip and have hairy bodies that easily pick up pollen grains.
- Cucurbit flowers open shortly after sunrise and remain open until late afternoon or early evening, so each flower is open for only a few hours.
- The honeybee is the most common and effective cucurbit pollinator. Honeybee activity closely coincides with the period when the flower is open.
- Honeybees begin to visit flowers an hour or two after sunrise and continue to visit until mid-afternoon. If temperatures are very warm, bee activity may decline about noon. When bees are active, spraying pesticides should be avoided.
- Research on watermelon in Florida showed that the number of bee visits was more important than the length of time that each bee stayed on the flower. Well-shaped, fully-expanded fruit developed after eight bee visits to a female flower.
- Fruit set was significantly reduced when bees only visited two to four times. Misshaped or undersized fruit could be the result of poor or incomplete pollination.
- The formation of seed in more complex plants depends on sexual reproduction in the flower. This process can be quite complex and is of fundamental interest to plant breeders.
- It is important that plantings be sufficiently separated when crossings are not wanted.
- Home gardeners who want to save seed from vegetable plants can expect Group 1 to contain less than five percent mixed seed.

- When seeds are saved from Groups 2, 3 and 4 and more than one variety of the same species is grown in the garden, gardeners should use special care to prevent crossing.

Table 7.6: *Approximate time from pollination to market maturity under warm growing conditions*

Vegetable	Time to market maturity (days)
Bean	7-10
Corn	18-23
Cucumber, pickling (¾ - 1 in. diameter)	4-5
Cucumber, slicing	15-18
Eggplant (maximum size)	25-40
Muskmelon	42-46
Okra	4-6
Pepper, green stage (about maximum size)	45-55
Pepper, red stage	60-70
Pumpkin, Connecticut field	80-90
Pumpkin, Big Max	120
Pumpkin, Small Sugar	65-75
Squash, summer, Crookneck	6-71
Squash, summer, Early prolific straightneck	5-61
Squash, summer, Scallop	4-51
Squash, summer, Zucchini	3-41
Squash, winter, Banana	70-80
Squash, winter, Boston marrow	60-70
Squash, winter, Buttercup	60-70
Squash, winter, Butternut	60-70
Squash, winter, Golden delicious	60-70
Squash, winter, Hubbard	80-90
Squash, winter, Table Queen or Acorn	55-60
Tomato, mature green stage	35-42
Tomato, red ripe stage	45-60
Watermelon	42-45

Self-incompatibility

- Evolution seems to favour (and be favoured by) genetic variability. Genetic variability is promoted by outbreeding (sexual reproduction between genetically dissimilar parents).
- Sexual reproduction is popular throughout the world of living things. Plants, being anchored in position, have a special problem in this regard.

- Many employ the services of animals (e.g., insects, birds, bats) to transfer pollen from plant to plant.
- But if the flowers have both sex organs: anthers producing pollen (the source of the male gametes) and the pistil producing the egg (the female gamete).

Dioecy

- The imperfect flowers are present on separate plants
- Dioecy is the equivalent of the separate sexes of most animals. But it is rather rare
- Some examples:
  - Poplar
  - Holly.

Monoecy

- The imperfect flowers are present on the same plant
- But if they mature at different times, self-fertilization is avoided
- Cucurbits are a common example.

Heteromorphic flowers

The flowers are perfect but come in two structural types:

1. Long stamens with a short style
2. Short stamens with a long style

- A single plant has one type or the other.
- If the pollinator has a short tongue, pollination is favoured from the first type to the second - but not the reverse.
- Heteromorphic flowers are not common, and even in the angiosperm families that favour them (e.g., primrose, flax), the same biochemical mechanisms of self that we will find (below) in homomorphic flowers are usually present as well.

Homomorphic flowers

- All flowers have exactly the same structure.
- Avoidance of self-fertilization depends on genetic/biochemical mechanisms.

There are two quite different types of self-incompatibility

1. Sporophytic Self-Incompatibility (SSI)
2. Gametophytic Self-Incompatibility (GSI)

1. Sporophytic Self-Incompatibility (SSI)

- This form of self-incompatibility has been studied intensively in members of the mustard family (Brassica), including turnips, rape, cabbage, broccoli, and cauliflower.

- In this system, rejection of self-pollen is controlled by the diploid genotype of the sporophyte generation.
- The control lies in the "S-locus", which is actually a cluster of three tightly-linked loci.
  - SLG (S-Locus Glycoprotein) which encodes part of a receptor present in the cell wall of the stigma.
  - SRK (S-Receptor Kinase), which encodes the other part of the receptor. Kinases attach phosphate groups to other proteins. SRK is a transmembrane protein embedded in the plasma membrane of the stigma cell.
  - SCR (S-locus Cysteine-Rich protein), which encodes a soluble, secreted ligand for the same receptor.
- Because the plants cannot fertilize themselves, they tend to be heterozygous, i.e., carry a pair of different S loci (here designated S1 and S2).
- However, dozens of different S alleles may be present in the population of the species, i.e., the S-locus in the species is extremely polymorphic (analogous to the major histocompatibility locus of vertebrates).

Gametophytic Self-Incompatibility (GSI)

- This form of self-incompatibility is more common than SSI but not so well understood.
- It occurs in nearly one-half of all the families of angiosperms, including the Solanaceous crops (Potatoes, tomatoes [wild, not cultivated).
  1. Beets (*Beta vulgaris*)
  2. Buttercups (Ranunculus)
  3. Lilies
  4. Roses
  5. Many grasses

The rules

- The S loci are (as in SSI plants) extremely polymorphic; that is, there is an abundance of multiple alleles in the population.
- Incompatibility is controlled by the single S allele in the haploid pollen grain.
- Thus a pollen grain will grow in any pistil that does not contain the same allele (so, as shown here and in contrast to what happens in SSI, S2 pollen from an S1S2 parent will grow down on S1S3 style.
- This appears to be the mechanism in tobacco.
- All pollen grains — incompatible as well as compatible — germinate forming pollen tubes that begin to grow down the style.
- However, growth of incompatible pollen tubes stops in the style while compatible tubes go on to fertilize the egg in the ovary.
- The block within incompatible pollen tubes is created by an S-locus-encoded

ribonuclease (RNase), which is synthesized within the style, enters the pollen tube and destroys its RNA molecules halting pollen tube growth.

- The RNase molecules contain a hyper variable region which is the basis for each S specificity (S1, S2, S3, etc.).
- The pollen tube expresses a molecule that binds RNase. This targeting molecule also exists in different S specificities (S1, S2, S3, etc.).
- In compatible ("nonself") tubes, the targeting molecule sequesters the RNase in a vacuole, thus permitting RNAs in the cytosol to survive and growth to continue.
- In incompatible ("self") tubes, the binding of, for example, the S1 targeting molecule to an S1 binding site on the S1 RNase releases the RNase into the cytosol so the RNAs of the pollen tube are destroyed and growth is halted.

Gametophytic Self-Incompatibility (GSI)

- In gametophytic self-incompatibility (GSI), the SI phenotype of the pollen is determined by its own gametophytic haploid genotype.
- This is the more common type of SI existing in the families: Solanaceae, Rosaceae, Plantaginaceae, Fabaceae, Onagraceae, Campanulaceae, Papaveraceae and Poaceae.
- Two different mechanisms of GSI have been described in detail at the molecular level, and their description follows,

1. The RNase mechanism

- The female component of GSI in the Solanaceae was found in 1989. Proteins in the same family were subsequently discovered in the Rosaceae and Plantaginaceae. Despite some early doubts about the common ancestry of GSI in these distantly related families, phylogenetic studies and the finding of shared male determinants (F-box proteins) early established homology. Consequently, this mechanism arose approximately 90 million years ago, and is the inferred ancestral state of approximately 50% of all plants. In this mechanism, pollen tube elongation is halted when it has proceeded approximately one-third of the way through the style.
- The female component ribonuclease, termed S-RNase, probably causes degradation of the ribosomal RNA (rRNA) inside the pollen tube, in the case of identical male and female S alleles, and consequently pollen tube elongation is arrested, and the pollen grain dies. The male component was only recently putatively identified as a member of the "F-box" protein family. The members of this group typically function as ubiquitin ligases, and they may function by recognizing the matching S-RNase molecules and sending them to proteasomal degradation. Despite some fairly convincing evidence that it may be the male component, several features also make it an unlikely candidate.

2. The S-glycoprotein mechanism

- The following mechanism was described in detail in *Papaver rhoeas*. In this mechanism, pollen growth is inhibited within minutes of its placement on the stigma. The female determinant is a small, extracellular molecule, expressed in the stigma; the identity of the male determinant remains elusive, but it is probably some cell membrane receptor. The interaction between male and female determinants transmits a cellular signal into the pollen tube, resulting in strong influx of calcium cations; this interferes with the intracellular concentration gradient of calcium ions which exists inside the pollen tube, essential for its elongation. The influx of calcium ions arrests tube elongation within 1-2 minutes. At this stage, pollen inhibition is still reversible, and elongation can be resumed by applying certain manipulations, resulting in ovule fertilization.

- Subsequently, the cytosolic protein p26, a pyrophosphatase, is inhibited by phosphorylation possibly resulting in arrest of synthesis of molecular building blocks, required for tube elongation. There is depolymerization and reorganization of actin filaments within the pollen cytoskeleton within 10 minutes from the placement on the stigma, the pollen is committed to a process which ends in its death. At 3-4 hours past pollination, fragmentation of pollen DNA begins, and finally (at 10-14 hours), the cell dies.

Sporophytic self-incompatibility (SSI)

- In sporophytic self-incompatibility (SSI), the SI phenotype of the pollen is determined by the diploid genotype of the anther (the sporophyte) in which it was created. This form of SI was identified in the following families: Brassicaceae, Asteraceae, Convolvulaceae, Betulaceae, Caryophyllaceae, Sterculiaceae and Polemoniaceae. Up to this day, only one mechanism of SSI has been described in detail at the molecular level—in Brassica (Brassicaceae).
- Since SSI is determined by a diploid genotype, the pollen and pistil each express the translation products of two different alleles, i.e. two male and two female determinants. Dominance relationships often exist between pairs of alleles, resulting in complicated patterns of compatibility/self-incompatibility. These dominance relationships also allow the generation of individuals homozygous for a recessive S allele.
- Compared to a population in which all S alleles are co-dominant, the presence of dominance relationships in the population raises the chances of compatible mating between individuals. The frequency ratio between recessive and dominant S alleles reflects a dynamic balance between reproduction assurance (favoured by recessive alleles) and avoidance of selfing (favoured by dominant alleles).
- *The SI mechanism in Brassica.* As previously mentioned, the SI phenotype of the pollen is determined by the diploid genotype of the anther. In Brassica, the pollen coat, derived from the anther's tapetum tissue, carries the translation products of the two S alleles. These are small, cysteine-rich proteins. The male determinant is termed SCR or SP11, and is expressed in the anther tapetum (i.e. sporophytically), as well as in the microspore and pollen (i.e. gametophytically). There are possibly up to 100 polymorphs of the S-haplotype in Brassica, and within these there is a dominance hierarchy.
- The female determinant of the SI response in Brassica is a transmembrane protein

termed SRK, which has an intracellular kinase domain, and a variable extracellular domain. SRK is expressed in the stigma, and probably functions as a receptor for the SCR/SP11 protein in the pollen coat. Another stigmatic protein, termed SLG, is highly similar in sequence to the SRK protein, and seems to function as a co-receptor for the male determinant, amplifying the SI response.

- The interaction between the SRK and SCR/SP11 proteins results in autophosphorylation of the intracellular kinase domain of SRK, and a signal is transmitted into the papilla cell of the stigma. Another protein essential for the SI response is MLPK, a serine-threonine kinase, which is anchored to the plasma membrane from its intracellular side. The downstream cellular and molecular events, leading eventually to pollen inhibition, are poorly described.

Other mechanisms of self-incompatibility

- These mechanisms are less abundant and have received only limited attention in scientific research. Therefore, they are still poorly understood.

1. Heteromorphic self-incompatibility

- A distinct SI mechanism exists in heterostylous flowers, termed heteromorphic self-incompatibility. This mechanism is probably not evolutionarily related to the more familiar mechanisms, which are differentially defined as homomorphic self-incompatibility.
- Almost all heterostylous taxa feature SI to some extent. The loci responsible for SI in heterostylous flowers are strongly linked to the loci responsible for flower polymorphism, and these traits are inherited together. Distyly is determined by a single locus, which has two alleles; tristyly is determined by two loci, each with two alleles. Heteromorphic SI is sporophytic, i.e. both alleles in the male plant, determine the SI response in the pollen.
- SI loci always contain only two alleles in the population, one of which is dominant over the other, in both pollen and pistil. Variance in SI alleles parallels the variance in flower morphs, thus pollen from one morph can fertilize only pistils from the other morph. In tristylous flowers, each flower contains two types of stamens; each stamen produces pollens capable of fertilizing only one flower morph, out of the three existing morphs.
- A population of a distylous plant contains only two SI genotypes: ss and Ss. Fertilization is possible only between genotypes; each genotype cannot fertilize itself. This restriction maintains a 1:1 ratio between the two genotypes in the population; genotypes are usually randomly scattered in space. Tristylous plants contain, in addition to the S locus, the M locus, also with two alleles. The number of possible genotypes is greater here, but a 1:1 ratio exists between individuals of each SI type.

2. Cryptic self-incompatibility (CSI)

- Cryptic self-incompatibility (CSI) exists in a limited number of taxa (for example, there is evidence for CSI in *Silene vulgaris*, (Caryophyllaceae). In this mechanism, the simultaneous presence of cross- and self-pollen on the same stigma, results in higher seed set from cross pollen, relative to self-pollen. However, as opposed to 'complete' or 'absolute' SI, in CSI, self-pollination without the presence of competing cross-pollen, results in successive fertilization and seed set; in this way, reproduction is assured, even in the absence of cross-pollination.

- CSI acts, at least in some species, at the stage of pollen tube elongation, and leads to faster elongation of cross-pollen tubes, relative to self-pollen tubes. The cellular and molecular mechanisms of CSI have not been described. The strength of a CSI response can be defined as the ratio of crossed to selfed ovules, formed when equal amounts of cross- and self-pollen are placed upon the stigma; in the taxa described up to this day, this ratio ranges between 3.2 and 11.5.

3. Late-acting Self-Incompatibility (LSI)

- Late-acting Self-Incompatibility (LSI) is also termed Ovarian Self-Incompatibility (OSI). In this mechanism, self-pollen germinates and reaches the ovules, but no fruit is set. LSI can be pre-zygotic (e.g. deterioration of the embryo sac prior to pollen tube entry, as in Narcissus triandrus) or post-zygotic (malformation of the zygote or embryo, as in certain species of Asclepias and in *Spathodea campanulata*.
- The existence of the LSI mechanism among different taxa and in general is subject for scientific debate. *Critics* claim that absence of fruit set is due to genetic defects (homozygosity for lethal recessive alleles), which are the direct result of self-fertilization (inbreeding depression). Supporters, on the other hand, argue for the existence of several basic criteria, which differentiate certain cases of LSI from the inbreeding depression phenomenon.

Application in Crop Improvement Programmes

- In vegetable breeding, there are various breeding objectives for consumers, such as quality, colour and nutrient content. For producers there are breeding objectives such as high yields, early maturity and disease resistance. However, there is one more objective in vegetable breeding to consider.
- Even if a variety has a good flavour and high yields and is disease resistant, unless the seeds are stable and can be relied upon to produce the same characteristics every year, it cannot become a good variety. Seed is also an agricultural product, and a good variety must have the characteristic of good seed production.
- In recent years, most of the vegetables cultivated over a large part of the world have been F1 hybrid varieties that realize high yields and early maturity through heterosis. Heterosis breeding is only possible if there is an efficient method of producing F1 seed on a large scale. For this, some control of reproduction or pollination is needed.
- In controlling pollination, self-incompatibility (SI) has been used for the Cruciferous crops. These include many important kinds of vegetables such as cabbage, radish, chinese cabbage, turnip and broccoli. The study of SI in crucifer crops began in Japan, where it still continues. In 1949, a Chinese cabbage F1 hybrid variety, "Nagaoka Kohai I Go", was produced by Shojiro Ito, and in 1961 a radish F1 hybrid variety, "Harumaki Minowase", was produced by a commercial seed

company. Most cruciferous vegetables grown in Japan and in other countries are now F1 hybrid varieties, whose seeds are produced using SI.

- However, SI is not always stable. It can be easily overcome under various external and physiological conditions. Therefore, stable F1 seed production has been the subject of study for many years.
- SI is governed by a series of multiple alleles (referred to as the S-gene) (Bateman 1955). However, there are genetic variations in the level of SI (referred to as LSI). Therefore, it can be assumed that SI is also regulated by genes other than the S-gene.
- There are two major seed production methods using the SI system for crucifer crops. One method is a single cross, in which SI is overcome in parental seeds by treating them with CO_2 gas. The other method is a double cross, in which parental seeds are produced using a pair of isogenic lines for the S-gene.
- There are genetic variations in the Reaction Level of Self-Incompatibility (RLSI) to a 4% CO_2 gas treatment. Thus, the parental lines in the parental seed production in a single cross have to show a marked reaction to CO_2. In F1 seed production on both a single and a double cross, the parental lines have to show a high LSI. It is therefore important to know the genetic relationship between these characteristics.

Male Sterility (MS)

- It is characterized by the production of non-functional or sterile pollen grains while the female gamete functions normally. Male sterility occurs in nature due to natural mutation. It can also be induced through interspecific crosses or by the use of gametocides, which will suppress the production of pollen grains.

MS is classified into the following groups:

- Cytoplasmic Male Sterility (CMS)
- Genic Male Sterility (GMS)
- Cytoplasmic Genic Male Sterility (CGMS)
- Environmental Genic Male Sterility (EGMS)
- Chemical Induced Male Sterility

1. Cytoplasmic Male Sterility (CMS)

- This type of sterility is determined by the cytoplasm.
- The zygote gets cytoplasm from, mother or egg cell.
- Hence the progeny will also be male sterile.
- This type of male sterility is known in many crop plants such as onion.
- Genes conditioning male sterility is located in the mitochondria or plasmid.

2. Genetic Male Sterility (GMS)

- GMS is normally governed by nuclear recessive genes 'ms ms'
- The A line is sterile because it has an 'ms ms' gene
- The B line is similar to A line but for the heterozygous condition (Ms ms)

- The R line is completely fertile (Ms Ms)
- This type of male sterility is used in tomato for the production of hybrids
- Male sterile line will be having white translucent anthers, while fertile line will be having yellow, plumpy anthers
- At the time of flowering, each and every plant is to be examined for identification of sterile and fertile lines
- They are to be tied with different colour threads
- At the time of harvest, sterile line and fertile line are to be harvested separately.

The difficulties in the use of GMS are

1. Maintenance of GMS requires skilled labour to identify fertile and sterile lines
2. Labeling is time-consuming
3. In hybrid seed production, spot identification of fertile line and removing them is expensive
4. Use of double the seed rate of GMS line is also expensive
5. High temperature leads to break down of male sterility in some crops.

3. Cytoplasmic Genic Male Sterility (CGMS)

- This is a case of cytoplasmic male sterility where the dominant nuclear gene restores fertility.
- This system is utilized for the production of hybrids in onion, maize and many other crops.
- The A line which is male sterile is maintained by crossing it with the isogenic B line (maintainer line).
- The B line is similar to the A line in all characters (isogenic) except fertile cytoplasm.
- The male sterile A line when crossed with the R line (restorer) restores fertility in F1.

4. Environmental Genic Male Sterility (EGMS)

- EGMS is more popularly termed as "Two Line Hybrid Breeding" as against "Three Line Hybrid Breeding" in case of CMS system. Certain GMS lines are conditional mutants, meaning thereby, in particular environment male sterile mutant plants turn into male fertile.
- After determination of critical environment (usually temperature or photoperiod) for sterility and fertility expression, such GMS mutants are classified under Environmental Genic Male Sterility (EGMS) lines. EGMS lines have been reported in several vegetable crops.
- However, from practical viewpoint, it is necessary to identify critical temperature or photoperiod for the fertility/sterility expression in temperature and photoperiod sensitive genetic male sterility, respectively and may not be hundred per cent.
- Seeds of EGMS line can be multiplied in an environment where it expresses male

fertility traits, while hybrid seeds can be produced in other environments where it expresses male sterility traits.

- Initially EGMS lines were thought to be of less practical value, as they were unstable. But now they are considered to represent the most efficient system for hybrid seed production.

5. Chemical Induced Male Sterility

- The chemical which induces male sterility artificially is called as male gametocide.
- It is a rapid method but the sterility is non-heritable.
- In this system A, B and R lines are not maintained.
- Some of the male gametocides used are gibberellins (rice, maize), sodium methyl arsenate (rice) and maleic hydrazide (wheat, onion).

Application of Cytoplasmic Male Sterility in Plant Breeding

- It is applied in production of hybrids in ornamental crops and vegetatively propagated crops. But in those of crop plants where seed is the economic part, it is of no use because the hybrid progeny would be male sterile. Example – Onion.
- Application of Genic Male Sterility in plant breeding.
- It is applicable in production of hybrids in both vegetative propagated crops and crops important for fruit or seed yield.
- Examples – Tomato and cucurbits

Application of Cytoplasmic-Genic Male Sterility in Plant Breeding

- It is used in commercial production of hybrid seeds in maize, sorghum and bajra.
- Examples – Morghum, bajra, sunflower, rice and wheat.
- The one with cytoplasmic male sterility would be included in the cytoplasmic genetic system as and when restorer genes for it are discovered.

Achievements

- Male sterility in vegetables is a never-ending process due to rapid advancement in molecular techniques and their implementation.
- Substantial progress has been made in understanding the mechanism of male sterility in selected vegetable crops.
- On a global level, cytoplasmic male sterility (CMS) and cytoplasmic genetic male sterility (CGMS) are the most widely utilized in the majority of vegetables.
- In India vegetable hybrids based on CMS and CGMS system have been limited. In India, genetic male sterility (GMS) has been exploited commercially only in the case of chilli and muskmelon to develop F x hybrid seeds commercially.
- Punjab Agricultural University (PAU), Ludhiana, India has released two chilli hybrids (CH-1 and CH-3) and one muskmelon hybrid (Punjab Hybrid) based on the GMS system.

- Similarly, in tomato, work on GMS lines is in progress at PAU. The CGMS system has been commercially exploited in chilli, onion and carrot.
- In the recent past, chilli CGMS lines were introduced at the Indian Institute of Vegetable Research (IIVR) from AVRDC, which are utilized directly or indirectly to produce CMS-based hybrids, i.e Kashi Surkh (CCH-2) and Kashi Early (CCH-3).
- The Indian Institute of Horticultural Research (IIHR), Bangalore, India has also released chilli hybrids based on the CGMS system, i. e. Arka Meghna (MSH-172), MSH-149 and MSH-96.
- In carrot, the Indian Agricultural Research Institute (IARI) regional station, Katrain (HP), India has developed one hybrid, Pusa Nayanjvoti, which is based on petaloid CGMS and it was identified for release from Delhi State Seed Subcommittee in 2009. In the tropical group of carrot, IARI, New Delhi has also reported CGMS system in different genetic backgrounds and evaluation of different hybrid combinations is in progress.
- In onion, IIHR, Bangalore has released two hybrids based on the CGMS system, i.e. Arka Kirtiman and Arka Lalima IARI, New Delhi has developed two hybrids in onion (Hybrid-63 and Hybrid-35) which are based on the same system. The CMS system has been commercially exploited in cabbage, cauliflower and onion.
- The IARI Regional Station Katrain (H.P.) has evolved a few hybrids of cabbage by utilizing Ogura CMS system and one of the hybrids (KCH-5) is under field testing in All India Coordinated Research Project on Vegetable Crops.
- The Ogura-type CMS has been transferred into heat-tolerant Indian cauliflower from kale and broccoli through repeated back crosses; four lines (MS-91, MS-51, MS-11 and MS-110) from the former and five lines (MS-01, MS-04, MS-05, MS-09 and MS-10) from the latter were developed.
- In India, research on the transgenic male sterility system has been initiated in selected vegetables but our first priority should be the utilization of existing and established, but unexploited male sterility systems.
- This will not only promote adoption of hybrid vegetable technology by economizing the cost of hybrid seeds but also provide basic material and scope for the development of a more efficient male sterility system in respective vegetable crops.

Breeding Methods

1. Mass Selection

- In mass selection, seeds are collected from (usually a few dozen to a few hundred) desirable appearing individuals in a population, and the next generation is sown from a stock of mixed seed.
- This procedure, sometimes referred to as phenotypic selection, is based on how each individual looks.
- Mass selection has been used widely to improve old "land" varieties, varieties that

have been passed down from one generation of farmers to the next over long periods.

- An alternative approach that has no doubt been practised for thousands of years is simply to eliminate undesirable types by destroying them in the field.
- The results are similar whether superior plants are saved or inferior plants are eliminated: seeds of the better plants become the planting stock for the next season.
- A modern refinement of mass selection is to harvest the best plants separately and to grow and compare their progenies.
- The poorer progenies are destroyed and the seeds of the remainder are harvested.
- It should be noted that selection is now based not solely on the appearance of the parent plants but also on the appearance and performance of their progeny.
- Progeny selection is usually more effective than phenotypic selection when dealing with quantitative characters of low heritability.
- It should be noted, however, that progeny testing requires an extra generation; hence gain per cycle of selection must be double that of simple phenotypic selection to achieve the same rate of gain per unit time.
- Mass selection, with or without progeny test, is perhaps the simplest and least expensive of plant-breeding procedures.
- It finds wide use in the breeding of certain forage species, which are not important enough economically to justify more detailed attention

2. Pureline selection

Pureline selection generally involves three more or less distinct steps:

1. Numerous superior appearing plants are selected from a genetically variable population,
2. Progenies of the individual plant selections are grown and evaluated by simple observation, frequently over a period of several years, and
3. When selection can no longer be made on the basis of observation alone, extensive trials are undertaken, involving careful measurements to determine whether the remaining selections are superior in yielding ability and other aspects of performance.

- Any progeny superior to an existing variety is then released as a new "pureline" variety.
- Much of the success of this method during the early 1900s depended on the existence of genetically variable land varieties that were waiting to be exploited.
- They provided a rich source of superior pureline varieties, some of which are still represented among commercial varieties.
- In recent years, the pureline method as outlined above has decreased in importance in the breeding of major cultivated species; however, the method is still widely used with the less important species that have not yet been heavily selected.
- A variation of the pureline selection method that dates back centuries is the selection of single-chance variants mutations or "sports" in the original variety.
- A very large number of varieties that differ from the original strain in characteristics

such as colour, lack of thorns or barbs, dwarfness, and disease resistance have originated in this fashion.

3. Hybridization

- From the 20th century, planned hybridization between carefully selected parents has become dominant in the breeding of self-pollinated species.
- The object of hybridization is to combine desirable genes found in two or more different varieties and to produce pure-breeding progenies superior in many respects to the parental types.
- Genes, however, are always in the company of other genes in a collection called a genotype.
- The plant breeder's problem is largely one of efficiently managing the enormous number of genotypes that occur in the generations following hybridization.
- As an example of the power of hybridization in creating variability, a cross between hypothetical wheat varieties differing by only 21 genes is capable of producing more than 10,000,000,000 different genotypes in the second generation. At spacing normally used by farmers, more than 50,000,000 acres would be required to grow a population large enough to permit every genotype to occur in its expected frequency.
- While the great majority of these second generation genotypes are hybrid (heterozygous) for one or more traits, it is statistically possible that 2,097,152 different pure-breeding (homozygous) genotypes can occur, each potentially a new pure line variety.
- These numbers illustrate the importance of efficient techniques in managing hybrid populations, for which purpose the pedigree procedure is most widely used

Pedigree breeding

- If the two original parents do not provide all of the desired characters, a third parent can be included by crossing it to one of the hybrid progeny of the first generation (F1).
- In the pedigree method, superior types are selected in successive generations, and a record is maintained of parent–progeny relationships. The F2 generation (progeny of the crossing of two F1 individuals or selfed F1) affords the first opportunity for selection in pedigree programmes.
- Starts with the crossing of two genotypes, each of which has one or more desirable characters lacking in the other.
- In this generation, the emphasis is on the elimination of individuals carrying undesirable major genes.
- In the succeeding generations the hybrid condition gives way to pure breeding as a result of natural self-pollination, and families derived from different F2 plants begin to display their unique character.
- Usually one or two superior plants are selected within each superior family in these generations.
- By F5 generation the pure-breeding condition (homozygosity) is extensive, and emphasis shifts almost entirely to selection between families.
- The pedigree record is useful in making these eliminations.

- At this stage, each selected family is usually harvested enmasse to obtain the larger amounts of seeds needed to evaluate families for quantitative characters.
- This evaluation is usually carried out in plots grown under conditions that simulate commercial planting practices as closely as possible.
- When the number of families has been reduced to manageable proportions by visual selection, usually by the F7 or F8 generation, precise evaluation for performance and quality begins.
- The final evaluation of promising strains involves.
  1. Observation, usually in a number of years and locations, to detect weaknesses that may not have appeared previously;
  2. Precise yield testing; and
  3. Quality testing. Many plant breeders test for five years at five representative locations before releasing a new variety for commercial production.

The bulk-population

- At maturity the crop is harvested enmasse, and the seeds are used to establish the next generation in a similar plot. No record of ancestry is kept.
- During the period of bulk propagation, natural selection tends to eliminate plants having poor survival value.
- The bulk-population method of breeding differs from the pedigree method primarily in the handling of generations following hybridization. The F2 generation is sown at normal commercial planting rates in a large plot.

Two types of artificial selection also are often applied:

1. Destruction of plants that carry undesirable major genes and
2. Mass techniques such as harvesting when only part of the seeds are mature to select for early maturing plants or the use of screens to select for increased seed size. Single plant selections are then made and evaluated in the same way as in the pedigree method of breeding.

- The chief advantage of the bulk-population method is that it allows the breeder to handle very large numbers of individuals inexpensively.
- Often an outstanding variety can be improved by transferring to it some specific desirable character which it lacks.
- This can be accomplished by first crossing a plant of a superior variety to a plant of a donor variety which carries the trait in question, and then mating the progeny back to a plant having the genotype of the superior parent. This process is called backcrossing.
- After five or six backcrosses the progeny will be hybrid for the character being transferred but like the superior parent for all other genes. Selfing the last backcross generation, coupled with selection, will give some progeny pure breeding for the genes being transferred.

- The advantages of the backcross method are its rapidity, the small number of plants required, and the predictability of the outcome.
- A serious disadvantage is that the procedure diminishes the occurrence of chance combinations of genes, which sometimes leads to striking improvements in performance.

Hybrid varieties

- The development of hybrid varieties differs from hybridization and selection for superior individuals in the subsequent generation.
- The F1 hybrid of crosses between different genotypes is often much more vigorous than its parents.
- This hybrid vigour, or heterosis, can be manifested in many ways, including increased rate of growth, greater uniformity, earlier flowering, and increased yield, the last being of greatest importance in agriculture.

History of selection

- Crop improvement refers to the genetic alteration of plants to satisfy human needs. In prehistoric times, our human forebears in various parts of the world brought into cultivation a few hundred species from the hundreds of thousands available.
- In the process, they transformed elements of these species into crops through genetic alterations that involved conscious and unconscious selection, the differential reproduction of variants.
- Through a long history of trial and error, a relatively few plant species have become the mainstay of agriculture and thus the world's food supply.
- This process of domestication involved the identification of certain useful wild species combined with a process of selection that brought about changes in appearance, quality, and productivity.
- The exact details of the process that altered the major crops is not fully understood, but it is clear that the genetic changes were enormous in many cases.
- In fact, some crop plants have been so changed that for many of them, maize, for example, their origins are obscure, with no extant close wild relatives.
- The selection of naturally occurring variants is the basis of crop improvement
- Over thousands of years this technique resulted in the development of modern basic crops.
- The discovery of techniques for asexual (vegetative) propagation, such as by using natural offshoots, rooting stem cuttings, or various grafting techniques, made it possible to "fix" genetic variants.
- This was the technique used for many tree fruits, enabling identical plants to be cultivated in orchards. Naturally produced seedlings derived from intercrosses of these selected plants were then available for selection again. Many present-day fruit crops are similar to those cultivated in antiquity, and some ancient selections are still cultivated: dates, for example.

- As humans carried improved crop varieties to new locations, opportunities opened to increase genetic variation from natural intercrosses with new wild populations.
- The changes that occur can be dramatic over time, as seen in the proliferation of breeds of animals and especially the wide range of changes brought about by fanciers of dogs, chicken, and pigeons. The observation of these changes influenced the thinking of Charles Darwin to suggest that natural selection, the survival of the fittest, could lead to enormous genetic changes if carried out over a long enough time, and could lead to the origin of new species.
- In the 18th and 19th centuries, a conscious attempt was made to predict the performance of plants that could be expected from one seed generation to the next.
- The concept that ancestry was important in crop improvement led to refinement in the selection process, brought about by keeping records and the assessment of lineage. Furthermore, it became obvious that variation could be managed by controlling the mating process, an extension of what had long been known in animal breeding.
- This new type of selection, termed "pedigree selection", was found to increase the efficiency of the process. Progeny testing (evaluating the genetic worth by progeny performance) increased efficiency of this process.
- The origin of commercial plant breeding began in the second half of the 19th century among seed producers. It involved controlled crosses (hybridization) between selections to control genetic recombinations, followed by selection of improved types.
- This is still the basis of traditional plant breeding. Interestingly, much of this early type of plant breeding was carried out without a clear idea of the genetic mechanism involved in inheritance. Until the famous experiments with the garden pea by Gregor Mendel (1822–1884), a Catholic priest in Brünn, a Moravian town then in the Austro-Hungarian Empire, the basic theory of inheritance involved the concept of blending.
- Mendel unraveled the basic concept of inheritance and clearly showed that characters in the pea were due to elements, later called genes, that remained unaltered as they were inherited.
- Many characters in peas, such as tallness and dwarfness, were shown to be controlled by a pair of genes of which one member was not always expressed (the concept of dominance and recessiveness).
- Mendel demonstrated that the gametes of the pea contained one member of the gene pairs that controlled characters and that recombined randomly at fertilization.
- Mendel's paper was published in 1866, but it had no impact until the paper was 'rediscovered' in 1900, when it created a sensation.
- It was soon obvious that the differences in appearance among plants (phenotypes) could be explained by the interaction of various genes (genotypes) as well as interaction with the environment.

20th Century Developments

- In the 20th century plant breeding developed a scientific basis, and crop improvement was understood to be brought about by achieving favourable accumulations and combinations of genes. Taking advantage of known genetic diversity could facilitate this, and appropriate combinations were achieved through recombinations brought about by the sexual process (hybridization). Furthermore, it was possible to move useful genes by special breeding strategies. Thus, a gene discovered in a wild plant could be transferred to a suitable adapted type by a technique known as the backcross method.
- A sexual hybrid was made, followed by a series of backcrosses to the desirable (recurrent) parent, while selecting for the new gene in each generation. After about fivc or six back crosses, the offspring resembled the recurrent parent but contained the selected gene.
- In the early 20th century, it was demonstrated that the extra vigour long associated with wide crosses (called hybrid vigour or heterosis), particularly in naturally cross-pollinated crops, could be exploited in plant breeding. For maize, a new system of hybrid breeding was developed, using a combination of inbreeding and outbreeding.
- Inbreeding was accomplished by crossing the plant with itself. This led to a decline in vigour as the step was repeated over several generations. Outbreeding was achieved by intercrossing the inbred lines to restore vigour. The hybrid between inbreeds derived from divergent inbreeds (called a single cross or F_1 hybrid) was uniform (homogeneous), and some were superior to the original populations before inbreeding. During the process of inbreeding, the inbreeds became weak, but vigour was restored when these inbreeds were crossed to produce the F_1. To increase seed set from weak inbreeds, two hybrids were crossed; this was known as the double cross method.
- Hybrid breeding technique in a sense is similar to arranging a Rubic's cube, where contradictory steps need to be taken to achieve the appropriate reformulation. In hybrid breeding, the first step produces a series of weak inbreeds, followed by a series of specific combinations to produce a series of new hybrids. Hybrid maize breeding led to enormous increases in productivity, which were soon exploited in a wide variety of seed-propagated crops, including naturally self-pollinated ones, such as tomato.
- A number of genetic techniques were developed and refined in 20th century, such as improved techniques to search for and store increased genetic variability, different techniques to develop variable populations for selection, and improved methods of testing to separate genetic from environmental effects. The exact details of the process for crops necessarily differed among naturally cross-pollinated plants (such as maize) and naturally self-pollinated plants (such as soybean or tomato) as well as those plants in which vegetative propagation (usually cross-pollinated) permitted the fixing of improved types directly.
- Conventional plant breeding can be defined as systems for selection of superior genotypes from genetically variable populations derived from sexual recombination.

The system is powerful because it is evolutionary; progress can be cumulative, with improved individuals continually serving as parents for subsequent cycles of breeding. Genetic improvement by conventional breeding has made substantial changes when the efforts have been long-term.

- Characters improved include productivity, quality, and resistance to diseases, insects, and stress. There are, however, limits to the progress of conventional breeding. These are due to limitations of the sexual system, because it is usually not possible to incorporate genes from nonrelated species or to incorporate small changes without disturbing the particular combination of genes that make a particular type unique. Thus a useful gene in cabbage cannot be transferred to wheat.
- Limitations of conventional breeding are particularly apparent when a needed character (such as disease or insect resistance) is unavailable in populations that can be incorporated by sexual crosses. Mutations may be induced, but they are often deleterious or connected with undesirable effects.
- With conventional breeding, it is also not possible to improve a unique genotype, such as 'artlett' pears, by adding a single character, since the recombination that results from hybridization makes it impossible to reconfigure this cultivar exactly. Finally, conventional breeding has technical or economic limitations to detect infrequent or rare recombinants, the lack of sufficient time to generate cycles of recombination, space to grow necessary populations to recover superior recombinants, or resources to be able to select, identify, evaluate, and fix desired recombinants.

Effects of self-pollination on genotype

- Self-pollination increases homozygosity with a corresponding decrease in heterozygosity.
- Inbreeding also increases homozygosity and reduces heterozygosity. Inbreeding is mating between individuals related by descent, that is, having a common parent or parents in their ancestry.
- Some examples of inbreeding are: sib mating (brother-sister mating), half-sib mating (brother-stepsister mating), etc.
- Self-pollination is the most intense form of inbreeding, since in this case the same individual functions as the male as well as the female parent.
- The effect of self-fertilization on homozygosity and heterozygosity may be illustrated by an example. Suppose an individual is heterozygous for a single gene (Aa) and the successive generations derived from it are subjected to pollination, every generation of self-pollination will reduce the frequency of heterozygote Aa to 5% of that in the previous generation.
- There is a corresponding increase in the frequency of the two homozygotes AA and aa. As a result, after ten generations of selfing, virtually all the plants in the population would be only homozygous and heterozygosity would be 0.097%, which is negligible.
- It is assumed here that the three genotypes AA, Aa and aa have equal survival. If there is unequal survival, it may increase or decrease the rate at which homozygosity is achieved.

- If Aa is favoured, the rate of increase in homozygosity would be lower than expected. But if Aa is selected against, homozygosity would increase at a faster rate than expected.
- When a number of genes are segregating together, each gene would become homozygous at the same rate as Aa. Thus the number of genes segregating does not affect the percentage of homozygosity.
- The term "homozygosity" denotes the frequency of genes in homozygous condition in the population. Similarly, linkage between genes does not affect the percentage of homozygosity in the population.
- Another way of visualizing the effect of self-pollination is to consider the frequency of plants which are homozygous for all the genes.
- In case of a single gene, the frequency of completely homozygous plants in a generation is the same, as the proportion of homozygosity increases at a much faster rate than that of completely homozygous plants.
- The proportion of completely homozygous plants is given by the following formula:

  Proportion of completely homozygous plants = $[(2m - 1)/2m]n$

  Where, m is the number of generations of self-pollination and n is the number of genes segregating.

  The proportion of completely homozygous plants under self-pollination is depicted.
- It is clear that as the number of genes increases, the proportion of completely homozygous plants after a given number of generation of selfing, decreases. But the effect of selfing is so strong that even if 100 genes are segregating, more than 95% of the population would be completely homozygous after only 12 generations.
- Linkage would increase the proportion of homozygous plants since it would reduce the number of genes segregating independently. Unequal survival of different genotypes would have the same effect as in the case of homozygosity.
- Thus, continued self-pollination has two main effects on the population: first, all the plants in the population become completely homozygous, and second, the population of a self-pollinated species represents a mixture of several homozygous genotypes.

Genetic advance under selection

- Selection of necessity is made on the basis of phenotype, and phenotype is produced by the joint action of genotype and environment. Therefore, the phenotypic superiority of selected plants or families over the original population is not solely due to their genotypic superiority.
- Improvement in the mean genotypic value of the selected families over that of the base population is known as genetic advance under selection (*Gs*).

Genetic advance under selection depends upon

1. The phenotypic variability among different plants or families in the base population,
2. The ineritability of the character under selection, and

3. Genetic advance under selection may be calculated as follows.

$$Gs = (k)\ (p)\ (H)$$

where, Gs is genetic advance under selection, k is selection differential, p is the phenotypic standard deviation of the base population (the population, which is being subjected to selection), and H is the ineritability of the character under selection.

- The formula for genetic advance under selection, therefore, may be rewritten as $Gs = (k)\ (\sqrt{Vp})\ (Vg/Vp)$.
- The selection differential, k, is based on the mean phenotypic values of the selected lines and of the base population, the phenotypic standard deviation (p), and the intensity of selection, that is, the proportion of population selected.
- k is expressed in terms of standard deviation units. Therefore, the value of k varies with the intensity of selection.

Developments Using DNA

- It has been suggested that many of the limitations of conventional breeding can be overcome with advances in molecular biology that rely on DNA, the genetic material.
- Recombinant DNA technology, called transgene technology or genetic engineering, is the most powerful and revolutionary of the new genetics developed in the last half of the 20th century.
- It is possible to isolate stretches of DNA from one organism, store it in a bacterial host, select unique combinations, and then incorporate them into the DNA of another species, where it can be expressed.
- This technique, which relies on cell and tissue culture, is a truly marvelous process.
- Refinements in the technique make it possible to concentrate mutations in desired genes, further increasing variability.
- Other uses of molecular biology known as genomics involve the detailed mapping of the DNA and the identification of useful stretches of the molecule.
- This makes it possible to improve the efficiency of selection, because it is based directly on the genes rather than the phenotypic expression of the organism, where the effects may be confounded by environment and genetic interactions.
- The limitations of the new breeding methods include technical problems, such as the difficulty of transformation, problems of gene expression, or the lack of knowledge concerning suitable genes to transfer.
- There are also nontechnical issues, such as legal problems, since the techniques and the genes are usually patented.
- However, in the short run, the greatest impediment has been problems of consumer acceptability and fear of the unknown.
- The term **"Frankenfood"** has been coined to refer to food altered by the process of using exotic genes incorporated by transgene technology.

- No convincing evidence shows that genetic engineering has produced harmful products, and an abundance of evidence shows that many foods derived from traditional systems have inherent problems (consider the allergic reactions of many people to peanuts).
- Nevertheless, molecular techniques have incited fear of this new technology in many people.
- Moreover, the surplus of food in the West has reduced the imperative to make the case for the need for new technology to consumers.

Pureline selection–Definition

- Pureline selection is a method in which new variety is developed by selection of the single best plant progeny among traditional varieties or land races.

Procedure of Pureline Selection

- Selection of individual plants from a local / traditional variety, land races or some other mixed population.
- Visual evaluation of individual plant progenies.
- Yield trials.
- Release of new pureline variety.

Step number	Year	Details
1.	First year	Many individual plants (about 200-3000) are selected from a local variety or some other mixed population based on phenotype. Any number of plants can be selected but importantly they should be homozygous. Seeds are harvested separately.
2.	Second year	Seeds from each plant are grown separately with proper spacing. Evaluated for characters under consideration. Probably in this first season it will not be possible to judge the value of separate lines well, but surely visually poor, weak and defective progenies can be rejected. Number of progenies is drastically reduced in this step, fewer are carried to step three.
3.	Third year	Seeds of second year crop are planted in a preliminary yield trial. Standard plots for comparison are introduced, undesirable progenies are rejected.
	Four to seven years	Replicated yield trials conducted at several locations. Standard plots for comparison are introduced, undesirable progenies are rejected. Tests are done for characters under consideration.
	Eighth year	Best progeny is released as a new pureline variety. Seed is multiplied for distribution.

Improvement of Local Varieties

- Pureline selection is the most popular method for the improvement of local varieties that have considerable genetic variability.
- A large number of varieties have been developed by this method.
- Some example of such varieties are, T_1 cowpea (*Vigna unguiculata*) and T_1 mung (*V. radiata*); these are pureline selection from old local varieties.
- Pusa Sawani bhindi (Okra; *Abelmoschos esculentus*) is a selection from a local variety collected from Bihar; it showed an acceptable level of field resistance to yellow mosaic virus.

Improvement of Old Pureline Varieties

- Purelines become variable with time. Often selection is used to isolate new pureline varieties from such genetically variable purelines.
- Sometimes off-type plants appear in purelines. Such off-type plants may be selected to develop new pureline varieties.

Applications of Pureline Selection

- As pureline selection gives out uniformity in maturity, height, etc. suitable for cultivars in which machines are used for the Because of the same reason as above it can be used for varieties grown for the processing market, as uniformity in texture, canning qualities is important here.
- It is practised after hybridization within segregating populations.
- Though it can serve the above purposes, this method has limited practical use in the breeding of major cultivated species. However, the method is still widely used while breeding less important species that have not yet been heavily selected.
- Logically it was the first step in the development of a uniform variety for the newly domesticated crops, with some amount of variability.
- Steps in the development of pureline variety are given as Procedure for Pureline Selection.

Advantages of Pureline Selection

- It is an easy and cheap method of crop improvement.
- Rapid method, lines are usually genetically fixed and yield trials can be immediately conducted.
- Plants in such variety react in similar fashion to environmental conditions, means they are uniform in performance and at the same time in appearance too.
- Maximum possible improvement over the original variety can be achieved.
- Useful in improving low heritability traits as selection is based on progeny performance.

Disadvantages of Pureline Selection

- Purelines have poor adaptability due to narrow genetic base, just opposite to mass selected variety.

- Superior genotypes can only be isolated from the mixed population. This selection is powerless to bring changes in hereditary factors, i.e. to develop new genotypes.
- Mostly popular or in fact limited to self-pollinated spp. only.
- Time and space consuming.
- More expensive yield trials have to be conducted than in mass selection.

Mass selection

- In mass selection, seeds are collected from a large number of phenotypically similar plants available in a population and are bulked to grow the next generation. In the simplest manner it is the creation of a mixture of different lines.
- While selecting, easily observable characters like plant height, leaf phenotype, branching habit, fruit colour, fruit size and number of fruits/plant, etc., should be observed and best the performing plants should be selected.

Procedure for Mass Selection

The general procedure is given below, however, steps may vary depending upon certain factors:

Year	Details
First year	Plants with similar but desirable traits are selected from source population and their seeds are bulked. Alternatively, undesirable population and the seeds of remaining pollination can be bulked.
Second year	Seeds carried from first year are planted in a preliminary yield trial along with standard varieties as checks. Selected plant types are critically evaluated for phenotype.
Third year to sixth year	Co-ordinated yield trials of the selected ones are conducted at several locations to confirm that selection would perform in a similar manner at different locations within an agro-climatic zone.
Seventh year	Mass selected variety is released if it is up to the expectation. Seeds are multiplied for distribution.

Methods of mass selection

There are different methods of mass selection

Hallets Mass Selection (1869)

- In this case, the plants are raised in the most optimum environment and best performing plants from that population are selected.
- These selected plants are bulked and advanced for the next generation.

Rimpai's (1867)

- Here the plants are grown either in the normal conditions or in stress conditions or in an unfavourable environment and from these stress conditions plants with desirable economic characters are selected and bulked for the next generation.

Fanned method

- The seeds are blown with wind and the heavy seeds are collected and raised in the next generation.

Application of mass selection

- In case of self-pollinated crops, mass selection has two major applications:
  - Improvement of desi or local varieties, and
  - Purification of the existing pureline varieties

Improvement of Local Varieties

- Mass selection is useful for the improvement of land, desi or local varieties of self-pollinated crops.
- The local varieties are mixtures of several genotypes, which may differ in flowering or maturity time, disease resistance, plant height, etc.
- Many of these plant types would be inferior and low yielding.
- As a result, they would lower the performance of the local variety.
- Hence the elimination of poor plant types through mass selection would improve the performance and uniformity of the variety.
- The local varieties have been under cultivation for a long time.
- Therefore, they would generally be well adapted to the local environment and would be relatively stable in their performance.
- Mass selection would improve the local variety without adversely affecting its adaptability and stability because the new variety would be made up of most of the superior plant types present in the original local variety.

Purification of Existing Pureline Varieties

- Purelines tend to become variable with time due to mechanical mixtures, natural hybridization and mutation.
- It is, therefore, necessary that the purity of pureline varieties be maintained through regular mass selection.
- At present, mass selection is used precisely for this purpose, and the nucleus seed of pureline varieties is produced through mass selection.
- Thus, as a breeding method, mass selection has only a limited application for the improvement of self-pollinated crops.
- It is generally not used for the handling of segregating populations derived from hybridization.

- But it has been used for the selection of desirable traits in segregating populations, e.g., for seed size in oats.
- But this application is of limited scope where mass selection for the concerned character is rather easy and avoids extensive data collection necessary for single plant selection, e.g., number of days to mature, seed size, etc.
- However, in breeding of cross-pollinated species, mass selection has been very important.
- In such crops, inbreeding must be avoided since it leads to a loss in vigour and yield.
- In case of mass selection, several plants are selected and their seeds are bulked together to raise the next generation; therefore, inbreeding is avoided or kept to a minimum.
- Further, because of the heterozygous nature of the population, several cycles of mass selection may effectively be practised

Advantages and disadvantages of Mass Selection

Advantages of Mass Selection

- Varieties developed by mass selection have wide adaptability, a wide genetic base and are stable.
- Rapid, simple and least expensive of plant breeding procedures.
- Retains considerable genetic variability
- Of use for the improvement of land races.

Disadvantages of Mass Selection

- Variation exists in the varieties developed by mass selection. Phenotype of these varieties is not as uniform as pureline varieties.
- To test the homozygosity of plants, progeny test is always required, as without progeny testing, selected types will show segregation in the next generation.
- Less improvement as compared to pure line selection.
- Limited use in improvement of self-fertilizing crops.
- Varieties are difficult to identify in seed certification programmes compared to pure lines.
- Only effective for characters with high heritability.
- Only those varieties which show genetic variation can be improved by this method, as it uses variability present in population.

Heterosis

- Heterosis is the superiority of F_1 (first generation offspring from cross) in one or more characters over mid-parental or even better parental value.

Hybrid vigour

- Another term used for heterosis is hybrid vigour.
- Shull (1914) used this term for the first time. Whadely (1994) opined that the two terms, namely, heterosis and hybrid vigour, are erroneously used synonyms of each other.
- According to him heterosis refers to the developmental stimulation while hybrid vigour denotes the manifest effect of heterosis.
- In short 'heterosis' is the 'influence 'while hybrid vigour is its effect.

Theories of heterosis

- The term 'heterosis' was first used by Shull in 1914.
- Heterosis is often called 'true heterosis' or 'euheterosis'.
- Heterosis may be defined as the superiority of an F_1 hybrid over either its parents in terms of yield or some other character.
- Heterosis or euheterosis is classified as mutational and balanced heterosis.
- Mutational heterosis results from dominant gene action, i.e., masking of the deleterious effects of recessive mutant alleles by their dominant counterparts.
- Balanced heterosis, on the other hand, arises due to over-dominant gene action, i.e. superiority of the heterogygote over both the corresponding homozygotes.

Causes–Basis of Heterosis

Various basis or causes of heterosis can be listed as follows:

- Genetic basis
- Dominance hypothesis
- Overdominance hypothesis
- Epistasis
- Physiological basis
- Cytoplasmic basis
- Biochemical basis.

Dominance hypothesis

- The dominance hypothesis was proposed by Charles Davenport (1908), and is the most widely accepted hypothesis among the other explanations for heterosis.
- As per this hypothesis, heterosis results due to superiority of dominant alleles over the deleterious recessive alleles by masking their effect and heterosis is in direct proportion with the number of dominant genes contributed by the parent.

Overdominance hypothesis

- The overdominance hypothesis was independently developed by Edward M. East (1908) and George Shull (1908).

- As per this hypothesis, heterosis results due to superiority of the heterozygote over its both homozygous parents, due to complementation between divergent alleles. Here, heterosis is in proportion with heterozygosis

Utilization of heterosis

- Wehner and Jenkins (1985) measured a natural outcrossing rate of 53% (36% between row).Some have reported heterosis from specific hybrid combinations (Ghaderi and Lower, 1979), especially when crossing diverse parents (Hayes & Jones, 1916).However, others have found no heterosis when parents were of similar type (Hayes & Jones, 1916). In a study involving three years, two replications and eight locations in the USA (Wehner et al., unpublished data), the total once-over harvest yield of the popular gynoecious hybrid 'Calypso' was 60 fruits per plot, while the gynoecious inbred parent Gy 14 had 61 fruits per plot.Where breeders have worked to develop elite hybrids, heterosis may average only a 5% advantage.
- In cucumber gherkins and melons, hybrids are produced commercially using a gynoecious inbred as the female parent crossed with a monoecious inbred as the male parent, and honeybees as pollen vectors. Gynoecious inbreds can be developed by self-pollination of plants that have been treated with silver nitrate or other ethylene inhibitors (Tolla & Peterson, 1979).One advantage of producing F_1 with gynoecious female parent by crossing with monoecious male parentis that the resulting hybrid will be gynoecious, having pistillate flowers at every node. If the F_1 hybrid is not parthenocarpic for commercial crop, a monoecious pollenizer must be mixed with the hybrid. Gynoecious cultivars are earlier and sometimes give higher yield than monoecious cultivars since they have pistillate flowers at every node (Wehner & Miller, 1985).
- Melon has no inbreeding depression, but some crosses were reported to express 3% heterosis for earliness and 8% for yield (Lippert & Legg, 1972). Andromonoecious lines had only 20 to 35% natural outcrossing (Whitaker & Bohn, 1952), so the lack of inbreeding depression might be explained in the same way as for cucumber. Hybrids are produced by hand pollination of emasculated perfect flowers on the female parent using staminate flowers from the male parent. Gynoecious inbreds have been produced, but fruits from pistillate flowers are oval rather than round like those produced from perfect flowers. Also, gynoecy is controlled by several genes and is complex to work with. The use of cytoplasmic genic male sterility would reduce hand labour requirements for crossing in hybrid seed production blocks.
- Summer squash (*Cucurbita pepo* L.) has been shown to have heterosis for yield, with 11 to 84% hybrid advantage over open-pollinated cultivars in zucchini types and 0 to 82% advantage for yellow types (Elmstrom, 1978). Hybrids are produced by crossing two monoecious inbred lines using honeybees as the pollen vector. One of the inbreds is made gynoecious for the first two to three weeks of flowering by spraying the plants with ethephon at the two- and four-leaf stage. Hybrids are used in 52% of U.S. crop. On the other hand, winter squash (*Cucurbita maxima* Duch and *C. moschata* Duch) is difficult to cross using monoecious inbreds since ethephon does not work well to change sex expression. Heterosis for yield in hybrid cultivars

may be 40% higher than for the open pollinated ones, and an increasing proportion of the crop is planted to hybrids.

- Watermelon is a monoecious, annual outcrossing species with little inbreeding depression. Heterosis is expressed for yield in some parental combinations, perhaps averaging 10%; however, the main use of hybrids has been to protect the parental inbreds and to produce seedless fruits. Seedless hybrids are triploids produced by crossing a tetraploid female parent with a diploid male parent. Tetraploids are produced by doubling the chromosome number of elite diploid inbred lines. Once the triploid F1s are planted for commercial production of water melon fruits, a diploid cultivar must be planted in alternating rows with such triploid cultivar to provide the pollen required for fruit set, so seeded watermelons are produced along with the seedless ones.
- The benefits provided to the world by heterosis in the vegetable crops is dependent on the crop. In the case of bean, pea and lettuce, there has been no effect of heterosis. In other cases, such as tomato, onion, cabbage and asparagus, there have been significant savings of agricultural land and a large increase in our ability to feed people as a result of heterosis.

Hybridization

- Hybridization refers to the crossing of two individuals or plants or lines with dissimilar genotype. It is the most important among the methods of plant breeding.
- The outcome of crossing of two individuals or plants or lines with dissimilar genotype is called as hybrid.

History of hybridization

- In comparison to selection, hybridization is of more recent origin.
- There is evidence that Babylonians and Assyrians hand-pollinated date palm as early as 700 B.C. for metaxenic effects of pollen.
- Metaxenia is the effect of pollen on the maternal tissues of fruit. Clearly, the artificial pollination in this case was not for the purposes of crop improvement.
- Sex in plants was discovered by Camararious in 1694. In 1717, Thomas Fairchild produced the first artificial hybrid, the Fairchild's mule, by crossing sweet William (*Dianthus barbatus*) with carnation (*Dianthus caryophyllus*).
- Subsequently, a large number of scientists used hybridization for scientific studies as well as for crop improvement.
- Notable among such scientists are Joseph Kolereuter, who made many crosses in tobacco during 1760-1766, and emphasized hybrid vigour in F1. Thomas Andew Knight, who developed several varieties of apples, pears, peaches, grapes and currants during 1759-1835; and Goss, Sargaret, Gaertner and Naudin, who had noted uniformity in F1, dominance in F1, and segregation and appearance of parental types in F2.
- But it was left for Mendel (1865) to propose the clear-cut laws of inheritance of qualitative traits. These and subsequent discoveries in genetics have given hybridization a scientific basis.

- Towards the end of the 19th century, hybridization was widely used for crop improvement.
- With the help of genetic principles, the breeder is able to predict the kind of progeny he is likely to obtain from a given cross.
- Therefore, he can plan the crosses from which he is most likely to obtain the desired plant type.
- Today, hybridization is the most common method of crop improvement, and the vast majority of crop varieties have resulted from hybridization.

Objectives or Aims of Hybridization

- To create genetic variability.
- To bring together desired qualitative characters found in different plants or plant lines into one plant or plant line i.e. to transfer desired character/s from other varieties to the considered one.
- To make F1 useful as a hybrid variety.
- Improvement of one or more quantitative characters

Types of Hybridization – Hybrids

Plant hybridization can be of two types:

- Intervarietal hybridization
- Distant hybridization

Intervarietal Hybridization

In this type of hybridization, two parents from the same species (may be from two different varieties, strains or races of the same species) are crossed.

- Single cross hybrids
- Double cross hybrids
- Three-way cross hybrids
- Triple cross hybrids
- Top cross hybrids

Single Cross Hybrids

- Results from the cross between two pure bred lines and produces an F1 generation called an F1 hybrid (F1 is short for Filial meaning "first offspring").
- The cross between two different homozygous lines produces an F1 hybrid that is heterozygous; having two alleles, one contributed by each parent and typically one is dominant and the other recessive.
- The F1 generation is also homogeneous, producing offspring that are all similar to each other.

Double Cross Hybrids

- Result from the cross between two different F1 hybrids.

Three-way Cross Hybrids

- Result from the cross between one parent that is an F1 hybrid and the other is from an inbred line.

Triple Cross Hybrids

- Result from the crossing of two different three-way cross hybrids.

Top Cross Hybrids

- Cross between inbred line and OPV.

Distant Hybridization (Population hybrids)

- It includes
- Interspecific hybridization – crossing between parents from two different species

Hybridization Procedure

Hybrid production involves following main steps (i.e. steps in crossing).:

1. Choice of parents
2. Evaluation or testing of parents
3. Protection of male parent
4. Emasculation
5. Bagging
6. Tagging
7. Pollination and bagging, tagging
8. Aftercare
9. Harvesting.

Production of F1 hybrids

- Crossing two genetically different plants produces a hybrid seed (plant) by means of controlled pollination.
- To produce consistent F1 hybrids, the original cross must be repeated each season.
- As in the original cross, in plants this is usually done through controlled hand-pollination, and explains why F1 seeds can often be expensive.
- F1 hybrids can also occur naturally, a prime example being peppermint, which is not a species evolved by cladogenesisor gradual change from a single ancestor, but a sterile stereotyped hybrid of watermint and spearmint.
- Unable to produce seeds, it propagates through the vining spread of its own root system.
- In agronomy, the term "F1 hybrid" is usually reserved for agricultural cultivars derived from two different parent cultivars, each of which are inbred for a number of generations to the extent that they are almost homozygous.

- The divergence between the parent lines promotes improved growth and yield characteristics through the phenomenon of heterosis ("hybridvigour"), whilst the homozygosity of the parent lines ensures a phenotypically uniform F1 generation.
- Each year, for example, specific tomato "hybrids" are specifically recreated by crossing the two cultivars over again.
- Two populations of breeding stock with desired characteristics are subject to inbreeding until the homozygosity of the population exceeds a certain level, usually 90% or more. Typically this requires more than ten generations.
- After this happens, both populations must be crossed while avoiding self-fertilization.
- Normally this happens in plants by deactivating or removing male flowers from one population, taking advantage of time differences between male and female flowering or hand-pollinating.
- In 1960, 99% of all corns planted in the United States, 95% of sugar beet, 80 percent of spinach, 80% of sunflowers, 62% of broccoli, and 60% of onions were hybrid.
- Such figures are probably higher today. Beans and peas are not commercially hybridized because they are automatic pollinators, and hand-pollination is prohibitively expensive.

Hybrids

- Hybrid plants are the result of the deliberate cross-pollination of two selected lines of plants.
- Hybrids are commonly produced to make seeds that combine the best characteristics of the parents--productivity, uniformity, vigour, pest resistance.
- Seeds collected from hybrids usually do not grow true to their hybrid parents; rather they often revert and display characteristics from more distant progenitors.

F1 hybrid seed

- Many selectively bred vegetable varieties are F1 hybrids (F1, the frist filial generation).
- F1 hybrids are made by crossing two distinctively different parent lines or parent cultivars to produce an offspring that is different from the parental types but with specific desirable characteristics from either or both parents.
- F1 hybrids are created for exceptional vigour and improved growth, quality, uniformity, and yield.
- Seed from F1 hybrids will not breed true-to-type. F1 hybrids must be remade by controlled pollination every time seed is needed. (That is why F1 hybrids are expensive).
- This is a time consuming and exacting breeding process not usually undertaken by home vegetable gardeners.

Difficulties in hybridization

- Ordinarily intervarietal hybridization presents little problem, if any, while distant hybridization is often beset by many difficulties.
- In many crops, however, some varieties, particularly when used as the female parent, show a considerably lower seed set than other varieties.
- This problem is not a serious one, and can be easily resolved by pollinating a proportionately larger number of flower buds.
- A more serious problem of hybrid (F1) mortality due to lethal genes is encountered in several cross-combinations of some crops.
- An excellent example is furnished by wheat (*Triticum aestivum*) where F1 plants from many cross-combinations show hybrid necrosis.
- The affected areas of leaves first turn a darker green, then greyish brown as the chlorophyll degenerates and finally the leaf tissue dies.
- Hybrid necrosis in wheat is believed to be the result of two dominant complementary genes Ne_1 and Ne_2 located on the chromosome 5BL and 2BS, respectively. According to an estimate, about 18% of wheat varieties possess Ne_1, while more than 43% have $Ne_{2.}$
- There is considerable variation in the severity of hybrid necrosis in the different cross combinations of wheat.
- In severe cases, necrosis may develop in the first or second leaf stage and the plants may die by the 3-6 leaf stag. In mild cases of necrosis, only leaf tips may be affected and the plants may grow and reproduce normally.
- All the grades between these two extremes are known to occur and a 0-8 scale has often been used for scoring the intensity of necrosis is presumably due to multiple alleles of both Ne_1(Ne_1^w, Ne_1^m and Ne_1^S) and Ne_2^w, $Ne_2^{m,}$ Ne_2^{ms} and Ne_2^S) genes, which generate weak (w), moderate (m), moderately strong (ms) or strong (s) necrotic effects.

  The problem of hybrid necrosis (or other lethal genes) may be overcome by

  1. Keeping necrosis crosses out of the breeding programme;
  2. Using one parent having recessive alleles of both the Ne genes;
  3. Inducing mutation for the recessive allele of the concerned Ne gene, while allele of the concerned Ne gene, where available;
  4. Utilizing spontaneous mutants having the recessive allele of the concerned gene, such as praline;
  5. Using a chemical treatment, e.g. spray of an amino acid, where available;
  6. Growing the F1 plants in an environment which will prevent the expression of the lethal gene(s), e.g. temperatures over 22°C in the case of Ne genes of wheat.

Consequences of hybridization

- In self-pollinated species, it is the easier self –pollination in F1 is easier to produce, F2 generation. Segregation and recombination of genes would produce several new genotypes, in addition to the two parental types, in F2.

- The number of genotypes produced in F2 increases rapidly with the number of segregating genes.
- The number of phenotypes with and without dominance also increases rapidly with an increase in F2 and increases rapidly with the number increase in the number of segregating genes.
- If, for example 10 genes are segregating, the F1 would produce 1,024 types of gametes, 59,049 different genotypes in F2, 1,024 different phenotypes with full dominance, and 59,049 different phenotypes without full dominance and epistasis.
- The smallest size of a perfect F2 population (a population that includes at least one plant of every genotype) would be 1,084,576.
- Most of the intervaietal crosses would differ for many more genes.
- It is, therefore, impractical to raise a perfect F2 population for any cross. Thus the difficulty in recovery of rare recombinants from segregating generations can be easily appreciated.
- This difficulty would be great enough even if the phenotypic effects of the genes were easily recognizable.
- However, in the case of quantitative characters the effects of genes are masked by those of the environment as well as by gene interactions making the identification of rare recombination even more difficult.
- In F2, vast majority of plants would be heterozygous for one or more genes.
- Hetrozygosity reduce the effectiveness of selection because such plants segregate to produce variable progeny, i.e., the genetic value of the progeny is not the same as that of the parent plants.
- The genetic composition i.e., the frequency of plants heterozygous for different numbers of genes, in a segregating generation is obtained by expanding and solving the terms of the following binomial $[1 + (2m - 1)]$

  Where, m is the number generation of selfing and n is the number of genes that are segregating (independently). If we assume the values of m and n to be 3 and 4, respectively, the appropriate binomial and its expansion will be as follows:

  [1 + (23 – 1)] 4 = (1 + (8 – 1)]4 = (1 + 7)] 4 = 14 + 4 (1)3 (7) + 4(1) (7) 3(7) + 3
- In the expanded binominal, each term has a coefficient, e.g., 1,4,6,4 and 1 in the above binominal, the factor (1) having indices from 4 to 0 and the factor (7) (= $2m - 1$) having indices from 0 to 4. In each term, the index of (1) denotes the number of heterozygous genes, while that of (7) i.e., ($2m - 1$) signifies the number of homozygous genes.
- Further, the solution of a term gives the frequency of plants having the genetic constitution described by the term.
- In the above binominal, solution of the first term will give the frequency of plants having 4 heterozygous and zero homozygous genes, that of second term will give the frequency of those having 3 heterozygous and 1 homozygous genes, of third term will be for plants having 3 heterozygous 2 homozygous genes, of fourth term will be for planting while that of fifth term will be for plants having all the four genes in homozygous state. The solution of the terms of the above binomial gives the following values: = 1 + 28 + 294 + 1,372 + 2,401

- These value are interpreted as follows to give the frequencies of plants having different genetic constitutions.
- 4 heterozygous and 0 homozygous = 1 plant.
- 3 heterozygous and 1 homozygous = 28 plants
- 2 heterozygous and 2 homozygous = 294 plants

Table 7.7: *Frequency (in per cent) of completely homozygous plants*

Number of genes segregating					
	F2	**F3**	**F4**	**F5**	**F7**
1	50.0*	75.0	87.5	93.8	98.4
2	25.0	56.3	76.6	87.9	96.9
3	12.05	42.2	67.0	82.4	95.4
4	6.3	31.7	58.6	77.2	93.9
5	3.1	23.7	51.3	72.4	92.4
10	0.0097	5.6	26.3	52.5	85.4
20	0.00009	0.32	6.9	27.5	73.0

Heterozygous and 3 homozygous = 1,372 plants 0 heterozygous and 4 homozygous = 2,401 plants.

- Thus in F4 generation of a cross segregating for 4 genes, the frequencies of plants segregating for different numbers of genes will be as given above.
- Frequency of completely homozygous plants in F2 and the subsequent generations of crosses segregating for different number of genes.
- The number of generations of self-pollination will be 1 in F2, 2 in F4 and $n - 1$ in Fn generation since F1 is the result of hybridization and only F2 and the subsequent generations are due to selfing.
- The frequency of completely homozygous plants = $[(2m - 1) / 2m)]n$ where, m is the number of generations of self- pollination increases, the frequency of completely homozygous plants increases sharply.
- Consequently, the confusing effects of heterozygote are reduced and selection becomes more effective in the advanced selfed generations.
- Therefore, many breeders feel that selection on individual plant basis should be delayed till F5 or even F6 generation.
- These considerations have assumed equal survival of all the genotypes, and no linkage.
- There is some evidence that heterozygotes may be favoured in nature.
- The increase in homozygosity, therefore, could be less than theoretically expected. Linkage would decrease the frequency of recombinant types and increase the frequency of completely homozygous plants.
- The latter effect is produced because linkage reduces the number of genes segregating independently.

- Linkage between favourable genes makes selection easier because the two desirable characters tend to be inherited together populations would be required to break this linkage.
- In case of quantitative characters, recombinations may produce genotypes superior to the two parents.
- For example, if a variety with the genotype AA bb cc dd is crossed with another variety with the genotype AA BB CC DD and aa bb cc dd would be produced.
- These plants would be superior and inferior, respectively, to both the parents; such recombinants are known as transgressive segregants. Breeding for yield generally aims at the recovery of transgressive segregants.
- Frequency of such segregants in F2, which 2 out of $4n$ plants (where n is the number of genes segregating), it would be apparent that the chances of recovery of such genes from each parent, but rarely a plant in a small F2 population would have all the genes from both the parents.

Pedigree method

- Pedigree may be defined as a description of the ancestors of an individual and it generally goes back to some distant ancestor or ancestors in the past. The pedigree is helpful in finding out if two individuals are related by descent, i.e., have a common parent in their ancestry and therefore are likely to have some alleles in common.

Maintenance of pedigree record

- Pedigree record may be kept in several different ways, but it should be simple and accurate. Generally, each cross is given a number; the first two digits of this number refer to the year in which the cross was made, and the remaining digits denote the serial number of the cross in that year. For example, the number 7911 denotes the cross number 11 of year 1979. In the segregating generations, one of the following two systems of designation may be followed:

1. Designation based on the location of progeny rows in the field

- In the first system, individual plant progenies in each generation are assigned row numbers corresponding to their location in the plot. In addition, each progeny in F4 and the subsequent generations is assigned the row number of the progeny in the previous generation from which it was derived.
- Thus each F3 progeny, derived from individual F2 plants, is given a number corresponding to the row number at which it is located in the F3 plot.
- Plants selected from an F3 progeny are identified by the row number of that progeny. When individual plant progenies are grown in F4, each progeny is also given the row number of the F4 plot at which it is located in addition to the F3 row number.
- Similarly, the plants selected from a progeny in F4 are given the row number of that progeny in the F4 plot.
- After the F5 progenies are planted, the row numbers of progenies in the F5 plot are added to the F4 row numbers. The same procedure is followed in the subsequent generations.

- Thus each progeny can be traced back to the F_3 progeny (or the F_2 plant) from which it originated. But for determining the pedigree of a progeny, the breeder has to consult the records of the previous years as well.

2. Designation based on serial numbers of selected plants

- In the second system, in each generation, the selected plants are assigned serial numbers within individual progenies; each progeny or selected plant bears the serial numbers of all the plants in the previous generations related to it by direct descent.
- Thus each selected plant F2 is given a serial number; the F3 progenies derived from these plants are given the serial numbers of their parent F2 plants.
- The plants selected from a progeny in F3 are given the number of that progeny; in addition, each selected plant from that progeny is also given a serial number.

Table 7.8: *The system of designating individual plant progenies based on progeny row numbers of the concerned and the previous generations*

Generation	Progeny number	Description
F2	7911-43	Progeny 11 No. 43 (Individual plant of whole F2 population of cross 11 made in the year 1979
F3	7911-7	Progeny in the 7th row in the F3 plot.
F4	7911-7-4	Progeny in the 4th row in the F4 plot, selected from the progeny in the 7th row of the F3 plot.
F5	7911-4-14	Progeny in the 14th row in the F5 plot, selected from the progeny in the 4th row in the F4 plot.
F6	7911-14-3	Progeny in the 3rd row in the F6 plot, selected from the progeny in the 14th row in the F5 plot.

These two numbers together make up the progeny numbers in the F4 generation. Similarly, the plants selected from an F4 progeny are given the F4progeny number, i.e. the serial numbers of the F2 and the F3 plants from which the concerned F4 progeny was derived and a new number showing the serial number of the plant among those selected from the F4 progeny.

- In this system, the pedigree of a progeny is immediately known, and one does not have to refer to the previous year's records. But there is a greater chance of error in this system since more numbers are to be recorded. The breeder may, however, devise his own system of record keeping.

Table 7.9: *The system of designating individual plant progenies based on the serial numbers assigned to the selected plants*

Generation	Progeny number	Description
F3	7911-7	Progeny obtained from plant number 7 selected in F2
F4	7911-7-4	Progeny from plant number 4 selected from the F3 progeny derived from the plant number 7 selected in F2
F5	7911-7-4-2	Progeny form plant number 2, selected from the F4 progeny derived from plant number 4 selected from the F3 progeny obtained from the plant number 7 selected in F2
F6	7911-7-4-2-8	Progeny from plant number 8, selected from the F3 progeny derived from the plant number 2 selected form the F4progeny of the plant number 4 selected from F3 progeny of the plant number 7 selected in F2.

In both the systems, the progenies are assigned a different serial number when they become homozygous and are included in preliminary yield trials. This serial number is given only to those homozygous lines that are included in preliminary yield trials.

- In the pedigree record, a brief note is made about the distinguishing features of each progeny. In keeping a pedigree record, the following should be kept in mind:
  1. Only important characteristics should be recorded. If a large number of characters are noted, the record keeping would become a great burden.
  2. Only promising progenies should be included in the record. Post-progenies may be simply marked 'discard'. This would give more time for studying promising progenies.
  3. The pedigree record must be accurate. Keeping no record is better than keeping an inaccurate record, which will only create confusion.
- The pedigree record is often useful in the elimination of some progenies. In late generations, e.g. F5, F6 etc., if some lines originated from the same F4 or F5 progeny and are similar, only one of these lines need be maintained. Further, it may often be possible to obtain a general idea about the inheritance of characters by studying the pedigree record.

Applications of Pedigree method

- The pedigree method is the most commonly used method for selection from segregating generations of crosses in self-pollinated crops.
- This method is often used to correct some specific weaknesses of an established variety.
- Pedigree method is also useful in the selection of new superior recombinant types.
- It is almost hoped that some transgressive segregants would be recovered.
- Thus the method is suitable for improving specific characteristics, such as disease resistance, plant height, yield, quality and maturity time etc.

Procedure for Pedigree selection

- Development of hybrids derivatives through pedigree method of selection

Hybridization

- The selection of parents to be used in a cross is the most important step in a breeding programme based on hybridization.
- The selected parents are crossed to produce a simple or a complex cross.
- For convenience, we refer to the seed obtained from both simple and complex crosses as the F1 seed.

F1 Generation

- F1 seeds are space-planted so that each F1 plant produces the maximum number of F2 seed.
- Ordinarily, 15-30 F1 plants should produce enough seed for a good F2 population.

F2 Generation

- In F2, 2,000-10,000 plants are space-planted to facilitate selection.
- About 100-500 plants are selected and their seeds are harvested separately.
- Usually, the F2 population size should be 10-100 time the number of F2 plants that are to be selected.
- The number of F3 progenies may vary from 100-500 depending upon the facilities available and the objectives of the breeding programme.
- When closely related varieties are crossed, the number of F3 progenies maintained would be considerably smaller than when the parent varieties are unrelated by descent.
- When the objective of breeding is to improve yield, to recover transgressive segregants, or both, a relatively larger number of F2 plants would be selected.

F3 Generation

- Individual plant progenies are space-planted. Each progeny should have about 30 or more plants.
- Individual plants with desirable characteristics are selected particularly from superior progenies.
- Outstanding plants from inferior progenies may also be selected.
- Disease and lodging susceptible progenies and progenies with undesirable characteristics are eliminated.
- The number of plants selected in F3 should be preferably less than that of the F3 progenies.
- If the number of superior progenies is small, the whole cross may be rejected.

F4 Generation

- Individual plant progenies are space-planted. Again, desirable plants are selected mainly from superior progenies.
- The number of plants selected in F4 is generally much smaller than that of the progenies raised in F4.
- Progenies with defects and undesirable characteristics are rejected.
- If two or more progenies coming from the same F3 progeny are similar and comparable, only one of them may be saved and the others may be rejected.
- Thus the emphasis is on selection of desirable plants from superior progenies.

F5 Generation

- Individual plant progenies are generally planted according to the recommended commercial seed rate.
- Often three or more rows are grown for each progeny to facilitate comparison among progenies.
- Many families would have become reasonably homozygous and may be harvested in bulk.
- In families showing segregation, individual plants are selected.
- The breeders have to visually assess the yielding potential of progenies and reject the inferior ones.
- The number of progenies must be reduced to a size manageable in preliminary yield trials, which are usually 25-100 progenies.

F6 Generation

- Individual plant progenies are planted in multi-row plots and evaluated visually. Progenies are harvested in bulk since they would have become almost homozygous.
- Progenies showing segregation may be eliminated unless they are outstanding; in such progenies individual plants may be selected. Preliminary yield trial is usually planted in F6 for those progenies which are reasonably homozygous and have enough seeds.
- Inferior progenies are eliminated based on yield data from preliminary yield trials and visual evaluation.

F7 Generation

- Preliminary yield trials with three or more replications are conducted to identify a few superior lines.
- The progenies are evaluated for plant height, lodging, disease resistance, flowering time, maturity time, etc.; quality tests for produce, organoleptic evaluation may serve as additional criteria for selection.
- Standard commercial varieties must be included as checks for comparison. Two to five outstanding lines, if superior to the checks, would be advanced to the coordinated yield trials.

F8 to F10 Generations

- The superior lines are tested in replicated yield trials at several locations.
- The lines are evaluated for yield, disease and lodging resistance, maturity time and quality including organoleptic quality etc., line that is superior to the best commercial variety included in the trial as check in yield and other characteristics is identified for release as a new variety.

F11 Generations

- When a strain is likely to be released as a variety, the breeder usually multiples its seed during its last year in trial.
- The breeder has the responsibility to supply the breeder's seed to the National Seeds Corporation, which is responsible for the production of foundation seed.
- Thus in F11 and F12, the seed of the new variety will be multiplied for distribution to the farmers.

Merits & Demerits of Pedigree Method

Merits

- Excellent method for improvement of easily observable and high heritable characters.
- As pedigree record is maintained, information regarding inheritance pattern of characters can be obtained as and when required.
- Each plant can be traced back to its parent plant.
- Only those progeny lines which contain plants with desired characters are selected for next generation. So there is scope for plant breeder's skills.
- Progeny tests are done, thus it is based on genotypic value rather than phenotypic value alone.
- Increased breeding efficiency by early identification of superior heterogeneous populations.
- Scope for transgressive segregation to occur for the characters like yield.
- New variety development takes short period as compared to bulk method.

Demerits

- Costly.
- Labour intensive.
- Requires skilled person as selection is practised.
- Pedigree record maintenance is time consuming.
- Selection for yield or other characters in F2 and F3 is ineffective.
- One important point to note is genetic variation available for selection gets decreased in later generations due to the individual plant selection carried out earlier.

Mass Pedigree method

- This method consists of selection among individual plants and their progeny during inbreeding following crosses between selected donor parents.
- The parents are normally crossed in a single cross system.
- The F2 populations are subjected to selection for apparent desirable traits viz. days to maturity, plant type, plant height, resistance to diseases and other easily identifiable traits.
- The F3 progenies of the F2 plants are grown in rows.
- The most desirable F3 progeny rows are selected / marked followed by selection and harvesting of superior individual plants from each of the selected rows.
- This practice of selection among rows followed by selections within rows continued till F5/F6 generation in which desirable rows are bulked within themselves separately and these bulks/new breeding lines are evaluated in the replicated trial for further selection.
- In this method, best individuals with desired characters are selected on the basis of phenotypic performance in a source population. Open-pollinated seeds of the selected individual plants are divided into two halves. Second year replicated progeny row trial is conducted using one set of half seeds from each plant. On the basis of progeny performance, the best parental individuals are identified. The remnant half seeds from the superior parental plants are mixed and grown in isolation for random mating during the third year. This method is called line breeding when selection is based on progeny tests and a group of progeny lines is composited. The name, mass pedigree method was given by **S.S. Rajan.**

Merits & Demerits of Mass Pedigree Method

Merits

- This method takes 14-15 years to release a variety while bulk method takes much longer time (15-16 years).
- The breeding values of selected plants are ascertained by progeny test.
- Therefore, pedigree selection is based on genotypic value rather than phenotype valuc.
- Pedigree method provides information about the mode of inheritance of different oligogenic characters which is not possible by other breeding method.
- There is chance of recovering transgressive segregants by pedigree method.

Demerits

- The selected materials become so large that handling of the same becomes very difficult.
- Because large number of progenies is rejected in this method, there are chances of elimination of some valuable materials.
- Records have to be maintained for all the selected plants and progeny which take lot of valuable time of a breeder.

Bulk Method

- It is a method which can handle segregating generations, in which F_2 and subsequent generations are harvested in bulk to grow the next generation.
- At the end of bulking period, individual plant selection and evaluation is carried out in the similar fashion as in the pedigree method.

Procedure for Bulk Method

Bulk method of plant breeding comprises the following main steps:

Step	Details
Hybridization	Crossing among selected parents
F1 generation	F1 seeds (minimum 20) planted. Bulk harvesting is carried out.
F2 to F6 generation	F2 to F6 are planted, harvested in bulk. Number of plants should be as large as possible. Generally artificial selection is not carried out.
F7 generation	Generally 30 to 50 thousand F6 seeds are space planted. Selection is carried out based on phenotype and 1000 to 5000 selected ones are harvested separately.
F8 generation	Individual plant progenies are grown and inferior progenies eliminated. Harvested in bulk.
F9 generation	Preliminary yield trials with standard varieties as check. Selection is based on yield.
F10 to F13 generation	Multi location yield trials are conducted using standard varieties as check. Evaluation of performance of lines is done.
F14 generation	Seed multiplication for distribution.

Natural and artificial selection during bulk period

- A large number of F2 populations of several thousand plants are grown at conventional planting rates which are in contrast to the pedigree method.
- At maturity all the plants are harvested in bulk and total seed is collected out of which a random sample is taken to grow approximately in the same area in the next season, i.e. in F3 generation.
- Normally the same procedure is followed uninterrupted upto F5 or F6 generation till sufficient level of homozygosity is attained.
- During all these years, natural selection is allowed to exert its influence, if any, to eliminate agronomically undesirable genotypes. As such, there is no attempt for artificial selection or to grow the material in off-season nurseries which may lead to loss of certain genotypes only on account of abnormal environmental conditions in the off-season.
- If, however, the off-season environment is expected to either reduce fitness of agronomically undesirable traits or even leave the characters unaffected, the generation can be advanced by taking two or more crops per year.

- Artificial selection is not conducted but there is no reason to lose sight of opportunity to improve the economic worth of population especially through negative selection for qualitative defects like susceptibility to diseases, late maturity etc.
- Even Nilsson-Ehle, the founding father of this method, helped natural selection in his bulk method, by discarding plants with winter damage which would have continued to produce at least some quantity of seed for several years.
- Florell (1929) also suggested that artificial selection towards a definite type may be practised to eliminate weak and undesirable plants.
- The use of artificially created environmental conditions can be made to exert specific type of selection pressure like that of incidence of diseases and pests.
- The single plant selection may be initiated any time in between F5 - F8 generation but that generation must be grown as space planted for better expression of potential of the individual plants as well as sufficient genetic variation which is usually suppressed under influence of interplant competition in thick planting.
- The plants are selected and evaluated in next generation like those in the pedigree method.
- The seed of individual plant is not sufficient for replicated yield trials so the growing of progeny rows serves the twin purpose of seed increase and also to reject some potentially less productive progenies by visual observation.
- Hence after the procedure for yield testing, multiplication and release of selections as new cultivars is essentially the same as in pedigree method.

Applications of Bulk Method

- Homozygous lines can be isolated in less time, automatic increase in homozygosity up to F6 or F8 generation, after this individual plant selection can be carried out.
- Natural selection may improve considered character (yield), with minimum expenses.
- Suitable for crops which are generally planted at high planting densities, e.g. small grain crops.

Modification of the Bulk Method

- A modification of the bulk method based on artificial selection in F2 and the subsequent generations is outlines below.
- **F2 and F3 Generations.** F2 and F3 generations are space-planted and a large number (1,000-5,000) of desirable plants are harvested separately.
- **F4 Generation.** F4 is space-planted and 1,000 to 5,000 desirable plants are selected. Seeds from the selected plants are harvested separately.
- **F5 Generation**. Individual plant progenies are grown. Selection among progenies is based on plant height, disease resistance, lodging resistance, maturity date, fruit weight, fruit shape, number of fruits and other agronomic characteristics. Undesirable and inferior progenies are eliminated. Often only 10-30% of the progenies may be saved. Seeds from each of the selected progenies are harvested in bulk.

- **F6 Generation.** A preliminary yield trial is planted using the bulk seeds from the selected individual plant progenies. Observations are recorded on agronomic characteristics and yield. Quality tests may be done on superior progenies.
- **F7 Generation**. Superior progenies selected on the basis of yield trial are space-planted for further purification. Individual plants are selected from the superior progenies and their seed is harvested separately.
- **F8 Generation.** Individual plant progenies are grown. Inferior progenies and progenies showing segregation are ordinarily rejected. Each selected progeny is harvested in bulk.
- **F9 Generation.** Preliminary yield trial is conducted to identify few superior progenies for detailed yield tests. Quality test is done to eliminate undesirable progenies.
- **F10 – F13 Generations.** Replicated yield trials are conducted at several locations with standard varieties as checks. The line or lines that are found to be superior to the standard checks would be released as new varieties.
- **F13 Generation.** Seed of the newly released variety is multiplied for distribution among the farmers.
- This modification of the bulk method provides an opportunity for the breeder to exercise his skill and judgment in selection of plant types in the early generations (F2 to F4).
- At the same time, it does not involve laborious record keeping as in the case with the pedigree method. Therefore, this modification has the appeal that it provides for selection in the early segregating generations and yet it involves less time and labour than the pedigree method.

Advantages and Disadvantages of Bulk Method

Advantages

- Simple and inexpensive.
- Requires little record keeping.
- Easy to handle populations as bulk harvest is done.
- Natural selection is effective, possible to carry out artificial selection if required.
- Selection for some environment dependent characteristics like disease resistance, cold and high temperature tolerance can be carried out when environment is favourable for disease epidemic or not conducive for normal growth and fruit set due to very high and very low temperatures. Till that year, the bulk can be maintained easily.

Disadvantages

- Requires more time duration to develop a new variety.
- Pedigree record is not maintained, so we cannot trace back progeny to the parent plant.
- Not a suitable method for greenhouses.
- Large number of progenies has to be selected at the end of the bulking period.
- Natural selection may also work against desirable traits.

- Not much scope for skills.
- Maximum productivity is established in F2 generation of method.
- No recombination occurrence among superior lines.

Single Seed Descent (SSD)

- This method is a modification of the bulk method.
- First proposed by Goulden (1939) and later modified by Brim (1966).
- Instead of bulking whole seed lot of selected plants, a single seed is selected randomly from each selected plant to make bulk.
- This method involves less record keeping (pedigree method) and works better where the main focus is on improvement of quantitative traits or characters such as yield and earliness, rather than qualitative traits or characters such as flesh colour and disease resistance.

Important features

- No special efforts for artificial selections or even there is no possibility for natural selection till F5 or F6 when the population is reasonably homozygous.
- Raising of F3 and later generations from a bulk of one seed from each F2 and subsequent generation plant in order to ensure that each F2 plant is represented equally in the end population.
- As a result of the speed and economy, the single seed descent scheme is becoming increasingly popular with breeders.

Procedure for Single Seed Descent Method

Single Seed Descent (SSD) method includes following stages as described below:

Step	Details
Hybridization	Crossing of selected parents
F1 generation	F1 seeds grown and bulk harvested
F2 generation	F2 seeds grown. One seed from each plant is selected randomly and mixed
F3 generation	F3 seeds are grown and harvested as above
F4 and F5 generation	The similar procedure as above is carried out
F6 generation	F6 seeds are planted. Selection for superior plants is conducted and selected ones are harvested separately. Number of plants could range from150 to 500
F7 generation	The main point in this step is individual plant progenies are grown and selected progenies are harvested in bulk
F8 generation	Preliminary yield trials and quality tests are conducted
F9 to F10 or F13 generation	Co-ordinated yield trials and tests for resistance and quality are conducted
F11 of F14 generation	Seed multiplication for distribution

Advantages and Disadvantages of SSD Method

Advantages

- Requires little space
- Saves time. Extensive field trials are not required
- Little record keeping
- Fast especially with multiple generations / year
- Easy method to handle populations, handles large number of crosses
- Good for traits with low heritability
- Natural selection has little influence

Disadvantages

- Little scope for natural selection
- Maximum productivity is established in F_2 generations, superior plants may be lost
- No recombination occurrence among selected lines Must evaluate large numbers of inbred lines to find out superior ones.

Backcross Method

- Backcross method is the one whose main objective is to incorporate desired gene, either dominant or recessive, to the highly productive, commercially successful variety which lacks that specific gene.

What is Backcross Breeding?

- Backcrossing has been used by plant breeders for decades to transfer specific character into elite lines.
- The variety which receives gene is the 'recipient parent' and variety which is the source of gene is called as 'donor parent'.
- Recipient parent is used repeatedly hence also named as 'recurrent parent'.
- Donor parent is used only once and hence known as 'non-recurrent parent'.
- Repeated backcrossing leads to homozygosity at the same rate as in case of selfing.

Use of Backcross Breeding Method

- Backcross to transfer dominant gene

Requirements of a Backcross Programme

For the successful development of a new variety through the backcross method, the following requirements must be fulfilled:

- A suitable recurrent parent must be available, which lacks one or two characteristics.
- A suitable donor parent must be available.
- The donor parent should have the character to be transferred in a highly intense form.

- The character to be transferred must have high heritability.
- Preferably, it should be determined by one or a few genes.
- A sufficient number of backcrosses should be made so that the genotype of the recurrent parent is recovered in full.
- Ordinarily, 6-7 backcrosses are sufficient for the purpose. In practice, 4-6 backcrosses are usually made.

Selection of Parents

- It is clear that the backcross method of breeding changes the genotype of the recurrent parent only for the gene under transfer.
- That is, only that defect of the recurrent parent, which is aimed to be corrected is removed through the backcross method.
- But some unexpected changes in one or more characters may also occur due to genes tightly linked with the gene or genes being transferred.
- Therefore, the recurrent parent must be the most popular variety of the area, which has high yielding ability, desirable quality and high adaptability. In each crop, one or two varieties dominate, and they are very popular with the cultivators.
- Such a variety may have one or two defects, e.g., susceptibility to disease or undesirable seed size or colour.
- These defects may be removed through the backcross method.
- The non-recurrent parent is selected for the character that is to be incorporated in to the recurrent parent.
- The yielding ability and other characteristics of the non-recurrent parent are not important.
- The only important factor in its choice is that it must have the character to be improved in an intense form.
- The intensity of the character should preferably be more than that desired in the recurrent parent because some intensity of the character is generally lost during the transfer and in the new genetic background of the recurrent parent.

Backcross-Dominant gene transfer

How Dominant Gene is transferred through Backcross Breeding?

- Let the variety *X* is well adapted and high yielding.
- Variety *Y* is resistant to a specific disease, which is governed by a dominant gene. We wish to transfer this dominant gene from variety *Y* to variety *X*.
- So, variety *X* – recurrent parent ( ) Variety *Y* – donor parent ( )

The various steps in backcrossing are as follows:

- **Hybridization:** The two varieties *X* and *Y* are crossed. Generally recipient / recurrent variety is used as female parent.
- **F_1 generation:** Selection for disease resistance is not performed. Plants from F_1 seed are backcrossed with recipient variety *X* and seeds are collected to raise BC_1 generation.

- **First backcross generation (BC_1):** Selection is done for disease resistance and selected ones are back crossed with recurrent parent X.
- **Second to fifth backcross generation (BC_2 to BC_5):** Segregation for disease resistance occurs in every back cross generation. Plants are selected on the basis of similarity with recurrent parent and resistance to disease. Selected ones are back crossed with recurrent parent X.
- **Sixth backcross generation (BC_6):** Disease resistant plants are selected. They are self pollinated and harvested separately.
- **BC_6 F_2 generation:** Individual plant progenies are grown from seeds of BC6 generation. Plants are selected on the basis of similarity with recurrent parent and at the same time resistance to disease. They are harvested separately.
- **BC_6 F_3 generation:** Individual plant progenies are grown from seeds of above cross. As done in above step, plants are selected on the basis of similarity with recurrent parent and resistance to disease but harvested in bulk.
- **Yield trials:** Replicated yield trials are conducted with recurrent parent as a check. The newly constituted variety should be similar to variety X for most of the important characteristics with resistant to disease. Seeds are multiplied for distribution.

Backcross-Recessive gene transfer

- Let variety *X* is well adapted and high yielding variety.
- Variety *Y* is another variety resistant to a specific disease; which is governed by a recessive gene. We wish to transfer this recessive gene form variety *Y* to variety *X*.
- So, variety *X* – recurrent parent ( ) Variety *Y* – donor parent ( ).
- When the desired character i.e. disease resistance is governed by a recessive gene, backcrosses cannot be performed in continuation as in dominant gene transfer method. After first backcross and then after every two subsequent backcrosses F_2 must be raised to test rust resistance.
- Only F_2 is tested for resistance as all F1 and backcross progenies are heterozygous and susceptible to disease.

The various steps for this backcross breeding are as follows:

- **Hybridization:** The two varieties *X* and *Y* are crossed. Generally recipient / recurrent variety is used as female parent.
- **F_1 generation:** F_1 plants are back crossed with the variety *X*.
- **BC_1 generation:** As disease resistance is controlled by recessive gene all the plants will be susceptible to disease, so disease resistance is not tested for this generation. Plants raised from seeds of above cross are selfed.
- **BC_1 F_2 generation:** Test for disease resistance is conducted. Resistant plants which are similar in plant characteristic to recurrent parent are selected and back crossed with recurrent parent.
- **BC_2 generation:** Plants are grown from seeds of above cross, plants those are similar in plant characteristic to recurrent parent are selected and backcrossed with recurrent parent. Resistance test is not conducted.

- **BC_3 generation:** Plants are selfed to grow F_2.
- Resistance test is not conducted for the sake of selection; it is done on the basis of resemblance to variety.
- **BC_3 F_2 generation:** Selection for resistance is conducted and selected plants are back crossed with variety *X*.
- **BC_4 generation:** Back cross of plants selected above with recurrent parent. Disease resistance test is not conducted.
- **BC_5 generation:** Plants are selfed to raise F_2. Disease resistance test is not conducted in this generation.
- **BC_5 F_2 generation:** Disease resistant plants similar in plant characteristics to recurrent variety are selected. Selfed and seeds are harvested separately.
- **BC_5 F_3 generation:** Individual plant progenies are grown. Selection is done for disease resistance and resemblance to variety *X*. Seeds of selected plants are bulked to constitute the new variety.
- **Yield trials:** Replicated yield trials are conducted with variety *X* as a check. The newly constituted variety should be similar to variety *X* for most of the important characteristics. Seeds are multiplied for distribution.

Genetic Consequences

- Repeated backcrossing to the recurrent parent accelerates homozygosity equals to selfing and improving the recovery of recurrent parent type progenies.
- Heterozygous F_1 will have genes in ratio of 50:50 from non-recurrent and recurrent parents.
- In subsequent backcross generations, this ratio will tilt towards the recurrent parent with the proportion of genes from donor parent being reduced by one–half and genes from recurrent parent increasing by the same rate.
- Contents of the donor parent in any backcross are given by the formula (1/2)*n* where *n* is number of crosses and backcrosses attempted with recurrent parent so in BC4 the genes from donor parent would be (1/2)5 = 1/72 or 1/32 × 100 = 3.125% indicating that contribution of recurrent parent by 96.875%.
- The proportion of homozygosity in any generation is given by the formula.

  $[(2m - 1/2m)]n$

  m = number of generation
- So after 3 generation of backcrossing, homozygosity will be 3-1/3 = 8-1/8 = 7/8 or 7/8 × 100 = 175/2 = 87.5%.
- For *n* number of genes, the proportion of homozygous genotypes would be (2 – 1/2)*n*
- So if we consider 4 genes, then after 3 generation of backcrosses, the proportion of homozygotes would be = (23 – 1/23)4 = (7/8)4 = 76.5%
- **Objective in backcrossing is to recover recurrent phenotype along with target gene.** Briggs and Allard (1953) suggested a complete the gene transfer through six backcrosses. Some important considerations in backcrossing program are: minimum of 53 plants from backcrossed seeds, 96 F_2 plants and 68 F_3 rows are required to

- If selection is not rigid for recurrent parent, the number of backcrosses required may increase to ten.
- Smaller populations necessitate more number of backcrosses.
- If the parental lines resemble agronomically then even fewer backcrosses are sufficient.
- If the target gene is linked to an undesirable gene then more number of backcrosses need to be attempted.
- Normally six backcrosses that follow strict selection for recurrent genotype are adequate for effective gene transfer.

Advantages and Disadvantages of Backcross Method

Advantages

- Gives predicted results. Extensive tests are not necessary as recurrent parent is usually a well established variety.
- We can obtain same results again and again.
- While eliminating defects of well established variety there is no much change in its adaptability, performance and genotypic background.
- Only method for interspecific gene transfer.
- Independent of environment.
- This method offers considerable control over the genetic variation in the segregating population under consideration.
- Little record keeping.
- Requires small populations.
- Backcrossing is applicable for both dominant gene transfer and recessive gene transfer.

Disadvantages

- Newly developed variety is superior to the recurrent parent only for the character which is transferred, not for others.
- Ineffective for traits with low heritability viz., quantitative traits.
- If undesirable genes are closely linked to the desired gene, the undesirable genes are probably transferred along with the desired gene to the new variety.
- Little recombination.
- It is a lengthy process. By the time we transfer a gene to the variety, that variety may have been replaced by other superior varieties.

Applications of Backcross Method

- Intervarietal transfer of qualitative characters and quantitative characters with high heritability.
- Interspecific transfer of traits with high heritability.
- To improve well established variety with respect to one or a few characters.

- Isogenic lines production.
- Cytoplasm transfer esp. with cytoplasmic male sterility (CMS) involved, used to transfer male sterility into purelines, which is a prerequisite for many hybrid production breeding programs.
- It may be modified so that transgressive segregation may occur for quantitative characters.

Ploidy breeding

- Ploidy refers to the number of sets of chromosomes in a biological cell.
- Plant germ cells (pollen and ovule) have one complete set of chromosomes from the male or female parent.
- Germ cells, also called gametes, combine to produce somatic cells.
- Somatic cells therefore have twice as many chromosomes as that of germ cells. The haploid number (n) is the number of chromosomes in a gamete.
- A somatic cell has twice that many chromosomes ($2n$).

Euploidy

- Euploidy is the state of a cell or organism having an integral multiple of the monoploid number, possibly excluding the sex-determining chromosomes.

Aneuploidy

- Aneuploidy is the state of not having euploidy. In humans, examples include having a single extra chromosome (such as Down syndrome), or missing a chromosome (such as Turner's syndrome).
- Aneuploid karyotypes are given names with the suffix 'somy' (rather than ploidy, used for euploid karyotypes), such as trisomy and monosomy.

Haploid and Monoploid

- As stated above, the haploid number (n) is the number of chromosomes in a gamete of an individual, and this is distinct from the monoploid number (x) which is the number of unique chromosomes in a single complete set.
- Gametes (sperm and ovary) are haploid cells.
- The haploid gametes produced by (most) diploid organisms are monoploid, and these can combine to form a diploid zygote.
- For example, most animals are diploid and produce monoploid gametes.
- During meiosis, germ cell precursors have their number of chromosomes halved by randomly "choosing" one homologue, resulting in haploid gametes.
- Because homologous chromosomes usually differ genetically, gametes too usually differ genetically from one another.
- All plants including algae and even many fungi switch between a haploid and a diploid state (which may be polyploid), with one of the stages emphasized over the other.

- This is called alternation of generations. Most fungi and algae are haploid during the principal stage of their life cycle.
- Male bees, wasps, and ants are haploid organisms because of the way they develop from unfertilized, haploid egg cells.

Polyploidy in Plants

- Polyploidy is pervasive in plants and some estimates suggest that 30-80% of living plant species are polyploid, and many lineages show evidence of ancient polyploidy (paleopolyploidy) in their genomes.
- Huge explosions in angiosperm species diversity appear to have coincided with the timing of ancient genome duplications shared by many species.
- It has been established that 15% of angiosperms and 31% of fern speciation events are accompanied by ploidy increase.
- Polyploid plants can arise spontaneously in nature by several mechanisms, including meiotic or mitotic failures, and fusion of unreduced ($2n$) gametes.
- Both autopolyploids (e.g. potato]) and allopolyploids (e.g. Okra, canola, wheat, cotton) can be found among both wild and domesticated plant species.
- Most polyploids display heterosis relative to their parental species, and may display novel variation or morphologies that may contribute to the processes of speciation and eco-niche exploitation.
- The mechanisms leading to novel variation in newly formed allopolyploids may include gene dosage effects (resulting from more numerous copies of genome content), the reunion of divergent gene regulatory hierarchies, chromosomal rearrangements, and epigenetic remodeling, all of which affect gene content and/or expression levels. Many of these rapid changes may contribute to reproductive isolation and speciation.
- *Lomatia tasmanica* is an extremely rare Tasmanian shrub which is triploid and sterile, and reproduction is entirely vegetative with all plants having the same genetic structure.
- There are few naturally occurring polyploid conifers. One example is the giant tree Coast Redwood (*Sequoia sempervirens*) which is a hexaploid (6x) with 66 chromosomes ($2n = 6x = 66$), although the origin is unclear.

Polyploid Crops

- Polyploid plants tend to be larger and better at flourishing in early succession habitats such as farm fields.
- In the breeding of crops, the tallest and best thriving plants are selected.
- Thus, many crops (and agricultural weeds) may have unintentionally been bred to a higher level of ploidy.
- The induction of polyploidy is a common technique to overcome the sterility of a hybrid species during plant breeding.
- For example, Triticale is the hybrid of wheat (*Triticum turgidum*) and rye (*Secale cereale*).

- It combines sought-after characteristics of the parents, but the initial hybrids are sterile.
- After polyploidization, the hybrid becomes fertile and can thus be further propagated to become triticale.
- In some situations polyploid crops are preferred because they are sterile.
- For example many seedless fruit varieties are seedless as a result of polyploidy. Such crops are propagated using asexual techniques such as grafting.
- Polyploidy in crop plants is most commonly induced by treating seeds with the chemical colchicine.

Examples of Polyploid Crops

- **Triploid crops:** banana, apple, ginger seedless watermelon, citrus.
- **Tetraploid crops:** durum or macaroni wheat, cotton, potato, cabbage, leek, tobacco, peanut, kinnow, Pelargonium.
- **Hexaploid crops:** chrysanthemum, bread wheat, triticale, oat, kiwifruit and sweet potato.
- **Octaploid crops:** strawberry, dahlia, pansies, sugarcane.
- Some crops are found in a variety of ploidy. Apples, tulips and lilies are commonly found as both diploid and as triploid. Daylilies (Hemerocallis) cultivars are available as either diploid or tetraploid. Kinnows can be tetraploid, diploid, or triploid.

Terminology

Autopolyploidy

- Autopolyploids are polyploids with multiple chromosome sets derived from a single species.
- Autopolyploids can arise from a spontaneous, naturally occurring genome doubling, like the potato.
- Others might form following fusion of $2n$ gametes (unreduced gametes). Bananas and apples can be found as autotriploids.
- Autopolyploid plants typically display polysomic inheritance, and are therefore often infertile and propagated clonally perfect

Allopolyploidy

- Allopolyploids are polyploids with chromosomes derived from different species. Precisely it is the result of doubling of chromosome number in an F_1 hybrid.
- Amphidiploid is another word for an allopolyploid. Some of the best examples of allopolyploids come from the Brassicas, and the
- Triangle of U describes the relationships among the three common diploid Brassicas (*B. oleracea, B. rapa,* and *B. nigra)* and three allotetraploids (*B. napus, B. juncea,* and *B. carinata*) derived from hybridization among the diploids.
- Current day cultivated okra is an example of an allopolyploid.

Types of Changes in Chromosome Number

- The somatic chromosome number of any species, whether diploid or polyploid, is designed as 2*n*, and the chromosome number of gametes is denoted as *n*. An individual carrying the gametic chromosome number, *n*, is known as haploid.
- A monoploid, on the other hand, has the basic chromosome number, *x*. In a diploid species, $n = x$; one *x* constitutes a genome or chromosome complement.
- The different chromosomes of a single genome are distinct from each other in morphology and/ or gene content and homology; members of a single genome do not show a tendency of pairing with each other.
- Thus a diploid species has two, a triploid has 3 and a tetraploid has 4 genomes and so on.
- The terminology of heteroploidy is summarized.
- Individuals carrying chromosome numbers other than the diploid (2*x*, and not 2*n*) number are known as heteroploids, and the situation is known as heteroploidy.
- The change in chromosome number may involve one or a few chromosomes of the genome; this is known as aneuploidy.
- The aneuploid changes are determined in relation to the somatic chromosome number (2*n* and not 2*x*) of the species in question. Therefore, the terminology for aneuploid individuals arising from diploid and polyploidy species is the same.
- Heteroploidy that involves one or more complete genomes is known as aneuploidy.
- By definition, therefore, the chromosome numbers of euploids are an exact multiple of the basic chromosome number of the concerned species, while those of aneuploids are not. Aneuploid individuals from which one chromosome pair is missing (2*n*-2) are termed as nullisomic

Variation in Chromosome Number

Table 7.10: *A summary of the terms used to describe heteroploidy (variation in chromosome number)*

Term	Type of change	Symbol*
Heteroploid	Having a chromosome number that is neither the haploid nor the diploid number normal in the species	-
Aneuploid	Organism has a partial set of chromosomes due to addition or deletion of specific chromosome(s). or one or more chromosome sets	2*n* ± few
Nullisomic	A genetic condition involving the lack of one of the normal chromosomal pairs for a species .	2*n*-2
Monosomic	One chromosome missing	2*n*-1
Double trisomic	One chromosome from each of two different chromosome pairs missing	2*n*+1+1
Trisomic	One chromosome extra	2*n*+1

Term	Type of change	Symbol*
Tetrasomic	One chromosome pair extra	$2n$+2
Euploid	Individual has one or more whole sets of chromosomes with each chromosome represented once in each set	
Monoploid	One copy of a single genome	x
Haploid	Gametic chromosome complement	n
Polyploid	More than 2 copies of one genome or 2 copies each of 2 or more genomes**	
Autopolyploid	Genomes identical with each other	
Autotriploid	Three copies of one genome	$3x$
Autotetraploid	Four copies of one genome	$4x$
Autopentaploid	Five copies of one genome	$5x$
Autohexaploid	Six copies of one genome	$6x$
Autooctaploid	Eight copies of one genome**	$8x$
Allopolyploid	Two or more distinct genomes (Generally each genome has two copies)	
Allotetraploid	Two distinct genomes	($2x1$+$2x2$)**
Allohexaploid	Three distinct genomes	($2x1$+$2x2$+$2x3$)**
Allooctaploid	Four distinct genomes	($2x1$+$2x2$+$2x3$+$2x4$)**

*$2n$ = Somatic chromosome number (and complement) and n = gametic chromosome number (and complement) of the species, whether diploid or polyploid.

x = The basic chromosome number (and complement) or genomic number. x1,x2, x3, x4 = Distinct genomes from different species.

** In general, this condition occurs; other situations may also occur.

- A diploid in which 1 chromosome is missing from the total complement ($2n$-1) is known as monosomics. A double monosomic individual has two chromosomes missing, but the two chromosomes belong to two different chromosome pairs ($2n$-1-1). An individual having one extra chromosome ($2n$+1) is known as trisomic, and that having two extra chromosomes each belonging to a different chromosome pair is called double trisomic ($2n$+1+1). When an individual has an extra pair of chromosomes, it is known as tetrasomic ($2n$+2). The same is true about the various features of and the facts accumulated about aneuploidy as well as euploidy. The breeder is generally concerned with monosomics and trisomics, and, in some situations, with nullisomics and tetrasomics.
- In euploids, the chromosome number is an exact multiple of the basic or genomic number. Euploidy is more commonly known as *polyploidy*. When all the genomes present in a polyploidy species are identical, it is known as *autopolyploid* and the situation is termed as *autopolyploidy*. In the case of allopolyploids, two or more distinct genomes are present. Euploids may have 3 (triploid), 4 (tetraploid), 5 (pentaploid), 6 (hexaploid), 7 (heptaploid), 8 (octaploid) or more genomes making up their somatic chromosomes number.

- In case of autopolyploidy, they are known as autotriploid, autotertaploid, autopentaploid, autohexaploid, autoheptaploid, auto octaploid and so on, while in the case of allopolyploidy they are termed as allotriploid, allotetraploid, allopentaploid, allohexaploid, alloheptaploid, allooctaploid etc, Amphidiploid is an allopolyploid that has two copies of each genome present in it and, as a consequence, behaves as a diploid during meiosis. A segmental allopolyploid contains two or more genomes, which are identical with each other, except for some minor differences.

Application of Allopolyploidy in Crop improvement

- Alloploypolidy has three major applications in crop improvement : (1) as bridging species in the transfer of characters from one species into another, (2) in the production of new crop species, and (3) for widening the genetic base of existing allopolyploid crop species.
- The promising alloplyploids are Raphanobrassica, the triploid (AAC) obtained by crossing *B. napus* (AACC) with *B. campestris* (AA), allopolyploid clovers, Festuca-Lolium hybrids and some species hybrids in Rubus and Jute (*Corchorus sp.,*). In Raphanaobrassica, the breeding objectives are to combine the hardiness of *B. oleracea* with quick growth and disease resistance of fodder radish. The problems of Raphanobrassica are the same as those of triticale. i.e., low fertility, cytogenetic and genetic instability and leafy rape-like plants that do not produce bulbs. There is evidence that hybridization and selection at the polyploidy level would be effective in improving Raphanobrassica.
- In cassava the triploids developed by crossing tetraploids and diploids, have been reported to be promising and it therefore seems possible to improve cassava by producing new chromosomal lines, in which the chromosome number does not go beyond an optimum level. The triploid plants derived from OP-4 (2x) X S-300 (4x) and OP-4(2x) X H-2304 (4x) consistently produced roots with high dry matter in the seedling and succeeding clonal generations, which ranged from 34 to 43%. Some of the triploids recorded high starch content from the eighth month onwards, being significantly higher than that of the control. Among these, the triploid 76-9 had a yield similar to that of H- 2304, the released cultivar, at CTCRI.
- The amphidiploids *B. napus* (AACC) crosses very easily with *B. campestris* (AA) to produce the triploid (AAC), which has some desirable features. The triploid is produced so easily that it may be used as a hybrid variety, a special case of hybrid varieties produced by crossing two different species.

Application of Autopolyploidy in Crop Improvement

- **Triploids.** Triploids are produced by hybridization between tetraploid and diploid strains. They are generally highly sterile, except in a few cases. This feature is useful in the production of seedless watermelons. In certain species, they may be more vigorous than the normal diploids, e.g. in sugarbeets.
- Seedless watermelons are grown commercially in Japan. They are produced by crossing tetraploid (4*x*, used as female) and diploid (2*x*, used as male) lines,

since the reciprocal cross ($2x \times 4x$) is not successful. The triploid plants do not produce true seeds; almost all the seeds are small, white rudimentary structures like cucumber (*Cucumis sativus*) seeds. But a few normal sized seeds may occur, which are generally empty. For good fruit setting, pollination is essential. For this purpose, diploid lines are planted in the ratio 1 diploid; 5 triploid plants. There are several problems, viz, genetic instability of 4x lines, irregular fruit shape, a tendency towards hollowness of fruits, production of empty seeds and the labour involved in triploid seed production (by hand-pollination).

- Triploid sugarbeets (*B. vulgaris*) produce larger roots and more sugar per unit area than do diploids, while tetraploids produce smaller roots and lower yields than diploids. Apparently, $3x$ is the optimum level of ploidy in sugarbeets. Triploid sugarbeet varieties have been grown commercially in Europe and Japan, but their popularity is declining rapidly. The triploid varieties are mixtures of triploid, diploid and other ploidy level plants. Seed production of triploid sugarbeet is difficult because the beet flower is small. Triploid seed may be produced in one of the following two ways: (1) using $4x$ plants as female and $2x$ as male or (2) using $4x$ as male and $2x$ as female. The first combination gives lower seed yield but a higher proportion of triploids, while the second gives higher seed yield but a lower proportion of triploids. Commercial triploid sugarbeet seed is produced by interplanting $4x$ and $2x$ lines in the ratio 3:1, and seed from both $4x$ and $2x$ plants is harvested. This seed consists of about 75% triploid ($3x$) seeds. Triploids sugarbeet may give 10-15% higher yields than diploids.
- Pusa Giant Berseem is the first autopolyploid variety released for general cultivation in India. It yields 20-30% more green fodder than the diploid berseem varieties.
- Autoployploidy is more likely to succeed in species with lower chromosome numbers than in those with higher chromosome numbers.
- Cross-pollinating species are generally more responsive than self-pollinating species.
- Crops grown for vegetative parts are more likely to succeed as polyploids than those grown for seeds.

Application of Aneuploids in Crop Improvement

- Aneuploids are useful in the studies on effects of loss or gain of an entire chromosome or a chromosome arm on the phenotype of an individual.
- Aneuploids are useful in locating a linkage group and a gene to a particular chromosome. By using a secondary or tertiary trisomic, the gene may be located to one of the two arms of a chromosome, or even to a part of the chromosome arm. The most important application of aneuploids is in locating genes on particular chromosomes; this will be considered in some detail.
- Aneuploids are useful in identifying the chromosome involved in translocation.
- They are useful in the production of substitution lines. Chromosome substitution may be desirable for studying the effects of individual chromosomes of a variety or for the transfer of the genes carried by specific chromosomes or a variety into another one.

Mutation

- It refers to the sudden heritable change in the phenotype of an individual. In other words, mutations arise due to change in DNA bases.

Mutation Breeding

- Hugo de Vries introduced the term "Mutation" in 1990. In 1927, Muller showed that x-ray induce genetic deviants in *Drosophila* In 1928, Stadler-reported high mutation rates in x-rayed maize and barley.
- In 1910 Morgan discovered white eye mutant in Drosophila by.

Transition

- Changes in the nucleotide involves replacement of one purine by another purine viz.,
- Adinine substituted by Guanine or vice versa. G or one pyrimidine by another pyrimidine viz., thymine substituted by cytosine or uracil.

Transversion

- Changes in the nucleotide involves substitution (adenine by thymine, cytocyine by thymine or vice versa)of a purine by a pyrimidine .

Mutations occur in two main ways:

1. By alteration in nuclear DNA (Point mutations)
2. By change in cytoplasmic DNA (Cytoplasmic mutations)

Types of Mutations

1. Spontaneous mutations- Mutations occur in nature
2. Induced mutations – Artificially induced by various mutagenic agent
3. Macro mutations – Mutations with distinct morphological changes in the phenotype. It is also called oligogenic mutations
4. Micro mutations – Mutations with invisible phenotypic changes .It is also called as polygenic mutations. Depending upon the effect on the survival of an individual, induced mutations are of types namely,
   - Lethal (all the individuals carrying such mutations are killed)
   - Sub-lethal (mortality is more than 50%)
   - Sub-vital (mortality is less than 50%)
   - Vital (all mutants survive)

Mutagens and their Mode of Action

- It refers to various physical or chemical agents which greatly enhance the frequency of mutations. Two types of mutagens.
- Physical mutagens: X rays, Gamma rays, Alpha particles, Beta particles, Fast and thermal neutrons, UV rays

Gamma Rays

- Gamma rays have generally a shorter wavelength and hence possess more energy per photon than X-rays. Gamma radiation in usually obtained from radioisotopes in contrast to X-rays. A gamma radiation facility can be used essentially in the same manner as an X-ray machine for acute or semi-acute exposures.
- The most distinct advantage of gamma radiation source for prolonged treatments is that it can be placed in a greenhouse or field so that plants can be exposed as they develop over long periods of time. Cobalt-60 and Cesium -137 are the main sources of gamma rays used in mutation breeding. They are stored in lead containers when not in use and operated by remote control mechanisms to irradiate plant material.

UV Rays

- Ultraviolet light has limited penetrating ability, therefore its use is limited to treating spores, pollen grains cells and cultured tissue. Wavelengths in the range of 2,500 to 2,800 nm are biologically most effective because this is the region of maximal light absorption by nucleic acids.

Beta Particles

- Beta particles such as those from Phosphorus-32 and Sulfur- 35 produce effects in tissues similar to those of X- or gamma rays. The penetrating ability of beta particles is lower than that of X- and gamma rays.
- The lower penetrating ability of beta particles can be overcome by putting the radioisotope in a solution and administering them to the plant material. P-32 and S-35 may then be incorporated directly into the cell nucleus giving a somewhat greater localization of the site of action. But because of the variability from tissue to tissue and cell to cell, it is difficult to determine the exact dose given by an internal emitter in plant material.

Neutrons

- Neutrons have been shown to be highly effective for the induction of mutation in plants, but a certain degree of confusion exists concerning the results of early experiments and lack of adequate dosimetric techniques.

Ion Beams

- Ion beams can give a large amount of energy with high LET (Linear Energy Transfer) to the localized position in tissues. Therefore, we can expect different biological effects will be given to plant material compared with low LET radiations, such as, gamma- and X-rays.
- Also ion beams can produce large structural changes in chromosomes and DNA. So we can expect to induce different kinds of mutations in plants with high frequency than gamma- and X-rays. But further basic and practical researches should be made to use ion beams efficiently in the future.

Table 7.11: *Mutagens and their mode of action*

Radiation	Main properties	Mode of action or changes caused
X rays	S.I., penetrating and non particulate	Induce mutations by forming free radicals and ions. Cause all changes.
Gamma rays	S.I., very penetrating and non particulate	Induce mutations by ejecting atoms from the tissues. Cause all changes.
Alpha particles	D.I., Particulate, less penetrating & +ve charged	Act by ionization and excitation. Cause chromosomal breakage and gene mutations
Beta particles	S.I., particulate, more penetrating and ⁻ve charged	Act by ionization and excitation. Cause chromosomal breakage and gene mutations
Fast and thermal neutrons	D.I., Particulate, neutral particles, highly penetrating	Cause chromosomal breakage and gene mutations
UV rays	Non-ionizing, low penetrating	Cause chromosomal breakage and gene mutations

Chemical Mutagens

- The number of chemical mutagens is numerous and continuously increasing. But for mutation inductin in cultivated plants only a few are readily very useful and most of them belong to the special class of alkylating agents such as, ethyl methane sulhonate (EMS), diethyl sulfate (dES), ethyleneimine (EI), ethyl nitroso urethane (ENU), ethyl nitroso urea (ENH), and methyl nitroso urea (MNH).

Group of mutagen	Name of chemical	Mode of action
Alkylating agents	Ethyl methane sulphonate. Methyl methane sulphonate. Ethyl ethane sulphonate	AT GC transitions Transitions GC AT transitions
Base analogues	5 bromo uracil 2 amino purine	AT GC transitions GC AT transitions
Acridine dyes	Acriflavin, proflavin	Deletion and addition
Others	Nitrous acid. Hydroxylamine Sodium azide	AT GC transitions GC AT transitions Transitions

Base Analogues and Related Compounds

- True base analogues are closely related to DNA bases, adenine, guanine, cytosine or thymine and can be incorporated into DNA without affecting its replication. But analogue differs from the normal base in certain substitutes hence its electronic structure is modified and it can be expected that occasional pairing errors will occur at the time of DNA replication after the analogue has been incorporated.
- The most frequently used analogues are 5-bromo-uracil (BU) and 5-bromo-deoxyuridine (BUDR), which are analogues of thymine. Apart from true analogues,

it has been found that N-methylated oxypurines have a chromosome effect.The most efficient compounds are 8-ethoxy caffeine (EOC) and 1,3,7,9 tetramethyluric acid (TMU), but their mode of action is still unknown. Maleic hydrazide (MH), an isomer of uracil, induces chromosome breaks in cell and aberrations are localized in heterochromatic regions near the centromere of the chromosomes.

Antibiotics

- Antibiotics such as azaserine, mitomycin C, streptonigrin and actinomycin D have been found to have chromosome breaking properties, but their usefulness are limited.

Alkylating Agents

- This is the most important group of mutagenic chemicals for mutation induction in cultivated plants.
- They have one or more reactive alkyl groups which can be transferred to other molecules.
- They react with DNA by alkylating the phosphate groups as well as the purine and pyrimidine bases. One should be extremely careful in using them because most are potential carcinogens, such as EI, EMS, MNH and should be used in small units.

Azide

- Azide is an effective mutagen under certain treatment conditions.
- It is possible to obtain high mutation frequencies with azide.
- Most mutations are gene mutations with some minor frequencies of chromosome aberrations.
- It is relatively safe, non-persistent and inexpensive but are also potential carcinogens.

Procedure of Mutation Breeding

Step 1. Choice of Material

- The best adapted variety of a crop should be chosen.
- Only one or two features of such variety have to be altered through mutagenesis depending upon the objectives.
- Suppose a variety is high yielding, but susceptible to a particular disease, the objectives of mutation breeding would be to induce resistance to that particular disease in the variety.

Step 2. Choice of Mutagen

- It depends upon the plant parts to be treated. Generally, chemical mutagens are more preferred for seed treatment and radiations for the treatment of vegetative parts.
- The penetration of chemical mutagens can be enhanced dissolving the mutagen in solvents like dimethyl sulphoxide.

Step 3. Mutagenic Treatment

- Plant species- Seeds, Pollens, Buds, Cuttings or Suckers.
- Dose of mutagen –LD 50 refers to a dose of mutagen that kills 50% of the treated individuals.
- Duration of treatment- depends on the intensity of radiations or concentration. The seeds are water soaked before treatment. After treatment the seeds or cuttings are immediately planted and pollens are used for pollination. Plants obtained from treated seeds or cuttings are called M1 plants.

Step 4. Handling of Treated Material

Seed propagated species

M1 Generation

- Several hundred (500 or more)treated seeds are space planted. All the M1populations are grown using wider spacing for easy identification. Generally the mutants are recessive. All the plants will be chimeras for the mutation present in heterozygous state. About 20 seeds from each M1 plant are harvested separately

M2 Generation

- About 2000 progeny rows are grown using wider spacings. Oligogenic mutants with distinct features are identified and selected. Only 1-3% of M2 rows may be expected to have beneficial mutations

M3 Generation

- Progeny rows form individual selected plants. Inferior mutant rows are eliminated. If the mutant progenies are homozygous, two or more M3 progenies containing the same mutation. Mutant M3 rows are harvested in bulk

M4 Generation

- A preliminary yield trial is conducted with promising mutant lines are selected for replicated multilocation trials

M5-M8 Generation

- Selected lines are tested in coordinated multilocation trials.
- The best performing line is released as a variety.
- In case of polygenic traits, identification of character is not possible through visual observations.
- The material is tested in replicated trial and screening is done for the character under improvement using appropriate statistical methods.
- Inferior plants are rejected in M3 and M4 generation based on screening tests and superior plants are bulked to raise the next generation.
- The homozygous progeny are tested in coordinated trial from M5 to M9 and the best line is released as a variety.

Advantages and Disadvantages

Advantages

- It is a cheap and rapid method of developing new varieties as compared to backcross, pedigree and bulk breeding methods.
- It is more effective for the improvement of oligogenic characters such as disease resistance than polygenic traits.
- This is the simple, quick and best way when a new character is to be induced in vegetatively propagated crops.

Disadvantages

- Most of the mutations are deleterious and undesirable.
- Identification of micro mutations, which are most useful to a plant breeder, is usually very difficult.
- Since useful mutations are produced at a very low frequency (0.1%), a very large plant population has to be screened to identify and isolate desirable mutants.
- It has limited scope for the genetic improvement of quantitative or polygenic characters.

Applications in Crop Improvement

- Development of improved varieties—All over the world, 1542 improved varieties have been developed in different crops (IAEA, 1991).
- Induction of male sterility—Use of GMS &CMS lines helps in reducing the cost of hybrid seed production.
- Production of haploids—Use of X ray irradiated pollens has helped in the production of haploids in many crops.
- Creation of genetic variability—In asexually propagated crops like potato, so- iatic mutations may be useful.
- Overcoming self-incompatibility—Mutation of S gene by irradiation leads to restoration of self-fertility in self-incompatible species.
- Improvement in adaptation—Induced mutations helped in improving the adaptation of some crop species.

Principles and Importance of breeding of Cross—Pollinated Crops

Principles

The most important methods of breeding cross-pollinated species are:

1. Mass selection,
2. Development of F1 hybrids, and
3. Development of synthetic varieties.

- Since cross-pollinated species are naturally hybrid (heterozygous) for many traits and lose vigour as they become purebred (homozygous), a goal of each of these breeding methods is to preserve or restore heterozygosity.

Hardy-Weinberg Principle

- The Hardy-Weinberg principle (also known by a variety of names such as HWP, Hardy-Weinberg equilibrium, Hardy-Weinberg Theorem, HWE, or Hardy-Weinberg law) states that both allele and genotype frequencies in a population remain constant—that is, they are in equilibrium—from generation to generation unless specific disturbing influences are introduced.
- These disturbing influences include non-random mating, mutations, selection, limited population size, 'overlapping generations', random genetic drift, gene flow and meiotic drive.
- It is important to understand that outside the lab, one or more of these 'disturbing influences' are always in effect. That is, Hardy-Weinberg equilibrium is impossible in nature.
- Static allele frequencies in a population across generations assume: random mating, no mutation (the alleles don't change), no migration or emigration (no exchange of alleles between populations), infinitely large population size, and no selective pressure for or against any traits.
- The Hardy-Weinberg model, named after the two scientists who derived it in the early part of this century, describes and predicts genotype and allele frequencies in a non-evolving population.
- The model has five basic assumptions.
  - The population is large (i.e., there is no genetic drift).
  - There is no gene flow between populations, from migration or transfer of gametes.
  - Mutations are negligible.
  - Individuals are mating randomly.
  - Natural selection is not operating on the population.
- Given these assumptions, a population's genotype and allele frequencies will remain unchanged over successive generations, and the population is said to be in Hardy-Weinberg equilibrium. The Hardy-Weinberg model can also be applied to the genotype frequency of a single gene

Importance

- The Hardy-Weinberg model enables us to compare a population's actual genetic structure over time with the genetic structure we would expect if the population were in Hardy-Weinberg equilibrium (i.e., not evolving).
- If genotype frequencies differ from those we would expect under equilibrium, we can assume that one or more of the model's assumptions are being violated, and attempt to determine which one(s).
- How do we use the Hardy-Weinberg model to predict genotype and allele frequencies? What does the model tell us about the genetic structure of a population?

Mass Selection

- Mass selection in cross-pollinated species takes the same form as in self-pollinated species; i.e., a large number of superior appearing plants are selected and harvested in bulk and the seed used to produce the next generation.
- Mass selection has proved to be very effective in improving qualitative characters, and, applied over many generations, it is also capable of improving quantitative characters, including yield, despite the low heritability of such characters.
- Mass selection has long been a major method of breeding cross-pollinated species, especially in the economically less important species.

Factors affecting equilibrium in populations

Migration

- Migration is the movement of individuals into a population from a different population. Migration may have introduced new alleles into the population or may change the frequencies of existing alleles. In plant breeding programmes, migration is represented by intervarietal crosses, polycrosses etc., wherein the breeder brings together into a single population two or more separate populations.

Mutation

- Mutation is a sudden and heritable change in an organism and is generally due to a structural change in a gene. Mutation may produce a new allele not present in the population or may change the frequencies of existing alleles.

Random Drift

- Random drift or genetic drift is a random change in gene frequency due to sampling error. Random drift occurs in small populations because sampling error is greater in a smaller population than in a larger one.
- Ultimately, the frequency of one of the alleles becomes 0 and that of the other allele becomes 1.
- The allele with the frequency of 1 is said to be fixed in the population because there would be no further change in its frequency.
- It may be expected that in a small populations all the genes would become homozygous, or would be fixed in due course of time. Breeding populations are generally small, hence a certain amount of genetic drift is found to occur in them.
- The breeder cannot do anything to prevent this genetic drift, except to use very large populations, which is often not practicable. Alternatively, he may resort to phenotypic disassortative mating, which would again require time, labour and money.

Inbreeding

- Mating between individuals sharing a common parent in their ancestry is known as inbreeding.
- In small populations, a certain amount of inbreeding is found to occur.

- Inbreeding reduces the proportion of heterozygotes or heterozygosity and increases the frequency of homozygotes or homozygosity.

Selection

- Differential reproduction rates of various genetopes are known as selection.
- In crop improvement, selection is very important because it allows the selected genotypes to reproduce, while the undesirable genotypes are eliminated.
- In a random mating population, if plants with AA or aa genotypes are selected, the frequency of A allele in the selected of population would be 1 or 0, respectively.
- Thus selection in a random mating population is highly effective in increasing or decreasing the frequency of alleles, but it is unable to either fix or eliminate them.
- However, in combination with a system of inbreeding, selection is highly efficient in the fixation and elimination of alleles.

Mutation Breeding

Physical Mutagens

- Physical mutagens include various types of radiation, viz X-rays, gamma rays, alpha particles, beta particles, fast and thermal (slow) neutrons and ultraviolet rays.
- A brief description of these mutagens is presented below:

Chemical Mutagens

Chemical mutagens can be divided into four groups, *viz.*

1. Alkylating agents
2. Base analogues
3. Acridine dyes
4. Others.

A brief description of some commonly used chemicals of these groups is presented below.

Variables

Methods

- The Hardy-Weinberg model consists of two equations: one that calculates allele frequencies and one that calculates genotype frequencies. Because we are dealing with frequencies, both equations must add up to

1. The equation $p + q = 1$

describes allele frequencies for a gene with two alleles. (This is the simplest case, but the equation can also be modified and used in cases with three or more alleles.) If we know the frequency of one allele (p) we can easily calculate the frequency of the other allele (q) by 1 ó $p = q$.

- In a diploid organism with alleles A and a at a given locus, there are three possible genotypes: AA, Aa, and aa.

- If we use p to represent the frequency of A and q to represent the frequency of a, we can write the genotype frequencies as $(p)(p)$ or p^2 for AA, $(q)(q)$ or q^2 for aa, and $2(p)(q)$ for Aa. The equation for genotype frequencies is $p^2 + 2pq + q^2 = 1$.

Systems of Mating

- The breeder has two basic tools to change the genetic composition of a population: Selection and Mating system.
- We have discussed the effects of selection in some detail. Here we would examine the consequences of various systems of mating.

There are five basic mating schemes

1. Random mating
2. Genetic assortative mating
3. Genetic disassortative mating
4. Phenotypic assortative mating
5. Phenotypic disassortative mating.

Random Mating

- In random mating, each female gamete is likely to unite with any male gamete equally and the rate of reproduction of each genotype is equal, i.e. there is no selection.

Genetic Assortative Mating

- In genetic assortative mating, the mating is between individuals that are more closely related by ancestry than in random mating.
- Thus in this mating system, it is not important that the genotypes of the plants selected for mating be correctly identified.
- This mating system is more commonly known as inbreeding.

Genetic Disassortative Mating

- In genetic disassortative mating, such individuals are mated, which are less closely related by ancestry than would be under random mating. Thus in this system, totally unrelated individuals are mated.
- These individuals often belong to different population.
- Example of such a mating are intervarietal and interspecific crosses. The effects of genetic disassortative mating are similar to those of migration.
- It may be expected that this mating system would reduce homozygosity and increase heterozygosity.

Phenotypic Assortative Mating

- Mating between phenotypically dissimilar individuals belonging to the same population is referred to as phenotypic disassortative mating. The consequences of this mating are as follows:

- It leads to the maintenance of or even some increase in heterozygosity.
- There is some reduction in population variance since it tends to produce intermediate phenotypes.

- There is a reduction in correlation between relatives or in prepotency due to the increase in heterozygosity.
- This mating system is very useful in making a population stable, i.e., in maintaining variability. Suitable parents may be selected to remove their weaknesses.
- The progeny from such a mating would be more desirable than the parents. It is also useful when the desirable type is an intermediate one and the available parents have the extreme phenotypes.
- But the most notable use of this mating system is in maintaining variability in relatively smaller populations as it reduces inbreeding.

Achievements

- Synthetics and composite approaches have been often used in improvement of cross-pollinated vegetables like onion, cauliflower, carrot, cabbage and spinach. Some examples of synthetic varieties are Pusa Early Synthetic 78-1, Pant Gobi -3 and Pusa Synthetic in cauliflower. Composites in some vegetables have been developed at SKUAST-J, Chatha, Jammu. Super Lal in carrot and C-13 in spinach beet are in advanced stage of testing.

Mass Selection

- In mass selection, seeds are collected from a large number of phenotypically similar plants available in population and are bulked to grow the next generation. In the simplest manner it is the creation of a mixture of different lines. While selection, easily observable characters like plant height, grain colour, grain size, tillering ability, disease resistance, etc. are considered.
- Mass selection increases frequencies of desirable genes. This type of selection method of plant breeding can be carried out multiple times as mass selected variety may contain considerable genetic variation. But, new genetic variability is not created; only the present one is used just to improve base population performance.

Genetic basis and Types of Mass Selection

Genetic basis of Mass Selection

- Mass selection is used in both self-pollinated and cross-pollinated species.
- In case of cross-pollinated species the mass selected varieties are heterozygous and heterogeneous.
- In self-pollinated species, these varieties are a mixture of several pure lines, means homozygous but heterogeneous.

Types of Mass Selection

It can be of two types as given below.

1. Positive mass selection 2. Negative mass selection

Procedure for Mass Selection

- Mass selection method of plant breeding is carried out in the manner explained below. It is the general procedure; steps may vary depending upon certain factors

Year	Details
First year	Plants with similar but desirable traits are selected from source population and their seeds are bulked. Alternatively we can rouge out the undesirable population and bulk the remaining.
Second year	Seeds carried from first year planted in a preliminary yield trial along with standard varieties as checks. Selected plant types are critically evaluated for phenotype.
Third year to sixth year	Co-ordinated yield trials of selected ones are conducted at several locations just to confirm that selection could perform in a similar manner at different locations within an agro-climatic zone.
Seventh year	Mass selected variety is released if it is up to the expectation. Seeds are multiplied for distribution.

Advantages and Disadvantages of Mass Selection

Advantages

- Varieties developed by mass selection have wide adaptability, wide genetic base and stability.
- Rapid, simple and least expensive of plant-breeding procedures.
- Retains considerable genetic variability.
- Of use for the improvement of land races.

Disadvantages

- Variation exists in variety developed by mass selection. Phenotype of these varieties is not as uniform as pure line varieties.
- To test the homozygosity of plants, progeny test is always required, as without progeny testing selected types will show segregation in the next generation.
- Less improvement as compared to pureline selection.
- Limited use in improvement of self-fertilizing crops.
- Varieties difficult to identify in seed certification programmes than purelines.
- Only effective for characters with high heritability.
- Only those varieties which show genetic variation can be improved by this method, as it uses variability present in population.

Applications of Mass Selection

- To maintain purity of existing variety simply by roguing of unwanted ones.
- In breeding of crops like forage crops, which are less considered from economic point of view.
- To preserve the identity of established cultivars.

Mass Selection in Cross-Pollinated Crops

- It is the simplest, easiest and oldest method of selection, where individual plants are selected based on their phenotypic performance, and bulk seed is used to produce the next generation by mixing it.
- Mass selection proved to be quite effective in maize improvement at the initial stages but its efficacy especially for improvement of yield, soon came under severe criticism that culminated in the refinement of the method of mass selection.
- The selection after pollination does not provide any control over the pollen parent as a result of which effective selection is limited only to the female parents.
- The heritability estimates are reduced by half, since only parents are used to harvest seed whereas the pollen source is not known after the cross-pollination has taken place.

Recurrent Selection and Principles

Recurrent Selection

- A method of breeding designed to concentrate favourable genes scattered among a number of individuals by selecting in each generation among progeny produced by matings inter se of the selected individuals or their self progeny of the previous generation.

Principles

- The probability of isolating an outstanding inbred line depends primarily on two factors:
  1. The proportion of superior genotypes present in the base population from which inbreds are isolated, and
  2. The effectiveness of selection during the inbreeding process in increasing the frequency of desirable genes or gene combinations.
- Recurrent selection is effective in increasing the frequency of desirable genes in the selected population. It is most suited for characters with high heritability. Based on the ways in which plants with desirable characters are identified, recurrent selection has been classified into four types.

Types of Recurrent Selection

1. Simple Recurrent Selection or Recurrent Selection for Phenotype

- In this a number of plants are self-pollinated in a source population in first year.
- At maturity, superior plants based on phenotypic performance are selected.
- In the second year, seeds produced by self-fertilization of the selected plants are planted and crossed in all possible combinations and the produce is bulked. This completes original selection cycle.
- Since selection is based on the phenotype of the plant, it is useful only for characters with high heritability.

- In case of cabbage and cauliflower, etc. where it is not possible to identity the desired selection before flowering, inter-crosses of selections may be made in the first year of each cycle and the second year may be eliminated from each cycle.
- Thus, strictly speaking, selfing is not an integral component of simple recurrent selection, rather it is done to prevent crossing from the inferior pollen grains before the plants reach selection stage.

2. Recurrent Selection for General Combining Ability (GCA)

- Here a three year cycle-involved.
- In the first year, a number of plants are self-pollinated as well as crossed to a broad-based heterozygous tester stock to identify the so-called plants with good general combining ability.
- In the second year, the crosses are evaluated to identify those that are superior. Selfs of first year are kept in reserve.
- In the third year, the reserve selfed seeds of those superior genotypes which exhibited good performance in the evaluation of GCA during the second year are grown out, inter-crossed in all possible combinations, and a composite of inter-crossed seed is used to establish an improved population for further selection.

3. Recurrent Selection for Specific Combining Ability (SCA)

- This method is the same as that of recurrent selection for general combining ability except that the tester selected is a narrow base inbred line.
- The recurrent selection for general and specific combining ability is equivalent to half sib progeny.

4. Reciprocal Recurrent Selection

- Aims at simultaneous improvement of two heterozygous and heterogeneous population, one acts as a tester for another and vice versa.
- This method is as effective as recurrent selection for SCA when non-additive effects are of major importance.

Role in Crop Improvement Programme

- Recurrent selection has been used successfully in corn, sweet clover bromegrass, strawberries and onion.
- Genetic patterns can be modified by control of light and heat. Since Borthwick and Hendrikes discovered phtyochorme in 1952, numerous studies have shown that controlled lighting and chemical treatment can modify flowering and fruiting in plants.
- This has not been tested in autogamous vegetable crops because of difficulties of producing enough seed for testing and necessary intercrossings in each cycle.
- Phenotypic recurrent selection has been tested in an elegant study in cantaloupe (muskmelon) by Andrus and Bohn (1967).
- Nine cycles of mass selection were practised in a heterozygous base population using an index representing 16 fruit characters.

Heterosis Breeding

- Hybrid vigour or heterosis, usually refers to the increase in size or rate of growth of offspring over parents. Heterosis or hybrid vigour is the increased or decreased vigour exhibited by the hybrid over their better parent value or their mid-parental value or to a standard variety.

Objectives

- Hybrid vigour in crop plants can be increased in yield of fruits/tubers/bulbs, or reduction in number of days to flower.
- Heterosis is credited for large increases in production per unit area.

Types of Heterosis

Types of heterosis on the basis of two different criteria are given below:

On the basis of Origin and Nature

1. Euheterosis or true heterosis
2. Mutational heterosis
3. Balanced heterosis
4. Pseudoheterosis or luxuriance

On the basis of types of estimation

1. Average or relative heterosis
2. Heterobeltiosis
3. Useful or standard or economic heterosis

Mutational Heterosis

- It is the simplest type of heterosis. Lethal (mostly), recessive, adaptively unfavourable mutants are either eliminated or sheltered by their non-lethal, dominant and adaptively superior alleles in cross pollinated crops. This is termed as mutational heterosis.
- Naturally occurring mutants are generally of recessive type and less adaptive to environmental conditions, hence risk of their elimination by natural selection process is higher.
- Its mutational heterosis which gives them to chance to be in population by sheltering them.

Balanced and Pseudo heterosis

Balanced Heterosis

- Well balanced gene combinations which are more adaptive to environmental conditions and useful from the agriculture point of view result in balanced heterosis. It has application in hybrid production.

Pseudoheterosis

- Also termed as luxuriance. Progeny possesses superiority over parents in vegetative growth, but not in yield and adaptation, usually sterile or poorly fertile. This concept cannot be utilized in hybrid varieties production.

Average or Relative Heterosis

- When heterosis is estimated over mid-parental value, i.e. average of two parents, it is referred to as average or relative heterosis.

  Formula

  Average heterosis = [(F1 – MP) / MP] × 100

  Where, F1 = value of F1, MP = mean value of two parents

Heterobeltiosis

- When heterosis is estimated over better parent it is called as heterobeltiosis.

  Formula

  Heterobeltiosis = [F1 – BP] / BP × 100

  Where, F1 = value of F1, BP = value of better parent

Standard Heterosis

- When heterosis is estimated over standard commercial hybrid it is called as standard heterosis. It has practical importance in plant breeding. It is also referred to as *useful* or *economic heterosis*.

  Formula

  Standard Heterosis = [(F1 – SH) / SH] × 100

Use of Heterosis in Crop Improvement Programmes

- Gains in yield and yield stability offered by heterosis have prompted use of hybrids in several crops.
- Genetic yielding ability has been increased greatly, and thus total production has been increased, with minimal dependence on chemical inputs and maximum use of biological power.
- Enthusiasm and funds have been directed to hybrid breeding, in part because of the proven efficiency of the inbred/hybrid method for producing products that farmers need and want, and in part because private capital was attracted to the profit potential of hybrid breeding and sales.
- The inbred/hybrid method has given breeders greater precision in developing, identifying, and multiplying the best hybrid genotypes in cross-pollinated crops.

Exploitation of Heterosis

- Private firms are attracted to the hybrid seed business because of the built-in plant variety protection of hybrids.

- Customers need to buy new seeds for every planting season. But for the breeding, production, and sale of hybrid seed the **"commercialization of heterosis"** can be successful only if it meets the following criteria:
  - The hybrids must satisfy the needs of the customer for all important traits. Simply to be "hybrid", or simply to exhibit "heterosis", is not enough.
  - The price of hybrid seeds must be low enough to enable the customer to make substantial profits from annually recurring investments in expensive hybrid seed.
  - A rule of thumb is that a first time use of hybrid seed should enable the farmer to earn an extra profit equal to at least three times the added cost of the seed.
  - The price of hybrid seed must be high enough to enable the seed company to make substantial profits from its investments in research, production, and sales.
  - A successful seed company needs to realize a 10-15% return on equity.
  - Its investments in research are one of the essential business expenditures for a research-based seed company and should be equivalent to 5-10% of sales income.

Two other criteria

1. Farmers will risk investment in improved seed only when they have some assurance of a fair price in a dependable market for their crop.
2. Government regulations, formal and informal, must give minimal hindrance to honest and prudent business operations. These two requirements apply to all seed firms and not just hybrid seed companies. They have particular significance in many developing countries.

To satisfy the three primary criteria for success in the hybrid seed business, companies must integrate a host of variables such as:

1. The pollinating system of the crop
2. Options for manipulation of the pollinating system
3. Supply and cost of labour for emasculation or other requirements for hybridization
4. The yield of the crop in the farmer's field
5. The commercial value of the crop per unit of land area
6. The seeding rate of the crop
7. The seed yield in the seed production field
8. The extra yield to be expected from heterosis
9. The implications of hybrid uniformity
10. The most important traits to improve in the crop, and their genetics
11. The ease of demonstrating improvements in new hybrids
12. Availability of inbred parents and other breeding materials in either public or private institutions. The following examples illustrate, for three different crop species, some of the many ways in which these twelve variables can be integrated

Single Cross

- Two inbreds are crossed A × B = AB (Hybrid). A hybrid progeny from a cross between two unrelated inbreds

Double Cross or Double Hybrid

- Two inbreds are crossed, similarly another two are crossed and then their F1 hybrids are crossed e.g. (A × B) × (C × D) = (AB × CD) = ABCD.

Poly Cross

- In case of inbred lines which cannot be crossed easily. Planting is done in such a way to get equal chance to cross with each other.
- Maintain the lines by vegetative propagation or interse mating of progenies of same generation if vegetative propagation is not possible.

Synthetic Variety

- The use of synthetic varieties for commercial cultivation was first suggested in maize(Hayes and Garber(1919).
- A synthetic variety is produced by crossing in all combinations a number of inbreds(4-6) that combine well with each other.
- The inbreds are tested for GCA. Synthetic is maintained by open pollination.
- The lines that make up a synthetic variety may be usually inbred lines but open pollinated varieties, or other populations tested for GCA also used. Synthetic varieties are common in grasses, clover, maize and sugarbeet.
- The normal procedure is equal amounts of seeds from parental lines is mixed and planted in isolation.
- Open pollination is allowed.

Features of Synthetic Variety

- Heterogeneous
- Synthetic variety can be developed by using clones, inbreds or OPV i.e. open pollinated variety
- Cross-pollination is a must
- Maintained by open pollination
- Unlike composite variety, exact reconstitution of synthetic variety is possible
- More adaptive to varying growing conditions as compared to hybrids
- Less uniform as compared to hybrids
- Less attractive as compared to hybrids
- Show some amount of heterosis as compared to OPV
- Have better disease resistance.

Steps Involved in Synthetic Variety

Development of synthetic varieties consists of four major steps

- Isolation of inbred lines
- Evaluation of inbred lines for GCA
- Intermating of good general combining inbreds in all possible combinations
- Mixing the seed of all F1 crosses in equal quantity

Procedure of Developing Synthetic Variety and Role in Crop Improvement

Procedure of Developing Synthetic Variety

Year	Main Breeding Activity
First	Selection of inbred lines, crossing with a common tester, harvesting top crossed seed separately.
Second	Evaluation of top crosses in replicated trial using standard hybrid or open-pollinated variety as check. Identification and selection of good general combining inbred lines based on top cross performance.
Third	Making all possible single crosses among inbred lines selected for good GCA and harvesting crossed seed of single crosses separately.
Fourth and fifth	Mixing seed of single crosses in equal quantity and seed multiplication by open pollination in isolation for one or two generations.

Role in Crop Improvement

- Synthetic varieties have been widely used in forage crops and in crops where pollination control is difficult. The maize breeding programme at CIMMYT, Mexico and the pearl millet breeding programme at ICRISAT, Hyderabad (to some extent) are based on synthetic varieties generally developed through population improvement.
- The maize breeding programme in India is placing increasingly greater emphasis on the production of synthetic or composite varieties.
- Pusa Early Synthetic cauliflower, a synthetic variety adapted to many environments, was developed at IARI.

Merits and Demerits of Synthetic Variety

Merits

- Less costly compared to hybrids.
- Farmer can maintain his synthetic variety for more seasons which is not possible in hybrids.
- Because of wider genetic variety base, the synthetics are more stable over years and environments.
- Seed production required more skilled operation in hybrids whereas it is not so in synthetics.

- Synthetic varieties have vast genetic variability which provides better protection from the infestation of new races of a disease.

Demerits

- Performance is little bit lower compared to single or double cross hybrids because synthetics exploit only GCA while hybrids exploit both GCA and SCA.
- The performance of synthetic is adversely affected by lines with relatively poorer GCA. Such lines have to be included to increase the number of parental lines making up the synthetics as lines with outstanding GCA are limited in number.
- The performance of synthetic varieties is less uniform and less attractive.
- Synthetic varieties are utilized in cross-pollinated species only whereas hybrids can be developed in both cross- and self-pollinated species.

Composite Varieties

Composite Varieties

- In cross-pollinated crops, the mixture of genotypes from several sources that is maintained in bulk from one generation to the next is referred to as composite variety.
- Composites are constituted by seed mixture of advanced generation material of intervarietal or inter-racial cross.

Steps In Development of Composite Variety

1. Selection of base or parental material.
2. Intermating of selected genotype.
3. Evaluation of crosses.
4. Mixing of parental material of superior crosses in equal quantity.

1. Selection of Base Material

- In developing composite cultivars, open-pollinated varieties or other heterozygous sources serve as the base material.
- Several open-pollinated varieties with similar maturity duration, height, and seed colour are selected for this purpose.

2. Evaluation of Crosses

- All single crosses are evaluated in replicated trials for yield performance in F1, F2, F3 & F4 generations using standard hybrid or open-pollinated variety as a check.
- Crosses exhibiting little or no inbreeding depression in F2, F3 and F4 generations are identified and selected for the development of composite cultivars.

3. Intermating Selected Genotype

- Selected genotypes are crossed in all possible combinations.
- Total crosses to be made are equal to $n(n-1)/2$, where n is the no of varieties selected for intermating.

- If there are 10 varieties there would be 10(9)/2 = 45 single crosses.
- Seeds of each cross are harvested separately.

4. Mixing Parental Seed of Superior Crosses

- Parental seeds (of the superior crosses) is mixed in equal quantities to develop a composite variety.
- Seed multiplication is carried out by open-pollination in isolation for one or two generations.
- Composite cultivars also exhibit more of additive gene action, because composites are constituted from the parents of only those crosses which exhibit little or no inbreeding depression in F2, F3 and F4 generations.

5. Features of Composite Variety

- Heterogeneous.
- Relevant to cross-pollinated species only.
- Can be developed from open-pollinated variety or any other heterozygous variety.
- Farmer can use his own saved seeds for 3 to 4 years, after that the seeds should be replaced.
- There can be two or more constituent genotypes.
- Evaluation for general combining ability (GCA) as in synthetic variety production, is not carried out.
- Exact reconstitution of a composite variety is not possible.

Procedure of Developing Composite Variety

Year	Main Breeding Activity
First	Selection of base material, intermating in all possible combinations and harvesting crossed seeds separately
second	Evaluation of F1 crosses in replicated trial using hybrid or open pollinated variety as check.
Third to fifth	Evaluation of F2, F3 and F4 generations in replicated trial using standard check, identification of crosses exhibiting little or no inbreeding depression.
Sixth and seventh	Mixing parental seeds of superior crosses in equal quantities to a constitute composite variety, seed multiplication by open-pollination in isolation.
Eighth	Release of new composite variety, distribution of the seeds to the farmers for commercial cultivation.

Differences between Synthetic and Composite Cultivars

Particulars	Synthetic cultivars	Composite cultivars
Base material	Inbreds or clones	Open-pollinated varieties or any other eterozygous source
Component genotypes	Two to eight	Two to many
Evaluation of GCA	Always done	Usually not done
Seed replacement	After 4-5 years	After 3-4 years
Exact reconstitution	Possible	Not possible

Achievements: Synthetic Varieties and Composite Varieties

Achievements: Synthetic Varieties

- Synthetic varieties have been developed in cross-pollinated crops like cauliflower, cabbage, beetroot, onion and carrot and several other crops gram in the USA.
- In India, synthetic varieties have been developed in pear millet at ICRISAT and in sugarbeet at GBPUAT, Pantnagar.
- Cauliflower synthetic 3 has also been developed.

Achievements: Composite Varieties

- In India, the first composite varieties were released in 1967: the six maize composites were, Ambar, Jawahar, Kisan, Vikram, Sona and Vijay. Subsequently, a number of other composites have been released by both thc central and provincial variety release committees.
- At present there are more than 25 maize composites recommended for cultivation in the various parts of the country.
- These composites range from those based on local materials (Makki Safed 1 and Moti), through those involving both Indian and exotic sources (Mansar, Chandan 3, Diara Composite, etc) to those developed from exotic lines only (Chandan Safed 2, Amber Pop, Chandan 1, etc).
- Some of the composites involves as many as 18 (composite C1) or even 22 (Agaity 76) lines, while certain others involving only two (Jawahar, Kisan, Sona) or even one (Chandan Safed 2, A-de-Cuba) population.
- Some of the recently released maize composites are: Co 1 (full season, resistant to downy mildew), NLD (full season, white seeded), Renuka (very early), Kanchan (very early), and Diara 3 (developed by three cycles of full-sib selection in Diara Composite).
- These composites yield as much as 90% of the best hybrid varieties.
- Three opaque 2 composites, viz., Shakti, Ratan and Protina, have been released.

They have twice the amount of lysine and tryptophan as compared to the normal maize hybrids and composites.

- Thus the nutritional value of these composites is superior to the other composites and hybrids.
- A composite variety, Composite I, has been evolved in *Brassica campestris var.* toria.
- It was developed by compositing 10 elite toria strains; it matures in 100 days, exhibits profuse branching, yields about 11 q/ha of seed, which contain about 40% oil.

Biometrics

Elements of Biometrics

- Biostatistics (a contraction of biology and statistics; sometimes referred to as biometry or biometrics) is the application of statistics to a wide range of topics in biology.
- The science of biostatistics encompasses the design of biological experiments, especially in medicine and agriculture; the collection, summarization, and analysis of data from those experiments; and the interpretation of, and inference from, the results.

Biometrics and the History of Biological Thought

- Biostatistical reasoning and modeling were of critical importance to the foundation theories of modern biology.
- In the early 1900s, after the rediscovery of Mendel's work, the conceptual gaps in understanding between genetics and evolutionary Darwinism led to vigorous debate between biometricians such as Walter Weldon and Karl Pearson and Mendelians such as Charles Davenport, William Bateson and Wilhelm Johannsen.
- By the 1930s statisticians and models built on statistical reasoning had helped to resolve these differences and to produce the neo-Darwinian modern evolutionary synthesis.
- The leading figures in the establishment of this synthesis all relied on statistics and developed its use in biology.
- Sir Ronald A. Fisher developed several basic statistical methods in support of his work *The Genetical Theory of Natural Selection*.
- Sewall G. Wright used statistics in the development of modern population genetics.
- J. B. S Haldane's book, *The Causes of Evolution*, re-established natural selection as the premier mechanism of evolution by explaining it in terms of the mathematical consequences of Mendelian genetics.
- These individuals and the work of other biostatisticians, mathematical biologists, and statistically inclined geneticists helped to bring together evolutionary biology and genetics into a consistent, coherent whole that could begin to be quantitatively modeled.
- In parallel to this overall development, the pioneering work of D›Arcy Thompson *Growth and Form* also helped to add quantitative discipline to biological study.

- Despite the fundamental importance and frequent necessity of statistical reasoning, there may nonetheless have been a tendency among biologists to distrust or deprecate results which are not qualitatively apparent.
- One anecdote describes Thomas Hunt Morgan banning the Friden calculator from his department at Caltech, saying Well, I am like a guy who is prospecting for gold along the banks of the Sacramento River in 1849. With a little intelligence, I can reach down and pick up big nuggets of gold. And as long as I can do that, I'm not going to let any people in my department waste scarce resources in placer mining. Educators are now adjusting their curricula to focus on more quantitative concepts and tools.

Applications of Biometrics

- The biometric principles and tools can be applied in the following areas:
- Public health, including epidemiology, health services, research, nutrition, and environmental health.
- Design and analysis of clinical trials in medicine.
- Population genetics and statistical genetics in order to link variation in genotype with a variation in phenotype. This has been used in agriculture to improve crops and farm animals (animal breeding).
- In biomedical research, this work can assist in finding candidates for gene alleles that can cause or influence predisposition to disease in human genetics.
- Analysis of genomics data, for example from microarray or proteomics experiments. Often concerning diseases or disease stages.
- Ecology, ecological forecasting.
- Biological sequence analysis.
- Systems biology for gene network inference or pathways analysis.

Statistical Tools in Crop Improvement

- A study of DNA polymorphism has become an active area of research in all important crops and several model plant species like *Arabidopsis thaliana* and *Brachypodium distachyon*.
- This involves development and use of molecular markers, which have proved useful not only for marker-assisted selection during plant breeding, but also for understanding crop domestication and plant evolution.
- To resolve the pattern of DNA polymorphism in any crop, the ultimate approach would be to sequence/resequence the entire genome (or a part of it) in a large number of accessions.
- This was, however, unimaginable during the 1980s and still remains cost ineffective, therefore DNA-based molecular markers (for example, Restriction Fragment Length Polymorphisms(RFLP), Random Amplified Polymorphic DNAs(RAPD), Simple Sequence Repeats (SSRs) and Amplified Fragment Length Polymorphisms

(AFLPs)) have largely been employed for the study of DNA polymorphism (Collard et al., 2005).

- Most of these molecular markers are based on the use of restriction digestion of genomic DNA, followed by hybridization of electrophoresed DNA, and/or visualization of the products of PCR carried out using suitably designed PCR primers.
- More recently, however, single nucleotide polymorphisms (SNPs), whose discovery was largely based on sequence information, became the markers of choice due to their abundance and uniform distribution throughout a genome.
- Once discovered, SNP genotyping can be done using any of the dozens of available methods. For SSR/SNP genotyping, some efforts in the past were made to provide for the desired high throughput and cost effectiveness through the use of PCR tetrad machines (handling 384 PCR reactions at a time), multiplexing, multiple loading of the gels and the use of automatic sequencers.
- However, this appeared inadequate and there has been an increasing demand to develop ultra-high-throughput low-cost assays for a variety of novel marker systems including SNPs.
- These new methods will allow ultra-high throughput genotyping of either one or few individuals for hundreds of thousands of markers, or that of thousands of individuals for one or few markers.
- High-density oligonucleotide arrays, which are now becoming available in several crops, provide a means for achieving this goal of low-cost ultra-high-throughput genotyping.
- These arrays may also be custom-made according to specific needs and, therefore, also allowed for the development of novel marker systems like single feature polymorphisms (SFPs) (including gene-specific hybridization polymorphisms and gene expression markers), diversity array technology (DArT) and restriction site-associated DNA (RAD) markers, which have now become the markers of choice.
- Technologies have also been developed, which make use of tag arrays for detection of the products of genotyping reactions.
- These novel array- or chip-based markers are useful for a variety of purposes including genome-wide association studies, population studies, bulk segregant analysis, quantitative trait loci (QTL) interval mapping, whole genome profiling and background screening and so on (Steinmetz et al., 2002; Winzeler et al., 2003; Wenzl et al., 2004, 2007a; Hazen et al., 2005; Kim et al., 2006).
- A brief account of the development and use of these high-throughput array-based molecular markers in plants is presented in this study.

Crop Biotechnology

Biotechnology is the use of living systems and organisms to develop or make useful products, or "any technological application that uses biological systems, living organisms or derivatives thereof, to make or modify products or processes for specific use" (UN Convention on Biological Diversity).

Genetically modified crops (**GMCs**, **GM crops**, or **biotech crops**) are plants, the DNA of which has been modified using genetic engineering techniques. In most cases the aim is to introduce a new trait to the plant which does not occur naturally in the species. Examples include resistance to certain pests, diseases, or environmental conditions, or resistance to chemical treatments (e.g. resistance to a herbicide), or the production of a certain nutrient or pharmaceutical agent.

GM Crop Status in India

A. Commercial Release of Transgenic Crops

Bt cotton is the only transgenic crop approved for commercial cultivation in India. Bt technology was deployed in cotton crops through genetic engineering techniques for control of bollworms, the major pest, thereby reducing the risk of crop failures and use of pesticides. Bt cotton producing a natural insecticide that comes from the ubiquitous **soil bacterium** known as ***Bacillus thuringiensis*** was approved by the Genetic Engineering Approval Committee (GEAC) for introduction in India in 2002 in the nine cotton growing states, namely, Andhra Pradesh, Gujarat, Haryana, Karnataka, Madhya Pradesh, Maharashtra, Punjab, Rajasthan and Tamil Nadu after extensive biosafety and agronomic testing.

As of date, the GEAC has approved Bt cotton hybrids expressing Cry 1 Ac gene (MON 531 event) and stacked genes Cry1 Ac and Cry 2Ab (MON 15985 event) BG-II developed by M/s Mahyco, encoding fusion genes (cry 1Ab+Cry Ac) GFM developed by M/s Nath Seeds and cry 1Ac gene (Event-1) developed by M/s J. K. Agrigenetics Ltd during 2002 to 2007.

In addition the GEAC also approved the commercial release of BN Bt (variety) developed by CICR, Nagpur. For the first time Bt technology has been introduced in a varietal background whereby the farmers can save seeds. During the period 2002 to 2008 the GEAC has approved 267- Bt cotton hybrids of which 132 were approved during 2008.

B. Status of Approval of GM Food Crops in India

The large-scale field trials of Bt brinjal developed by M/s Mahyco expressing cry 1Ac gene from *Bacillus thuringiensis* tolerant to the fruit and shoot borer is under progress at 11 locations within the research farms of Indian Institute of Vegetable Research(IIVR) /State Agricultural Universities/ Indian Council of Agriculture Research. The trials are being conducted under the supervision of Director IIVR, Varanasi. Details of other GM crops under various stages of field testing are as follows:

Table 7.12: *Field trials of gm crops (containing new genes/events) approved by the GEAC in year 2008*

S. No.	Crop	Company Name	Trial	Trait	Gene/Event
1.	Cauliflower	Sungro Seeds Research Ltd.	BRL-I	Insect resistant	*cry1Ac*
		Nunhems India Pvt. Ltd	Event selection	Insect resistant	*cry1b and cry 1C*
2.	Cotton	Dow Agro Sciences India Pvt. Ltd.	BRL-I	Insect resistant	cry1Ac& cry1F (Wide Strike = Event 3006-210-23 and Event 281-24-236)
		JK Agrigenetics Ltd.	BRL-I	Insect resistant	cry1Ac (Event-1) and *cry1EC (Event-24)*
		MAHYCO	BRL-I	Insect resistance and herbicide tolerant	cry1Ac & cry2Ab (MON 15985) and CP4EPSPS (MON 88913)
		Metahelix Life Sciences	LST	Insect resistant	cry1C (MLS9124 event)
		Central Institute for Cotton Research	LST	Insect resistant	cry 1Ac
3.	Rice	Bayer Bioscience Pvt. Ltd.	Event selection	Insect resistant	cry 1 Ab, cry 1C & bar
		Avesthagen Ltd.	Event Selection	Oxdative stress	*Oryza sativa* taipae 309
4.	Tomato	Avesthagen Ltd.	Event selection	Increased lycopene content	unedited NAD9
5.	Groundnut	ICRISAT	Event selection	Tobacco streak virus against peanut stem necrosis disease.	Coat protein gene
6.	Cabbage	Nunhems India Pvt. Ltd	Event selection	Insect Resistant	cry1b and cry 1C
7.	Potato	Central Potato Research Institute.	Event selection	Late blight resistant	RB gene
8.	Corn	Monsanto India Ltd.	BRL-I	Insect resistant and herbicide tolerant	Cry 1A.105 (chimeric gene) and cry2Ab2

BRL-I: Biosafety Research Level-I

BRL-II: Biosafety Research Level II

LST: Large-Scale Trial

Source: Website of Ministry of Environment & Forests, Government of India

- Biotechnology is a rapidly developing area of contemporary science.
- It can bring new ideas, improved tools and novel approaches to the solution of some persistent, seemingly intractable problems in food crop production.
- Given the pressing need to enhance and stabilize food production in response to mounting population pressures and increasing poverty, there is an urgent need to explore novel terminologies that will break traditional barriers.
- Biotechnology is usually defined as "any technique that uses living organisms, or substances from those organisms, to make or modify a product, to improve plants or animals, or to develop microorganisms for specific uses".
- It comprises a continuum of terminologies, ranging from traditional biotechnology such as plant tissue culture to modern biotechnology such as genetic engineering of plants and animals and represents the latest front in the ongoing scientific progress of this century.
- However, its increasing importance, at least in plant improvement, should not obscure the fact that traditional plant breeding based on hybridization followed by selection and evaluation of a large population in the field, accounts for over 50% of the global increase in agricultural productivity.
- Not only have particularly important new genotypes been bred in Asia through the so-called Green Revolution but also, worldwide, new varieties have been bred in response to the changing needs of agriculture.

Key areas in which biotechnology has application

- Multiplication of selected/elite clones from hybrid or specific parental lines and rapid production of disease-free propagates through conventional micro-propagation or somatic embryogenesis.
- Development of plant varieties, in conjunction with whole plant breeding through genetic engineering of plant species.
- Application of agricultural microbiology to produce microorganisms beneficial to crop cultivation.
- Development of techniques, based on the use of monoclonal antibodies and nucleic acid probes, for the diagnosis of plant pests and diseases and the detection of foreign chemicals in food.
- Application of genetic mapping techniques, based on the use of molecular markers, as an aid to conventional plant breeding programmes.

Current Role of Biotechnology in Crop Improvement

Some of the applications of biotechnology that are currently being used in crop production are given below:

- The fact that a whole plant can be regenerated from a single cell, explant, or organ makes tissue culture a valuable technique to proliferate genetically identical material and select interesting variants for commercial purposes.
- Totipotency, which states that cells are autonomic and, in principle, capable of regenerating to give a completely new plant, also allows a genetic change made at the cellular level, to become an established traits of a whole plant.

- The newly introduced or selected trait can, subsequently, be passed on to future generations of the species by conventional breeding methods.
- Tissue culture is particularly important for vegetatively propagated crop species since it reduces the labour associated with line maintenance and germplasm conservation.
- Micropropagation is also used for the exchange of disease-free plant material. Anther or pollen culture also facilitates in early achievement of homozygosity, and is particularly useful to breed for recessive traits.

Genetic Engineering in Plants

Genetic Engineering Appraisal Committee

The Genetic Engineering Appraisal Committee (GEAC) is the apex body constituted in the Ministry of Environment and Forests under 'Rules for Manufacture, Use, Import, Export and Storage of Hazardous Micro-organisms/Genetically Engineered Organisms or Cells 1989', under the Environment Protection Act, 1986. The Rules of 1989 also define five competent authorities, i.e. the Institutional Biosafety Committees (IBSC), Review Committee of Genetic Manipulation (RCGM), Genetic Engineering Approval Committee (GEAC), State Biotechnology Coordination Committee (SBCC) and District Level Committee (DLC) for handling of various aspects of the rules

Table 7.13: *Landmark discoveries leading to development of transgenics*

Year	Discoveries
1900	De Vries, Carl Correns, and Enrich von Tschermak- Rediscovered Mendel's law of inheritance
1902	Walter Sutton and T. Boveri- proposed the chromosome theory of heredity
1909	Wilhelm Johannsen- Coined the term 'Gene'
1910	Thomos Hunt Morgan- Demostrated linkage between genes
1941	G.Beadle and E. Tatum- Proposed one gene one enzyme hypothesis
1944	Avery, Macleod and Mc Carty- Demonstrated DNA as the genetic material
1953	Francis Crick and James Watson- Discovered double helical structure of DNA
1954	Skoog and Miller- Plant regeneration from cell cultures
1961	S.Brenner, F. Jacob, and M. Meselson- Discovered mRNA
1966	Nirenberg, Khorana and Ochoa- Cracked the genetic code
1968	M. Meselson and H.C. Smith, K.W. Wilcox and T.J. Kelley- Described first restriction enzyme
1972	Stanely Cohen and H. Boyer- Produced the first recombinant DNA molecule
1973	Stanley Cohen cloned the first animal gene- in to *E.coli*
1975-77	Sanger, Maxam and Gillbert- Developed DNA sequencing methods
1976	Genentech- First genetic engineering company established
1983	E.L. Herrera-Estella, A. Depicker, M. Van Montagu J.S. Chel- Demonstrated Agrobacterium mediated transformation in plants

Year	Discoveries
1987	Burke et al- Developed Yeast Atificial Chromosome (YAC) as vectors in plants
1990	N.Sternberg- Developed Bacterial Artificial Chromosome (BAC) as vectors in plants
1994	Calgene- Developed Flavr Savr, first transgenic tomato showing extended shelf life.

Engineering Plant Metabolic Pathways through RNAi

RNAi has been used to modify plant metabolic pathways to enhance nutrient content and reduced toxin production (Summarized in Table 1). The technique takes advantage of the heritable and stable RNAi phenotypes in plants.

Table 7.14: *Examples of novel plant traits engineered through RNAi.*

Trait	Target Gene	Host	Application
Enhanced nutrient content	Lyc	Tomato	Increased concentration of lycopene (carotenoid antioxidant)
	DET1	Tomato	Higher flavonoid and b-carotene contents
	SBEII	Wheat, Sweet potato, Maize	Increased levels of amylose for glycemic management and digestive health
	FAD2	Canola, Peanut, Cotton	Increased oleic acid content
	SAD1	Cotton	Increased stearic acid content
	ZLKR/SDH	Maize	Lysine-fortified maize
Reduced alkaloid production	CaMXMT1	Coffee	Decaffeinated coffee
	COR	Opium poppy	Production of non-narcotic alkaloid instead of morphine
	CYP82E4	Tobacco	Reduced levels of the carcinogen nornicotine in cured leaves
Heavy metal accumulation	ACR2	*Arabidopsis*	Arsenic hyperaccumulation for phytoremediation
Reduced polyphenol production	s-cadinene synthase gene	Cotton	Lower gossypol levels in cottonseeds for safe consumption
Ethylene sensitivity	LeETR4	Tomato	Early ripening of tomatoes
	ACC oxidase gene	Tomato	Longer shelf life because of slow ripening
Reduced allergenicity	Arah2	Peanut	Allergen-free peanuts
	Lolp1, Lolp2	Ryegrass	Hypo-allergenic ryegrass
Reduced production of lachrymatory factor synthase	lachrymatory factor synthase gene	Onion	"Tearless" onion

Meristem and bud Culture

- Zygotic embryo culture, anther and microspore culture, cell and tissue culture, chromosome engineering, protoplast culture, genetic engineering, molecular markers, monoclonal antibodies, micro propagation for commercial purposes, genetic conservation, and exchange of material, inter-specific crosses, haploid production, in vitro selection, somaclonal variation, somatic embryogenesis, artificial seeds $2n$ gametes for inter-specific crosses, fusion for somatic hybridization gene transfer, aid to breeding programmes to develop resistant varieties to plant diseases.
- In vitro culture of zygotic embryos (embryo rescue) has enabled us to overcome barriers to a number of inter specific crosses from zygotic failure. Several interspecific hybrids have been obtained in this way—for example, between Lycopersicon species and Solanum species, among cucurbits, and among legumes. In vitro culture of anthers allows the regeneration of large number of haploid plants from which by chromosome doubling via colchicine treatment or via further in vitro culture, diploid plants, homozygous at all loci can be obtained.
- Lines produced from anther culture of hybrids are obtained in less time and show greater variability than those obtained by self-pollination; several varieties produced using this method have already been released. Direct culture of microspore has proved to be efficient in Brassica species.

Agrobacterium-mediated Genetic Engineering

- Recombinant DNA procedures can be used to make discrete changes in the genetic makeup of plants.
- Transgenic plants of over 50 species that contain genes from other plant species, bacteria, viruses, and animals, are currently available.
- The foreign genes expressed in transgenic plants confer on them a variety of important agronomic traits including: insect resistance, virus resistance, microbial resistance, herbicide resistance, altered macromolecular composition, modified reproductive capacity, and delayed senescence.
- Alien genes from organisms that cannot hybridize with AVRDC mandate crops can be introduced through genetic engineering. Such gene transfers are impossible with conventional breeding methods.
- Genetic engineering also allows introduction of only one or two well-characterized genes at a time, unlike hybridization where extensive backcrossing is needed to remove undesirable genes.
- Efforts have been made at AVRDC in producing transgenic mungbean, hotpepper (chilli), tomato, and vegetable soybean plants via the Agrobacterium-mediated uptake of marker genes.
- Bacterial genes that conferred antibiotic resistance 'marked' transformed cells.
- These selectable marker genes permitted the transformed cells to be grown in antibiotic that would kill non-transformed cells. Such marker genes have been used to optimize the transformation protocol and to study the inheritance and stability of introduced genes.

- Cucumber mosaic virus (CMV) causes large yield losses in pepper and tomato.
- Since there are no resistant sources in germplasm and no effective way to prevent the infection, the novel approach of genetic engineering becomes a new solution to problems.
- The truncated CMV RNA-2 replicase and coat protein genes were transferred to pepper and tomato.
- The detection of certain nucleotides had resulted in a truncated translation product about 75% of the full length protein.
- The AVRDC scientists have observed a high degree of CMV resistance in the RO and R1 lines of the transgenic chilli pepper and tomato plants.
- Once a gene is successfully transferred, transgenic plants need to be grown and tested in a newly-constructed, special confinement facility — a greenhouse for transgenic plants — in order to establish the stability and effectiveness of the gene, and to further produce transgenic plants.
- The movement of transgenic plant materials from the laboratory or greenhouse to the field is highly regulated in many countries.
- The regulations state that field tests of transgenic material can only be done under a permit that is granted on a case by case basis.
- However, the results from many studies in the last five years indicate that the use of genetic engineering does not impose extraordinary risks on environment and society.
- It is obvious from the level of activity in this area that the pipeline is full with a large variety of potential products.

Isolation of useful Genes

- Before the means to transfer the gene (trait) into a recipient plant being established, the trait or gene of interest has to be identified and isolated.
- AVRDC scientists have identified several strains of bacteria isolated from field soils and plant roots, which have strong inhibitory effects of *Fusarium* wilt pathogen and other fungi.
- Genomic libraries have been constructed from the two bacteria, and transformants from these libraries are presently being screened for antifungal genes.
- Another bacteria has also been identified to possess an inhibitory effect on *fusarium* wilt pathogen as well as chitinolytic activity, i.e., the digestion of the major component of fungal cell wall, chitin.
- The chitinase gene from this bacterium is also in the process of being cloned.
- Although both mungbean and vegetable soybean are rich in proteins, their nutritional values are low due to low sulfur-containing amino acids.
- Among AVRDC's mungbean and soybean germplasm collection, there are no accessions with significant levels of sulfur-rich proteins. In cooperative project with the University of Hawaii, AVRDC is attempting to introduce the cloned cDNA encoding the 2S proteins from paradise nuts for protein quality improvement in mungbean and vegetable soybean.

Biosafety

- It is prevention of large-scale loss of biological integrity, focusing both on ecology and human health.
- Biosafety is related to several fields.
- In ecology (referring to imported life forms from beyond ecoregion borders).
- In agriculture (reducing the risk of alien viral or transgenic genes, or prions such as BSE/"MadCow", reducing the risk of food bacterial contamination).
- In medicine (referring to organs or tissues from biological origin, or genetic therapy products, virus; levels of lab containment protocols measured as 1, 2, 3, 4 in rising order of danger).
- In chemistry (i.e., nitrates in water, PCB levels affecting fertility)
- In exobiology (i.e., NASA's policy for containing alien microbes that may exist on space samples - sometimes called "biosafety level 5").
- The international Biosafety Protocol deals primarily with the agricultural definition but many advocacy groups seek to expand it to include post-genetic threats: new molecules, artificial life forms, and even robots which may compete directly in the natural food chain.
- Biosafety in agriculture, chemistry, medicine, and exobiology and beyond will likely require application of the precautionary principle, and a new definition focused on the biological nature of the threatened organism rather than the nature of the threat.
- When biological warfare or new, currently hypothetical, threats (i.e., robots, new artificial bacteria) are considered, biosafety precautions are generally not sufficient. The new field of biosecurity addresses these complex threats.
- Biosafety level refers to the stringency of biocontainment precautions deemed necessary by the Centers for Disease Control and Prevention (CDC) for laboratory work with infectious materials.

Embryo Culture

- Removal of developing young embryos from seeds and their cultivation in vitro. Exchange of genes between cultivated and wild species and transfer of resistant gene(s) from wild species to cultivated varieties is hampered due to certain barriers in sexual hybridization. Therefore, the in vitro method has been successfully used to overcome such barriers in distant hybridization. In this case, hybrid embryo are rescued prior to their abortion and cultured directly into artificial media. The embryo culture technique has been practically utilized for proper development of fertilized embryos in the following cases:
- *Solanum lycopersicum* syn (*Lycopersicon esculentum) x L.peruvianum*
- *Capsicum pendulum x C.annuum*
- *Phaseolus vulgaris x P.acutifolius*
- *Cucurbita pepo x C.ecuadorensis*
- *Cucurbita metuliferus x C.zeyheri*
- *Abelmoschus esculentus x A. Moschatus*
- *Abelmoschus esculentus x A. ficulneus*

Meristem Culture

- Regeneration of a plant from tissues of an actively dividing organ like stem tip, root tip or vegetative bud.
- Aseptic culture of shoot apical meristem on nutrient media for the production of complete plant.
- Apical meristems are excised from the shoot tip and placed on media.
- Meristem is not capable of growing independently unless some leaves primordial are retained below it.
- By this method plants can be rapidly multiplied, regenerated plants are genetically similar to the donar plants and pathogen free plants especially plants free from viruses can be produced.

Anther Culture

- Culture of anthers (or pollen grains) on a suitable medium (in vitro) for the production of callus and/or haploid plants.
- This technique can be used to obtain haploid in many vegetable crops e.g.potato, brinjal, tomato, capsicum, pea, French bean, muskmelon, and sprouting broccoli.

Somatic Hybridization

- Somatic hybridization allows fusion of complete cytoplasm of two parents. The cytoplasm mix, obtained offers the opportunity of producing hybrids.

Their role in Crop Improvement Programme

- Somatic cell fusions have been successfully employed for resynthesis of *Brassica napus* by fusing protoplasts of *B. oleracea* and *B. compestris.*
- Cytoplasmic male sterile gene(s) can also be transferred in vegetables like onion and carrots to produce hybrid seed cheaply.
- Hexaploid hybrid resulting from protopast fusion of *Solanum brevidem x S. tuberosum*, which show resistance to leaf roll virus and race 'O' of late blight.
- Also protoplast fusion of *Solanum melongena x S. sisymbrifolium* show resistance to phomopsis blight.
- Fusion between sexually incompatible members of the same family has produced hybrid plants that retain some chromosome numbers of both parents. The best known example is the **'pomato'** which is potato – tomato somatic hybrid.

Molecular Markers

Molecular markers are mainly divided into two categories

1. Hybridization based markers
2. PCR based markers

Two types of PCR based Markers

1. PCR with arbitrary primers
2. PCR with sequence specific primers

Hybridization based Markers

- Hybridization based marker technology use cDNA, cloned DNA elements or synthetic oilgonucleotides as probes, which are labeled with radioisotopes or with conjugated enzymes that catalyze a coloured reaction to hybridize DNA.
- The DNA, either cleaved with restriction enzyme or amplified by PCR, are separated by gel electrophoresis and transferred to a solid support matrix by Southern hybridization and hybridized mainly with labeled probe of known origin and sequence.
- RFLP is representative of this type of technology. Restriction polymorphism occurs when mutation removes an existing restriction site or creates a new restriction site. These alterations are detected by using a labeled probe.

Restriction Fragment Length Polymorphism (RFLP)

- RFLP was the first technique which enables the detection of polymorphism at the DNA sequence level. In this method, DNA is digested with restriction enzymes that cut the DNA at specific sequences, electrophoreed, blotted on a membrane and probed with a labeled oligonucleotide.
- The DNA sequence variation detected by this method was termed restriction fragment length polymorphism.

Advantages

- RFLP are co dominant markers, enabling heterozygotes to be distinguished from homozygotes.
- The method is simple and no sequence specific information is required.
- They are reliable markers in linkage analysis and can be easily determined if linked.
- Trait is present in a homozygous or heterozygous state in an individual, an information highly desirable for recessive traits.
- This technique has been successfully employed for tagging genes of economic value.

Limitations

- Utility of RFLP has been hampered due to the large amount of DNA required for restriction digestion and southern blotting.
- The requirement of radioactive isotopes makes the analysis relatively expensive and hazardous.

PCR based Markers

- It is a PCR method, whereby a specific sequence of nucleotides within a double stranded DNA is amplified.
- The sequence is identified by the use of short synthetic oligonuclcotides that are complementary to the terminal regions of the DNA sequence to be amplified; these oligonucleotides are extended by the thermostable Taq DNA polymerase on the DNA template.

PCR with Arbitrary Primers

- In this category of PCR based markers, the random or arbitrary primers are used for DNA fingerprinting. Random primers of different lengths and sequences are mainly used for this purpose.
- Primer sequence is mainly selected in such a way that it binds with most probability to their complementary sequence on the template DNA.
- The molecular markers of this category are divided on the basis of their primer lengths like RAPD (Random Amplified Polymorphic DNA), AP-PCR (Arbitrary Primed-PCR), DAF (DNA amplification Fingerprinting), and AFLP (Amplified Fragement Length Polymorphism).

Random Amplified Polymorphic DNA (RAPD)

- A PCR based molecular marker technique in which a single type of random decamer primers are used to amplify a set of DNA segment distributed randomly throughout the genome.
- In this technique, a single species of decamer (10 bp) primers is used to exploit random sites from the genome and result in polymorphism between two distinct species.

Advantages

- Only a small amount of DNA is required for PCR amplification.
- No need of radioactivity as RFLP.
- There is no need of prior genomic information for designing primer.
- The primers are random and can be used across a large variety of species.

Disadvantages

- It is a dominant type of marker system and this causes a loss of information relative to the marker which shows codominance.
- It is not a reproducible method because of short primer length that binds to nonspecific sites in the genome and makes the results inconvenient for data interpretation.

DNA Amplification Fingerprinting (DAF)

- DNA Amplification Fingerprinting (DAF) is one of the recent amplification based nucleic acid scanning techniques. DAF utilizes primers of 5-12 nts of length of arbitrary sequence.
- Primers as short as 5 nts in length can produce complex banding patterns that are resolved by polyacrylamide gel electrophoresis and silver staining.
- DAF uses low stringency amplification condition, so that primers can anneal arbitrarily at multiple sites on template DNA strand and initiate DNA synthesis.
- DAF does not depend on cloning or DNA sequence information and can generate fingerprints from DNA of viral, bacterial, fungal, plant and animal origins.
- DAF can be used to assess genetic diversity, to identify genotype and also for segregation and linkage analysis.

Amplified Fragment Length Polymorphism (AFLP)

- It is based on the PCR amplification of genomic restriction fragments generated by specific restriction enzymes and oligonucleotide adaptors of a few nucleotide bases.
- AFLP is a combination of RAPD and RFLP methods, but detects a tenfold greater number of loci than those detected by RAPD analysis, thus the AFLP have the capacity to rapidly screen thousands of independent genetic loci.
- It is a novel DNA fingerprinting technique in which no prior sequence knowledge is required for amplification.
- The number of DNA fragments in single reaction can be tuned by selection of specific primer sets.
- AFLP technique uses stringent reaction condition for primer annealing and combines the reliability of RFLP technique with the power of PCR technique.

Advantages

- AFLP is highly reproducible over RADP and RFLP.
- AFLPs are faster, less labour intensive and provide more information than RFLPs.
- Compared to RAPD, fewer primers are needed to screen all possible sites.
- AFLPs can be a codominant marker like RFLPs. So it can be used for the discrimination of heterozygotes from homozygotes. AFLP analysis is especially useful in screening backcross individuals.
- AFLPs are extremely useful as a tool for DNA fingerprinting and also for cloning and mapping of a variety specific genomic sequences.
- AFLP has widespread application including assessment of genetic diversity studies, construction and saturation of linkage maps and tagging of genes.

Disadvantages

- AFLP generates a huge quantity of information which may need automated analysis therefore computer technology.
- AFLP markers display dominance.

PCR with Sequence Specific Primers

- An approach opposite to the PCR with arbitrary primers is to design primers to target specific genome. These primers are based on sequence specific knowledge of the genome and bind to a particular complementary sequence because of their long length and specificity.
- Sequence Tagged Sites (STS).
- Sequence tagged sites are primers that are based on some degree of sequence knowledge of genome under study. Sequence tagged site primers target and amplify a specific region of the genome. A sequence tagged site is a general term given to a marker defined by its primer sequences amplified from the arbitrary primers.

- Examples of STSs are"
  - Sequence Tagged Microsatellites (STMs).
  - Sequence Characterized Amplified Region (SCAR).
  - Cleaved Amplified Polymorphic Sequence (CAPS).

Advantages

- Only a limited quantity of DNA is required for this process.
- Highly polymorphic in nature.
- Good analytical resolution and high reproducibility because of the longer primer sequences.
- Codominant in nature that is useful in distinguishing between homozygotes from heterozygotes.
- Evenly distributed throughout the genome.

Disadvantage

- One obvious drawback of SSR marker is the development of polymorphic SSR marker because time, cost and skill is required for the construction and screening of library and then sequencing of the hybridized clone to determine the flanking region of the repeats for SSR primer designing.

Sequence Characterized Amplified Regions (S CAR)

- A technique in which RAPD marker termini are sequenced and longer primers are designed (15-30bp long) for specific amplification of a particular locus.

Advantages

- PCR products of SCAR produce a more clear banding pattern than RAPDs.
- Usually dominant but can be converted into codominant markers that can be used for genetic mapping.

Disadvantages

- There is no need to design and screen a library to pick up a sequence for primer synthesis, but sequencing efforts are required to find out the end sequences of RAPD products to synthesize specific primers for amplification.
- Skill, efforts and expenses are required to design specific primers for each locus.

Cleaved Amplified Polymorphic Sequence (CAPS)

- If specific primers designed for STMS, EST and STMS do not show polymorphism, then the amplified products are digested with a particular restriction enzyme and scored for polymorphism which is termed as Cleaved Amplified Polymorphic Sequences.

Advantages

- Codominant in 'nature' so the information can be effectively used for genetic mapping.
- CAPSs are reproducible because of specific and long primer sequences.

Disadvantages

- To design a primer some genomic sequence specific knowledge is required.
- Efforts and expenses are required to design specific primers for each locus.

Molecular Breeding for Crop Improvement

- The process of developing new crop varieties can take almost 25 years. Now, however, biotechnology has considerably shortened the time to 7-10 years for new crop varieties to be brought to the market.
- Molecular breeding is an addition to the existing science of plant breeding. It involves studying the results of breeding with the techniques of molecular biology.
- The tools of molecular breeding include:
  - Molecular markers
  - QTL mapping
  - Marker Assisted Selection
  - Genetic transformation

Molecular Markers

- Molecular markers may be defined as DNA sequences with a known location on a chromosome and associated with a particular gene or trait. Their inheritance can be easily mapped. These are the signs of identification of any trait of an organism.
- The use of molecular markers is based on naturally occurring DNA polymorphism. One of the main aspects of this technology is that linkage between molecular marker and trait of interest can be detected in a single cross.
- An ideal marker should be highly polymorphic i.e. it must exist in different forms so that the chromosome carrying the mutant gene can be distinguished from the chromosome carrying normal gene by the marker; it should show codominant inheritance i.e. should be able to distinguish the homozygous and heterozygous states of diploid organisms; it should have frequent occurrence in genome; should be easily available; should have selective neutral behaviour; should be easy and fast to assay; should possess high reproducibility and should permit easy exchange of data between laboratories.

Applications of Molecular Markers

DNA based markers find applications in a wide array of fields:

- Assessment of genetic variability and characterization of germplasm
- DNA fingerprinting and identification of cultivars

- Detection of monogenic and quantitative trait loci (QTL)
- Marker-assisted selection

Mapping of Quantitative Trait Loci in plants

- The term Quantitative Trait Loci (QTL) was coined by Gelderman in 1975. These may be defined as the regions of the genome associated with an effect on a quantitative trait.
- A quantitative trait is any attribute of an organism that can be measured and quantified. Many of these traits are continuously distributed and are affected by multiple genes. Thus, QTL mapping is a process of locating genes with effects on quantitative traits using molecular markers.
- The basic idea in mapping studies is to look for markers that are inherited with the trait. Physical linkage will result in co-inheritance, while recombination and independent assortment will disrupt these associations.

Mapping of QTL Requires

- Suitable mapping population generated from parents highly contrasting phenotypically for the target trait and genetically divergent so as to enhance the possibility of identifying a large set of polymorphic markers well-distributed throughout the genome Saturated linkage map based on molecular markers i..e. markers are available for each chromosome from one end to the other and adjacent markers are spaced sufficiently close such that recombination events occur only rarely between them.
- Practically, this is considered to be less than 10 recombinations per 100 meioses, or a map distance of less than 10 centiMorgans (cM).
- Reliable phenotypic screening of mapping population for the trait under study. The target quantitative trait has to be measured as precisely as possible and limited amount of missing data are required. Data is pooled over locations and replications to have a better understanding of QTL X environment' interaction, if any.
- Appropriate statistical packages to analyze the genotypic information in combination with phenotypic information. Different programs are available such as:
  - MAPMAKER/QTL
  - QTL Cartographer
  - PLABQTL
  - QTL Mapper
  - Qgene

Statistical Methods of QTL Mapping

- Single marker approach
- Simple Interval mapping
- Composite Interval Mapping
- Multiple Interval Mapping

Single Marker Approach

- Also called single factor analysis of Variance (SF-ANOVA) or single point analysis, is carried out for each marker locus independent of information from other loci. The population is divided into subpopulations based on allelic segregation of individual loci.
- Phenotypic trait and average in each subpopulation is measured and statistically significant differences determined by t-test, ANOVA or linear regression (R^2).

Simple Interval Mapping

- This method takes full advantage of a linkage map. It involves analysis of intervals between adjacent markers instead of single markers.
- The presence of a putative QTL is estimated if the log of odds ratio (LOD) exceeds a critical threshold.
- It eliminates problems of recombination between marker and QTL by using tightly linked markers.

Composite Interval Mapping

- It carries out analysis of intervals between adjacent markers + additional single markers unlinked to the interval markers so that *n-1* tests for interval QTL associations are performed on a chromosome with *n* markers.
- It combines interval mapping for a single QTL and a given interval with multiple regression analysis on a marker associated with other QTLs.

Multiple Interval Mapping

- It is an extension of interval mapping to multiple QTLs. It allows one to infer the locations of QTLs to positions between markers, makes proper allowance for missing genotype data and can allow interactions between QTLs.
- The simplest MIM method involves use of multiple regression. This method is relatively new and computationally intensive.

Utility and Prospects of QTLs

- QTLs are becoming an essential tool of molecular breeding approaches.
- QTL mapping is being combined with methods in functional genomics, developed for the study of gene expression, which can be utilized to develop markers from genes themselves – the candidate gene approach.
- Comparative mapping will become more widely practiced with the availability of high-density consensus maps greatly facilitating the construction of new maps and mapping specific chromosomal regions.
- The development of high resolution maps will also facilitate the isolation of actual genes (rather than markers) via 'map based cloning' (also 'positional cloning').
- Plant scientists will continue to use QTL maps and markers that tag genes of interest for many years to come.

Marker Assisted Selection (MAS)

- MAS refer to the use of DNA markers that are tightly-linked to target loci as a substitute for or to assist phenotypic screening.
- By determining the allele of a DNA marker, plants that possess particular genes or quantitative trait loci (QTL) may be identified based on their genotype rather than their phenotype. QTL mapping thus represents the foundation of the development of markers for MAS.
- MAS is a technique that does not replace traditional breeding, but can help to make it more efficient.
- It does not include the transfer of isolated gene sequences such as genetic engineering, but offers tools for targeted selection of the existing plant material for further breeding.
- MAS is also referred to as **SMART breeding - "Selection with Markers and Advanced Reproductive Technology"**

Steps followed in MAS

1. Crossing of parents contrasting for the desired trait
2. Raising of F1 populations
3. Selfing of F1 to raise F2 populations
4. Genotyping of parental DNA to obtain markers linked to the desired trait
5. Genotyping of F2 plants to screen plants carrying the desired trait by looking for the presence of markers linked to trait of interest, selected in the parental genotyping step.
6. Screening out plants carrying the marker and further raising of only those plants

Targeting Induced Local Lesions in Genomes (TILLING)

- This method was given by Craig McCullum (McCallum *et al.*, 2000) at the Fred Hutchinson Cancer Research Center.
- It allows directed identification of mutations in a specific gene.
- The method combines a standard and efficient technique of mutagenesis with a chemical mutagen such as Ethyl methanesulfonate (EMS) with a sensitive DNA screening technique that identifies single base mutations (also called point mutations) in a target gene.
- The TILLING method relies on the formation of DNA heteroduplexes that are formed when multiple alleles are amplified by PCR and are then heated and slowly cooled.
- A "bubble" forms at the mismatch of the two DNA strands, which is then cleaved by a single stranded nuclease. The products are then separated by size on several different platforms.

Applications of TILLING

- Identifies mutation events in predetermined genome regions

- Allows use of induced mutations to broaden the genetic base of crops
- Can be used for gene discovery

Advantages of TILLING

- It is a non-transgenic, high throughput reverse genetics approach.
- Naturally occuring SNPs can be detected by eco-TILLING.
- It is applicable to any self-fertilized plant since it does not require transgenic and tissue culture tools.
- Rare SNPs can also be detected.
- Pooling strategy is used which is time saving

Disadvantages of TILLING

- Creation of mutant population difficult.
- Different species respond differently to chemical mutagens.
- High quality DNA of all plants is required to equally represent the pool.
- Occurance of false-positives and false-negatives.
- Screening of large mutagenized population.

Transgenic Breeding

- It is a modern breeding technique done with the help of genetic engineering. An elite genotype is selected having one undesirable character eg.
- Disease susceptibility for which resistant gene construct is prepared and the gene construct is then introduced in elite genotypes by genetic engineering. More than one gene can be introduced. This is called pyramiding.

Advantages of transgenic breeding over conventional breeding

- **Gene construct:** It can be prepared for a particular character which is then introduced into the elite genotype.
- **Gene pyramiding:** It is done with the help of genetic engineering and the presence of genes is confirmed by using molecular markers which reduces the time.
- **Transfer of a particular known gene:** A known gene can be transferred to the genotype eliminating the transfer of other undesirable genes.
- **Evaluation:** Evaluation for a particular trait can be done at seedling stage with the help of molecular markers.
- **Time period:** It takes 4-5 years to develop a variety.
- **Gene transfer in distantly related species:** Genes from distantly related species can be easily transferred to others which otherwise offer many barriers during crossing.

Disadvantages in Transgenic Breeding

- Introduction of gene is not easy as it requires processes of genetic transformation, tissue culture and molecular techniques.

- **Position effect:** The introduced transgene may interrupt the endogenous gene. These can be reduced by the following approaches.
  - **MARs (Matrix Attachment Regions):** Flank the transgene with MARs which will lead to increased expression level of transgene without altering the endogenous gene.
  - **Site specific transgene integration system:** It facilitates the integration of trangene where low or no deletirious position effect has been observed.
- **Transgene silencing:** It is a genetic variation in which when two or more copies of a transgene or transgene with homology to an endogenous gene are integrated into a plant genome, some or all of the homologous copies are inactivated or silenced.
- It is also called homology dependent gene silencing because it results from interaction between homologous nucleic sequences.

It occurs at Two Levels:

A. **Transcriptional gene silencing (TGS):** Is associated with heavily methylated and inactive promotor sequences. Multiple copies of closely linked transgenes have a strong tendency to become methylated and inactivated, which can also silence the other unlinked locus sharing DNA sequence homology with it.

B. **Post-Transcriptional Gene Silencing (PTGS):** It occurs when a promoter is active but mRNA species fail to accumulate. Double stranded RNA derived either from a transgene locus or from copying an aberrant RNA , can trigger methylation of homologous DNA and RNA degradation.

Transgene Selection: Whether or not transformants possc desired transgenic trait of interest. Whether or not a transgenic trait has a predictable inheritance and expression pattern over generations, genetic background and environmental conditions.

Containment Area: Transgenics are always grown in a containment area where they do not disturb other plants. Unwanted plants called refuge are grown around the transgenic plants.

Breeding of Tomato

- **Centre of origin**: Peru, Equador and Bolivia
- **Ploidy level**: $2n = 2x = 24$
- Tomato *Solanum lycopersicum* Mill. (syn. *Lycopersicon esculentum*) is one of the most important warm-season fruit vegetables grown throughout the world. Among the vegetables, maximum attempts have been made to improve this crop.
- Short duration of crop, easiness in cultivation and a large number of seeds per fruit has made it an ideal crop for many research works.

Origin

- Tomato is a native of Peru in South America. The crop spread to North America primarily by migrating birds.
- The largest concentration of wild tomatoes is present in Mexico.

Floral Biology

- Dehiscence of anther occurs 1-2 days after the opening of the corolla.
- It occurs from base to top and is longitudinal. It is mainly self-pollinated but certain percentage of cross-pollination also occurs.
- If the pollen is shed as the style grows up through the anther tube, self-fertilization is the rule.
- Self-pollination also takes place when the style is short and the stigma is not extroverted beyond the connivent anther.
- A certain degree of cross-pollination happens when the stigma protrudes outside the level of anther (extrovert).

Breeding Goals

- Earliness
- Increased fruit yield
- Fruit quality
- Indeterminate cultivars for greenhouse production
- Resistance to disease (Wilt, Tospo virus, TLCV, blight, anthracnose, mosaic and root knot nematode)
- Resistance to insects (Fruit borer, whitefly)
- Resistance to abiotic stresses
- Cold set varieties
- Hot set varieties
- Drought tolerance
- Salt tolerance
- Low temperature germination and growth
- Chilling injury tolerance
- Herbicide tolerance

Cytogenetics

- Tomato is a true diploid with $2n = 24$. Haploids, tetraploids, trisomics and monosomics have also been produced in the tomato.

Advances in Breeding of Tomato

- Tomato is one of the most important and widely grown vegetable in the world.
- It is rich source of vitamins and minerals and it can be consumed either raw or in the form of various processed products.
- Tomato seed contains 24% oil, which after extraction is used as salad oil or in the manufacture of margarine.
- Tomato is adapted to a wide range of soils and climates and its cultivation extends from the tropics to a few degrees within the Arctic Circle.

- Thus, there is vast scope for its improvement which can be achieved through breeding high yielding varieties/hybrids, developing diseases and insect-pests and stress-tolerant varieties along with improved cultivation technology such as raising them in disease-free nursery, optimum fertilization, proper irrigation, effective weed control, plant protection measures to enhance the yield, fruit quality and maximum economic returns to the growers.

Improvement in tomato crop has been made keeping in view the following objectives:

1. To develop early and high yielding varieties/hybrids
2. To develop varieties/hybrids resistant to important diseases and insect pests of the areas
3. To develop stress resistant varieties/hybrids
4. To increase the shelf life
5. To breed varieties/hybrids suitable for processing

- Germplasm nature and magnitude of genetic variability available in the gerrmplasmare basic requirement for the genetic improvement of the crop.
- Cultivated and wild species which provide valuable genes for disease, insect pest resistance and abiotic stress. Lot of information on genetic studies in tomato are available which ultimately helps in selecting some elite lines for achieving desired objectives.
- Tomato breeders have succeeded in developing varieties which are superior in yield, horticultural traits and resistant attributes made through different breeding methods.

Table 7.15: *Varieties of tomato*

Breeding methods	Varieties
Introduction	Sioux, Marglobe, Rutgers and Labonita
Selection	Arka Saurabh, Arka Abha and Pant T-3
Hybridization	Hisar Arun, Hisar Anmol, Pusa Ruby, Punjab Chuhara
Varieties developed by Dr Y S P UHF, NAUNI, SOLAN, HIMACHAL PRADESH	
Selection	Solan Vajar , Solan Lalima, Solan Gola
Hybridization	Solan Sindhur, Solan Shagun, Yashwant, and Solan Garima

The increased interest towards F1 hybrid breeding is due to the possibility of combining a complex of valuable attributes in a genotype for example high fertility, earliness, uniformity of plants and fruits, complex resistant to diseases, adaptability to different environmental conditions, increases shelf life and higher yield.

- Bulk of the hybrid seed in tomato is produced by hand emasculation and pollination.
- There is possibility of decreasing expenditure of hybrid seed production if self-emasculating line i.e. male sterile lines are used as the female parent.
- Different types of sterility such as pollen sterility (non-functional pollens), functional sterility with indehiscent anthers, structural male sterility and positional

male sterility (experted stigma) are available but their use in hybrid seed production is negligible, because the female line segregates in 1:1 ratio for fertile and sterile plants, detection and removal of fertile plants before anthesis not possible. The rouging is possible only after flowering starts.

- Tomato is susceptible to many diseases and pests. The losses due to these are considerable.
- Disease resistant varieties not only bring down the cost of cultivation but also helps to reduce the pollution occurring due to pesticides. Bacterial wilt *Pseudomonas solanacearum* is a serious obstacle in tomato production.
- The resistance is controlled by polygenes. The resistant genes have been identified in *Lycopersicon pimpinellifolium* (PI 127805 A). Commercial varieties are resistant to bacterial wilt viz., Venus, Saturn, Arka Abha and Arka Alok.
- Immunity of fusarium wilt is controlled by a single dominant gene reported in *Lycopersicon pimpinellifolium* Missouri Acc 160 (PI 7932).
- Other sources of resistance are SI 676 (*Lycopersicon hiustum*) and two accessions of *Lycopersicon peruvianum* viz. EC 128897 and EC 128898.
- *Alternaria solani*, red cherry and some accessions of *Lycopersicon hisutum* possess resistance to this disease but no commercial variety has been developed which is highly resistant.
- In buckeye rot, Rattan and Saini (1979) reported that resistance was under control of a single dominant gene and the source was EC 54725 (*Lycopersicon pimpinellifolium*).

Quality Improvement

- IIHR Banglore scientists have developed a purple coloured genetically engineered tomato which contains two genes from the flowers of the snapdragon plant.
- The fruits are rich in anti-oxidants and may protect people from a host of diseases including cancer and heart ailments.
- The level of anthocyanin as well as purple colour has gone up by 70-100% in transformed fruits.

Breeding of Chilli

- Chilli (*Capsicum* spp.) is an important commercial spice and vegetable crop for small and marginal farmers in Asia, Africa and South America.
- Among the five cultivated species of the genus Capsicum, *C. annuum* is the most widely cultivated in India for its pungent (chilli syn. hot pepper) and non-pungent (sweet pepper syn. capsicum, bell pepper) fruits.
- The cultivation of *C. frutescens, C. chinense,* and *C. baccatum* is limited and usually restricted to homestead gardening in different regions.
- Pepper (chilli and sweet) market types prevalent in India can broadly be grouped into the following four categories: (i) fresh market (green, red, multi-colour whole fruits), (ii) fresh processing (sauce, paste, canning, pickling), (iii) dried spice (whole fruits and powder), and (iv) industrial extracts (paprika oleoresin, capsaicinoids and carotenoids).

- Chillies are also called as pungent pepper grown all over the world except in colder climates. Bell peppers are constituents of many foods, add flavour, colour, vitamin C and pungency.
- **Origin:** Tropical America
- **Distribution:** Mainly cultivated in Brazil, Mexico, Spain South and Central America China and India. In India, Andhra Pradesh, Maharashtra, Karnataka, Tamilnadu and H.P etc.

Five major cultivated species in the Genus Capsicum

1. *Capsicum annuum*
2. *C. frutescens*
3. *C. chinense*
4. *C. pendulum*
5. *C. pubescens.*

Classification of chilli type based on fruit characters

1. *C. annuum* var *acuminatum* (Nepal pepper): Fruits are long, thin, pendulous and pungent
2. *C. annum* var *longum*: (Long chilli) the fruits are long and stout with a very broad base
3. *C. annuum* var *grossum:* Big chilli Sweet pepper, bell pepper. Fruits are large and bell shaped, turn bright red on ripening, little pungent, used as vegetable
4. *C. annuum* var *ceraciferma:* Very small chilli fruits, round, slightly pungent.
5. *C. fruitiscens* var *minima:* Bird pepper with white and long pedicel, fruits small and highly pungent *e.g., Golconda Mirapa, Seema Mirapa*

Breeding Objectives

1. Earliness
2. Desirable fruit shape and size (obovate and round fruit in bell pepper and long fruits in chilli).
3. Superior fruit quality (pleasing flavour, high sugar/acid ratio, high pigment content and vitamin C in bell pepper and high capsaicin.
4. Resistance to diseases such as fruit rot, Cercospora leaf spot, powdery mildew, bacterial leaf spot, phytophthra root rot, root knot, common TMV.
5. Resistance to insects (thrips, mites, aphids, fruit borers).
6. Resistance or tolerance to abiotic stress (heat, water stress, salinity, etc.).

Breeding Methods

1. **Pureline selection:** This method is applicable to land races or local cultivars being grown by farmers. *e.g.,* G1 , K 1, Co1, Sindhu.
2. **Pedigree method:** Involves selection of superior plants following hybridization between superior cultivars. *e.g.,* Andhra Jyothi, Pusa Jwala.

3. **Backcross method:** Used to transfer a single gene or a few genes from primitive cultivars or wild forms to leading cultivars.
4. **Heterosis breeding:** F1 hybrids are popular in the USA and Europe and are gaining popularity in India. The first hybrid in India was **Bharat** developed by Indo-American Hybrid Seed Company, Bangalore (1973) followed by several companies.
5. **Mutation breeding:** Found to be an effective and efficient breeding tool in pepper.
6. **Biotechnological methods**

Capsicum Diversity and Characterization

- The National Bureau of Plant Genetic Resources (NBPGR), New Delhi facilitates collection, regeneration, characterization, conservation and distribution of chilli germplasm to researchers in India.
- However, at NBPGR, indigenous collections constitute only 18% of the total *Capsicum* collections, while the majority of the accessions (2412) are exotics (Kalloo *et al.,* 2005).
- The NBPGR currently has ~3000 *Capsicum* accessions under *ex situ* conservation, but online passport and characterization data are not available.
- Phylogenic analysis based on seed protein has revealed that *C. annuum* accessions from Kerala might have been introduced from Portugal (Anu and Peter, 2003). Random amplified polymorphic DNA (RAPD) based cluster analysis of 7 chilli landraces from Manipur separated *C. annuum* genotypes (1 cluster) from *C. frutescens* and *C. chinense* (another cluster). Genetic variation among *C. annuum* genotypes was higher compared with *C. frutescents* genotypes and *C. chinense* genotypes, most of which were closely related (Sanatombi *et al.*, 2010).

Discovery of Novel Genetic Resources

- The northeast region consisting of states like Arunachal Pradesh, Assam, Manipur, Meghalaya, Mizoram, Nagaland, Sikkim and Tripura is recognized as the hot-spot for chilli diversity.
- During the last decade, *Capsicum* landraces from this region have received renewed attention due to the discovery of India's hottest chilli (naga jolokia) from Assam (Mathur *et al.*, 2000).
- Later, a variant of this landrace (bhut jolokia) was reported as the world's hottest with possible origin through natural crossing between *C. chinense* and *C. frutescens* (Bosland and Baral, 2007).
- A number of variants of these landraces probably derived from natural inter-specific crosses are known in the region (Kumar *et al.,* 2011) under different local names (Sanatombi *et al.,* 2010; Verma *et al.,* 2013).
- In general, these landraces have highly pungent fruits (Sanatombi *et al.,* 2010) and are potential sources of capsaicinoids with broad-spectrum ethno-pharmaceutical uses (Meghvansi *et al.,* 2010).

- Landraces like BS35, GKC29 and bhut jolokia are resistant to *pepper leaf curl virus*- Varanasi strain (Kumar *et al.*, 2006b, Table 1) and naga chilli is susceptible to *chilli veinal mottle virus* (Banerjee *et al.*, 2014).

Table 7.16: *Identified resistant/tolerant source against major pests of chilli in India, since 2000*

Biotic stress and *causal organism*		**Name of line and reference**
Anthracnose (*Colletotrichum* spp.)	:	Bhut Jolokia (Gerg *et al.*, 2013); PBC80 (VI046804), PBC81 (VI046805), PBC932 (VI047018); LLS, Breck-1, Breck-2, Jaun (Kaur *et al.*, 2011)
Bacterial wilt (*Psedomonas solanacearum*)	:	AVPP0102 (PP0107-7011), PBC66 (VI037518), PBC67 (VI037519), PBC384 (VI037548), PBC385 (VI039374), PBC535 (VI037556), MC-4
Phytophthora blight (*Phytophthora capsici*)	:	GKC29, PI201234, IC364063
Chili/pepper leaf curl virus (ChiLCV)	:	BS35, GKC29, Bhut Jolokia (Kumar et al., 2006b; Kumar et al., 2011), CHUH-4 (Mondal, 2013), Pepper Line Nos. 59 & 86
Chilli veinal mottle virus (ChiMoV)	:	Individual plant selections from: PBC495 (VI037455), PBC521, PBC370 (VI037453), PBC569 (VI046889), PBC371 (VI039369), Tiwari (Erect), 9852-131 (AVPP9807), Punjab Gucchedar, Perennial, Punjab Surkh, Pusa Sadabahar, Pant C1, Perennial HDV (Reddy and Reddy, 2010)
Cucumber mosaic virus (CMV)	:	Perennial, PBC495 (VI037455), VC246, VR42, VR55 (Reddy and Reddy, 2010), AVPP9812 (PP9852-10)
Peanut bud necrosis virus (PBNV)	:	EC121490, IC119611 (Kalloo *et al.*, 2005)
Nematode (*Meloidogyne javanica)*	:	EC402105, EC402113, EC405253, NIC19969, IC214965, IC214985, IC215012, EC391083, EC391087, EC378632, EC378688 (Pandravada *et al.*, 2010)
Powdery mildew (*Leveillulataurica*)	:	Arka Harita, Arka Suphal, PBC167 (VI046819)
Thrips	:	Calcapin Red, Chamatkar, P46-A, X1068, X743, X1047, BG4, X226, X230, X233 (Kalloo *et al.*, 2005)
Yellow mites (*Polypagotarsonemus latus*)	:	Jwala, RHRC Erect, AEG77 (Desai *et al.*, 2007)

Distinctness of variants of bhut jolokia from *C. frutescens* and *C. chinense* has been demonstrated by phylogenetic analysis based on ribosomal RNA (rRNA) gene-internal transcribed (ITS) region sequences (Purkayastha *et al.*, 2012a).

- The existence of unique diversity in the region has been further strengthened with the discovery of a naturally occurring allotetraploid chilli in the region (Jha *et al.*, 2012).

- In a molecular marker diversity analysis, the landraces from the northeast region clustered with non-*annuum* (*i.e. C. frutescens, C. chinense, C. baccatum*) genotypes (Rai *et al.,* 2013).
- Crossing attempts involving three of these landraces to *C. annuum* revealed that they were reproductively isolated (pre-hybridization barrier and post-hybridization barrier of hybrid breakdown) from *C. annuum* genotypes (Rai *et al.,* 2014) and are believed to have originated from sympatric domesticated species rather than introduction from elsewhere (Rai *et al.,* 2013).
- This hypothesis (origin of landraces from natural crossing between sympatric populations) is further strengthened by the results of an independent study involving diversity analysis of 53 landraces collected from the northeast region of India (Yumnam *et al.,* 2012).
- The study revealed (i) the presence of novel single nucleotide polymorphisms (SNPs) across *Pun1* locus and (ii) that out of the total variation, 58% of the variations occurred among plants within populations and 34% among populations within the same group.
- This points to possible gene-flow among populations in the recent past and indicates that populations have not been isolated from each other long enough (Yumnam *et al.,* 2012). It seems that these landraces have not yet reached the species differentiation level; it is premature to propose them as separate species (Verma *et al.,* 2013) in contrast to a proposal to treat some of these landraces from Assam as a separate sixth domesticated species (*C. assamicum*) by Purkayastha *et al.* (2012b).

Genetic Resources against Biotic Stresses

- Commercial chilli production is adversely affected by attack of insect pests and diseases. Host plant resistance is generally considered a cost-effective and safe method of controlling insect pests with regard to the environment and health of producers and consumers.
- A detailed account of global sources of resistance to major chilli pests and diseases, including those reported from India was given by Babu *et al.* (2011).

Characterization of Chilli Pathogens

- Understanding of prevalent species and the pathotype/biotype or strain of a given pathogen in the targeted production region is crucial for breeding resistant cultivars.

Colletotrichum spp.

- Different species of *Colletotrichum*, *viz*., *C. capsici*, *C. gloeosporioides* and *C. acutatum* are known to cause anthracnose disease of chilli in India and many other countries in Southeast Asia and Oceania. *Colletotrichum capsici* and *C. gloeosporioides* were the two predominant species in Southern India (Rao *et al.,* 2007). *C. capsici* was prominent in Karnataka, Tamil Nadu and Maharashtra (Rao *et al.,* 2007), Himachal Pradesh (Sharma *et al.,* 2005), Andhra Pradesh (Rao *et al.,* 2007; Pratibha, 2009; Nanda, 2011) and in Arunachal Pradesh (Selvakumar, 2007).

- Isolates have been characterized based on mycelium and conidia morphology (Pratibha, 2009), conidial shape, internal transcribed spacer (ITS) of ribosomal DNA and amplified length polymorphism (AFLP) markers (Nanda, 2011), level of aggressiveness (Sharma *et al.,* 2005), differential reactions against given host genotypes (Khirbhat *et al.,* 2004; Deshpande and Rawal, 2007; Garg *et al.,* 2013). Recently, a real time polymerase chain reaction (PCR) protocol for specific and efficient detection of *C. capsici* has been reported (Srinivasan *et al.,* 2014).

Viruses

- More than 20 viruses are known to infect chilli and sweet peppers in India. Among these, *Chilli veinal mottle virus* (ChiVMV) of the *Potyvirus* genus, *Cucumber mosaic virus* (CMV) of the *Cucumovirus* genus, chilli infecting begomovirus causing leaf curl diseases (referred here as *Chilli leaf curl virus*, ChiLCV) and *Groundnut bud necrosis virus* (GBNV) or *Peanut bud necrosis virus* (PBNV) of the *Tospovirus* genus are currently considered to be the most important (Reddy and Reddy, 2010).
- ChiVMV is transmitted by aphids in a non-persistent manner and is the most prevalent potyvirus in India. Isolates of ChiVMV were identified from Karnataka and Tamil Nadu based on host range, serological relationship, electron microscopy and phylogenetic analysis of coat protein sequences. Nucleotide sequence analysis of these isolates revealed > 90% similarity with already known ChiVMV sequences. Symptom variation and severity also varied among the ChiVMV isolates (Reddy and Reddy, 2010).
- CMV is transmitted mainly by aphid species in a non-persistent manner. Host range, electron microscopy and partial molecular characterization studies have confirmed CMV incidence and distribution on chilli from the northern Telengana region of Andhra Pradesh (Reddy and Reddy, 2010).
- ChiLCV transmitted by whitefly has emerged as a serious problem throughout India. *Tomato leaf curl Joydebpur virus* (reported earlier from tomato in Joydebpur, Bangladesh) has also been found to be associated with chilli leaf curl disease in Punjab (Shih *et al.,* 2006).
- A complete begomovirus sequence associated with chilli leaf curl disease from Varanasi, Uttar Pradesh shared ~95% sequence identity with ChiLCV-PK and infectivity of the cloned molecule was demonstrated in the natural host for the first time (Chattopadhyay *et al.,* 2008). Complete sequencing of DNA-A from different samples indicates association of 3 distinct begomoviruses with this disease.
- An isolate from Kalyani (West Bengal) shared maximum nucleotide identity with *pepper leaf curl Bangladesh virus* (96.3%) while 3 isolates from Rajasthan showed nucleotide identity of 88.9 to 91.1% with *Chilli leaf curl virus*-Multan, and 8 isolates (2 each from Delhi, Haryana, Punjab and Uttar Pradesh) shared nucleotide identity of 98.5% with *tomato leaf curl New Delhi virus* (Reddy and Reddy, 2010). Based on *cp* sequence analysis, emergence of new variants of ChiLCV has been reported (Rai *et al.,* 2010).
- For the first time, the virus-vector relationships and host range of ChiLCV isolate from Rajasthan have been described (Senanayake *et al.,* 2012). The beta-satellite

of this virus resembled most closely (97.3%) with tomato leaf curl Bangladesh beta-satellite previously reported from chilli and tomato in India (Senanayake *et al.*, 2012).

- In recent years, tospoviruses (PBNV/GBNV) transmitted by thrips have emerged as a major threat for chilli cultivation in India. Tospovirus, (PBNV) was widely recorded on tomato and chilli peppers across 14 Indian states representing southern, north-western, north-eastern and central regions (Kunkalikar *et al.*, 2011).

Cytogenetics

- *In situ* nuclear DNA contents varied significantly between 23 *C. annuum* genotypes with somatic chromosome number $2n = 24$ (Mukherjee and Sharma, 1990). *Capsicum* cytogenetic analyses in India were carried out by a few groups, especially by Aneil Kumar O and his associates at Andhra University, Vishakhapatnam.
- During the 1980s they studied meiotic behaviour of cytogenetic stocks such as accidentally induced interchange (Rao and Kumar, 1983), induced octaploid (Panda *et al.*, 1984) and inter-specific crosses (Kumar *et al.*, 1987; 1988). Besides some minor chromosome structural differences, the *C. annuum* accession used differed from *C. chinense* by 2 translocations and from *C. baccatum* accession by 2 translocations and 1 inversion (Kumar *et al.*, 1987). In the past decade, progress has been limited to meiotic analysis of male sterile lines (Dash *et al.*, 2001; Kumar *et al.*, 2001), study on meiotic consequences of chromosomal aberrations (Dhamayanthi and Reddy, 2000), meiotic analysis to determine chromosome number and structure in the process of polyploid induction (Ranganathan and Jagatheeshwari, 2013), and to study meiotic behaviour and karyotype of an allotetraploid (Jha *et al.*, 2012).

Tissue Culture and Transgenics

- Peppers (chilli and sweet pepper), unlike tomato, tobacco and petunia have an inherent problem associated with *in vitro* regeneration. The severe recalcitrant morphogenic nature of peppers, formation of rosette shoots and ill-defined shoot buds and genotypic dependent tissue culture response in peppers (Kothari *et al.*, 2010) have hampered progress in tissue culture advancements and genetic transformation of chilli. Progress on transformation of *Capsicum* spp.

Table 7.17: *Reports on transformation of capsicum in India* (*Since* 2000)

Genotype; transformation method	Vector, reporter gene/s (selection agents)	Remarks	Reference
Pusa Jwala; *Agrobacterium* (C58)	pGV1040:: *phosphotransferase-II* (*nptII*) + -glucuronidase (*gus*) (kanamycin and cefotaxime)	The transformed shoots were selected, rooted and transgene presence was confirmed through GUS assay, PCR and southern blotting of *nptII*	Shivegowda *et al.* (2002)

Genotype; transformation method	Vector, reporter gene/s (selection agents)	Remarks	Reference
K1, K2, PLR1; *Agrobacterium* (EHA 105)	T-DNA with CaMV 35S::*gus* (cefotaxime and anamycin)	The integration of *gus* gene in genome was confirmed by histochemical *gus* staining, PCR and southern analysis	Sobhakumari and Lalithakumari (2005)
Byadagi Kaddi, *Agrobacterium* (EHA105)	pBinBt3::*cry*I*(Ac)* + *nptII* under *CaMV35S* promoter and *nos* terminator (kanamycin)	Pre-cultured explants were more suitable for transformation. 0.5 × 10^8 bacterial cells/ ml was optimum for transformation	Channappagoudar (2007)
C. frutescens; Pollen transformation; *Agrobacterium* (EHA 101)	pCAMBIA::hygromycin phosphotransferase (*hpt II*) under CaMV *35S* promoter and terminator + - glucuronidase (*uid* A) under CaMV *35S* promoter and *NOS* terminator (kanamycin)	Successful transformation of pollen from *in vitro* flowers induced through application of silver nitrate (AgNO3) and cobalt chloride (CoCl2)	Sharma *et al.* (2008)
Arka Gaurav, Arka Mohini; *Agrobacteriumin planta* (EHA105)	pCAMBIA1301:: - glucuronidase (*uid* A) + *hptII* reporters; (kanamycin)	Genotype independent protocol for *in planta* transformation of sweet pepper, identified stable transformants based on PCR analysis	Kumar *et al.* (2009)
California Wonder; *Agrobacterium* (LBA 4404)	pBI121::*npt*-II + *gus* reporters (cefotaxime and kanamycin)	Standardization of pre-culture and co-cultivation time	Verma *et al.* (2013)
Pusa Jwala *Agrobacterium* (EHA105)	pBinAR C1::*nptII* + C1 (kanamycin and rifampicin)	Transgenic plants with C1 gene (satellite DNA molecule of *chilli leaf curl virus*, useful in understanding C1-mediated pathogenesis	Kumar *et al.* (2012)
Sweet pepper; *In planta Agrobacterium*	Pathogenesis-related gene (*NPR1*) for resistance to powdery mildew	Technology can be used as a versatile tool in economically important crop species	Arthikala *et al.* (2014)

In India is summarized in Table 2. Parthasarathy *et al.* (2010) and Kothari *et al.* (2010) have reviewed *Capsicum* biotechnology. Recently, embryo rescue was successfully attempted to bypass pre-zygotic barriers in inter-specific crosses (Debbarama *et al.*, 2013).

- Research attempts to examine the potential of capsaicinoids synthesis under *in-vitro* culture by chilli pepper cells, tissues and organs has been also made (Kehie *et al.*, 2014).

Molecular Markers

- In *Capsicum*, markers have been used to tag desirable traits including QTLs (Ramchiary *et al.*, 2013).
- However, the in-past progress of marker development and its application in marker-assisted selection in chilli has been slow due the larger genome size of chilli and non-availability of the complete pepper genome sequence till recently.
- Now pepper genome sequences have become available (Kim *et al.*, 2014; Qin *et al.*, 2014), which will substantially accelerate the process of markers application in chilli breeding. In India, markers have mostly been used for *Capsicum* germplasm diversity analysis and DNA fingerprinting of varieties (Singh *et al.*, 2008).
- The IIHR has completed marker based fingerprinting of all its released OPVs and inbreds of hybrids and at IIVR attempts were made to validate RAPD marker associated with *Rf* gene (Kumar *et al.*, 2007). Marker validation and utilization for screening CMS trait has been demonstrated (Kumar *et al.*, 2009). Morphological (floral characteristics) and molecular markers have been demonstrated to be useful for delineation of species specificity and identification of genetic stocks (Thul *et al.*, 2012).
- Cleaved amplified polymorphism (CAPS) analysis of populations derived from ChiVMV-resistant Perennial indicated that resistance was tightly linked with *pvr11* locus on chromosome 4. Further, comparison of nucleotide sequence of the coding region indicated 99.1 to 99.5% protein and 99.1 to 99.8% nucleotide identity with genotype carrying *pvr11* gene of the eukaryotic translation initiation factor *elF4E* family (Reddy and Reddy, 2010).
- Ten QTLs for yield-related traits were mapped on 4 linkage groups (LG) out of which 9 were stable. The phenotypic contribution of these QTLs ranged from 8 to 51% (Dwivedi *et al.*, 2013).

Varietal Development

- Considerable progress has been made in the development, release and commercialization of chilli cultivars in India.
- In India, the chilli varietal release procedure is facilitated by All India Co-ordinated Research Project (AICRP) on vegetable crops (responsible for coordination of multi-location evaluation and recommendation of better performing OPVs or hybrids for release and cultivation) and the Central Variety Release Committee (CVRC) on Horticultural Crops (responsible for official release notification of cultivars).
- In addition, different state governments also release or notify varieties through State Varietal Release Committees (SVRCs) in respective states.
- According to the Protection of Plant Variety and Farmers Right Act (PPV&FR Act 2001), it is up to the breeder, farmer or institution to decide to register notified cultivars with the PPV&FR authority (responsible for executing PPV&FR).
- If a plant breeder (a person or group of persons, a farmer or group of farmers, or an institute as defined in the PPV&FR Act) wish to protect the cultivar, it must

go through a distinctiveness, uniformity and stability (DUS) test implemented by ICAR on behalf of the PPV&FR authority.

Table 7.18: *List of chilli cultivars recommended for national release in India* (1975-2014)

Cultivar type (developed by)	:	Name (recommended zone/s)
Open pollinated (all 24 by public sector)	:	G-4 or Bhagyalakshmi, G-5 or Andra Jyothi (I, IV, V, VI, VII, VIII), K-2 (IV, VI), J-218 (I, IV, V, VI, VII), Bhaskar or X-235 (IV, V, VI, VIII), Muslawadi (V), Arka Lohit/ Sel-1 (V, VII, VIII), LCA206-B (V, VI, VII, VIII), Phule C-5(VII), AKC-86-39 (VII), BC-14-2 (V, VI), RHRC-Cluster Erect (VII), PMR57/88-K (VII), LCA-334 (III, IV, V, VII), ASC-2000-02 (VII), KA-2 or Kashi Anmol (IV), LCA-353 (IV, V, VII),BC-25 (V, VI, VII), PC-7 (V), HS- HP-154 (VIII), IVPBC-535 or Kashi Sinduri-Paprika (VIII), PC-56 (I, IV, V, VII, VIII), VR- 338 (II), ACS-06-2 (VIII)
Hybrid (10 by private and 5 by public sector)	:	HOE-888 (IV, VIII), ARCH-236 (IV), Sungrow-86-235 (IV, VIII), ARCH-228 (IV, V, VI), Arka Meghana or MSH-172 (IV, V, VI & VIII), Arka Sweta or MSH-149 (IV, VI, VIII), CCH-2 or Kashi Surkh (II, IV, V, VI), KCH-3 (IV), CCH-3 or Kashi Early (IV, V, VIII), BSS-453 (II), NCH-587 (IV, VII), VNR-332 or Rani (IV, VIII), IIH-41786 (VII), BSS-378 (VII), VNR-Vidya (IV)

Zones: I = Humid western Himalayan region, II = Humid Bengal-Assam basin, III = Humid eastern Himalayan and Bay islands, IV = Sub-humid Satlej Ganga Alluvial plains, V = Sub-humid eastern and south eastern plains, VI = Arid western plains, VII = Semi-arid plateau and central highlands, VIII = Humid, semi-arid Western Ghats and Karnataka plateau.

- Since its inception in 1971, AICRP has recommended 24 open pollinated varieties (OPVs) and 15 hybrid chilli cultivars through multi-location evaluations for different agro-climatic zones (Table 3).
- The Central Variety Release Committee (CVRC) has notified 57 chilli cultivars including 7 hybrids (notified after 2005) for cultivation in different zones. In the beginning, centers like the Regional Research Center, Lam, Andhra Pradesh were pioneers in developing and releasing several popular chilli cultivars (e.g., G-4, G-5, Sindur, etc.).
- Likewise Pusa Jwala released by IARI became very popular (Tiwari and Ramanujam, 1974) and has now became a well- recognized market segment (pod type) in the pepper seed industry worldwide.

Cost-effective Hybrids using Male Sterility

- Male sterility systems are widely used to produce cost-effective hybrids and their seeds.

- In India, both nuclear male sterility (GMS) and cytoplasmic male sterility (CMS) systems have been developed and characterized in chilli and are being used for the development of experimental crosses, potential hybrids and production of hybrid seeds.
- It is expected that use of male sterility system in chilli would facilitate to reduce the cost of producing hybrids seeds by 40% (Lin *et al.*, 2013).

Nuclear/Genic Male Sterility (GMS)

- MS-12, a monogenic, recessive gms line, has been developed and commercially exploited in India.
- Thanks to pioneer efforts at Punjab Agricultural University (PAU), Ludhiana, where 2 hybrids (CH-2, CH-3) based on MS-12 were developed, farmers were trained to produce hybrid seeds and were provided with male sterile female parents (MS-12) and male parents.
- This resulted in the adoption of hybrid seed production and hybrid production technologies by farmers in Punjab (Dhaliwal, 2010). Seed producers also market hybrid seeds in nearby regions of Uttar Pradesh, Haryana, and Rajasthan states. Since MS-12 line possess *ms- 10* mutatnt gene induced through mutagenensis in France, the commercial exploitation of MS-12 is a classic example of indirect use of mutagenesis in crop improvement and PAU continue to use *ms-10* gene-based GMS line in heterosis breeding program (Singh *et al.*, 2014)

Cytoplasmic Male Sterility (CMS)

- The area under chilli hybrid production in India has increased from 2.4% in 1997-1998 to about 25% at present. Recognizing the role of CMS lines in labour-saving, cost-effective hybrid seed production, IIHR developed 4 pairs of CMS and maintainer lines from a locally isolated male sterile cytoplasm (Reddy *et al.*, 2002). The male sterile cytoplasm was genetically similar to the most widely used Petersons cytoplasm (Kumar *et al.*, 2009).
- CMS lines were used to develop a number of experimental crosses. Based on promising performance (yield and quality attributes), 4 hybrids (Arka Meghana, Arka Harita, Arka Sweta and Arka Kyathi) were identified for further promotion through different approaches.
- More than dozen private seed companies have acquired CMS lines developed by IIHR and used them in commercial hybrid development and seed production. Eight seed companies acquired parents of hybrids and directly commercialized them.
- The exact area under commercial hybrid production by all seed companies is not known.
- Since 2010, government sector undertakings such as NSC and NHRDF also started producing hybrid seeds by engaging small seed producers with backstopping support from IIHR scientists.
- According to an estimate (based on amount of hybrid seeds produced and made available to the farmers by the NSC), Arka Meghna and Arka Harita were grown

on about 2500 and 1500 ha, respectively in 2012. Recently, PAU has developed numerous potential crosses based on CMS line introduced from AVRDC (Singh *et al.*, 2014).

Present Progress and Outlook

- In the current scenario of developmental efforts being undertaken in northeast India, a regional *Capsicum*-specific exploration that recognizes the role of local communities for *in situ* conservation should be conducted before important genetic resources get replaced by commercial cultivars due to profitable intensification.
- Frequent and simultaneous occurrences of pests are a major constraint for commercial chilli production in India. Continued emergence of new pathogen races and strains against available resistant cultivars necessitates the search and use of new resistant sources. Following breeding progresses are underway at various institutions:
- The identified resistant sources to leaf curl virus are used to transfer resistance into desirable breeding lines. Progress is underway to study the genetics of resistance to the leaf curl complex causing begomovirus and to identify genes involved in plant defense against the leaf curl complex.
- Combining resistance to ChiVMV, CMV and thrips.
- Gene expression studies, development of functional markers and QTL mapping for anthracnose disease resistance.
- Diversification of CMS lines and incorporation of ChiVMV and *Phytophthora* root rot resistance into male sterile, maintainer and restorer lines.
- Development of paprika varieties to meet high demand of non-pungent pods with high colour value for oleoresin extraction industries.
- Increasing industrialization, migration of populations to urban areas, a shortage of agricultural labour, shrinking cultivable land, increasing risks of crop failure due to unpredictable climate, emerging global markets and networks and increasing demand (domestic and export) for more nutritious and safer foods will affect future chilli breeding, production and marketing.
- In general, chilli research in India should aim to deliver technologies pertaining to the following scenarios by integrating breeding research with multidiscipline:
- To address agricultural labour shortages, the development of cost-effective technologies pertaining to chilli seed production, chilli production, harvesting, processing and marketing will require close collaborative research between agricultural engineers and production scientists.
- Initial progress towards mechanization already has been made at several ICAR institutes, including IIHR (e.g. low-cost hand operated dibbler to facilitate sowing in protrays, tractor drawn bed former cum transplanter); these technologies should be evaluated and refined through participatory pilot projects in major chilli producing regions.
- To provide safe chilli products for domestic consumers and also to ensure smooth exports, agricultural policy must support research to control the presence of aflatoxin in dry chilli and related products.

- The share of improved chilli seeds (especially hybrids) within India will increase and there is also potential for seed export. It is imperative to provide support to the pepper seed industry while at the same time ensuring that hybrid seeds are available to small-scale growers at affordable prices. The approaches undertaken by IIHR in the past decade on improved seed dissemination are realistic, as they not only support seed companies but also empower marginal farmers through community-based hybrid seed production. Similar approaches should be considered by other ICAR institutes and SAUs engaged in chilli breeding research to multiply the development impact from research.
- Organizing periodical consultation meetings of active stakeholders including private sector industries (seed, oleoresin, capsaicin extraction, and export) is critical to predict and identify future commercial domestic and international trade opportunities. Specific breeding research traits can then be identified and breeding plans can be executed by the interested stakeholders.
- These advocacy actions would sustain and boost the Indian chilli industry by facilitating mutual benefit sharing among the stakeholders.

Breeding of Capsicum

- Capsicum, bell pepper or sweet pepper is one of the important vegetable crops grown all over the world.
- Bell pepper fruits are of bulky shape having 3-4 lobes and rich in vitamin A and C content.
- The crop originated in New World Tropics and subtropics (Greenleaf, 1986) and was introduced in India by Britishers in Himachal Pradesh and Tamil Nadu (Nilgiri).
- Bell pepper is an annual or short lived perennial herb upto 1.5 m in height. It has well developed tap root system with many laterals.
- Stems are branched, erect and often woody, growth determinate or indeterminate. The fruits have many seeded berries with 2 or more locules.
- Bell pepper is a solanaceous vegetable having-diploid chromosome with 2n equal to 24. Genetic resources for the improvement of peppers in tropical regions are far from exhausted by the capsicum breeders.
- Germplasm collection and conservation is the major step in the improvement of bell pepper. This material can be used in hybridization programme.
- Promising germplasm may directly be introduced from other countries. Quality specifications such as size, shape, colour and non-pungency make the task of developing new hybrids/cultivars very challenging.

Breeding Objectives in Bell Pepper

1. Earliness
2. Higher yield
3. Good average fruit weight

4. More number of fruits per plant
5. Fruit shape
6. Wider adaptability
7. Resistance to diseases like leaf blight and fruit rot, anthracnose, cercospora leaf spot, viruses, powdery mildew and bacterial wilt etc.

Improvement Methods in Bell Pepper

Introduction

- It is the easiest and quickest method of breeding. Germplasm may directly be introduced from abroad. Some promising direct introductions in bell pepper are Yolo Wonder, California Wonder, Chinese Giant, World Beater, Ruby King and King of North.

Pureline Selection

- When selection is carried out in heterozygous lines substantial improvement can be done. Arka Mohini from Titan (USA), Arka Gaurav from Golden DalWonder (USA), Arka Basant from Soroksari (Hungary).

Pedigree Method

- This method involves selection of plants in segregating generations following hybridization between superior cultivars along with maintenance of pedigree record.
- Selection of superior parental lines is crucial step in this method. Spartan Garnet is cross between California Wonder and Dwarf Pisniento, Spartan Enerold between Morgold and California Wonder, Sannette between (Morgold x California Wonder) and Keystone Resistant Giant.
- In pedigree method both parental lines should be promising and superior.

Back Cross Method

- This method is used to transfer a single or few desired genes from wild species i.e. incorporation of disease resistance into commercial cultivars from wild species/ landraces.

Heterosis Breeding

- Hybrid vigour in plants has long been noted and probably the first suggestion that it be exploited in vegetables (Hayes and Jones, 1916) in cucumber. Heterosis in bell pepper has been reported for plant height, days to flowering, average fruit weight, number of fruits per plant, early yield and total yield.
- Different hybrids developed by different institutions and firms have been presented in table 2. Hand emasculation and pollination is most common method used for hybrid seed production in bell pepper.

- Use of genetic male sterility and chemical hybridising agents have also been reported in the literature. Flower buds one day before anthesis are selected on the female parent and emasculated.
- Pollen from the desirable male parent is collected and dusted on the stigma of emasculated flowers. Crossing work is done in the morning.
- After emasculation and pollination, bagging is done by perforated butter paper bags and then tagging is done indicating parentage and date of emasculation and pollination.
- After 3-4 days, the bags are removed from the plants. 5-8 crossed fruits are kept on each plant and other fruits are removed to facilitate the vigorous growth of crossed fruits.
- After 40- 55 days of pollination fruits get mature and ready to harvest. Crossed fruits are harvested and the hybrid seed extracted.

Table 7.19: *Varieties of bell pepper*

Varieties	Institution
California Wonder, Yolo Wonder	IARI, New Delhi
Arka Mohini, Arka Gaurav, Arka Basant	IIHR, Bangalore
Solan Bharpur	UHF, Solan
Punjab-27	PAU, Ludhiana

Breeding for Resistance to Diseases and Insect Pests

- Sweet pepper is highly susceptible to a number of fungal, bacteria, viral diseases and insects.
- However besides the availability of resistant sources, the knowledge of pattern of inheritance of various resistance genes is a must.
- The review of available literature suggests that most of the resistance is controlled by single dominant and single recessive genes.
- However, many of the sources available are pungent types which are difficult to utilize because of the dominant nature of the pungency.
- Despite its economic importance bell pepper remains a neglected crop in terms of research barring a few isolated effects in India.
- However, despite so many constraints the potential for crop improvement is great. Instead of looking for exotic germplasm sources, we can locate the locally adapted material for useful genes as in most of the major production areas where farmers prefer to save their own seed which resulted in evolution of new types having wider adaptability and probably a matter of resistance genes as well.

Table 7.20: *Resistant /tolerant sources to diseases/insects-pests*

Disease/insects	Resistant/tolerant source
Fruit rot (*Colletotrichum* sp.)	*C. chinense*, Accr 1555, 1554, 906, Chinese Giant, Hungarian Yellow Wax, Spartan
Cercospora leaf spot	California Wonder, Hungarian Wax
Bacterial leaf spot	PI 163192, PI 260435, PI 163189, PI 163192, PI 271322, PI 32219
Phytophthora root rot	PI 201234
Phytophthora nicotianae var. *nicotianae*	PBG 631, UHF-1
Root knot nematode	Santaba
TMV	Sonnette, Keystone Resistant Giant, YW, Yolo Y

Breeding for Quality Attributes

- *Capsicum annuum* L. is a diverse group of vegetables. Bell group forms a small group of peppers.
- The group is clearly distinguished due to the bulky three to four lobed fruits and non-pungency. The major breeding objective in bell pepper group is fruit shape and non-pungency.
- Besides this, breeding for fruit colour is another important objective. However, fruit colour is a trait entirely dependent upon consumer preference.
- In some regions, people prefer dark green coloured fruits while in other cream or yellow and red coloured fruits may be prepared.
- Here, it should be clearly understood that sweet pepper is a rather broad group having various shapes, sizes and colours. Thus, to breed a true to type bell pepper having superior quality traits such as bell shape having three to four lobes, pericarp thickness, desired colour and most importantly non-pungency is a challenging task.

Breeding of Brinjal / Egg Plant

- Brinjal is an important commercial vegetable crop grown in India.
- **Origin:** Indo-Burma
- **Distribution:** India, Japan, Indonesia, China, Bulgaria, Italy, France, USA and African countries. In India all the states grow brinjal.
- **Wild species:**
  1. *Solanum torvum*
  2. *S. nigrum*
  3. *S. indicum*
  4. *S. mamosum*

Breeding Objectives

1. High yield
2. Earliness
3. Fruit shape, size and colour as per consumer's preference
4. Low proportion of seed
5. Soft flesh
6. Lower olanine content
7. Upright study plant free from lodging.
8. Resistance to diseases like bacterial wilt, blights.
9. Resistance to insects like shoot and fruit borer, jassids, etc.

Breeding Methods

1. **Introduction:** This method may be useful in other countries but not to India.
2. **Pureline selection:** *e.g.,* Pusa Purple long, Co.1.
3. **Pedigree selection:** Many varieties have been developed through hybridization and subsequent pedigree selection. e.*g.,* Pusa Kranti.
4. **Heterosis breeding:** Many F1 hybrids have been developed and released for commercial cultivation. *e.g.,* Arka Navneet.
5. **Hybridization for resistance breeding:** Resistance is not available in the cultivated varieties. Wild varieties like *S. incanum, S. anthocarpum* are reported to be resistant and used in hybridization programme. However the inferior quality of the fruits of wild species associated with resistance are often expressed in the hybrid progenies.
6. **Backcrossing:** is normally follow to transfer genes conferring resistance to diseases like bacterial wilt resistance which has been shown to be under a single dominant gene.
7. **Mutation Breeding**
8. **Biotechnological Methods:** *e.g..* Bt. Brinjal

Breeding for Disease Resistance in Solanaceous Crops

- The solanaceous vegetables include important crops like tomato, sweet pepper as well as chilli and eggplant.
- Solanaceous vegetables are major significant sources of pro-vitamin A carotenoids, Vitamin C and other micronutrients. Breeding for disease resistance in solanaceous crops thus offers golden opportunities to substantially boost income and alleviate poverty.
- Unfortunately, diseases of solanaceous crops caused by fungi, bacteria and viruses take away a significant proportion of farm productivity thus leading to huge losses to the farming community.
- Many single genes and quantitative trait loci (QTL) that confer resistance to all major classes of plant pathogens have been mapped and cloned from solanaceous vegetables.

- R gene refers to phenotypically defined single gene or quantitative resistance loci (QRL) that function to confer disease resistance to a particular pathogen for which sequence information may or may not be available.
- In breeding for disease resistance, usually four methods i.e. introduction, selection, hybridization and mutation are followed. It is important to identify a source of resistance before undergoing breeding process.
- To cite an example in tomato, *Solanum lycopersicum* is crossed with *S. pennellii* against *Alternaria alternata* f. sp. *lycopersici*, *S.pimpinellifolium* against *Ralstonia solanacearum* and *Cladosporium fulvum, S.peruvianum against Cladosporium fulvum* and *Clavibacter michiganensis* against potato *Solanum tuberosum* is crossed with *S.berthaultii* and *S. phureja* against viruses and with *S. demissum* and *Solanum tuberosum* subsp. *andigena* against PVY and *Phytophthora infestans* and viruses while in pepper *Capsicum annuum* is crossed with *C. chinense* against viruses.
- Meeting breeding challenges against biotic stresses require robust protocols to screen and identify useful alleles from exotic germplasms and efficient selection methods to facilitate incorporation of multiple stress tolerance genes into new varieties.
- Development of disease resistant varieties in tomato and pepper should be taken as a high priority area.
- Tomato production in the tropics has been devastated by increased incidence and severity of tomato yellow leaf curl diseases vectored by whitefly begomoviruses.
- Development of resistant varieties is complicated because the pathogen is highly diverse and almost all resistance genes originated from wild tomato species.
- Application of marker assisted selection (MAS) has facilitated development of tomato varieties homozygous for multiple begomovirus resistance genes that offer higher levels of resistance to a wider range of begomoviruses.
- Early efforts by the worldwide tomato breeding community to identify TYLCD resistance in cultivated tomato were unsuccessful but high levels of resistance were found in some wild tomato species, especially some accessions of *S. chilense, S. peruvianum* and *S. habrochaites.* Interspecific crosses and backcrosses have let to the introgression of *Tomato yellow leaf curl virus* (TYLCV) resistance genes *Ty*-1, *Ty*-2, *Ty*-3, *Ty*-4, *Ty*-5. Early TYLCD resistance breeding focused on *Ty*-2 , which offered excellent resistance to begomoviruses in Taiwan, south India and parts of the Middle East.
- However, the failure of *Ty*-2 to provide resistance to *tomato yellow leaf curl Thailand virus,* first detected in Taiwan in 2005 and other major begomoviruses in the tropics prompted to adopt a strategy of pyramiding multiple Ty genes into its lines to increase the chances of broad-based resistance to multiple begomoviruses.
- Screening for TYLCD resistance by exposing seedlings to viruliferous whiteflies and scoring plants for symptom severity was limited by several factor such as seasonal fluctuations of whitefly populations affecting TYLCD incidence.
- Even though multiple Ty lines demonstrate effective resistance to many begomoviruses, the diversity of tomato-infecting begomoviruses increases the likelihood of virus forms that will overcome resistance.

- Use of cultural practices for TYLCD control such as pre- or post-transplant nets to exclude whiteflies, application of imidacloprids, and crop-free periods coupled with resistance can augment TYLCD resistance. Varieties resistant to both begomoviruses and their vector offer great potential for durable TYLCD resistance.
- In tomato, sources of genes to improve abiotic traits include wild relatives, landraces, or other exotic germplasm.
- The problems of working with unadapted germplasm with poor horticultural traits and linkage drag, and traits conditioned by many small quantitative trait loci (QTLs) present tremendous challenges to plant breeders.
- Success will require strong interdisciplinary teams working to design effective protocols that identify trait sources, efficient high throughput marker protocols that effectively screen large populations for many markers, and basic research to understand the mechanisms of crop adaptation to diverse group of pathogens.
- A successful breeding programme is one that introduces a new resistant cultivar over a large area, giving the pathogen a great scope for survival and does so without incurring epidemiological mutation.
- Under environments where disease occurs in varying intensities, disease resistance breeding provides a far greater protection to yield enhancement.
- When resistance to a disease is race specific, some varieties of solanaceous vegetables are resistant to certain diseases in one area while susceptible in other area.
- The expression of resistance to a disease can also be affected by the age of the plant. The nutrition of the host plant can also affect the expression of resistance to diseases.
- Availability of adequate level of field resistance to some diseases in plants, severe losses of yield are not obtained.
- Recently, DNA microarray technology helps the breeders to learn more about different diseases and to further classify them on the basis of the patterns of gene activity in the plant system.
- Molecular breeding including both marker assisted breeding and genetic engineering has experienced significant innovations and advances during the past three decades.
- Starting with Mendelian genetics, plant breeders have progressed from using morphological markers to protein isozymes and then ultimately to DNA-based markers, which have enabled routine genome-wide analysis.
- As a result, thousands of genes and quantitative trait loci (QTL) have been mapped across the major crop species, laying the foundation for marker-assisted selection (MAS) techniques, such as precise marker-assisted backcrossing, to transfer desirable loci into breeding lines.
- With the subsequent development of advanced MAS strategies, including QTL pyramiding, marker-assisted recurrent selection and genomic selection, marker-assisted approaches have led to rapid gains in selection for plant breeding programms.

- Breeding by design and genetic modelling have received great attention as designing a desirable plant based on marker and associated gene information becomes increasingly possible.
- Molecular breeding increases genetic gain per crop cycle, stacks favourable alleles at target loci and reduces the number of selection cycles.
- In the last decade, the private sector has been benefitted immensely from Molecular breeding, which demonstrates its efficacy.
- In contrast, molecular breeding adoption is still limited in the public sector and it is hardly used in developing countries. Major bottlenecks in these countries include shortage of well-trained personnel, inadequate high-throughput capacity, poor phenotyping infrastructure, lack of information systems or adapted analysis tools or simply resource-limited breeding programmes.
- The emerging virtual platforms aided by the information and communication technology revolution will help to overcome some of these limitations by providing breeders with better access to genomic resources, advanced laboratory services and robust analytical and data management tools.

Breeding of Onion

- Onion (*Allium cepa L.*) is an important vegetable crop grown throughout the world. It is a cool season vegetable. It is grown for its bulbs.
- In India, the most important onion-growing states are Maharashtra, Tamil Nadu, Andhra Pradesh, Bihar and Punjab. Onion is one of the crops which India exports.
- A global review of area and production of major vegetables shows that onion ranks second in area and third in production in the world.
- About 25.40 million tonnes of onion are produced in the world from 1.79 million hectares of land. India ranks first in area and second in production (0.29 million hectares and 2.45 million tonnes).

Origin

- The Russian scientist Vavilov reported Central Asia as the primary centre of origin and near east as the secondary centre of origin. Onion originated from the region comprising north-west India, Afghanistan, the Soviet Republic of Tazik and Uzbek and Western Tiensens. Western Asia and the area around Mediterranean sea are its secondary centres of origin. It is now cultivated throughout the world.
- Ploidy level = $2n = 2x = 16$

Floral biology

- The flower structure is called an umbel which is an aggregate of many small inflorescences (cymes) of 5 to 10 flowers, each of which opens in a definite order causing flowering to be irregular and to last for two or more weeks.
- Each individual flower contains 6 stamens, 3 carpels united into one pistil and 6 perianth segments.
- The pistil contains 3 locules each of which has 2 ovules.

- The flower also contains nectarines which secrete nectar to attract insects for cross-pollination.
- The flowers are protandrous and anthers shed pollen over a period of 3-4 days prior to the time when full length of style is attained. Anthesis occurs in early morning (6-7 a.m.).
- Anther dehiscence is between 7.00 a.m. and 5.00 p.m. and on next day also with peak between 9.30 a.m. and 5.00 p.m. Pollen fertility is maximum on the day of anthesis.
- Thus, stigma becomes receptive 3-4 days after shedding of pollen grains and protandry leading to favour cross-pollination.

Breeding Goals

- Long bulb storage life
- Resistance to diseases (purple blotch, basal rot, stemphyllium blight, bacterial storage rot)
- Resistance to insect pests (thrips)
- Resistance to abiotic stresses (moisture stress, high temperature, salinity, alkalinity)
- Bulb quality (size, shape, colour, pungency, firmness, dormancy, amount of soluble solids)

Cytogenetics

- The *Allium cepa* species are diploid with basic chromosome number of $x = 8$ ($2n = 16$). Occasional tetraploids have also been reported.
- There are morphological and cytological similarities between the species of section cepa, but still strong crossing barriers exist between them.
- This prevents gene flow between the two even where sympatric distribution of two species occurs.
- Introgression of genetic material from wild to cultivated species is also difficult. Low success has been recorded in several interspecific hybridizations.

Breeding Methods

- Introduction, mass selection, selfing and massing, inbreeding followed by hybridization to exploit heterosis have been used for improvement of onion.

Plant Introduction

- Early Grano is an introduction into India. In long-day types, Brown Spanish was also a successful introduction.

Mass Selection

- Common in cross-pollinated crops. Most of the onion varieties in India have been developed by mass selection.

Selfing and Massing

- Suggested by Jones and Mann (1963). This method is very good for improvement in a crop, where inbreeding depression is common. Improvement in cultivar can be effected by selfing followed by massing. The procedure is as follows:
- **First year (bulb crop):** Select 100 best bulbs of desired type.
- **Second year (seed crop):** Grow selected bulbs, self one or more umbels per plant to initiate a separate line each.
- **Third year (bulb crop):** Grow the progenies of each inbred line separately. Discard the poor performing lines during the growing season, at harvest or in storage. Select at least 25 best lines and keep 15-20 bulbs of each for selfing and open pollination for next year.
- **Fourth year (seed crop):** Self-pollinate 1-2 umbels in each plant and allow others to open pollinate. Mass open pollinated seeds and increase for large-scale production for more than two generations to avoid much inbreeding depression.
- **Fifth year (bulb crop):** Grow the selfed progenies separately. Select again the best 25 lines and 15-20 bulbs of each lines as above in the third year.
- **Sixth year (seed crop):** Composite and plant bulbs of all selections in a field or in a cage for free open pollination in between the unrelated lines. The open pollinated seed can be massed and increased as foundation seed.

Inbreeding

- In any onion improvement programme, a considerable amount of inbreeding or selfing is necessary. Brown paper bags and three ring muslin cloth bags are used for selfing.

Hybridization

- Used when we want to introduce characters from other varieties.
- Intervarietal : Very common
- Interspecific : Very rare
- Utilized in *A. cepa* x *A. fistulosum.*

Heterosis Breeding

- The F1 hybrids are high yielding with uniformity in bulb size, the two most desired characters.
- One of the main components for exploitation of heterosis in onion is isolation of male-sterile lines.
- The male-sterile lines have been isolated in Pusa Red at IARI.
- Male-sterile lines have also been isolated at IIHR, Bangalore, but there is no hybrid in commercial production yet; work is in progress to isolate the best combiners.
- Heterosis breeding involves three important stages, namely:
  - Production of inbred lines
  - Testing combining ability of inbred lines

- Production of the seeds of F1 hybrids

- A line = male sterile line (S msms) caused by sterile cytoplasm (S) and nuclear recessive gene (msms).
- B line = maintainer line (N msms)
- C line = male parent (N MsMs or N Msms or N msms or S MsMs or S Msms)
- Maintenance AxB 1:1 F1 production AxC 4:1 or 8:2.

Salient breeding achievements and Future Prospects

- The varieties Italian Red and Local Brazilian are resistant to purple blotch.
- Granex and Excel 35 are reported to be resistant to pink rot.
- Some promising F1 hybrids identified are Abundance, Aristocrat, Banaza, Champion and Contender.
- F1 hybrids developed in India, namely, Purple x Red, Purple x White and Red x White have more weight of plants and increased bulb diameter.
- Yellow-red colour of bulb was linked with resistance to *Colletotrichum arienans.*
- Punjab Selection and N 2-4-1 were good for dehydration. Pusa Red and Punjab Red Round were best suited for storage at room temperature (up to 5 months).
- Developed CO ON.5 in aggregatum onion at TNAU which has profuse seed setting ability and can be propagated by seeds unlike the other varieties of aggregatum onion.

Future Prospects

- Development of varieties with improved quality and high yield suitable for different agro-climatic conditions.
- Breeding varieties resistant to major diseases and insect pests, such as purple blotch, basal rot, stemphyllium blight and thrips.
- Heterosis breeding for development of short-day, high yielding F1 hybrids in red, white and yellow types with longer storage life.
- Development of varieties of multiplier onion with bright red colour and higher bulblets.
- Standardization of improved package of practices particularly for new varieties and hybrids.

Breeding of Cowpea

- Cowpea (*Vigna unguiculata*) also called as southern pea and black eyed pea belongs to family leguminosae.
- Less attention has been given to the vegetable cowpea. It is a nutritious leguminous crop low in anti-nutritional factors.
- It has a wide range of ecological adaptations and can be grown more widely.
- In fact, it probably has the greatest potential among all food legumes in the semi-arid to sub-humid tropical areas.
- The typical vegetable cowpea is characterized by a long pod (>30 cm), stringless pods, fleshy pod pericarp, thin and long seeds and higher monosaccharide: polysaccharide ratio. Pole and bushy types are available in vegetable cowpea.

Origin

- Africa is considered as the primary centre of origin of 170 species of genus Vigna, 22 are found in India.
- The crop spread from Africa to Asia and Europe through Egypt. Now cowpea is widely distributed throughout the tropics and subtropics.

Floral Biology

- The flower opens early in the morning, closes before noon and falls down on the same day.
- The extra floral nectarines at the base of the corolla attract ants, flies and bees but a heavy insect is required to depress the wing petal and expose the stamen and stigma.
- Cowpea flowers bloom between 6.30 a.m. and 7.00 a.m.
- The process of opening of corolla takes 45-60 minutes.
- Anther matures and starts dehiscing about 78 hours before the time of opening corolla.
- Pollen remains viable for 42 hours at room temperature and relative humidity 91%. Stigma becomes receptive from 12 hours before blooming to 6 hours after blooming.

Breeding Goals

- High green pod yield (vegetable type varieties)
- High seed yield (dry-seed type varieties)
- High fodder yield (fodder type varieties)
- Dual purpose (seed and vegetable type and seed and fodder type)
- Earliness
- Appropriate plant type (erect, determinate for vegetable and seed type cultivars and spreading for fodder type cultivars)
- Wider adaptability
- Photo-insensitivity
- Short, tender pods for whole pod processing
- Long, tender and stringless pods for fresh consumption
- Varieties suitable for inter-cropping
- Resistance to diseases
- Resistance to insects
- Better seed quality.

Cytogenetics

- It is true diploid with 2n=22. Sen and Bhowal (1960) made cytotaxonomic studies on *Vigna sinensis* (L.) Savi, *V. catjang* (Bur.M) Waqlp and *V. sesquipedalis* (L.) Fruw, popularly known as cowpea, catjang bean and asparagus bean.

- The three cultivated species could be easily crossed among themselves. The hybrids among them were fertile like intervarietal hybrids.
- Chromosome number of each of three cultivated species was found 2n = 22. As mentioned elsewhere, all the above cultivated species are put under *Vigna unguiculata*

Breeding Methods

Pedigree Selection

- The pedigree system of breeding is the most common method used by cowpea breeders.
- This method has been successful in developing cowpea cultivars with new combinations of characteristics and resistance to diseases.
- Single plant selections are carried out within larger F2 / F3 populations. Individual plant progenies are planted in one or more rows, 4-6 m in length and 1.5 m apart.

Backcross Method

- The backcross breeding procedure has been found efficient for transferring single-gene resistance to specific diseases into cowpea cultivars.
- For example, this procedure has been used to transfer Cls gene, which provides resistance to Cercospora leaf spot, into the susceptible cultivar Colossus in USA.

Heterosis Breeding

- Seven *Vigna unguiculata* var. *sesquipedalis* varieties were crossed in diallel. Of the 42 hybrids, only three exceeded the higher parent in seed protein content and pods / plant.
- The methods of F1 hybrid production are hand emasculation and pollination, use of 'protruded stigma' types female and use of male sterile line.

Mutation Breeding

- Mutation breeding in cowpea has been utilized on a limited scale through irradiation (9-40 Krad) by gamma rays to isolate mutants with increased yield and earliness.

Salient Breeding Achievements

- Varieties Pusa Phalguni suited for spring season, Pusa Barsati for rainy season and Pusa Dofasli for both the seasons have been evolved.
- K 1552 (Pusa Komal) evolved is an early variety which is resistant to bacterial blight.
- Variety Grant is resistant to Fusarium wilt, Chinese Red is resistant to Phytophthora stem rot, Mississipi 57-1 and Iron are resistant to root-knot nematodes have been evolved.
- Variety Goit, Dixies-Cream and Alabunch are resistant to cowpea mosaic disease.

Future Prospects

- Many valuable genes like male sterility, high protein and methionine content, resistance to thrips, aphids, pod-borers, root-knot nematodes, anthracnose and cowpea mosaic virus are available which need further utilization.
- Bush types of yard-long bean have been developed (Los Banos Bush Sitao No.1).
- There is a need to evolve high essential amino acid lines in cowpea. The extent of damage due to various diseases has not been estimated except for a few diseases.

Breeding of Amaranthus

Floral Biology

- Most *Amaranthus species* are monoecious. The flowers can be terminal or axial but are always organized in to glomerules within the inflorescence. Within the glomurules, the first flower is generally staminate, and the later flowers are pistillate (Pal 1972; Joshi and Rana 1991).
- Emasculation in *A.spinosus* is restricted to the topping off of the upper part of the panicle carrying male flowers above the female flowers.
- After emasculation, the flower heads are bagged with butterpaper bags in *A.tricolor, A.lividus and A.dubius*, emasculation process is very tedious due to the intermixed nature of the male and female flowers within the same glomerule. The time of anthesis is 8.30 to 9.30 am.

Breeding Objectives

- To identify suitable vegetable amaranth types with high yield potential coupled with nutritional qualities and without anti-nutrient factors.
- To develop an 'erect' and 'clipping' (cutting) types, suited for early harvest for 'pulling' and continued harvest by periodical cutting respectively to suit different regions of Tamil Nadu
- To develop a dual amaranth types i.e., grain-cum-leafy amaranth.

Breeding Methods

- Selection is the only method through which a number of improved varieties have been developed.
- Improved selections of Amaranthus developed at TNAU

Name of variety	Pedigree of the variety	Chief characteristics
CO.1 amaranthus *A.dubius* Mart.ex.-Thellyung	Selection from a local type	Seeds–black, small (about 4000 seeds/g); lack initial vigour but rapid growth rate after 25-30 days; leaves–dark green, broad with ridged surface; stem–round, glossy green and succulent even at mature stage; specially suited for late harvest at 30-35 days after sowing as whole plant green (Mulakkerai) with green yield 7-8 t/ha.
CO.2 amaranthus (*A.tricolor* L.)	Selection from a local type	Seeds–black; and bolder in size (1200-1400 seeds/g); good early germination and vigorous growth; leaves – green and long lanceolate; stem-green; highly suited for 'tender green' harvest at 20-25 days after sowing as the stems are tender (Thandukeerai) with yield–10-11 t/ha.
CO.3 (*A.tricolor*L.var.*tristis*)	Selection from a local type	2800 seeds/g; grows to a height of 90-100 cm with 12-15 basal branches; lends itself for periodical cuttings (clippings- Arukeerai or Arakeerai) and hence can be retained for more than 3 months. 1st clipping commences 20 days after sowing and thereafter at weekly intervals. A total of 10 clippings can be had, yields 10-12 t/ha.
CO.4 amaranthus (*A. Hypochondriacus* L.)	Selection from a local type	Seed–pale yellow, medium sized (2000-2300 seeds/g); leaves–green, lanceolate; stem–green; inflorescence–green,long, terminal and generally unbranched; flowers in 40-45 days after sowing and matures in about 90 days; yields 8 t/ha by thinning at 25 days and about 2-2.5t. of grains per ha at maturity. The grains are rich in protein (15.9%) and other essential amino acids; popping rate of grain is about 2.5 times by volume.
CO.5 amaranthus	Single plant selection	The leaves are double coloured (green and pink). It gives a rosette growth in early stages. First harvest starts 25 days after sowing and in duration of 55 days it produces 40t/ha.

Breeding of Cole Crops

Genetic Mechanisms for Commercialization of Heterosis Breeding in Cole Crops

- Cole crops are one of the most important groups of cool season vegetables. Initially, these crops were confined to Europe and other temperate regions.
- But, after the Second World War, their cultivation has spread to tropics and sub tropics as well. All the members of this group have originated primarily from the common ancestor wild cabbage (*Brassica oleracea* var. *oleracea*) and are fully cross-compatible with each other.
- In general, these are **highly cross-pollinated crops** and have shown preponderance of non-additive gene action for most of the economic traits, thereby suggesting the relevance of heterosis breeding.
- In the absence of appropriate genetic mechanism for the production of hybrid seed with ease at relatively lower cost, heterosis in a crop is of no practical value.
- Fortunately, the genetic phenomena of sporophytic self-incompatibility and male sterility (particularly cytoplasmic male sterility) exist in cole crops and have proved instrumental in commercialization of hybrid seed production.
- Self-incompatibility
- The term self-incompatibility refers to the partial or complete incapacity of a plant, producing functional gametes, to set viable seed on self-pollination.
- Alternatively, it is the failure of the genetically identical gametes to unite in fertilization.
- In cole crops, **sporophytic self-incompatibility (SSI)** exists wherein the incompatibility reaction is governed by the genotype of the pollen producing plant and it is controlled by a series of alleles at a single locus 'S'.
- SSI system has been confirmed in kale (Thompson, 1957), sprouting broccoli (Sampson, 1957), cabbage (Adamson, 1965) and cauliflower (Hoser-Krauze, 1979). However, its use in cabbage hybrid seed production had been demonstrated much earlier (Odland and Noll, 1950).

Basic Steps in the use of SSI

1. Identification of self-incompatible plants in diverse population/genotypes.
2. Development of homozygous self-incompatible lines.
3. Identification of S-alleles in the homozygous self-incompatible lines.
4. Establishment of inter-allelic relationships among the S-alleles.
5. Identifying the best combining lines.
6. Maintenance of parental self-incompatible lines.
7. Commercial hybrid seed production.

- Identification of self-incompatible plants is carried out by making different types of pollination viz., self-pollination in freshly opened flowers (OP), compatible pollination with unrelated S-allele pollen (CP) and bud pollination to ascertain whether the plant in question is male and female fertile and also to get selfed seeds to raise its progeny.

- The plants are categorized as self- incompatible/compatible based on seed set in OP in relation to CP (Watts, 1963). For quicker (within 24-48 hours of pollination) identification of self-incompatible plants, fluorescence microscopy can be used (du Crehu, 1968; Vidyasagar and Chatterjee, 1984).
- Wallace (1979) suggested the use of seed set as well as pollen tube penetration (fluorescence microscopy) data.
- Homozygous self-incompatible lines are developed by making intra-progeny crossings/pollinations in freshly opened flowers in the progeny of a self-incompatible plant in full diallel by utilizing at least 7 progeny plants (Mackay, 1977).
- The intra-progeny crossings/pollinations reveal heterozygous nature of the original self-incompatible plant. After grouping the progeny plants on the basis of compatible/incompatible reaction, we get at least two (one homozygous and one heterozygous) or all the three (both homozygous and one heterozygous) phenotypic groups.
- By selecting plants from the homozygous groups, homozygous lines could be developed in the next 2-3 generations.
- Information about the inter-allelic relationships among the S-alleles is important to ascertain their level of dominance and S-allele interactions.
- The S-alleles high in the dominance series are likely to give minimum selfs and sibs in hybrid seed.
- S-allele interactions (in pollen and stigma) are worked out by using the data (seed set as well as pollen tube penetration) obtained from the pollinations carried out in freshly opened flowers reciprocally between a heterozygote (SxSy) and its two corresponding homozygotes (SxSx and SySy).
- Four types of S-allele interactions viz., type I (same S-allele dominant over the other in both pollen and stigma), type-II (one S-allele dominant over the other in pollen but codominant in stigma), type-III (one S-allele dominant over the other in stigma but codominant in pollen) and type-IV (both S-alleles codominant in pollen as well as stigma) have been reported.
- In order to have heterotic hybrids, it is essential to identify the best specific combining S-allele lines.
- This can be ascertained from the SCA studies and *per se* performance. Self-incompatible lines can also be combined with self-compatible lines but the hybrid seed quantities shall be lesser since the hybrid seed will be harvestable on the plants of self-incompatible lines only.
- Maintenance of self-incompatible lines is a costly affair. The various methods being followed are large-scale seed production through manual sib mating in bud stage, carbon dioxide gas treatment (3-5% conc.) for 8-24 hours at 100% relative humidity in air-tight growth chambers and tissue culture by using meristem as explant.
- Of late, sprays of sodium chloride (3-5%) have been reported to be effective in temporary breakdown of self-incompatibility (Kucera, 1990; Yang *et al.*, 1995 and Kucera *et al.*, 2006). Common salt (5%) sprays made 15 minutes after pollination with self pollen in opened flowers in all the flowering regimes on the plants of

self incompatible lines of cabbage have proved effective at CSK HPKV Palampur (Singh and Vidyasagar, 2012).

- This method is being followed at IARI to produce the seeds of the parental self-incompatible line ScSc which is one of the parents of Pusa Hybrid -2 of cauliflower (personal communication).
- Commercial hybrid seed production is done by way of developing single cross, three-way cross or double cross.
- In single cross, two self- incompatible but cross-compatible best combiners are planted in alternate rows in isolated plots. The hybrid seed is harvested on both the lines.
- In three-way cross, one single cross and a self-incompatible line are planted in alternate rows. Similarly in double cross, two single crosses are used. In USA, for hybrid seed production of cabbage, top cross is being used.
- For every 2 or 3 rows of a self-incompatible line, one row of a good open pollinated (OP) cultivar as a pollen parent is provided.
- However, the hybrid seed is harvested from the self incompatible plants only.
- The main problems being faced in hybrid seed production are depression in S-allele lines by continuous inbreeding, pseudo-compatibility, the effect of environmental factors on the level of self-incompatibility and higher proportion of selfs/sibs in hybrid seed due to lack in proper synchronization of flowering. These could be managed by resorting to vegetative propagation, using the S-allele lines which behave stable under diverse environments and selecting the parental lines which have perfect synchronization in flowering.

Male Sterility

- Male sterility is defined as the deviant condition in normally bisexual plants (monoecious as well as hermaphrodites) when no viable pollen is formed. Genic male sterility which is governed by recessive nuclear genes has been reported in cabbage (Rundfeldt, 1960), cauliflower (Nieuwhof, 1961) and other members of the cole group as well.
- In view of the problems associated with the maintenance of genic male sterile lines, its commercial use has been rather limited. In cauliflower, it has been utilized for hybrid production by maintaining the male sterile lines clonally (Grout, 1988).
- Pearson (1972) reported male sterility controlled by *Brassica nigra* nuclear-cytoplasm interaction but this could not be used on commercial scale due to partial opening of petals and less developed nectar glands leading to poor bee-activity.
- Ogura male sterility (Ogura, 1968) was reported in Japanese radish. No restorer nuclear genes could be found in Japanese genotypes but the same could be obtained in European radish.
- Bannerot *et al.* (1974) could introduce cabbage nucleus into the cytoplasm of Ogura male sterile radish (R- cytoplasm) by repeated backcrossing.
- This CMS in cabbage had the problem of chlorosis at low temperature (below 12°C). This drawback was overcome by replacing radish chloroplasts with *Brassica oleracea* chloroplasts through protoplast fusion by Robertson *et al.* (1987).

- Sigareva and Earle (1997) standardized the protoplast fusion procedure to transfer desirable male sterility cytoplasm from broccoli into cabbage much more rapidly than conventional backcross method.
- No fertility restorer genes have been found for R- cytoplasm (Ogura) induced male sterility in cole crops (*Brassica oleracea*). Ogura-CMS is stated to be a mitochondrial DNA encoded male sterility.
- This implies that all *Brassica* spp. including cole crops act as maintainers of R-cytoplasm (Ogura) induced male sterility.
- Fang *et al.* (1997) have also reported the presence of dominant male sterile gene in cabbage. Its usage may not be preferred over R-cytoplasm (Ogura) induced male sterility because of the possible problems associated in maintaining the male sterile lines in homozygous dominant condition. R-cytoplasm (Ogura) induced male sterility already present/introduced in any genotype of a cole crop can be transferred into the desired genetic background through back cross method.

Basic steps in the use of CMS

1. A-line (female parent) of desired genetic background (genetically S- ms ms).
2. B-line (maintainer): This is an isogenic line in the genetic background as that of A-line and its function is in the maintenance of A-line (genetically N-ms ms).
3. C-line (male parent): This is a male parental line and also the best specific combiner with A line. Genetically this could be N/S-Ms/- (fertility restoration) or N-ms ms since the economic product in cabbage and cauliflower is not the true seed.
4. Maintenance of A, B and C lines.
5. Hybrid seed production.

- The first increase of the seeds of the lines A, B and C must be in insect-proof cages/structures followed by an increase in the open but not more than twice to obtain a high degree of purity in the stock seeds of the parental lines used for the production of F1 hybrid seeds.
- The seeds of lines A and B can be produced in the same cage while the C line must be grown separately.
- Hybrid seed production on commercial scale is carried out in the open by providing recommended isolation distance of at least 1000 m.
- Usually for every 2 or 3 rows of A-line, one row of C-line is planted. The lines A and C must have perfect synchrony in flowering for good pollination and seed set. The hybrid seed is harvested from the plants of A-line only.
- The main problem being faced is the lower quantity of hybrid seed on account of honeybees preference for the pollen fertile C-line.
- This could be overcome to a certain extent with appropriate flower morphology and manipulating the ratio of plants of A and C lines, spacing and planting design.
- Inbreeding depression is another problem which results in low seed production of inbred parental lines and also that of single cross hybrids. Three- and four-way (double cross) crosses may prove economical.

Self-incompatibility versus Cytoplasmic Male Sterility

- Both the genetic phenomena have their own merits and demerits. The use of self-incompatible lines ensures the harvest of hybrid seed on all the plants in the hybrid seed production block.
- In CMS system, the hybrid seed will be harvested only from 67-75% of the plants in the hybrid seed production block.
- The latter phenomenon (CMS) has been considered better in avoiding selfs in the hybrid seed. However, in CMS lines also, some pollen production has been noticed especially at relatively higher temperatures nearing end of flowering stage thereby creating conditions for the possible selfs in the ultimate hybrid seed in this system as well.
- In my opinion, both the genetic systems (SSI and R/Ogura-CMS) will prove equally good provided the S-allele and CMS lines are tested over space and time in variable weather conditions for their stability and their consequent use in the hybrid seed production programme.

Breeding for Resistance in Cole Crops

- Cabbage, cauliflower, kale, Brussels sprout, broccoli and knol-khol are the important cole crops.
- During cultivation, these crops are severely affected by various fungal and bacterial diseases with substantial yield reduction.
- These diseases can be kept under check by following elaborate chemical spray programmes, but they are cost prohibitive and not eco-friendly.
- The only sustainable and long lasting approaches to combat all this problem is the development of resistant cultivars.
- To develop a suitable resistant variety, a good resistant source is required which can only be obtained if the available material is properly screened with standard procedures and potential disease resistant sources are identified.
- In this section diagnostic characteristic of important diseases affecting cole crop and screening techniques for important diseases of cole crops has been discussed.

1. Sclerotinia Rot

- The disease is caused by *Sclerotinia sclerotiorum* (Lib.) deBary.
- Leaves lose their shine and drop down, basal portion of the leaf petioles start rotting and turn dark brown to black with a white cottony tuft of mycelium on the surface.
- Stalks rot from inside, become hollow and filled with black sclerotia. Curds lose compactness followed by development of characteristic soaked white rot.
- Water soaked lesions also appear on bolting shoots later exhibiting a white silvery appearance followed by complete wilting.
- There is no pod formation in the diseased plants. Sclerotia produced by the fungus can survive in soil for many years.
- Temperature range of 18-23°C and 95-100% relative humidity is congenial for the disease development.

Screening Technique

- Cauliflower plants are inoculated with approximately 100 mg culture of the fungus at the leaf axil near collar region just before the curd initiation.
- The inoculated plants are kept in humidity chambers having 95±5% relative humidity at 25±1°C. The symptoms of the disease on inoculated plants are recorded after 25 days of inoculation.

Under Laboratory Conditions

- Forty days old seedlings are transplanted in pots containing sterilized soil. The plants are allowed to establish for 5 days in the pots.
- These are then inoculated with 10 days old mycelium (10 mg of mycelium in corn sand meal) one cm away from the main stem. The inoculum is covered with sterilized soil and watering is done at regular intervals of time (3 days interval). Symptoms usually appear after 15 days.
- Data on disease reaction is recorded on the basis of presence of dark brown lesions at the collar region of the plant after 25 days of inoculation.

Sources of Resistance

- Early Winter Adam's White Head, EC 162587, EC 103576, Kn-81, Janavon, EC 131592, RSK 1301.

2. Alternaria Leaf Spot

- The disease is characterized by distinct spots with concentric rings on the lower leaves; the dark dusty fungus growth develops on these spots during moist periods.
- During storage the spots enlarge. Soft rot bacteria may enter through dead leaf spots. The disease is caused by *Alternaria brassica* (Berk.) Sacc. and *A. brassicicola* (Sachow.) Wiltsh.
- The fungus overwinters in seed and in residue from diseased plants. Wet conditions promote disease development.

Screening Technique

- Spray plants with an atomiser @ 1×10^4 spores/ml.
- Provide optimum required temperature and humidity conditions for pathogenesis.

Source of resistance: Kt-9

3. Downy Mildew

- The disease is caused by *Peronospora parasitica* (Pers.) Fr. an obligate parasite.
- The disease is most serious in seed beds and appears as small leaf spots which first are yellow and later turn brown with bluish black markings.
- In moist weather, a white downy mold develops on the underside of the leaf spots. Damage to heads curds may occur under conditions of high humidity, warm days and cool nights followed by heavy dew. The causal fungus overwinters on seed, in crucifer weeds and in soil.

- Disease is promoted by cool weather and occurs at 8-14°C and 90- 95% relative humidity. Downy mildew can predispose plants to bacterial soft rot.

Screening Technique

- Spray plants with 1×10^5 spores/ml after 3 days of seedling emergence and provide proper temperature and humidity conditions.
- Inoculum in diseased leaves can also effectively result in disease development.

Source of resistance:

- **Cauliflower lines:** PI 208474, CC3-5-1-1, PI, S, PI 231210, PI 246077 and Early Winter Adam's White Head
- **Cabbage lines:** PI 263056, PI 253057, PI 357374, PI 418984, PI 418986, PI 418987, PI 418988.

4. Black Rot

- The disease is caused by the bacterium *Xanthomonas campestris* pv. *campestris* (Pam.) Dowson; affects young as well as mature plants.
- Affected seedlings turn yellow and die. On older plants, the tissue turn yellow at leaf margins and chlorosis progresses toward centre forming 'V' shaped lesions toward the centre of the leaf. Affected areas later turn brown and die.
- The veins and veinlets of the chlorotic tissue become dark. The infection on head results in brown to black discolouration of the surface which at later stages leads to rotting.
- The black rot bacterium overwinters on seed and in residue from diseased plants; it persists in residue for 1 to 2 years. The disease is also spread by movements of contaminated water.

Screening Technique

- Inoculate the plants at 4-5 leaf stage by cutting the margins of leaves with scissors dipped in 48 hours old bacterial culture (1×10^8 cfu/ml). Keep the plants at high relative humidity (>90%) for 10 days and record disease progress after 14-days of inoculation.

Sources of Resistance

- 294 bc, Pua Kea, MGS-23, Sn 445, Sn 401, Ivan's Glory,Brunswick Pie Corto, PI 436606, Pusa Mukta, RBSI, EC 162587, Lawyana, EC 191025, EC 191095.

5. Soft Rot

- The disease is caused by *Erwinia carotovora* (Jones) Holl. Affected areas appear to be water-soaked, develop a soft decay and have a distinctive foul smell.
- Affected cabbage and cauliflower heads decay rapidly and turn dark.
- The bacteria usually infect plants through surface areas injured by insects, cold, or mechanical means. Warm wet conditions promote disease development.

Sources of Resistance

- *Brassica rapa* accessions viz.; CC-23, CC-1, Majica and Arcadia are having partial tolerance. Cabbages line MR-1 carries high degree of resistance.

Breeding of Cabbage

- All the cole crops including cabbage have originated primarily from the same wild diploid species viz., *Brassica oleracea* var. *oleracea* (syn. *sylvestris*).
- The non-heading leafy cabbage was the first to evolve in eastern Mediterranean region and Asia Minor.
- The compact headed white cabbage got evolved at a later date in North Europe through mutation and introgression from wild species, human selection and adaptation.
- In India, cabbage was introduced by the Portuguese during the Mughal Period but the crop became popular during the British rule.
- It occupies slightly lesser area as compared to cauliflower in our country but it tops the cole group in area and production in the world.
- In view of the temperate nature of the old European introductions, its seeds used to be imported from Europe.
- After the interruption of seed supplies during the Second World War, seed production technology was standardized indigenously and quality seeds were produced at Quetta (now in Pakistan), Kashmir valley, Kullu valley and Kalpa in Himachal Pradesh.

Genetic diversity

- After kale, cabbage possesses the maximum genetic diversity among the cole crops.
- There are distinct groups on the basis of leaf colour and texture (white, red and savoy), head shape (round, pointed and flat/drum), maturity (early, mid and late) and temperature requirements for entering into reproductive phase/seed production (temperate and tropical).
- Within the same botanical variety of *Brassica oleracea,* there exists a quite good amount of genetic variability.

Gene action

- The nature of gene action plays a deciding role in selecting the most appropriate breeding method/approach to effect genetic improvement in a crop.
- Net weight of head and head maturity are the two most important economic traits besides head compactness, shape and colour.
- The genetic studies carried out in India and abroad have reported the preponderance of dominance (non-additive gene action) for these traits besides appreciable amount of additive x additive (i) type of epistasis as well.
- This justifies the importance and development of hybrids in cabbage. Commercialization of hybrids would have been impossible in the absence of

genetic mechanisms ensuring free cross-pollination in the production of hybrid seeds in panmictic plots.

Genetic Mechanisms

- **Self-incompatibility** and **male sterility** have proved instrumental in commercialization of hybrids in cabbage followed by in other cole crops.
- Although **sporophytic self-incompatibility system** was confirmed in cabbage by Adamson (1965), but the methods of seed production by utilizing self-incompatible parental lines had been suggested by Odland and Noll (1950) much earlier.
- The use of self-incompatibility involves the basic steps of identification of self-incompatible plants in diverse genotypes, development of homozygous self-incompatible S-allele lines, studying inter- allelic relationships of S-alleles, ascertaining stability of S-allele lines in diverse environments, pinpointing thc most heterotic S-allele parental lines combination, maintenance of S-allele lines and their utilization in the production of hybrid seed on commercial scale.
- Recessive nuclear male sterility has been reported by Rundfeldt (1960) and Nieuwhof (1961) but not commercialized due to the problems encountered in its maintenance and the absence of effective seedling markers.
- **Pearson (1972)** reported male sterility controlled by *Brassica nigra* nuclear-cytoplasm interaction but this could not be used on commercial scale due to partial opening of petals and less developed nectar glands leading to poor bee- activity.
- **Ogura male sterility (1968**) was reported in Japanese radish. No restorer nuclear genes could be found in Japanese genotypes but the same could be obtained in European radish.
- Bannerot *et al*. (1974) could introduce cabbage nucleus into the cytoplasm of Ogura male sterile radish (R- cytoplasm) by repeated backcrossing. This CMS in cabbage had the problem of chlorosis at low temperatures (below 12°C).
- This drawback was overcome by replacing radish chloroplasts with *Brassica oleracea* chloroplasts through protoplast fusion by Robertson *et al*. (1987).
- Sigareva and Earle (1997) standardized the protoplast fusion procedure to transfer desirable male sterility cytoplasm from broccoli into cabbage much more rapidly than conventional backcross method.
- No fertility restorer genes have been found for R- cytoplasm (Ogura) induced male sterility in cole crops (*Brassica oleracea*).
- This implies that all *Brassica* spp., including cole crops, act as maintainers of R-cytoplasm (Ogura)-induced male sterility. Fang *et al*. (1997) have also reported the presence of dominant male sterile gene in cabbage.
- Its usage may not be preferred over R-cytoplasm (Ogura)-induced male sterility because of the possible problems associated in maintaining the male sterile lines in homozygous dominant condition.

Commercialization of hybrid seed production

- Initially, it was only the sporophytic self-incompatible system which was commercialized for hybrid seed production. In this system, two heterotic but cross-compatible S-allele lines are grown in alternate rows in a panmictic plot to allow for free cross pollination and the hybrid seed is harvested from all the plants of both the parental S-allele lines.
- IARI, Regional Station Katrain has developed cabbage **hybrid KGMR-1** by using SSI and the same has been identified through AICRP on vegetable crops.
- With the successful transfer of R-cytoplasm (Ogura) induced CMS system in cole crops (including cabbage), there has been a lot of interest in its use for hybrid development over a decade now.
- This is seen as an alternative approach capable of overcoming the problems (selfs in hybrid seed due to breakdown of self-incompatibility and decline in vigour of S- allele lines due to continuous selfing and sib mating) experienced while using sporophytic self-incompatibility system.
- In CMS system, two or three rows of CMS male sterile lines are planted alternatively with one row of pollen fertile parental line (giving heterotic performance in combination with CMS male sterile line) in a panmictic plot and the hybrid seed is harvested from the plants of male sterile lines only thereby sacrificing about 25-33% plants.
- This does not mean that there are no problems associated with R-cytoplasm (Ogura) induced CMS system.
- It has been observed at CSKHPKV Palampur that a few plants in the backcross progenies of CMS genotypes do exhibit the production of some pollen especially when the flowering is near ending and prevailing temperatures are invariably higher, giving ample chance of sibs/selfs in hybrid seed.
- As of now, a few hybrids based on R-CMS system may be available in the market. IARI Regional Station Katrain (Kullu Valley) is engaged in the utilization of CMS system in the development of hybrids. In my opinion, both the genetic systems (SSI and R-CMS) will prove equally good provided the lines are tested over locations and years in variable weather conditions for their stability and consequential use in hybrid seed production.

Resistance Breeding

- Screening for biotic stresses has been carried out primarily at IARI Katrain. Resistance to cabbage yellows (*Fusarium oxysporum* f.sp. *conglutinans*) has been noticed in Ac238, Spitzkool and EC93559 whereas resistance to black rot (*Xanthomonas campestris*) has been reported in EC10109, Sel.8, Spitzkool, and Greenland Hammer.
- MR-1, an introduction from USA carries multiple disease resistance to *Sclerotinia* rot, *Rhizoctonia solani*, downy mildew, black rot and soft rot (*Erwinia carotovora*).
- As regards the genetics of resistance there are variable reports in literature. Type A resistance (stable at wider temperature range) to cabbage yellow disease is controlled by a single dominant gene whereas Type B (unstable at temperature

above 25°C) is polygenic. Resistance to *Sclerotinia* rot is under the control of a major recessive gene with modifiers. For downy mildew (*Peronospora parasitica*), the resistance is monogenic dominant against each race. Resistance to black rot is polygenic dominant.

- Red type cabbage is resistant to caterpillars (*Pieris brasssicae*) but susceptible to aphids (*Brevicoryne brassicae*) and its vice-versa is true in case of green/white cabbage.
- The **hybrid KCH-5** is tolerant to both aphids and caterpillars.

Biotechnology

- Micropropagation of plants through tissue culture has been successful in the maintenance of elite inbreds, self-incompatible and male sterile lines. Homozygous inbreds have been developed through anther/microspore culture (haploids) followed by doubling of chromosomes by colchicine treatment.
- Cybrids have been produced between *Brassica oleracea* and other *Brassica* spp. and Crucifers (*Raphanus sativus*).
- Recombinant DNA technology has been used in producing transgenics including Bt cabbage but the latter is yet to reach the farmer's fields for commercialization.

Breeding Achievements in India

- Only a limited genetic improvement research work has been carried out in India primarily at IARI Regional Station Katrain (established in 1949) and Kalpa in Himachal Pradesh as the earlier introductions of cabbage were temperate in nature requiring a chilling period of about six to eight weeks to enter into the reproductive phase.
- With the development of tropical cabbages in late 1970's in Japan, it has been made possible to carry out breeding work as well as seed production of cabbage even in sub-tropical (IARI New Delhi) and sub-temperate (CSKHPKV, Palampur) climate.
- Tropical cabbages have also proved instrumental in enhancing acreage as well as quality of cabbage heads in Peninsular India.
- The breeding methods used in the genetic improvement of cabbage include plant introductions, mass selection, inbreeding, hybridization, recurrent selection, disruptive selection, development of synthetics and heterosis breeding.

Table 7.21: *The improved varieties, synthetics and hybrids developed in India*

Variety/hybrid	Pedigree/breeding method	Institution
Golden Acre	Introduction	IARI, Katrain
Pusa Drum Head	Selection from EC 6774 (Japan)	IARI, Katrain
Pusa Mukta	Hybridization of EC24855 × EC 10109	IARI, Katrain
Pusa Ageti	Selection (Tropical)	IARI, New Delhi
Pusa Sambandh	Synthetic-	IARI, Katrain

Variety/hybrid	Pedigree/breeding method	Institution
Copenhagen Market	Introduction	IARI, Katrain
Pride of India	Introduction	Deptt. Agri. HP (Solan)
September	Introduction	Deptt. Agri. Tamilnadu
Drumhead Savoy	Introduction	Deptt. Agri. Tamilnadu
Red Acre	Introduction	Deptt. Agri. Tamilnadu
KGMR-1	F1 hybrid	IARI, Katrain
KTCBH-81	F1 hybrid	IARI, Katrain

There are a large number of cabbage hybrids from multinational seed companies as well as from the Indian seed companies having tie-ups with foreign counterparts especially after the implementation of new policy on seed development (NPSD) in October 1988.

- Currently at IARI Regional Station Katrain, sporophytic self-incompatibility (SSI) and cytoplasmic male sterility (CMS) are being used in the development of hybrids.
- KGMR-1 and KTCBH-81 based on SSI have already been identified for release during 2005 and 2013 respectively and KTCBH-84 based on CMS is being evaluated in trials of All India Coordinated Research Project on Vegetable Crops.
- At CSKHPKV Palampur, mild chill requiring genotypes viz., KGAT-I, II and III have been developed through hybridization of temperate and tropical genotypes.
- Besides, strains of Golden Acre, Glory and Savoy Cabbage and some other elite inbreds capable of bolting and producing seeds satisfactorily at Palampur have been developed. Homozygous self-incompatible lines have been developed in mild chill genetic background.
- Cytoplasmic male sterility has also been introduced in these genotypes and BC6 CMS progenies have been obtained.

Breeding of Cucumber

- Cucumber (*Cucumis sativus* L.) is one of the most important cucurbitaceous vegetable crops grown extensively in tropical and sub-tropical parts of the country.
- Cucumber is one of the Asiatic species and a member of the cucurbitaceae which has 90 genera and 750 species.
- **Origin:** India – It is considered as the home of cucumbers.
- **Distribution:** China, USA, Africa, Europe. In India it is grown in north and south and lower as well as higher hills.
- It is grown for its tender fruits, which are consumed either raw as salad, cooked as vegetable or as pickling in its immature stage.
- It is a rich source of vitamin B and C, carbohydrates, Ca and P (Yawalkar, 1985). The gynoecious hybrids are produced commercially.
- Gynoccious expression affects yield and earliness of fresh market cucumber. The gynoecious hybrid has 99-100% pistillate flowers on the first ten nodes of the main stem as compared to 3-5 for monoecious hybrid. Monoecious hybrids have a higher

percentage of quality fruits but genoecious yielded high and were precocious (Wehner and Miller, 1985).

- Cucumber has been the subject for production of multiple disease resistance. In the hybrids Slicer and Pickle, there is resistance to downy mildew, powdery mildew, anthracnose, angular leaf spot, cucumber mosaic and scab.

Breeding Objectives

1. Early fruiting.
2. High female to male sex ratio.
3. Attractive green or dark green fruits with smooth surface and without prominent spines or prickles.
4. Uniform long cylindrical shape without crook neck.
5. Fruits free from carpel separation without hollow spots.
6. Fruits free from bitterness.
7. Less seeds at edible maturity.
8. Resistance to diseases (powdery mildew, downy mildew, anthranose, cucumber mosaic virus).
9. Resistance to insect pests.
10. Resistance to abiotic stresses.

Genetics of Sex

Table 7.22: *Different sex types in cucumber*

Monoecious plants	Staminate and pistillate flowers
Androecious plants	Only staminate flowers
Gynoecious plants	Only pistillate flowers
Hermaphrodite plants	Only hermaphrodite flowers
Andromonocious plants	Staminate and hermaphrodite flowers.

Chemical Regulation of Sex Expression

Increasing Female Tendency

(*i*) Auxin

(*ii*) Ethylene

(*iii*) Acetylene

(*iv*) 2-chloroethylphosphonic acid

Increasing Male Flower Promotion

(*i*) GA_3

(*ii*) GA_4

(*iii*) GA_7

(*iv*) Silver nitrate

(*v*) Silver thiosulfate

Breeding Methods

1. **Pedigree method:** This includes selection of single plant segregates in segregating generations F2, F3 etc. derived from crosses between desirable parents.
2. **Backcross method:** This is useful for transferring characters governed by single genes, *e.g.*, disease resistance or quality traits from donor lines to more stable recurrent parents.
3. **Use of sex inheritance and chemicals in breeding methods :** In this system gynoecious lines are treated with $AgNO_3$ to stimulate production of staminate flowers and such lines are allowed inter-pollination to build up a source population with wide genetic variation. The gynoecious lines treated with $AgNO_3$ pollinate a mixture of same lines which have been sown about 2-3 weeks later and rouged from any staminate flowers segregates. A gynoecious population emerges from the harvested seed and serves as a gene pool for carrying out further selection of desired genotypes and lines.
4. **Population improvement and extraction of inbred lines:** This method is based on recurrent selection and aims at long-term gain for the characters having low to moderate heritability.
5. **Hybridization:** Hybrid varieties of cucumber are becoming predominant day by day. Their number is continuously increasing in India also.
6. **Biotechnology methods**
   (*a*) Introduction of somaclonal variation and its application in breeding.
   (*b*) *In vitro* mutagenesis on haploids and diploids.
   (*c*) *In vitro* selections for disease resistance and cold resistance etc.
   (*d*) *In vitro* fertilization.
   (*e*) Production of transgenic plants.
   (*f*) Embryo rescue technique

Commercial Application of Heterosis

- Heterosis is observed in almost every crop species studied. Of course, the degree of heterosis should be considerably high to permit its commercial exploitation.
- Heterosis is commercially used in the form of hybrid or synthetic varieties. Such varieties have been most commonly used in cross-pollinated and in often cross-pollinated crop species.
- In several self-pollinated species also, hybrid varieties have been commercially used.
- The major difficulty in commercial exploitation of heterosis is in the production of large quantities of hybrid seeds.
- This is particularly difficult in those self-pollinated species where the number of seeds per fruit is small.

Breeding of Root and Tuber Crops

- Root and tuber crops are the mainstay of a winter diet. The most commonly grown root crops in India are carrot *(Daucus carota* L.), radish *(Raphanus sativus* L.), turnip (*Brassica rapa* L.) and beetroot (*Beta vulgaris* L.).

- The main tuber crops after potato are sweet potato (*Ipomoea batatas*), yam (*Dioscorea alata*), tapioca (*Manihot esculenta*), elephant foot (*Amorphophallus campanulatus*), colocasia (*Colocasia esculentum*) and coleus (*Solenostemon rotundifolius*).
- Majority of them are perennial in nature but agriculturally these are managed as an annual.
- Due to their short duration and high productivity, these crops fit well in sequential, inter and relay cropping which enable maximum use of enable land.
- These crops are generally storage organs which enlarge to store energy in the form of carbohydrates. These are highly efficient in producing the highest amount of nutrients and health promoting substances per unit area.

Improvement of Root Crops

- A clear cut objective of breeding is a useful pre-requisite for greater chances of success in the improvement of a vegetable crop.
- The breeding method depends upon breeding system, life cycle, mode of reproduction, sex forms and genetic architecture.
- The commonly used breeding methods for crop improvement are selection hybridization, backcrossing and heterosis breeding through utilization of hybrid vigour.
- The development of hybrid and hybrid seed production in root crops can be made by the following methods:
  1. Emasculation (by pulling out the male plants) and use of wind pollination (Garden beet).
  2. Free insect pollination.
  3. Using self-incompatibility: Radish and turnip.
  4. Using male sterility: Carrot, radish and beetroot.

The following steps are necessary for the production of hybrid seeds in these crops:

(*i*) Inbreeding and production of inbred lines.

(*ii*) Testing of combining ability.

(*iii*) Improvement of inbred lines/varieties.

(*iv*) Production of hybrid seeds.

Breeding of Carrot

Breeding Objectives

Main breeding goals in carrot are:

- Early and high root yield.
- Combining desirable characters of both European and Asiatic groups, especially high carotene content and ability to set seeds is plains.
- Developing F1 hybrids using cytoplasmic male sterility.

- Desirable uniform root size, shape (cylindrical), dark orange external and internal colour, i.e. uniform in xylem and phloem, top-root ratio, i.e., small tops and smooth heavy tender root.
  - Broad shouldered, cylindrical, uniformly tapering or stump rooted thin and self-coloured slow bolting carrots.
  - Resistance/ tolerance to defects such as excessive secondary root development, splitting, secondary growth and cavity spot.
  - Resistance to Alterneria blight *(A. dauci)* and cercospora leafblight *(Cercospora carotae)* and root knot nematode *(Meloidogyne hapla).*
  - Tolerance to environmental stresses and wider adaptability.
- The production of F1 hybrid seed by hand emasculation and pollination is not possible commercially as the flowers are very small and single pollination gives only one or two seeds. In carrot, the inflorescence is compound umbel.
- The flowers are arranged in umbellets, and the umbellets into an umbel. Flowers are protandrous. It produces the first order umbel called King umbel, second order (secondary) and third order (tertiary) and even umbels of fourth and fifth orders also.
- Pollen dehiscence may lasts one or two days per floret and 6 to 7 days per umbel.
- Stigma becomes receptive on the third or fourth day after anthesis and may remain receptive for about a week.
- Thus a plant may take about 4-6 weeks for complete blooming. Anthesis in a single umbel is completed in 7-9 days.
- Sufficient buds at peak stage of flowering are emasculated and the remaining young ones are removed.
- The pollen parent umbel is enclosed in the same bag along with the emasculated umbel of the female parent.
- Daily for a few days in the morning, the male umbel is gently rubbed over the female to ensure cross-pollination.
- Some times when the pollen parent possesses some dominant marker gene with the help of which the hybrids can be distinguished in the seedling stage, it is not necessary to emasculate the flowers.

Male Sterility

- Welch and Grimball (1947) reported the first male sterile plant in 1945-46. Now morphologically there are 3 types:
  1. **Wax coated and swollen anthers:** These anthers do not produce any fertile pollen.
  2. **Brown coloured anthers:** These anthers are shriveled and brown without any viable pollen.
  3. **Anthers replaced by petals:** These anthers are petaloid. This type of sterility is more stable.

Genic Male Sterility

- Hausche and Gabelman (1963) reported that male sterility is conditioned by two nuclear genes. The plants carrying ms5 ms5 or ms4 in heterozygous condition (MS4 ms4) or homozygous condition (MS4 MS4) result in male sterility.

Gene Cytoplasmic Male Sterility

- Jones (1950) and Lamprech (1951) reported that the male sterility is due to interaction between dominant nuclear gene and cytoplasm.
- It is not clear that how many genes are involved, but according to Timin *et al.* (1981), the brown anther male sterility is conditioned by dominant and recessive genes (MS1 MS1 MS2 MS2) in (s) cytoplasm, while the petaloid type is conditioned by three dominant genes (MS3 MS3 MS4 MS4 MS5 MS5) in (s) sterile cytoplasm.

Breeding of Radish

Major breeding objectives

- Early maturity and high root yield.
- Non-pithy blunt ended cylindrical roots.
- High pungency and nutrients.
- Slow and late bolting with less foliage.
- Tolerance to heat, drought and wet situations, and cold hardiness, etc.
- Resistance to alternaria blight, white rust, radish mosaic virus and aphids.

- The production of F1 hybrid seed in radish by hand emasculation and pollination is easy as compared to carrot as the flower buds are bigger, easy to emasculate and single pollination yields more number of seeds.
- However, the male sterility can be used for commercial hybrid seed production. In general, the system of breeding remains the same as followed in other brassicas.

Use of Male Sterility and Self-incompatibility in Hybrid Seed Production Male Sterility

- Ogura (1968) reported gene-cytoplasmic male sterility in radish which is due to interaction between recessive nuclear gene ms ms and sterile (s) cytoplasm.
- Later it has been reported that there are two pairs of ms genes in the nucleus (Kalloo & Bergh, 1993) hence the genotype of the male sterile line (A) could be s ms 1 ms1 ms2 ms2, maintainer line (B) Nms1 ms1 ms2 ms2; pollen parental line (C) NMS1 MS1 MS2 MS2. Ogura (1968) described first sterile cytoplasm in a Japanese population of radish. On the bases of flower morphology, radish CMS is classified into three types:
  - Degenerative corolla
  - Shriveled stamen
  - Abortive pollen

- Ogura cytoplasm has been found widely distributed among the wild Japanese radish plants and most of the early European radish population.
- Male sterility in line 64A has been found to be associated with tapetum abortion and serious vacillation of cytoplasm of tapetum cells which resulted in hypertrophy (Su *et al.* 1995).
- The successful transfer of Ogura CMS to three cultivars of radish (Japanese White, Pusa Chetki and Pusa Desi) has been reported by Sodhi, *et al.* (1993).
- These three cultivars producing flowers with pollenless anthers are now available as male sterile lines for hybrid seed production in radish.
- Hawaldar *et al.* (1997) crossed Ogura radish, a cytoplasmically genetic male sterile line, with four local and three Japanese cultivars to identify the maintainer lines.
- Out of seven F1 families, one or involving a local cv. Ayushi produced 100% male sterile (MS) progenies. The hybrid seed production is done by using A and C lines in 1:1 ratio.

Self-Incompatibility

- Radish is normally self-incompatible and insect-pollinated. The incompatibility system is saprophytic in type in which the papillae on the stigma surface form the barrier to pollen tube penetration.
- Hawaldar and Mian (1997) investigated the self-incompatibility mechanism through seed set analysis and pollen tube growth behaviour.
- Moderately strong self-incompatibility was observed in Red Kalpin and Kuni White.
- For the production of hybrid seed 1:1 ratio of two parental strains is advocated. Excess nitrogen causes poor pod set in radish.
- Spraying of 30 ppm boric acid or 0.2% multiplex at flowering is useful for seed setting. Radish is less preferred by bees in comparison to cabbage, turnip, carrot, etc. and necessitates keeping bee hives in the seed crop.
- For maintenance purpose, the self-incompatibility can be overcome by bud pollination or by CO_2 at relatively high concentration of 3-5% for 2 hours in the evening on other day to their opening by their own pollen.
- It has been observed that the seed is better when the buds are pollinated only a day prior to opening when there is at least 30% flowering on the plant than bud pollinating only a day prior to opening just at the commencement of flowering.

Breeding of Turnip

The main breeding objectives are:

- Earliness in attending marketable size.
- High yield.
- Stump rooted varieties with thin tap root and non-branching habit.
- Slow bolting and no pithiness.
- Resistance to club root, powdery mildew, turnip mosaic virus, while rust, cabbage root fly and turnip root fly.

- All the systems of hand emasculation and pollination, use of incompatibility and production of hybrid seed are similar to that of radish and other brassicas.
- Heterosis has not been exploited commercially though reported for yielded and dry matter content and other characters in turnip.
- Synthetic cultivars developed by using inbred lines of good combining ability have out yielded the best commercial cultivars in Holland.

Breeding of Beet Root

The main breeding objectives are:

- Breeding high yielding varieties having dark red, uniformly colored roots with absence of any internal whiterings.
- Uniform shape and size of roots, and slow bolting habit.
- Developing varieties of spherical, flattened spherical, cylindrical or conical root shapes depending upon regional preferences.
- The seed production is taken in the hills for European types and in the plains for Asiatic types. European types require high chilling of 4-7°C for a period of about 4-6 weeks.
- The mild summer and low rainfall in hills especially during flowering and seed setting stages are beneficial.
- The 'root to seed' and 'seed to seed' methods can be used seed production preferable 'root to seed'. In root to seed method, fully matured roots (before pith development) are harvested, true to the type roots are selected and after giving proper root and shoot cuts are transplanted in a well prepared field.
- The sowing time should be so adjusted that the roots become available and their stecklings could be set in before chilling months.
- In heavy snowfall areas where chilling period is long, the roots after uprooting are stored in trenches before the onset of winters and replanting is done in the month of March-April. In such case stecklings are prepared just before planting.
- The seed is ready for harvesting from July-August in the hills and from May to June in the plains depending upon the weather, crop and cultivar.

Tuber Crops

- Tuber crops are mainly replanted each year from tubers, rhizomes, corms, cloves, cuttings, etc. Some of these crops flower regularly and thus develop seeds while others do not flower and thus form no seeds.
- In general, irrespective of their reproduction (sexual/asexual) behaviours, these crops are invariably highly cross-pollinated and highly heterozygous in nature and thus exhibit severe inbreeding depression. These crops comprise agriculturally superior clones where the superiority has been protected by continuous vegetative reproduction over a long period of time as these clones presumably carry favourably interacting genes complexes of alleles of many different loci (much similar to the genotype of F1) have succeeded and survived in agriculture.

- In practical breeding once this superiority is recognized, it has been reproduced and protected vegetatively. Thus, these crops consisted of large number of clones (derived from single superior plant) and have progeny exactly identical to the parental genotype unlike sexually reproducing crops where segregation and recombination produce new gene combinations and thus each progeny differs from their parents both genotypically and phenotypically.
- It has also been widely recognized that when vegetatively propagated materials reproduce sexually (where flowering and seed setting takes place) whether by selfing or out crossing with other superior clones, the resulting progeny will be much inferior to the parental clones because of segregation and recombination.
- So the crops where flowering takes place, the general approach to the breeder is to self or hybridize the superior clones to generate population of genetically variable individuals where only the better plant/clones are identified and then propagated vegetatively.
- This will ensure that each vegetative descendent within each family will have identical genotype and the differences observed in phenotype might be due to environmental and genotype x environmental interactions.
- A common method of breeding in vegetatively propagated crops does not depend upon the method of reproduction (sexual or asexual) but rather rely on giving genetically identical materials unless degenerated, by the presence of systemic diseases (mostly viruses and bacteria), mutation, somatic degeneration mechanical mixture etc.
- Thus, clonal selection is considered an immortal, quick and easy method of breeding in these crops.
- All breeding methods like introduction, clonal selection, hybridization and selection, heterosis breeding, mutation breeding, polyploidy breeding, interspecific hybridization, inter-specific grafting, etc. depending upon the crop, its utility and type of reproduction, have been reported in the literature.
- Irrespective of methods of breeding used, it is very important to understand that breeding materials and commercial cultivars in these crops are maintained by asexual reproduction.
- This thus provides unique advantages and opportunities in their breeding. In these crops, a single outstanding plant selected from a population may become the source of a new variety.
- Thus, the breeding behaviour or genotype of the plant is not important since there will not be sexual reproduction. This valuable plant(s) may be selected strictly from asexual (no flower and seed set) or other (flowering and seed setting takes place), unimproved variety or population produced by crossing two or more clones.
- Moreover, these crops too follow the same pre-conditions as in sexually reproducible crops before selection, i.e. utilization of existing variability; creation of new variability where it does not exist; and selection and utilization of the best genotype to produce superior clone/variety.
- The asexually reproducible crops comprise large number of clones and all these clones are expected to have genotypic constitution as the parent plant.

- Consequently these are genetically identical to each other and the phenotypic differences observed within clones do not have genetic basis rather is purely environmental in nature.
- So the selection within clones will not be effective (as within pure lines) or rather useless. The selection procedure adopted for clones do not differ much from sexually reproducible crops, that is, single plant selection is followed in all those characters which are highly heritable whereas for low heritability characters like yield, the performance of the clone is judged by observing more number of plants in replicated trials.
- Thus, strict selection procedure should be followed in the early stage (generation) so that weak and undesirable plants/clones are eliminated and only superior ones are carried for replicated testing over time and space.

Breeding of Tapioca

Breeding Objectives

- To develop varieties with high tuber yield and starch content with low HCN content.
- To develop varieties with short duration and non-branching type.
- To develop varieties resistant to Cassava Mosaic Virus (CMV).

Hybridization and Selection

- Tapioca being propagated vegetatively, breeding programme is more time consuming due to its long production period.
- Induction of flowering in sparingly flowering varieties is a problem.
- It can be done by increasing day length, since long day condition favours induction of more flowering.
- Foliar application of IAA, NAA and ascorbic acid at 50 ppm is also recommended to increase the intensity of flowering.

Varieties	Pedigree	Yield (t/ha)	Duration (days)	Starch (%)	Year of release
H.97	Manjavella x Brazillian seedling selection	25,30	290-310	29	1971
H.165	Chadyamangalam vella x clone similar to Kalikalan	33-38	240-260	25	1971
H.226	Ethakkakaruppan x M4	30-35	290-310	29	1971
Sree Visakham (H.2304)	Multiple hybrid involving 5 parents, viz., Acc. 468 Ac. No.174.S.1310 (Madagascar) Ac.No.82 & M4	35-40	300-320	31	1977
Sree Prakash (S-856)	Selection from local accessions	35-40	220-240	29	1987

Clonal Selection

- In India, considerable amount of variability exists in crops and much more is available in other cassava growing regions, particularly in tropical south Latin America, which is the birth place of cassava.
- This breeding technique has been used to develop improved cultivars at TNAU such as CO 1, CO 2 and CO 3.

Varieties	Pedigree	Yield (t/ha)	Duration (months)	Starch (%)	Year of release
CO 1	Clonal selection (ME7) from a local type collected from Tiruchirapalli district.	29.97	8.5 to 9	35.0	1977
CO 2	Clonal selection (ME 167) from an open pollinated seedling progenies raised from the seeds collected from a Thiruvarur type in Thanjavur district.	35.37	8 to 8.5	34.50	1984
CO 3	Clonal selection (ME120-1) from seedling progenies of open pollinated seeds obtained from IITA, Ibandan, Nigeria.	42.58	8	35.60	1993
CO (TP)4	Clonal selection from one of the seedling progenies of the hybrid SM1679 (CM 2766-5 x CM 4843) obtained from CIAT	50.60	8.5	40.0	2002

Interspecific hybridization

- To enhance the protein content and to incorporate disease resistance in cultivated cassava, hybridization between cera rubber (*Manihot glaziovii*) and selected clones of cassava was attempted at CTCRI, Thiruvananthapuram.
- Repeated backcrossing of interspecific hybrid of cassava was carried out, but acceptable root quality was not recovered.

Heterosis, Polyploid and Mutation Breeding

Heterosis Breeding

- To exploit the hybrid vigour in cassava, work done at CTCRI, Thiruvananthapuram revealed that some hybrids are performing better than their parents and even existing varieties in terms of tuber yield, quality, etc.

Polyploid breeding

- Induced tetraploids by colchicine treatment was attempted at CTCRI, Thiruvananthapuram and found that they were poor yielding and less adapted to the normal field conditions.

- However, one of the triploid strains evolved after crossing tetraploids and diploids out-yielded both the parents and the scientists were of the opinion that there is a possibility of improving cassava by producing new chromosomal races in which the chromosome underdoes not go beyond an optimum level.
- Evolution of triploids by crossing diploids and the artificially induced tetraploids has been done.
- Though a few of the selected triploids showed relatively high dry matter content and extractable starch, they did not prove to be superior in yield to the released varieties namely Sree Visakham and Sree Sahya.
- A spontaneous triploid was identified from a cross involving the two released diploid varieties H-165 and Sree Sahya.

Mutation Breeding

- Mutation with gamma irradiation has been employed successfully on single node cutting of cassava but no economic mutant has been isolated so far.

Breeding of Ginger

Floral biology of ginger

- The floral biology of ginger was studied at Kasaragod (AICSCIP, 1975 and CPCRI, 1975). Flowering is not common in ginger.
- Cultivar differences exist with respect to flowering. It is a crop with very rare flowering ranging from 0.5 to 67.5 % and some species do not flower at all.
- Early planting i.e. March-April favoured flowering when the crop was grown with rhizome bits of standard size i.e. 15.0g.
- Maintaining plant as biennial using larger rhizome bits resister still earlier and more profuse flowering. NAA (50 ppm) induces early flowering with more number of flowers, whereas, kinetin (15 ppm) enhances duration of flowering and both together increase inflorescence production.
- Flowering starts in October and continues till December with a peak in November. Flower bud development takes 20-25 days from the bud initiation to flowering.
- A period of 23-28 days was required to complete blooming in an inflorescence. Anthesis takes place between 1.30 pm and 3.30 pm under South Kerala conditions. Anther dehiscence almost coincides with the flower opening or follows it immediately.
- The studies on stigma receptivity could not be followed up since hand pollination failed to set seeds. However, observation under hand lens indicated that stigma was receptive at the time of anther dehiscence.
- Pollen grains are spherical, size ranging from 90 to 100 micron. Pollen sterility ranged up to an average of 76%. Flowers fade and fall the next day.
- Natural seed set was not reported so far in ginger. Reasons for absence of seed set under natural conditions were investigated by many workers.
- Studies revealed presence of spiny stigmatic surface which prevents proper adherence and germination of pollen grains (Valsala, 1994).

- Pollen fertility in ginger reported by various workers ranged from 0.9 to 40 % (Valsala, 1994; Binbu, 1997 and Adaniya, 1998).
- Germination of pollen grains was very slow and they get coiled during advanced stages of pollen germination.
- Under *in vitro* conditions the pollen tube attained maximum size of 108 micron after 24 h on the style which was very large (39,000 micron).
- The flower abscises within 12 h and so the possibility of pollen tube reaching the ovule is remote.
- Self-incompatibility was reported by Das *et al.* (1999). He observed the hermaphrodite flowers with *pin* and *drupe* type incompatibility and dehisced pollen grains did not reach the stigma head.
- There was no seed set during selfing and crossing. Damayanthi *et al* (2003) reported that heterostyly may be a contributing factor to sterility in ginger.
- Different techniques tried for breaking incompatibility such as bud pollination, artificial rubbing, artificial cross pollination, mixed pollination, mentor pollination and use of growth regulators but failed to produce fruit and seed set (Nybe, 2001).

Genetic Resources in Ginger

- Largest variability of ginger is reported from China followed by India. Variability for yield and quality in cultivated ginger exists mainly in NE India and Kerala.
- Most varieties are named after their places of origin/ domestication and are locally adopted. Apart from cultivated ginger, the other important economic species of the genus are *Z. zerumbet, Z. odoriferum, Z. spectabile, Z. squarrosum* and *Z. casumunnar* which are used for medicine and aromatic purposes.

Exploration and Collection

- Germplasm of ginger includes land races, improved varieties, tetraploids, exotic collections/ introductions and related taxas of *Zingiber* sp available in the country.
- NBPGR, New Delhi has done an all India survey in collaboration with CPCRI, Kasaragod (Kerala) and Dr YSPUHF, Solan (HP) for collection of ginger germplasm during 1976-77.
- The survey brought to light the importance of collection of ginger germplasm because of the extinction of local cultivars due to the introduction of high yielding exotic ones like Rio de Janeiro, China, Sierra Leone, Jamaica etc.
- Subsequently, IISR, Calicut, NBPGR Regional/ Satellite Stations, different Centres under AICRP on Spices have also made extensive surveys in the ginger growing areas of the country for collection of germplasm. Exotic collections from Nepal, Jamaica, China, Nigeria, USA, Brazil, New Zealand and Fiji are also available in the germplasm conservatory of IISR, Calicut.

Characterization and Documentation

- Characterization and documentation of ginger germplasm are being carried out by the IISR, Calicut, NBPGR Regional and Base Centres located at Thrissur and

Shillong and the AICRP on Spices Centres located at Solan (Dr YSPUHF, HP), Pottangi (HARS OUAT, Odisha), Raigarh (RS IGKV MP), Pundibari (UBKV WB) and Dholi (TCA RAU Bihar).

Evaluation and Utilization

- The ginger germplasm when subjected to evaluation for various morphological characters, yield, quality attributes and pest as well as disease incidence revealed much variability for most of the traits.
- Subsequently, the promising germplasm shortlisted from such evaluations were utilized for further evaluation and clonal selection for developing elite cultivars having traits such as high yield, dry ginger recovery, oleoresin, essential oil, crude fibre, reaction to insect pests and diseases, etc.

Conservation

- Conservation of germplasm plays an important role in the maintenance of ginger since it is a vegetatively propagated crop. National Conservatory of ginger germplasm at IISR, Calicut has conserved exotic cultivars, indigenous collections, improved cultivars, mutants, tetraploids and related species and taxas.
- In addition to IISR, germplasm is being maintained at different centres under AICRP on Spices. These accessions are being maintained in storage pits in the fields as *ex situ* bank and *in vitro* gene bank at IISR for medium term storage of germplasm.

Improvement of Ginger

- In India cultivars so far developed are through introduction, selection and mutation. Polyploidy breeding was also attempted in ginger. Recombination breeding is not possible due to the lack of seed set.

Introduction

- **Local: within state:** Number of locally adopted material from one place of cultivation to other place has been utilized.
- **Outside state:** The adopted material has been introduced from one state to other e. g. Wynad, Maran, Gurubathan, Nadia etc.
- **Exotic:** Introduced from other countries e. g. China, Rio de Janeiro, Jamaica, etc.

Selection

- Ginger does not flower, therefore, no conventional breeding method is much effective. It sets no seeds, therefore, no variability is generated.
- But studies conducted at the NRCRI, Umudike, Umuhia and Nigeria showed that seed set is possible by hybridization. Fruit set has been obtained from pollination of stigmas of both the protruding *'pin type'* and the hidden *'thrum type'* styles but significantly more from the latter, however, seeds are not viable. The ginger is propagated through rhizomes.

- Local materials for years have developed variation. It is being exploited through selection for yield, resistance to diseases and quality attributes.
- Selection is also practised to the introduced material for adaptation to local market, climatic conditions and other characteristics.
- As a part of selection, collection of germplasm, their characterization and yield evaluation are being done at several research institutions and some cultivars have already been developed. Developed material through clonal selection is tested by SAU's, ICAR Institutions and AICRP on Spices Centres.
- Cultivar Himgiri is released through clonal selection by Dr YSPUHF Solan (HP), Suprabha and Suruchi by HARS, Pottangi (OUAT, Odisha), Varada, Mahima and Rejatha by IISR, Calicut (Kerala).

Mutation Breeding

- Due to improbability of employing recombination breeding, alternative to induce variability is through physical and chemical mutagens.
- The specific defects in the commercial cultivar or the incorporation of genetic variability in the materials can be accomplished by mutation breeding.
- This can also be involved for the purpose of inducing resistance to common diseases. The vegetative propagating nature of the crop may be an additional advantage for the propagation if desirable mutants are recovered.
- Number of research workers like Dutta and Biswas (1985) using chemical mutagens, Mohanty and Panda (1988) using physical mutagens, Mohanty and Panda (1992) using physical and chemical mutagens, Jayachandran and Mohankumaran (1992) using physical mutagens, Giridharan and Balakrishanan (1992) and Bhanu (1994) using physical and chemical mutagens have attempted mutation studies.
- In majority of cases, besides identifying LD50/ LR50 not much breakthrough has been highlighted. Some mutants showed variation in aerial parts by way of dwarf plants, branched tillers and chlorotic leaves.
- The chlorotic mutants will not be of much value except for ornamental purpose as they will affect the source sink relationship, ultimately affecting rhizome development and finally yield.
- The attempts to induce flowering and its utilization for recombinant breeding also didn't yield fruitful results. Giridharan and Balakrishanan (1992) observed very meagre number of flowers (1-2.6) but no viable pollen was noticed and thus no seed set.
- However, Mohanti and Panda (1992) from HARS, Pottangi (OUAT, Odisha) were able to release one variety, Surubhi, out of the materials treated with physical and chemical mutagens.

Polyploidy Breeding

- It is an alternative method to generate variations and to explore the possibilities of inducing flowering and seed set in ginger.

- The advantage of rhizome as commercial product can possibly be capitalized by pursuing polyploidy breeding in ginger as there will not be a problem of its maintenance and propagation unlike flowering crops.
- Ratnabal (1979) reported the induction of polyploidy in ginger cultivar Rio-de-Janeiro through colchicines treatment in which majority of tetraploids showed stunted growth than the diploids and had reduced length and breadth of leaves.
- There was slowing down of metabolic activities, resulting in no flowering. These lines later reverted to the diploid parental state in the subsequent generation.
- As cited by Dewey (1980) the crops most amenable to attempt polyploidy should have- i) low chromosome numbers ii) be harvested for vegetative growth iii) be cross pollinating iv) have ability to reproduce vegetatively and v) have perennial habit. Ginger fulfils most of these conditions and thus can be studied for polyploidy breeding. Ramachandran (1982) and Ramachandran and Nair (1992) reported successful induction of stable tetraploids in ginger.
- The polyploids were more vigorous than the diploids and flowered during the second year of induction.
- They could isolate stable tetraploids having $2n = 4x = 44$. These autotetraploids had larger rhizomes and high yield. However, oil content of these rhizomes was lower than the original diploid cultivars.
- These tetraploids are now available in the germplasm collection of IISR. The Buderim Ginger Co. in Queensland Australia released a tetraploid line 'Buderim Gold' for cultivation.

Biotechnological Techniques

- Crop improvement programmes in ginger are hampered by lack of seed set leading to limited variability.
- The techniques involve tissue culture, micro propagation, somaclonal variants, somatic hybrids, screening of germplasm against diseases and identification of genetically variable materials.
- Somaclonal variation could be an important source of variability that could be exploited to evolve high yielding, high quality lines and to develop lines resistant to rhizome rot and bacterial wilt.
- Tissue culture techniques could also be used for *in vitro* pollination, embryo rescue and possible production of 'seeds' in ginger.
- However, so far the possibility of tissue culture technique for the multiplication of ginger has been standardized by various workers using different explant materials e.g. stem cutting of three-month old plant, shoot tip of rhizomes kept in storage, meristem, leaf/aerial pseudo stem, rhizome buds, young leaf shoots, shoot tips, young leaf, young growing buds from stored rhizomes and callus of shoot primordia. However, the cost involved in raising *in vitro* plants, their number, timely availability, proper size and survival in the field on commercial scale are some questions which need deliberations.
- The development of micro rhizomes through tissue culture can be used for maintenance of germplasm and quality seed material. The biotechnological

techniques can also be used to delete the duplicates which will save time, energy, space and cost involved under *in vitro* conservation of germplasms. Besides, these techniques can also be used to screen the cell or callus or cell culture for their insensitivity to toxic metabolites of pathogen.

- The regenerated plantlets are tested for their reaction with the actual pathogens. Such studies can be ameliorated with multiple resistances for fungal, bacterial nematodes, and drought tolerance coupled with quality and productivity.

Breeding of Turmeric

- Turmeric (*Curcuma longa* L.) 2n=3x=63 belongs to the family Zingiberaceae and plays a vital role in the national economy.
- It has originated from Southern Asia, probably India, in the slopes of hills in the tropical forests of west coast of South India.
- About 18 species occur in India of which a few are important spice plants. *C. longa* is cultivated in large areas and *C. aromatica* popularly known as Cochin turmeric or Kasturi manjal, used for the preparation of *kum-kum*, is also grown in some parts. *C. amada*, the flesh of which has the taste and flavour of raw mango and therefore it is known as mango ginger. *C. caesia,* black turmeric, claimed to have high medicinal value has been regarded as endangered by the Central Forest Department of India. It is a principal ingredient in curry powder.
- **Turmeric oleoresin** is used in brine pickles and to some extent in non-alcoholic beverages, gelatins, butter and cheese, etc.
- The **colour curcumin** extracted from turmeric is used as a colorant. Turmeric is also used as a dye in textile industry. It is used in the preparation of medicinal oils, ointments and poultice.
- It is a stomachic, carminative, tonic, blood purifier and an antiseptic. It is also used in cosmetics. The aqueous extracts have bio-pesticidal properties.
- It is grown on a large-scale in India, China, Sri Lanka, Indonesia, East Indies, Fiji, Queensland and Pakistan. About 18 states in India cultivate turmeric; however, turmeric cultivation is largely confined to south and eastern India.
- AP followed by Odisha, TN and Maharashtra states in India constitute the lion's share in India's turmeric production. It is grown to a small extent in the lower and mid-hill altitudes in Meghalaya and HP.
- India has a prime position in the world and is the largest producer, consumer and exporter of turmeric and accounts for more than 50% of the world trade.
- The area under turmeric in India is 195,100 ha with an annual production of 992,900 t, i.e. productivity of 5.1 t/ ha.
- Out of the total turmeric produced in India 90% is consumed locally and remaining 10% is exported to various countries like USA, UK, Middle East, Japan, Singapore, Malaysia, South Africa, Australia and other countries.
- Although India is leading in turmeric production, the average productivity and quality are not satisfactory.

- Though few of the cultivars are capable of yielding 30-35 t/ha of raw turmeric, the actual realized yield is much less and this has been attributed to very little crop improvement work carried out.
- In HP, turmeric has not attained significant status among spice crops probably due to poor yield and being a long duration crop, however, some growers especially in the lower hills have started showing interest in the crop and the area is steadily increasing as the crop can be successfully planted under rainfed conditions under minimal care and attention.
- Moreover, increasing monkey menace and engagement of farmers in other occupations offer better opportunities for increasing acreage under this crop in the state.

Turmeric breeding

Flowering in turmeric

- Flowering varies depending upon cultivars and climate ranging from 1.2 to 95.8%.
- Majority of genotypes flower within a period of 110 to 155 days though some may flower earlier and later.
- Flowering season extends from July to October and inflorescence takes 8-12 days to bloom after their emergence.
- Duration of flower opening lasts for 8-10 days. Separate peduncle arising directly from the rhizome bears the flowers in cone-shaped spikes in the tuft of leaves. The spikes are shorter than the leaves and supported by stout peduncles.

Pollination and Seed Set

- In turmeric, pollination is brought about by insects. Fruit set and seed set are not generally observed in turmeric. Nazeem and Menon (1994) observed seed set in crosses involving *C. domestica* and *C. aromatica* types. Renijith (1999) and Vijayasree (2001) reported seed set in crosses involving short duration cultivars of *C. domestica* but failed to obtain success in crosses involving short and medium duration cultivars.
- Fruits matured within a period of 20-22 days. Seeds /fruit ranged from 4 to 6. Pollen pistil interaction studies after *in vitro* cross pollination of short and medium duration cultivars showed pollen germination on stigma and pollen tube passing through the style and reaching the ovules.
- Lack of seed set could be due to other factors (Vijayasree, 2001). Lad (1993) observed seed setting in turmeric cultivars Ca 92/11 and Ca 17/1.

Genetic Resources of Turmeric

- The genus *Curcuma* belonging to the family Zingiberaceae has a wide spread occurrence in the tropics of Asia, Africa and Australia.
- Apart from the *Curcuma longa* L. or C. *domestica* Val., the common culinary turmeric, there are several other species of *Curcuma* which are mainly used as colouring agents for production of arrowroot, cosmetics and for medicinal purposes.

Species such as *C. aromatica* (Kasturi manjal), *C. caesia* (Black turmaric), *C. amada, C. xanthorriza, C. aeruginosa*, *C. phaeocaulis* and *C. petiolata* are also cultivated in different places and regions.

- The genus *Curcuma* consists of about 117 species, whereas, from India around 40 species are reported (Velayudhan *et al*., 1999).
- Among the 40 species recorded in India, only two species viz., *C. longa* L. and *C. aromatica* Salisb. are commercially cultivated, the first for turmeric and the latter for 'Kasturi turmeric' used in toiletries, cosmetics and medicines. More than 50 turmeric types are known to exist in India belonging to *C. longa* and a few cultivars that belong to *C. aromatica*.
- Most of these cultivars go by local names mostly derived from the place of occurrence (Nair *et al*., 1980). The other species, *C. angustifolia, C. aeruginosa, C. caesia, C. mangga, C. purpurascens, C. xanthorrhiza* and *C. zeodaria* are cultivated to a limited extent in different areas and these are used locally in traditional medicines.

Improvement of Turmeric

- The first ever organized research on turmeric was initiated at Udayagiri in Odisha during 1944 under the auspices of ICAR. Presently, crop improvement work in turmeric is under progress at ICAR IISR, Calicut (Kerala) and under the aegis of ICAR AICRP on Spices in different centres such as Pottangi (Odisha), Coimbatore (Tamil Nadu), Jagtial (AP) and Solan (HP) besides various SAU's in the country.
- Various methods and tools of breeding employed are: Selection, Mutation, Hybridization and Biotechnology.

Selection

- The varietal improvement work so far attempted is only through clonal selection and hybridization could not be practiced due to sterility, since the commercial types are sterile triploids of *C. longa.* In a study for evolving high yielding cultivars, Shankaracharya and Natarajan recorded fresh yield of 16.8 to 22.4 t/ha under irrigated conditions.
- In Andhra Pradesh, Rao *et al.* (1975) collected more than 100 accessions within the country and the performance of these accessions was studied and broadly classified into three duration groups depending on their crop life.
- These are (a) Short duration "kasthuri" types belonging to *C. aromatica* maturing in 7 months, referred as "Ca" types. (b) Medium duration "Kesari" types and referred to an intermediate or CLi types. (c) Long duration types belonging to *C. longa* maturing in 9 months and referred to as Cll types.
- Clonal selections were made among the more popular cultivars in these duration groups and yield trials carried out for three years resulted in identification of 8 promising types in short duration, 6 types in medium duration and 8 in long duration group.
- The screening of germplasm revealed various sources of resistance for important pest and diseases of turmeric.

Mutation Breeding

- Since turmeric is vegetatively propagated, efforts were made to release improved varieties through mutation breeding. Induced mutations were also tested in turmeric.
- In Tamil Nadu, two mutants, Co-1 and BSR-1, identified from 'Erode Local', were released for large-scale cultivation in the name of Co-1 and BSR-1. They have a yield potential of 24.70 and 30.69 t/ha over 'Erode Local' (20.2 t/ha) (Balashanmugam *et al*, 1986).
- These lines are having moderate curing percentage i.e. 18.4 and 20.5% with 4.2% curcumin content. Chezhiyan and Shanmugasundaram (2000) reported a turmeric cultivar, BSR-2, derived by X-ray mutagenesis from the variety Erode Local.
- It produces 690-890 g fresh rhizome yield per plant, curcumin content 3.75% and curing percentage of 20.

Hybridization

- The crop improvement was concentrated on clonal selection as hybridization could not be practiced due to sterility, since the commercial types are sterile triploids of *C. longa* (Ramachandran, 1961).
- However, Nazeem *et al*., (1993) reported VK 5 of *C. longa* (2n=63) produced seeds when it was crossed with Kasthuri and Dindigam, which belong to *C. aromatica* (2n=84).
- They indicated the possibility of crossing high-yielding *longa* types with selected *aromatica* types having high curing percent.
- Viable seed set in turmeric reported by Sasikumar *et al.,* (1994) has taken shape for recombination breeding programmes.

Biotechnology

- Induction of somo-clonal variation is yet another important tool for breeders engaged in crop improvement programmes.
- Induced mutagenesis using single cell culture to isolate chimera free mutants for yield and quality parameters is a potential avenue in the case of turmeric as many workers have obtained various degrees of success in tissue culture techniques in turmeric.
- Organogenesis and plantlet formation were achieved from the callus cultures of turmeric. Variations with high curcumin content were isolated from tissue cultured plantlets (Nadgouda *et al*, 1982).
- *In vitro* microrhizome formation was reported in turmeric by Nirmalbabu *et al.,* (2003). Protoplasts could be isolated successfully from leaf tissues as well as from cell suspension cultures of turmeric (Nirmalbabu *et al.*, 2006).

Intellectual Property

- Traditionally, farmers in all nations saved their seeds from year to year. Allowing to follow this practice with genetically modified seeds would result in seed developers losing the ability to profit from their breeding work.

- Therefore, genetically-modified seeds are subject to licensing by their developers in contracts that are written to prevent farmers from following this traditional practice.
- Many objections to genetically modified food crops are based on this change.
- Enforcement of patents on genetically modified plants is often contentious, especially because of gene flow.
- In 1998, 95-98% of about 10 km^2 planted with canola by Canadian farmer Percy Schmeiser were found to contain Monsanto Company's patented Roundup Ready gene although Schmeiser had never purchased seed from Monsanto.
- The initial source of the plants was undetermined, and could have been through either gene flow or intentional theft. However, the overwhelming predominance of the trait implied that Schmeiser must have intentionally selected it.
- The court determined that Schmeiser had saved seeds from areas on and adjacent to his property where Roundup had been sprayed, such as ditches and near power poles.
- Although unable to prove direct theft, Monsanto sued Schmeiser for piracy since he knowingly grew Roundup Ready plants without paying royalties.
- The case made it to the Canadian Supreme Court, which in 2004 ruled 5 to 4 in Monsanto's favour.
- The dissenting judges focused primarily on the fact that Monsanto's patents covered only the gene itself and glyphosate resistant cells, and failed to cover transgenic plants in their entirety.
- All of the judges agreed that Schmeiser would not have to pay any damages since he had not benefited from his use of the genetically modified seed.
- In response to criticism, Monsanto Canada's Director of Public Affairs stated that "It is not, nor has it ever been Monsanto Canada's policy to enforce its patent on Roundup Ready crops when they are present on a farmer's field by accident...Only when there has been a knowing and deliberate violation of its patent rights will Monsanto act."

OOO

CHAPTER 8

Basic Information on Vegetable Crops

Ajit Singh
Senior Scientist and Head
Krishi Vigyan Kendra, Burhanpur (M.P.)

Table 8.1: *Common name, scientific name, family, origin, chromosome number, inflorescence type and fruit type of vegetables crops*

S. No	Common name	Scientific name	Family	Origin	Chromosome number (2*n*)	Inflorescence type	Fruit type
1.	Amaranthus	*Amaranthus blitum* (Chhoti chaulai) *Amaranthus tricolor* (Badi chaulai)	Amaranthaceae	Central and South America, India, South-East Asia	32	Spike	Utricle
2.	Beetroot	*Beta vulgaris* L.	Amaranthaceae	Mediterranean region	18	Spike	Aggregate
3.	Spinach/ Vilayati palak	*Spinacia oleracea* L.	Amaranthaceae	Persia (Iran)	12		
4.	Beet leaf/ Palak	*Beta vulgaris* var. *bengalensis* L.	Amaranthaceae	Indo-China	18		
5.	Onion	*Allium cepa* L	Amaryllidaceae (Liliaceae)	Central Asia	16	Umbel	
6.	Garden Leek	*Allium porrum*	Amaryllidaceae	Central Asia	16		
7.	Parsley	*Petroselinum crispum*	Apiaceae	Central Mediterranean region	22	Umbel	
8.	Coriander	*Coriandrum sativum* L.	Apiaceae (Umbelliferae)	Southern Europe and Asia Minor	22	Compound umbel	Schizocarp
9.	Celery	*Apium graveolens* L.	Apiaceae (Umbelliferae)	Mediterranean region	22	Compound umbel	Schizocarp
10.	Carrot	*Daucus carota* L.	Apiaceae (Umbelliferae)	South Western Asia especially Afghanistan	18	Compound umbel	Schizocarp
11.	Colocasia/ Taro	*Colocasia esculenta* (L.) Schott.	Araceae	Indo– Malayan region	28	Spadix	Berry
12.	Elephant foot yam	*Amorphophallus paeoniifolius* (Dennst.) Nicolson	Araceae	South east Asia	28	Spadix	

S. No	Common name	Scientific name	Family	Origin	Chromosome number (2*n*)	Inflorescence type	Fruit type
13.	Elephant ear/ Tannia	*Xanthosoma sagittifolium* (L.) Schott	Araceae	Tropical America and West Indies	26	Spadix	
14.	Asparagus	*Asparagus officinalis* L.	Asparagaceae	Temperate Europe and Asia	20		
15.	Globe Artichoke	*Cynara scolymus* L.	Asteraceae (Compositae)	Mediterranean region	34	Capitulum	Achenes
16.	Lettuce	*Lactuca sativa* L.	Asteraceae (Compositae)	Middle East (Egypt and Iran)	18	Capitulum	Achenes
17.	Ceylon spinach, Malabar spinach, Indian spinach	*Basella alba* L.	Basellaceae	South Asia	48		
18.	Sprouting broccoli	*Brassica oleracea* var. *italica* Plenck	Brassicaceae (Cruciferae)	Mediterranean region	18	Cyme	Silique
19.	Brussels sprouts	*Brassica oleracea* var. *gemmifera* DC.	Brassicaceae (Cruciferae)	Mediterranean region	18	Raceme	Silique
20.	Cabbage	*Brassica oleracea* var. *capitata* L.	Brassicaceae (Cruciferae)	Mediterranean region	18	Raceme	Silique
21.	Cauliflower	*Brassica oleracea* var. *botrytis* L.	Brassicaceae (Cruciferae)	Mediterranean region	18	Raceme	Silique
22.	Chinese cabbage	*Brassica rapa, subspecies pekinensis and chinensis*	Brassicaceae (Cruciferae)	China	20	Raceme	Silique
23.	Knol khol	*Brassica oleracea* var. *gongylodes* L.	Brassicaceae (Cruciferae)	Central countries of northern Europe	18	Raceme	Silique
24.	Radish	*Raphanus sativus* L.	Brassicaceae (Cruciferae)	Mediterranean region	18	Raceme	Silique
25.	Turnip	*Brassica rapa* L.	Brassicaceae (Cruciferae)	Mediterranean region	20	Raceme	Silique
26.	Sweet potato	*Ipomoea batatas* (L.) Lam.	Convolvulaceae	South America	90	Cyme	Capsule
27.	Ash gourd	*Benincasa pruriens* f. *hispida* (Thunb.) de Wilde & Duyfjes	Cucurbitaceae	Java, Japan	24		Pepo
28.	Bitter gourd	*Momordica charantia* **L.**	Cucurbitaceae	Tropical Asia	22		Pepo
29.	Bottle gourd	*Lagenaria siceraria* (Molina) Standl.	Cucurbitaceae	South Africa and India	22		Pepo
30.	Chow-Chow	*Sechium edule* Swaetz.	Cucurbitaceae	Central America	24		Pepo
31.	Cucumber	*Cucumis sativus* L.	Cucurbitaceae	India	14		Pepo
32.	Ivy gourd	*Coccinia grandis* (L.) Voigt	Cucurbitaceae	India	24		Pepo
33.	Longmelon	*Cucumis melo* var. *utillisimus*	Cucurbitaceae	Tropical Africa	24		Pepo

S. No	Common name	Scientific name	Family	Origin	Chromosome number (2*n*)	Inflorescence type	Fruit type
34.	Muskmelon	*Cucumis melo* L.	Cucurbitaceae	Tropical Africa	24		Pepo
35.	Pointed gourd	*Trichosanthes dioica* Roxb.	Cucurbitaceae	India	22	Recimose	Pepo
36.	Pumpkin	*Cucurbita moschata* (Duchesne) Duchesne ex Poir.	Cucurbitaceae	North western part of South America	40		Pepo
37.	Ridge gourd	*Luffa acutangula* (L.) Roxb.	Cucurbitaceae	Sub-tropical Asian regions including India	26		Pepo
38	Round gourd/ Tinda	*Praecitrullus fistulosus* (Stocks) Pangalo	Cucurbitaceae	Tropical Africa	22		Pepo
39.	Snake gourd	*Trichosanthes cucumerina*	Cucurbitaceae	India	24		Pepo
40.	Snapmelon	*Cucumis melo* var. *momordica*	Cucurbitaceae	Tropical Africa	24		Pepo
41.	Spine gourd	*Momordica dioica* Roxb. ex Willd.	Cucurbitaceae	Indo –Malayan region	28		Pepo
42.	Sponge gourd	*Luffa cylindrica* (L.) Roem.	Cucurbitaceae	Asian regions including India	26		Pepo
43.	Summer squash	*Cucurbita pepo* L.	Cucurbitaceae	North America	40		Pepo
44.	Watermelon	*Citrullus lanatus* (Thunb.) Matsumura & Nakai	Cucurbitaceae	Southern Africa	22		
45.	Winter squash	*Cucurbita maxima*	Cucurbitaceae	Mexico	40		Pepo
46.	Yams	*Dioscorea* Spp.	Dioscoreaceae	South-East Asia	40,60,80	Recimose	Capsule
47.	Cassava (Tapioca)	*Manihot esculenta* Crantz	Euphorbiaceae	Brazil	20	Cyathium	
48.	Broad bean	*Vicia faba* L.	Fabaceae (Leguminoceae)	India	12		Legume
49.	Cluster bean	*Cyamopsis tetragonoloba* (L.)Taub.	Fabaceae (Leguminoceae)	India	14		Legume
50.	Cowpea	*Vigna unguiculata* (L.) Walp.	Fabaceae (Leguminoceae)	Africa	22	Raceme	Legume
51.	French Bean	*Phaseolus vulgaris* L.	Fabaceae (Leguminoceae)	Southern Mexico and Central America	22		Legume
52.	Lablab Bean/ Indian bean/	*Lablab purpureus* (L.) Sweet, *Dolichos lablab* L. (ambiguous synonym)	Fabaceae (Leguminoceae)	India or South-East Asia	22	Raceme	Legume
53.	Garden pea	*Pisum sativum* L.	Fabaceae (Leguminoceae)	Ethiopia and Western Asia	14	Raceme	Legume
54.	Winged bean	*Psophocarpus tetragonolobus* (L.)DC.	Fabaceae (Leguminoceae)	Madagascar in Africa	18		Legume
55.	Fenugreek	*Trigonella foenum-graecum* L	Fabaceae (Leguminoceae)	Mediterranean region and West Asia	16		Legume

S. No	Common name	Scientific name	Family	Origin	Chromosome number (2*n*)	Inflorescence type	Fruit type
56.	Coleus (Chinese potato)	*Plectranthus rotundifolius* (Poir.) Spreng.	Lamiaceae	Tropical Africa	64,84	Raceme	
57.	Okra	*Abelmoschus esculentus* (L.) Moench	Malvaceae	India and Africa	130		Capsule
58.	Arrowroot	*Maranta arundinacea* L.	Marantaceae	Andean region of South America	18		
59.	Drumstick	*Moringa oleifera* Lam.	Moringaceae	African tropics and North-West India	28		
60.	Chekkurmanis	*Sauropus androgynus* (L.) Merr.	Phyllanthaceae	India and Burma region			
61.	Rhubarb	*Rheum rhabarbarum* L.	Polygonaceae	Temperate Asia, probably Siberia	44		
62.	Curry leaf	*Murraya koenigii* (L.) Spreng.	Rutaceae	India	18	Cyme	
63.	Brinjal	*Solanum melongena* L.	Solanaceae	Indo-Burma region	24	Scorpioid cyme	Berry
64.	Capsicum	*Capsicum annuum* L.	Solanaceae	Central and South America	24	Cyme	Berry
65.	Chilli	*Capsicum annuum* L.	Solanaceae	Mexico	24	Cyme	Berry
66.	Potato	*Solanum tuberosum* L.	Solanaceae	South America	48	Cyme	Berry
67.	Tomato	*Solanum lycopersicum* L. (Syn. *Lycopersicon esculentum* Mill.)	Solanaceae	Central and South America	24	Cyme	Berry

Note: For latest botanical and zoological names site www.sp2000.org may be consulted

Table 8.2: *Technical information on vegetable crops*

Particulars	Vegetables
Self-pollinated crops	Beans, cowpea, peas, fenugreek, tomato, potato, lettuce
Cross-pollinated crops	Cucurbitaceous vegetables, cole crops, coriander, onion, turnip, radish, carrot, cassava, spinach, *Amaranthus* spp, Beet leaf/ Palak, Chekkurmanis, sugarbeet, colocasia/taro, asparagus
Often cross-pollinated crops	Brinjal, capsicum, chilli, okra
Monoecious	Cassava, sweet potato, ash gourd, cucumber, bitter gourd, bottle gourd, sponge gourd, water melon, pumpkin, summer squash, winter squash, chow-chow, Chekkurmanis
Andromonoecious	Muskmelon
Dioecious	Asparagus, pointed gourd, ivy gourd, spine gourd, spinach, yams,
Monoecious and Dioecious	*Amaranthus* spp.
Dicliny	Cucumber, watermelon, pumpkin, squash, asparagus
Dichogamy	Sweet corn, sugarbeet
Protandry	Carrot, leek, onion, coriander, globe artichoke
Protogyny	Cole crops, amaranthus, colocasia/taro
Heterostyle	Brinjal
Chasmogamy	Lettuce
Herkogamy	Lima bean
Cleistogamy	Lettuce
Sporophytic self-incompatibility	Cole crops, radish, turnip, sweet potato
Gametophytic self-incompatibility	Tomato [wild, not cultivated], potato, sugarbeet
Genic male sterility	Cole crops, tomato, muskmelon, watermelon and okra
Cytoplasmic male sterility	Onion, carrot, cucumber, sweet pepper, chilli
Gene-cytoplasmic male sterility	Radish

CHAPTER 9

Climate and Soil Requirements for Vegetable Crops

Vijay Agrawal
Deputy Director Horticulture
Department of Horticulture & Food Processing
Govt. of Madhya Pradesh

Crop	Climate	Soil
Amaranth	It is also grown successfully in hot summer season and humid conditions of kharif season.	Amaranth can be grown on a wide variety of soils. However, sandy loam soil is best suited for its successful cultivation.
Ash gourd	It grows well in warm, humid tropical climate. The temperature of 22-35°C is ideal.	Though a deep loamy soil is best suited, it can be grown on heavy clay as a rainfed crop. During summer, the crop can be grown in tank slopes. The optimum pH is 6.5-7.5.
Beetroot	It can grow in almost every climate. Being fairly hardy, it is generally considered as a cool weather crop. It grows best during winter in plains. It requires abundant sunshine for proper development. Beet is most productive at 20-22°C.	Beetroot thrives best in a light to medium heavy soil. Loamy sand, deep, well-drained and fertile soils are preferred. Good beet is produced on a wide variety of soils, but slightly alkaline (with a pH up to 8) is ideal.
Bittergourd	A long period of warm, dry weather with 30°-35°C temperature is optimum. The minimum temperature should not go below 18°C. If temperature goes above 40°C, more male flowers are produced and plants become prone to mosaic disease.	A well-drained, loamy soil with a pH of 6.5-7.0 is ideal for its cultivation.

Crop	Climate	Soil
Bottle Gourd	Slightly wet to semi-dry ecological condition is suitable for this crop. Night and day temperature of 18-22°C and 30-35°C respectively is optimum for its proper growth and high fruit set.	Bottle gourd can be grown on all types of soils if these are not too much acidic (pH less than 5.5) or saline and alkaline. Loam or sandy-loam soil is most suitable.
Brinjal	Brinjal is susceptible to frost. A long and warm growing season with a mean temperature of 20-30°C is most favourable for its successful production. In northern plains, it is adversely affected during December-February due to low night temperature.	Well-drained and fertile sandy loam soils with a pH of 5.5-5.6 are best suited for its cultivation. However, several cultivars are grown successfully under high pH level with a rich application of farmyard manure or green manuring practiced before transplanting.
Broccoli	It is a cool season crop resistant to mild frost. The temperature of 20-25°C is optimum for its proper growth, while 15-20°C for heading stage. The heads become loose with rise in temperature.	Broccoli can be grown in a wide variety of soils but deep loamy soil is best suited. The pH of 5.0-6.5 is optimum.
Brussels sprout	Brussels sprout thrives best in cool and humid climate with a temperature of 15-25°C. It can withstand frost as well as snowfall. Generally, good quality sprouts are obtained when temperature is 10-20°C. The rise in temperature results in loose and elongated sprouts with poor flavour.	It can be grown in a wide range of soils but sandy loam is best-suited. The pH of 6.0-6.8 is optimum.
Cabbage	It thrives best in a relatively cool moist climate. It can withstand extreme cold and frost relatively better than cauliflower. The growth in most of the cabbage varieties is arrested when temperature rises above 25°C. With the introduction of heat tolerant hybrids and development of tropical lines/ breeds, the temperature barrier has been removed.	It can be grown on all types of soils. In clay loam or silt, higher yield may be obtained from the late crop. It does not grow well on a highly acidic soil and is best grown between pH 5.5 and 6.5.
Capsicum	It cannot withstand heavy rains during flowering or fruit set. Very cold temperature (<12°C) during night also hampers growth and fruit set. Ideal temperature for flowering is 26-28°C day and 16-18°C night.	Although a wide range of soils are suitable, red or sandy loam soil with a pH of 5.5-6.8 is ideal

Crop	Climate	Soil
Carrot	Carrot can be grown on all types of soils.	It thrives best on a deep, loose, loamy soil. For early crop, a sandy loamy soil is preferred but for high yield, deep, loose and loamy soil is desirable. Soil with a pH of 6.5 is ideal.
Cauliflower	It can grow at an average temperature of 5-8°C to 25-28°C. The optimum temperature for growth of young plants is around 23°C and 17-20°C in later stages.	Fairly deep loamy soils are ideal for growing cauliflower. Soil with a pH of 5.5- 6.5 is optimum.
Chilli	During the growing period which extends over 4 months maximum temperature ranging from 20-30°C and a minimum temperature not below 10°C is ideal. The fruit colour is affected at temperatures below 15°C.	Chilli is grown in many types of soils—sandy to heavy clay. A well-drained, fairly light fertile loam with a fair moisture-holding capacity is ideal. The light soils produce better quality fruits than heavy soils. Chilli crop prefers a soil reaction ranging from pH 6-7.
Cowpea	The germination is better at 12-15ºC temperature and the crop thrives best between 21-35ºC temperature. It can tolerate drought to some extent.	Cowpea can be grown almost in any type of soil but well-drained loamy or slightly heavy soils are better. Saline or alkaline soils are not good.
Cucumber	The optimum temperature favouring anthesis and dehiscence is 13-18°C. Nevertheless for optimum vegetative growth and yield, the temperature requirement is 18-24°C.	A well-drained loamy soil is preferred by cucumber. Lighter soil which warm quickly in spring area usually utilized for early crop. Cucumber prefers a soils with pH of 5.5-6.8.
Frenchbean	A mean air temperature of 20-25°C is optimum for its growth and high pod yield.	French bean is grown over a wide range of well-drained, alluvial friable soils, but is cannot withstand extreme acidic and alkaline soils. The optimum soil pH is 5.5-6.8.
Knol khol	Knol khol thrives best in a relatively cool, moist climate. It grows well with a monthly average temperature of 15-20°C, maximum and minimum average being 24°C and 4.5°C.	Knol-khol can be grown on all types of soils. A soil rich in manures and fertilizers produces excellent knobs. Sandy loam is ideal for an early crop and clay or silt loam for higher yield and late crop. The optimum soil pH is 5.5-6.8.
Muskmelon	The optimum temperature for germination of seeds is 27-30°C.	A well drained, loamy soil is ideal. Muskmelon is sensltive to acidic soils. The crop cannot be grown successfully below a pH of 5.5. A soil pH between 6.0 and 7.0 is ideal. Alkaline soils with high salt concentration are not suitable.

Crop	Climate	Soil
Okra	Okra requires a long, warm and humid growing period. It can be successfully grown in hot humid areas. It is sensitive to frost and extremely low temperatures. For normal growth and development a temperature between 24°C and 28°C is preferred.	It is grown on sandy to clay soils but due to its well-developed tap root system, relatively light, well-drained, rich soils are ideal. A pH of 6.0-6.8 is ideally suited.
Onion	Onion is a cool season crop. However, it can be grown under a wide range of climatic conditions. The ideal temperature requirement of the onion crop is 12.8-12°C before bulbing and 15.5-25°C for bulb development. Very low temperature in the early stages favours bolting, whereas a sudden rise in temperature favours early maturity in *rabi,* resulting in small-sized bulbs.	Onion can be grown on all types of soils. However, deep friable loamy and alluvial soils are best for its successful production.
Pea	Temperature of 15-25°C is favourable for its cultivation.	It can be grown on many types of soils—light sandy loam to clayey soil. Pea needs well-drained soils. The pH of 6.0-7.5 is ideal.
Pumpkin	A temperature of 18-30°C is optimum. Short days, low temperatures (18-30°C), high relative humidity, bright sunshine are ideal for its cultivation.	Pumpkin can be grown in various kinds of soils but sandy loam to loamy soils are ideal. The pH of 5.5-6.8 is optimum.
Radis	The temperature of 10-15°C is optimum for proper development of roots, texture and flavour. Maximum root growth occurs initially at 20-30°C but later at temperature of 10-14°C.	Radishes can be grown almost in all kinds of soils but friable loamy soil with high humus content is most suited.
Ridge gourd	It grows very well in warm hot climate, the optimum temperature being 25-30°C. Very high temperature especially in the early crop growth stage helps produce more male flowers, reducing the yield. Very low temperature also affects growth of vines.	Sandy loamy soil rich in organic matter, is most suited for higher yield.
Round melon	It grows well under warm and dry climate but can be grown both under hot as well as mild climates. The temperature of 27-30°C is optimum for seed germination.	Sandy-loamy soil rich in organic matter is ideal for its cultivation. A pH between 6.0-7.0 is ideal its.

Crop	Climate	Soil
Tomato	Tomato, a warm season vegetable, is grown extensively in cool season also. The optimum temperature required for its cultivation is 15-27°C. At high and low temperatures, there is low germination of seeds, poor plant growth, flower drop, poor fruit set and ripening. Under extreme high and low temperature conditions, the yield and quality of fruits is reduced. Mild winter condition in northern plains is ideal for seed germination, plant growth, fruit set, fruit development and ripening.	Tomato is grown in varied types of soil-sandy loam to clay, black soil and red soil-having proper drainage. However, sandy loam, rich in organic matter is ideal for its cultivation. The pH of soil should be 7-8.5. Tomato can tolerate moderate acidic and saline soils. In red and black soils of Karnataka, Maharashtra and Madhya Pradesh, tomato hybrids are cultivated commercially.
Turnip	The roots develop best flavour, texture and size at a temperature of 10-15°C.	Turnip can be grown in all kinds of soils but it grows well in loamy soil having sufficient humus. The well-drained, sandy loam soil is ideal for its cultivation.
Watermelon	Plant growth is optimum under 28-30° C, while fruiting is better at 24-27° C. higher temperatures are beneficial during ripening. Arid regions of Rajasthan are best suited for production of quality fruits.	Watermelon may be grown on a wide variety of soils. Sandy loams are best for early crop, while loamy soils have high yielding potential. Alluvial river-beds are also good for watermelon. It is generally cultivated in river-beds by making trenches and sowing in hills or pits. A pH of 6.5-7.0 is ideal.

CHAPTER 10

Improved Varieties and Hybrids of Vegetable Crops

S.K. Tyagi
Scientist (Horticulture)
Krishi Vigyan Kendra, Khargone (M.P.)

Table 10.1: *Crop-wise improved varieties/hybrids released from ICAR/SAUs/other Organizations along with their characteristic features*

Name of variety/hybrid	Salient features
Tomato	
Varieties developed at IARI, New Delhi	
Pusa Divya	It is a cross between Long style x Roma. Plants indeterminate, profusely branched; fruits thick skinned, round to oval; first picking 80 days after transplanting. Average yield 350q/ha. It is recommended for Punjab, U.P. and Bihar.
Pusa Gaurav	Plants are dwarf, bushy with moderate foliage cover. Fruits are smooth, elliptical (egg-shaped) and borne in clusters. The unripe fruits are firm with thick flesh (0.6cm) with two well-filled locules facilitating easy transportation over long distances. Whole fruit is suitable for processing and canning as it is without neck constriction and has higher TSS (6%) and better keeping quality at room temperature. It has average yield potential of 330-350 q/ha. It is recommended for Punjab, UP, Bihar, Rajasthan, Gujarat, Haryana and Delhi.
Pusa Hybrid-2	Plants are compact, semi-determinate with good foliage cover having prolific bearing. Fruits are round to flattish round, firm, smooth and attractive with uniform red at maturity. Plants are highly tolerant to root knot nematode. Average yield 600-625 q/ha. It is recommended for Humid western Himalayan regions, sub-humid Sutluj- Ganga alluvial plains and Western plains, semi-arid plateau.

Name of variety/hybrid	Salient features
Pusa Hybrid-4	Plants are determinate, compact with good dark green foliage. Fruits are attractive, round, smooth, average fruit weighing 70-80 g with three locules, thick pericarp and uniform ripening. It has field resistance to root-knot nematode. It has average yield potential of 425-450 q/ha. It is recommended for semi-arid plateau and Central high lands.
Pusa Hybrid-8	Plants are determinate, with green compact foliage and heavy fruit bearer. Fruits are round, medium (75-80g) with uniform ripening. Average yield 430-450 q/ha. It is recommended for Punjab, U.P., Bihar and Jharkhand
Pusa Rohini	Fruits with thick pericarp, suitable for long distance transportation. Maturity 80 days. Average yield 415 q/ha. It is recommended for NCR.
Pusa Ruby	This is an early maturing variety. Plants are indeterminate with a height of 80-85cm and spreading habit with few hardy branches. Fruits are flat to round, slightly lobed (4- 5 locules), medium-sized, uniform red at maturity with slightly acidic pulp, suitable for fresh consumption and processing. It is ideal for autumn, winter and spring summer seasons in north India. Average yield 280-300 q/ha. It is recommended for all over India.
Pusa Sadabahar	Determinate, thermo-insensitive, fruit set takes place almost round the year, both hot and cold set, Maturity 60 days, average yield 350 q/ha. It is recommended for NCR.
Pusa Sheetal	Plants are determinate; fruits set successfully under low night temperature up to 8°C and suitable for early spring season; fruits flattish round with yellow stem end, smooth, attractive, medium size, red colour and uniform ripening. Harvesting starts from early March in northern plains. Average yield 350 q/ha. It is recommended for north Indian plains.
Pusa Uphar	Plants indeterminate, prolific bearer, upright, compact with dark-green foliage; fruits in bunches with 4-6 fruits per bunch, attractive, round, medium sized, thick-skinned and uniform in ripening. Average yield 370 q/ha. It is recommended for Punjab, U.P., Bihar, Rajasthan, Gujarat, Haryana and Delhi.
Pusa-120	Plants are semi-determinate, spreading, late maturing with dark green foliage. Fruits are flattish-round, attractive, medium-large, uniform red, less-acidic, less seeded, resistant to nematode and suitable for winter and summer season cultivation. Average yield 300-320 q/ha. It is recommended for all over India.
Varieties developed at IIHR, Bangalore	
Arka Abha	Fruits oblate, with light green shoulder and have stylar end scar with average fruit weight of 75g. Develops deep red colour on ripening. Resistant to bacterial wilt caused by *Ralstonia solanacearum*. Bred for fresh market and average yield 43 t/ha. It is recommended for all over India.

Name of variety/hybrid	Salient features
Arka Abhijit	High yielding F1, hybrid with bacterial wilt resistance. Bred for fresh market. Average yields 65 t/ha in 140 days. It is recommended for Chhattisgarh, Orissa, Arunachal Pradesh, Rajasthan, Gujarat, Haryana, Delhi, Karnataka, Tamil Nadu and Kerala.
Arka Alok	Fruits on the lower clusters square round, large (120g) and in later cluster oblong, medium (80g) firm fruits with light green shoulder. Resistant to bacterial wilt. Bred for fresh market average yield 46 t/ha. It is recommended for H.P., J&K, Hills of U.P., Punjab, Tarai region of UP, Bihar, Jharkhand, Karnataka, Tamil Nadu and Kerala.
Arka Ananya	High yielding F1, hybrid with combined resistance to Tomato Leaf Curl Virus (TLCV) and bacterial wilt. Fruits oblate-round, medium (65-70g), Deep red, medium firm fruit Suitable for fresh market. Average yields 65-70 t/ha in 140 days. It is recommended for Punjab, Tarai region of U.P., Bihar, Jharkhand, Chhattisgarh, Orissa, Arunachal Pradesh, Rajasthan, Gujarat, Haryana, Delhi, M.P. and Maharashtra.
Arka Saurabh	Fruits medium large, (70-80g), round with light green shoulder, deep red firm fruits with nipple tip. Bred for both fresh market and processing. Average yield 30-35 t/ha. It is recommended for H.P., J&K, Hills of UP, Punjab, Tarai region of U.P., Bihar, Jharkhand, Chhattisgarh, Orissa, A.P., Rajasthan, Gujarat, Haryana, Delhi, M.P., Maharashtra, Karnataka, Tamil Nadu and Kerala.
Arka Vardan	High yielding F1, hybrid with root knot nematode resistance. Bred for fresh market. Average yields 75 t/ha in 160 days. It is recommended for H.P., J&K, Hills of U.P., Punjab, Tarai region of U.P., Bihar, Jharkhand, Karnataka, Tamil Nadu and Kerala.
Arka Vikas	Fruits medium large (80-90 g), oblate with light green shoulder, which develops deep red on ripening. Bred for fresh market Adapted to both rainfed and irrigated conditions. Average yield 35 t/ha. It is recommended for H.P., J&K, Hills of U.P., Punjab, Tarai region of U.P., Bihar, Jharkhand, Chhattisgarh, Orissa, A.P., Rajasthan, Gujarat, Haryana, Delhi, M.P., Maharashtra, Karnataka, Tamil Nadu and Kerala.
Arka Vishal	High yielding F1, hybrid for fresh market. Average yields 75 t/ha in 160 days. It is recommended for Punjab, Tarai region of U.P., Bihar, Jharkhand, Karnataka, Tamil Nadu and Kerala.
Varieties developed at IIVR, Varanasi	
Kashi Amrit	Fruits are round, attractive, red and fleshy with an average weight of 108 g. Suitable for cultivation during TOLCV infested period. Gives an average yield of 620 q/ha. It is recommended for U.P., Bihar and Jharkhand.
Kashi Anupam	Plants are determinate, fruits large, flattish round (slightly indented at blossom end of fruit), attractive red with 5-6 locules, medium maturity 75-80 days after transplanting. Average yield 500-600 q/ha. It is recommended for Rajasthan, Gujarat and Haryana.

Name of variety/hybrid	Salient features
Kashi Hemant	The plants are determinate, fruits attractive, red and round, weight varies from 80 to 85 g. Average yield 400- 420 q/ha. It is recommended for Chhattisgarh, Orissa, A.P. and M.P.
Kashi Sharad	Plants are indeterminate, leaves broad, fruits attractive red, slightly oval, firm, thick pericarp, longer shelf life, weight 90 to 95 g. Average yield 400-500 q/ha. It is recommended for J&K, H.P. and Uttaranchal.
Kashi Vishesh	This variety is resistant to TLCV and has been developed using *L. hirsutum f glabratum* B'6013' as donor parent following backcross pedigree selection method. Plants are determinate, dark green, fruits red, spherical, size medium to large, weight of 80 g, first harvest at 70-75 days after transplanting. Average yield 400-450 q/ha. It is recommended for J&K, H.P., Uttaranchal, Punjab, U.P., Bihar, Jharkhand, Chhattisgarh, Orissa, A.P. and Karnataka, Tamil Nadu and Kerala.
Varieties developed at HAU, Hisar	
Hisar Arun	Plant determinate, heavy bearer, early and high yielding. Average yield is 275-300 q/ha. It is recommended for all over India.
Hisar Lalima	Plant determinate, extremely early, medium to large fruited. Average yield is 275-300 q/ha. It is recommended for Haryana.
Hisar Lalit	Plants determinate, fruit maturity in 65-70 days after planting, average fruit weight 50 g, resistant to root knot nematodes. Average yield is 250-300 q/ha. It is recommended for all over India.
HS-101	Plant determinate sturdy, multi-branched, fruits in clusters of 2-3, round, small to medium in size, red on ripening and suitable for winter season. Average yield 250-275 q/ha. It is recommended for all over India.
HS-102	Plant determinate, fruit in clusters of 3-4, small to medium in size, uniform in ripening and suitable for winter and summer season. Average yield 250-275 q/ha. It is recommended for all over India.
HS-110	Plant determinate, potato leaf type, late, fruits large fleshy, suitable for table and processing. It is recommended for Haryana.
Varieties developed at AAU, Anand, Gujarat	
Anand Tomato- 3	The fruits of this variety are big in size with high pulp to juice ratio and very good firmness. The fruits are round in shape, dark red in colour and possess long shelf life. Average yield is 326 q/ha. It is recommended for middle Gujarat.
Gujarat Tomato-2	Plant growth habit of this variety is determinate type. Immature fruits colour is green with dark green shoulder and at mature stage; fruits are attractive red colour with hard pericarp. Average yield is 340 q/ha. It is recommended for middle Gujarat.

Name of variety/hybrid	Salient features
Varieties developed at G.B.U.A&T, Pant Nagar	
Pant Bahar	The plants are bushy and profusely branched. Fruits are flattish round, medium in size with 5-6 locules slightly ridged and become red on maturity. The first picking can be had in 75-80 days after transplanting. The yield potential of this variety is about 250q/ha. It is recommended for H.P., J&K, hills of U.P., Punjab, Tarai region of U.P., Bihar and Jharkhand.
Pant poly house tomato-1	It is suitable for poly house cultivation. No. of fruit per bunch is 5-6. Average fruit weight is 202 g and yield potential is 1400 q/ha. It is recommended for H.P., J&K, hills of U.P., Punjab, Tarai region of U.P., Bihar and Jharkhand.
Pant poly house tomato hybrid-1	It is suitable for poly house cultivation. No. of fruit per bunch is 7-8. Average fruit weight is 265 g and yield potential is 1800 q/ha. It is recommended for H.P., J&K, hills of U.P., Punjab, Tarai region of U.P., Bihar and Jharkhand.
Pant T-3	The fruit weight is about 70 g. After ripening the fruits attain uniform red colour. It is suitable for processing also. It is suitable for cultivation in northern hills and plains of the country. The yield potential of this variety is about 300 q/ha. It is recommended for H.P., J&K, hills of U.P., Punjab, Tarai region of U.P., Bihar and Jharkhand.
Varieties developed at MPKV, Rahuri	
Dhanashree	Fruits are medium small and round with orange red colour, suitable for kharif and rabi season, average yield 600-800 q/ha. It is recommended for Maharashtra.
Phule Raja	Fruits are orange red in colour, medium size, and attractive thick skin. Suitable for kharif and rabi season, average yield 650 q/ha. It is recommended for Maharashtra.
Varieties developed at TNAU, Coimbatore	
CO 1	Round fruits, light green when unripe and red at ripening, bears in clusters. Average yield 36 t/ha. It is recommended for TN.
CO 2	Fruits ovate, large and flat. Average yield 42 t/ha. It is recommended for TN.
CO 3	Dwarf plants fruits globular and red; bears in clusters of 4-5 fruits. Suitable for summer. Average yield 50 t/ha. It is recommended for TN.
COTH 2	F1 hybrid developed by crossing LCR2 x CLN 2123A. Semi determinate. Fruits are smooth, round to slightly oblong, medium sized. Resistant to leaf curl virus. Average yield 90 t/ha. It is recommended for TN.
COTH1	F1 hybrid developed by crossing IHR 709 x LE 812. Plants are semi-determinate. Fruits are medium sized, bright red, smooth, round to slightly oblong and born in clusters of 4 - 5. Average yield 95 t/ha. It is recommended for TN.

Name of variety/hybrid	Salient features
Paiyur1	A hybrid derivative suitable for irrigated and rainfed conditions of Dharmapuri district. Fruits are round, smooth, fleshy and suitable for long distance transport. Average yield 32 t/ha. It is recommended for TN.
PKM1	Determinate plants; fruits flattish with ridges. Suited for long distance transport. Average yield 32 t/ha. It is recommended for TN.
TNAU Tomato Hybrid Co3	F1 hybrid developed by crossing HN2xCLN 2123A semi-determinate, Fruits are medium-sized, smooth round borne in clusters of 3-5. Resistant to leaf curl virus and moderately resistant to root knot nematode. Average yield 95 t/ha. It is recommended for TN.
Varieties developed at CSAUAT, Kanpur	
Azad T-5	Indeterminate, round, red, bold, better shield life. Average yield 400-450 q/ha. It is recommended for Punjab, Tarai region of U.P., Uttarakhand, Bihar and Jharkhand.
Azad T-6	Determinate, round red, medium, smooth. Average yield 450-500 q/ha. It is recommended for Punjab, Tarai region of U.P., Uttarakhand, Bihar and Jharkhand.
KTH-1 (Hybrid)	Oval shaped fruit, red in colour. Average yield 500-550 q/ha. It is recommended for Punjab, Tarai region of U.P., Uttarakhand, Bihar and Jharkhand.
KTH-2 (Hybrid)	Round shape, red colour. Average yield 550-600 q/ha. It is recommended for Punjab, Tarai region of U.P., Uttarakhand, Bihar and Jharkhand.
Brinjal	
Varieties Developed at IARI, New Delhi	
DBHL-20	Fruits are long, dark purple, glossy, weighing 90-100g and it takes about 55 days from transplanting to first harvest. Its average yield is 524.0 q/ha. It is recommended for Punjab, Delhi, Uttar Pradesh and Bihar.
DBL-02	Plants are non-spiny with erect branches, having light purple pigmentation partially on younger leaves, mid-rib and veins. Its fruits are long, violet-purple with round distal end, each fruit weighing 80-90g. Maturity 55 days from transplanting. Average yield 40 t/ha. It is recommended for Delhi, Punjab and Haryana.
Pusa Ankur	Plants are semi-upright (semi-erect stem), medium-tall (65cm) with 6-8 branches. Fruits are small, round and purple. Average yield 560-580 q/ha. M.P. and Maharashtra.
Pusa Anupama	Plants medium tall with slightly violet-tinged stem and foliage; leaves with purple rib; fruits cylindrical, tender, purple, borne in clusters with 3-5 fruits. Resistant to bacterial wilt; first picking 65-70 days after transplanting. Average yield 370 q/ha. It is recommended for Kerala, Tamil Nadu and Karnataka.

Name of variety/hybrid	Salient features
Pusa Bindu	Plants are 55-60 cm tall with green leaves. Fruits are borne in cluster of 4-5 with each fruit weighing 125g. This variety has yield potential of 250-300q/ha. It is recommended for northern central plains, areas of Gujarat and Maharashtra.
Pusa Hybrid-5	Plants are vigorous, non-spiny with semi-erect branches. Fruits are long, glossy, attractive, dark purple with partially pigmented peduncle, each fruit weighing 100 g. It takes 80-85 days from sowing to first picking and further pickings continue up to December in the Central and northern plains, whereas in Kerala, Tamil Nadu, Karnataka and Andhra Pradesh from November to May. Average yield 520-540 q/ha. It is recommended for northern plains, central India, non-coastal areas of Kerala, Tamil Nadu, and Karnataka.
Pusa Hybrid-6	Plants are non-spiny, semi-erect, branched, vigorous, partial pigmentation on younger leaves and growing tips. Fruits are round, glossy, attractive, violet purple, partially pigmented peduncle with each fruit weighing 200-250g. Being an early bearing hybrid it takes 85-90 days from sowing to first picking and further pickings continue up to December. It has average yield potential of 450-470 q/ha. It is recommended for northern plains, central India, non-coastal areas of Kerala, Tamil Nadu, Karnataka and Andhra Pradesh.
Pusa Hybrid-9	Plants have green, non-spiny, strong, erect branches with light pigmentation on younger leaves. Fruits are oval, glossy, attractive, dark purple, peduncle partially pigmented with average weight of 300g. It takes 90 days from sowing to first pickings and further picking continue up to December. Average yield 675-700 q/ha. It is recommended for Gujarat and Maharashtra.
Pusa Kranti	This variety has been derived from a three-way cross (PPL x Hyderpur) x Wynad Giant. Fruits are oblong, 15-20 cm long, dark purple with shining green calyx and less seeded. Average yield 275-300 q/ha. It is recommended for Punjab, U.P. and Bihar.
Pusa Purple Long	Fruits are long (20-25 cm), purple, glossy and tender. It is suitable for summer and autumn season cultivation. Average yield 280-310 q/ha. It is recommended for Punjab, U.P., Bihar, Rajasthan, Gujarat, Haryana, Delhi, M.P., Maharashtra, Karnataka, Tamil Nadu and Kerala.
Pusa Purple Cluster	Stem purple; foliage purplish green, non-spiny; fruits 10-12 cm long, deep purple, borne in clusters with 4-9 fruits per cluster. Most suitable for hills and southern states; field-resistant to bacterial wilt *(Pseudomonas solanacearum)*; first picking 60-65 days after transplanting. Average yield 250 q/ha. It is recommended for all over India.
Pusa Shyamla	Fruits are long, glossy, attractive dark purple, weighing 80-90g. Maturity 50-55 days. Average yield 390 q/ha. It is recommended for Punjab, Uttarakhand and Bihar.

Name of variety/hybrid	Salient features
Pusa Upkar	Plants are semi-upright, medium-tall (70 cm) with 5-6 branches and greenish purple leaves bearing -15 fruits. Fruits are round (9cm diameter), dark purple and each fruit weighing -300g. Average yield 650-670 q/ha. It is recommended for Punjab, U.P. and Bihar.
Pusa Uttam	Plants are semi-upright, vigorous, branched and free from spines. Mature plants appear green with occasional light pigmentation also present on growing shoots. Flowers appear in clusters having bluish-violet corolla. Fruits are pendent, oval, medium, large-sized, and glossy with dark purple skin and green peduncle. The bearing habit is solitary. Single fruit weight 250-300g. It has yield potential of 425-450 q/ha. It is recommended for northern and central plains, areas of Gujarat and Maharashtra.
Varieties Developed at IIHR, Bangalore	
Arka Kusumakar	Fruits green, small, borne in clusters. Soft texture with good cooking quality. Average yield 40 t/ha. It is recommended for Punjab, Tarai region of U.P., Bihar, Jharkhand, Rajasthan, Gujarat, Haryana, Delhi, Karnataka, Tamil Nadu and Kerala.
Arka Anand	High yielding F1 hybrid with resistance to bacterial wilt. Fruits are green, medium long with excellent cooking quality. Average yields 55-60 t/ha in 140 days. It is recommended for Rajasthan, Gujarat, Haryana, Delhi, M.P., Maharashtra, Karnataka, Tamil Nadu and Kerala.
Arka Keshav	Red purple glossy fruit skin with green calyx. Green leaves with purple leaf base and purple veins when young. Light purple green stems. Fruits tender with slow seed maturity with no bitter principles. Resistant to bacterial wilt. Average yield 45 t/ha. It is recommended for Karnataka.
Arka Navneeth	High yielding F1 hybrid with large (450 g) purple oblong fruits. Average yields 65-75 t/ha in 160 days. It is recommended for H.P., J&K, Hills of U.P., West Bengal, and Assam, Punjab, Tarai region of UP, Bihar, Jharkhand, Rajasthan, Gujarat, Haryana, Delhi, M.P., Maharashtra, Karnataka, Tamil Nadu and Kerala.
Arka Neelkanth	Fruits tender with slow seed maturity with no bitter principles. Resistant to bacterial wilt. Average yield 43 t/ha. It is recommended for Karnataka.
Arka Nidhi	Fruits purple, tender with slow seed maturity with no bitter principles. Resistant to bacterial wilt. Average yield 48.5 t/ha. It is recommended for Karnataka, Tamil Nadu and Kerala.
Arka Sheel	Fruits medium sized with deep shining purple skin. Calyx fleshy and green. Solitary bearing habit average yield 38 t/ha. It is recommended for Karnataka, Tamil Nadu and Kerala.
Arka Shirish	Fruits green, extra-long. Solitary bearing habit. Average yield 39 t/ha. It is recommended for Karnataka, Tamil Nadu and Kerala.

Name of variety/hybrid	Salient features
Varieties Developed at IIVR, Varanasi	
Kashi Komal	Semi-upright, height 90-100 cm with light green stem and leaves; first flowering takes in 35-40 days after transplanting. Fruits are light purple, long, soft textured, average length 13 cm and diameter 3 cm. The picking starts in 65-70 days after transplanting and gives an average yield of 800q/ha. It is recommended for all over India.
Kashi Prakash	Plants are semi-upright, stems and leaves are green. Fruits are attractive with light green spots, calyx spiny, average weight 190 g. The picking starts in 80-82 after transplanting and gives an average yield of 650-700 q/ha. It is recommended for U.P. and Jharkhand.
Kashi Sandesh	Green stems, purplish green leaves and purple flowers appear in 45 days after transplanting. Fruits are purple, medium sized, round shaped, fruit length 12.4 cm, diameter 10.2 cm and weight 225.3 g. The picking starts in 76 days after transplanting. Gives an average yield of 780q/ha. It is recommended for all over India.
Kashi Taru	Plants are tall and erect, height 120-130 cm, leaves and stem dark green; first flowering starts 45-50 days after transplanting. Fruits are long, purple, length 31 cm and diameter 5 cm. The picking starts 75-80 days after transplanting and gives an average yield of 700-750 q/ha. It is recommended for U.P. and Jharkhand.
Varieties Developed at HAU, Hisar	
BR-112	Plants are bushy, fruits round with bright purple colour and fleshy. Average yield is 250 q/ha. It is recommended for all over India.
Hisar Jamuni	Fruits oblong, dark bright purple colour and high yielding. Average yield is 260-280 q/ha. It is recommended for Haryana.
Hisar Pragati	Suitable for summer season, fruit long bright dark purple colour. Average yield is 325-350 q/ha. It is recommended for Haryana.
Hisar Shyamal	Fruits are round, bright, dark purple colour. Average yield is 300-325 q/ha. It is recommended for all over India
HLB-12	Tolerant to high temperature and shoot and fruit borer. Average yield is 200 q/ha. It is recommended for Haryana.
PH-4	Plants are bushy, fruit medium long to long and thin, deep purple in colour, flesh light green. Average yield is 250- 300 q/ha. It is recommended for all over India.
Varieties Developed at JAU, Junagadh	
GJB-3	The fruits are medium to big in size and oval in shape with green colour and good shine. Suitable for *late kharif* (15th Aug to 15th Sept). Average yield is 394 q/ha. It is recommended for Gujarat.

Name of variety/hybrid	Salient features
Varieties Developed at AAU, Jorhat, Assam	
JC-1	Plants are tall, pedicel non spiny, fruit long, medium size, purple colour with pointed apex. Tolerant to bacterial wilt, phomopsis blight, little leaf virus and borer. Maturity 130 days. Average yield 336 q/ha. It is recommended for entire state of Assam and North East region.
JC-2	Plant tall, pedicel spiny, fruit large, oblong, purple colour with blunt apex. Tolerant to bacterial wilt, phomopsis blight, little leaf virus and borer. Maturity 130 days. Average yield 350 q/ha. It is recommended for entire state of Assam and North East region.
Varieties Developed at AAU, Anand, Gujarat	
Anand Brinjal Small Round-2	The fruits are small, round, light pinkish, shiny with non-spiny calyx. Average yield is 408 q/ha. It is recommended for M.P. and Maharashtra.
GBH-1	The fruits are dark purple with single green styler spot at apex. This hybrid is tolerant to shoot and fruit borer. Average yield is 578 q/ha. It is recommended for Punjab, Tarai region of U.P., Bihar, Jharkhand, Rajasthan, Gujarat, Haryana, Delhi, M.P. and Maharashtra.
Oblong Brinjal -1	Fruits are attractive due to shape, size, black colour and shiny. It is moderately tolerant to leaf disease. Average yield is 486 q/ha. It is recommended for middle Gujarat.
Varieties Developed at GBUA & T, Pant Nagar	
Pant Brinjal Hybrid1	The fruits are long, bright deep purple in colour and have soft flesh. The fruiting is in clusters. It has field resistance against bacterial wilt and phomopsis blight. It is less infested by shoot and fruit borer. This variety has wider adaptability. It takes 70-75 days for first picking and yields about 600 q/ha fresh fruits. It is recommended for hills and plains both in U.P and Uttarakhand.
Pant Rituraj	Fruits are purple colour almost round with slight tapering towards the bottom. It takes about 60 days for first picking after transplanting. It is suitable for planting both in winter and summer seasons and is a prolific bearer unlike Type 3 which is the parental cultivar and a shy bearer. The yield potential is 300 q/ha. It is recommended for H.P., J&K, hills of U.P., Punjab, Tarai region of U.P., Bihar and Jharkhand.
Pant Samrat	It is a long fruited variety which produces dark purple, medium long fruits in clusters. Plant is tall (80-100cm) and robust. Young leaves are purplish green. It takes about 70 days for first picking after transplanting. It has resistance against phomopsis blight and bacterial wilt under field conditions. It has been reported to be tolerant against fruit and shoot borer. The yield potential is about 300 q/ha. It is recommended for throughout the country including the hilly regions.

Name of variety/hybrid	Salient features
Varieties Developed at MPKY, Rahuri	
Krishna	Plants are tall and hardy, resistant to lodging. Egg shaped, spiny purple coloured fruits with white stripes, suitable for Kharif and summer seasons. Average yield 480 q/ha. It is recommended for Maharashtra.
Phule Arjun	Fruits are green with purple and white stripes, oval in shape and spiny. Suitable for Kharif and summer seasons, average yield 450 q/ha. It is recommended for Maharashtra.
Phule Harit	Fruits are green in colour with a white patch at the apex. Suitable for Kharif and summer seasons. Average yield 200 q/ha. It is recommended for Maharashtra.
Varieties Developed at TANU, Coimbatore	
CO 1	Fruits are light green with a white base and oblong. Medium-sized fruits (50-60g); good keeping quality. Average yield 24 t/ha. It is recommended for T.N.
CO 2	Fruits are oblong with brown stripes. Suitable for Periyar and Coimbatore districts. Average yield 35 t/ha. It is recommended for T.N.
COBH 1	Fruits are slightly oblong, glossy violet. Average yield 56 t/ha. It is recommended for T.N.
COBH 2	F1 hybrid developed by crossing EP65xPusa Uttam. Fruits are medium sized, oblong in shape, dark violet, weighing 55 to 60g. Average yield 60 t/ha. It is recommended for T.N.
KKM1	Fruits egg-shaped, milky white, borne in clusters of 2 - 4. Average yield 37 t/ha. It is recommended for T.N.
MDU1	Fruits round, bright purple and weigh 280 g. Average yield 34 t/ha. It is recommended for T.N.
PKM1	Fruits are medium-sized, 45 - 65g; drought tolerant and can withstand long distance transport. Average yield 34 t/ha. It is recommended for T.N.
PLR (BR) 2	Single plant selection from Sevanthampatti local. Fruits are oval in shape, dark purple with pink tinge and glossy. Average yield 38 t/ha. It is recommended for T.N.
PPI 1	Fruits are oblong, pale green and moderately resistant to shoot and fruit borer and wilt. Average yield 45 t/ha. It is recommended for T.N.
TNAU Brinjal VRM 1	Pure line selection. Fruits are oval in shape, glossy pink with green tinge at the distal end. Average yield 45 t/ha. It is recommended for T.N.
Varieties Developed at CSAUAT, Kanpur	
Azad B-2	Round, bright purple, resistance to phomopsis fruit rot. Average fruit 350-400 q/ha. It is recommended for Punjab, Tarai region of U.P., Uttarakhand, Bihar and Jharkhand.

Name of variety/hybrid	Salient features
Azad B-3	Long, cylindrical, shining purple, suitable for both seasons. Average fruit 400-450 q/ha. It is recommended for Punjab, Tarai region of U.P., Uttarakhand, Bihar and Jharkhand.
Azad B-4	Round fruited, light purple in colour, calyx green. Average fruit 300-350q/ha. It is recommended for Punjab, Tarai region of U.P., Uttarakhand, Bihar and Jharkhand.
Azad Hybrid	Long, bright purple, calyx green. Average fruit 500-550 q/ha. It is recommended for Punjab, Tarai region of U.P., Uttarakhand, Bihar and Jharkhand.
Chilli	
Varieties Developed at IARI, New Delhi	
Pusa Jwala	Plants dwarf, bushy, light green; fruits 9-10 cm long, light green, ripe fruits light red, highly pungent; fairly tolerant to thrips and mites. Average yield 85 q/ha (green) and 18 q/ha (dry). It is recommended for all over India.
Pusa Sadabahar	Plants erect, perennial (2-3 years), 60-80 cm tall; fruits 6- 8 cm long, borne in clusters with 6-14 fruits per cluster, ripe fruits dark red, highly pungent, resistant to CMV, TMV and leaf curl complex; first picking in 75-80 days after transplanting. Average yield 95 q/ha (green) and 20 q/ha (dry). It is recommended for all over India.
Varieties Developed at IIHR, Bangalore	
Arka Harita	F1 hybrid, fruits light 10 cm and width 1cm. Fruits are dark green and turn to red on maturity. Resistant to powdery mildew and viruses. Average yield 38.2 t/ha green chilli or 5.86 t/ha dry chilli. It is recommended for Karnataka, Tamil Nadu and Kerala.
Arka Lohit	Fruits dark green turning deep red on maturity (Capsanthin 0.21%); Fruits highly pungent (Capsaicin 0.71%); suitable for irrigated and rainfed cultivation. Average yield 25 t/ha (Green chilli) and 3 t/ha (Dry chilli). It is recommended for Chhattisgarh, Orissa, Arunachal Pradesh, M.P., Maharashtra, Karnataka, Tamil Nadu and Kerala
Arka Meghana	CMS-based high yielding F1 hybrid with resistance to powdery mildew and viruses. Fruit length 10.6 cm and width 1.2 cm. Very early types, Fruits are dark green and turn to red on maturity. Average yields 33.5 t/ha green chilli or 5 t/ha dry chilli. It is recommended for Punjab, Tarai region of UP, Bihar, Jharkhand, Chhattisgarh, Orissa, Arunachal Pradesh, Rajasthan, Gujarat, Haryana, Delhi, Karnataka, Tamil Nadu and Kerala.
Arka Suphal	Fruits green, smooth, medium long (6-7 cm x 1 cm), pendent, turning deep red on maturity. Suitable for irrigated and rainfed cultivation. Tolerant to powdery mildew. Average yield 25 t/ha (Green Chilli) and 3t/ha (Dry Chilli). It is recommended for M.P. and Maharashtra.

Name of variety/hybrid	Salient features
Arka Sweta	CMS based high yielding F1 hybrid for fresh market. Suitable for *kharif* & *rabi* seasons under irrigated conditions. Fruit length 11-12 cm, width 1.2 - 1.5 cm; smooth and medium pungent. Fruits light green and turn red on maturity. Tolerant to viruses. Duration 160 days. Yields 33 t/ha (fresh) & 5 t/ha (dry). It is recommended for Punjab, Tarai region of U.P., Bihar, Jharkhand, Rajasthan, Gujarat, Haryana, Delhi, Karnataka, Tamil Nadu and Kerala.
Varieties Developed at IIVR, Varanasi	
Kashi Anmol	Plants are determinate, dwarf (60-70 cm) with nodal pigmentation on stem and bear green attractive pendant fruits. First picking starts 55 days after transplanting; yield 200 q/ha. It is recommended for Punjab, Uttar Pradesh, Bihar and Jharkhand.
Kashi Early	Plants are tall (100-110 cm height) without nodal pigmentation on dull green stems and bear pendant fruits. Fruits are long (8-9 x 1.0-1.2 cm), attractive, dark green and turn bright red at physiological maturity, pungent with smooth surface. First picking of the green fruits starts at about 45 days after transplanting. The average yield of this hybrid is 250 q/ha (red ripe). It is recommended for Uttar Pradesh, Bihar, Madhya Pradesh, Jharkhand, Uttaranchal, Karnataka, Delhi, Punjab, Haryana, Andhra Pradesh and Chhattisgarh.
Kashi Surkh	Plants are semi-determinate (1-1.2 m), erect with nodal pigmentation on stem. Fruits are light green, straight, length 11-12 cm, suitable for green as well as red fruit production. First harvest starts after 55 days of transplanting. Green fruit yield is 240 q/ha. It is recommended for West Bengal, Assam, Punjab, Tarai region of U.P., Bihar, Jharkhand, Chhattisgarh, Orissa, Arunachal Pradesh, Rajasthan, Gujarat and Haryana.
Varieties Developed at HAU, Hisar	
Hisar Shakti	Fruits upright in clusters, pointed, thin, medium long, highly pungent, tolerant to leaf curl mosaic virus, fruit rot and powdery mildew. Average yield is 170-190 q/ha. It is recommended for Haryana.
Hisar Vijay	Early, fruit similar to Hisar Shakti but with blackish tinge on fruit surface under low temperature, tolerance like Hisar Shakti. Average yield is 150-170 q/ha. It is recommended for Haryana.
Varieties Developed at AAU, Anand, Gujarat	
Chilli 101	The fruits are medium pungent, long, compact and straight with green shining colour, more shelf life and high ascorbic acid content. Average green fruit yield of 145 q/ha. It is recommended for south Gujarat regions, M.P. and Maharashtra.
Chilli 111	The fruits are attractive, medium long, straight, pungent and slightly wrinkled surface with good shine. Average yield is 121 q/ha. It is recommended for Gujarat.

Name of variety/hybrid	Salient features
Chilli 112	The fruit of this variety are pungent, elongated straight, light green colour and semi wrinkled surface. Average yield is 127 q/ha. It is recommended for Gujarat.
Chilli 131	The fruits are elongated, straight, compact, short pedicel length, smooth shiny surface and pointed blossom end. The fruits having light green colour and non-pungent with good aroma. Average yield is 89 q/ha. It is recommended for Gujarat.
Chilli hybrid-1	The fruits of this hybrid are elongated and straight with pointed blossom end having semi-wrinkle surface and light green colour. Average yield is 273 q/ha. It is recommended for Gujarat.
Varieties developed at GBUA & T, Pant Nagar	
Pant C-1	The green pod yield of this variety is 75 q/ha. The dry pod yield is about 15 q/ha. This variety is easily distinguishable from the other varieties because of the fact that it has upright fruiting pods. The pods are highly pungent, relatively small in size, broader at the base and narrow towards the tip. It is moderately resistant to mosaic and leaf curl virus. The green pod yield of this variety is 75 q/ha. The dry pod yield is about 15 q/ha. It is recommended for all over India.
Varieties developed at MPKV, Rahuri	
Phule Jyoti	Plants are tall and spreading. Fruits are smooth and medium long. Tolerant to thrips, mites and wilt. Suitable for kharif season, average yield 250 q/ha. (Green). It is recommended for Maharashtra.
Phule Mukta	Resistant to powdery mildew and *fusarium* wilt. Fruits are small, dark green and smooth. Suitable for kharif season. Average yield 15 q/ha. (Red dry). It is recommended for Maharashtra.
Varieties developed at TANU, Coimbatore	
CO 1	Fruits long and red at ripening; seed content 55%; samba type. Average yield 2 t/ha. T.N.
CO 2	Fruits oblong and bright red on ripening; seed content 60%, gundu type. Average yield 2 t/ha. It is recommended for T.N.
CO 3	Fruits long, slender and dark green. Suitable for export. Average yield 4 t/ha. T.N.
CO 4	Vegetable chilli, fruits are dark green, less pungent and suitable for high density planting. Average yield 3 t/ha. (Green). It is recommended for T.N.
K1	Plants bushy; pods green; long samba type; dark red on ripening. Average yield 2 t/ha. It is recommended for T.N.
K2	Plants tall, compact, pods long and bright red on ripening; samba type; high seed content. Average yield 2 t/ha. It is recommended for T.N.
MDU.1	Determinate growth. Fruits borne in clusters of 4 - 9. Fruits are long, dark red and shiny. Average yield 2 t/ha. It is recommended for T.N.

Name of variety/hybrid	Salient features
PKM 1	Hybrid derivative of the cross between AC. No. 1797 x CO.1. It is suitable for cultivation under irrigated conditions. Bold pods, which are dark red in colour. The number of pods per plant is 204 with 85 seeds per fruit. Average yield 3 t/ha (Dry pod). It is recommended for TN.
PLR1	Suitable for green chilli. Average yield 19 t/ha. (Green). It is recommended for TN.
PMK.1	Suitable for semi-dry condition. Fruits contain high capsaicin0.36%, gundu type. Average yield 3 t/ha. It is recommended for TN.
TNAU Chilli Hybrid Co1	F1 hybrid developed by crossing Selection 1 x CA97. Fruits are long, smooth and dark green. It possesses 0.58% capsaicin Suitable for green and dry chilli. Moderately resistant to fruit rot. Average yield 28 t/ha. (Green) and 7 t/ha. (Dry). It is recommended for TN.
Capsicum	
Varieties developed at IARI, New Delhi	
California Wonder	Plants upright, fruits upright, fine flavoured with 3-4 distinct lobes, smooth, thick flesh, deep green, turns bright crimson at maturity; first picking 90-100 days after transplanting. Average yield 170 q/ha. It is recommended for all over India.
Pusa Deepti	Plants erect, medium and bushy; fruits yellowish green, conical, turns dark red at maturity; 9 11 cm long, very early; first picking in 60-70 days after transplanting. Average yield 225 q/ha. It is recommended for all over India.
Varieties developed at IIHR, Bangalore	
Arka Basant	Thick fleshed, 2-3 lobed conical fruits. Average fruit weight 50-80 g. Fruits erect, cream coloured, which turn orange red on ripening. Average yield 15 t/ha. It is recommended for all over India.
Arka Gaurav	Thick fleshed, 3-4 lobed green blocky fruits. Average fruit weight 130-150 g. Fruits erect, which turn orange-yellow on ripening. Average yield 16 t/ha. It is recommended for all over India.
Arka Mohini	Thick fleshed, 3-4 lobed dark green blocky fruits. Average fruit weight 180-200 g. Fruits pendent, which turn red on ripening. Average yield 20 t/ha. Seed yield 150 kg/ha. It is recommended for all over India.
Cauliflower	
Varieties developed at IARI, New Delhi	
Improved Japanese	This November maturity group variety has been introduced from Israel. It is erect with bluish green leaves. Curds are white, and compact. It is susceptible to heat and curd availability period is late November to December (16-20°C) from late July to early August sown crop. Average yield 110-125 q/ha. It is recommended for north Indian plains.

Name of variety/hybrid	Salient features
Pusa Deepali	Plants are medium tall. Leaves are erect, short, green and waxy. Curds are compact, self-blanching, white and medium in size. Uniform curds are well protected by leaves and resulting no riceyncss. Curds are ready in late October when the average temperature is around 20- 25°C. It is recommended for sowing from May end to early June. Average yield 150-175 q/ha. It is recommended for Delhi and Punjab.
Pusa Early Synthetic	Plants are erect with bluish green leaves. Curds arc small medium, flat, creamy white and compact. It is suitable for early planting and resistant to riceyness. Average yield 120-150 q/ha. It is recommended for Punjab, U.P., Bihar, Karnataka, Tamil Nadu and Kerala.
Pusa Himjyoti	Plants erect; leaves bluish green with waxy coating, self- blanched; retentive white curds, 500 600g in weight; takes 60-75 days from planting to harvesting. Only variety suitable from April to July sowings in the hilis above 1000 m height. Average yield 160 q/ha. It is recommended for all over India.
Pusa Hybrid-2	It is semi-erect with bluish-green upright leaves. Curds are creamy-white, highly compact with average curd weight of 907 g. It matures from mid-November to mid- December in the North Indian Plains. Plants are resistant to downy mildew. Average yield 250-270 q/ha. It is recommended for humid Bengal Assam basin and Sub-Humid Satluj Ganga Alluvial plains.
Pusa Kartik Sankar	Temperature requirement for curd formation is 22-25°C. Maturity is 96 days, which is 2 weeks earlier than Pusa Early Synthetic. Curds are compact, white, medium sized and each weighing about 475 g. It is resistant to downy mildew. Average yield 145-160 q/ha. It is recommended for West Bengal, Assam, Punjab, U.P., Bihar and Jharkhand.
Pusa Meghna	Extra early (Sept. maturity), white compact medium sized curd (350-400 g). Maturity 95 days. Average yield 125 q/ha. It is recommended for NCR.
Pusa Paushja	Dec.-Jan. maturity, curd compact white. Maturity 75days. Average yield 400 q/ha. It is recommended for Punjab, U.P., Uttarakhand and Bihar.
Pusa Sharad	Mid-early (Nov.) compact curd weighing 900 g. Maturity 80 days. Average yield 240 q/ha. It is recommended for NCR.
Pusa Shubra	Plants are erect with long stalk and bluish green leaves. Curd is compact and white each weighing 700-800 g. Plants are highly tolerant to riceyness and are resistant to black rot under field and artificial inoculation conditions. It takes 125-130 days for 50% curd formation and best temperature for curd is 12-16°C. The yield potential of this variety is 250-300 q/ha. It is recommended for Delhi, Punjab, Haryana, U.P. and Bihar.
Pusa Shukti	Dec.-Jan. maturity, curd compact white. Maturity 75 days. Average yield 440 q/ha. It is recommended for NCR.

Name of variety/hybrid	Salient features
Pusa Snowball K-1	Curds snow white, self-blanched, last to form curd in the snowball group extending its availability period by about a fortnight. Field resistant to black rot; matures in 90-95 days after transplanting. Average yield 300 q/ha. It is recommended for all over India.
Pusa Snowball Kt-25	It is suitable for cultivation in the *late rabi* season. Leaves are dark green waxy and slightly curly. Curd is white, very compact. It is a self-blanching type cultivars. Suitable sowing time under north Indian condition is Oct-Nov. It is tolerant to black rot. Average yield 17.5- 30.0 t/ha. It is recommended for all over India.
Varieties developed at IIVR, Varanasi	
Kashi Kunwari	This is an early maturity group variety. Suitable time of sowing is end of June to July under Indian plain conditions. It can tolerate high rainfall during its vegetative growth. Curds are semi-dome type, white compact; fine texture and weigh about 300-450 g; and yield of 300-350 q/ha. It is recommended for Punjab, U.P., Bihar and Jharkhand.
Varieties developed at HAU, Hisar	
Hisar-1	Suitable for mid-late season, curds medium-large size and white in colour. Average yield is 250 q/ha. It is recommended for all over India.
Varieties developed at GBUA & T, Pant Nagar	
Pant Gobhi-2	The marketable curds of this variety are available in October in the plains. The curds are yellowish and medium compact. The yield potential of this variety is about 100q/ha. It is recommended for northern plains.
Pant Gobhi-3	This is a synthetic variety. It is classified as September maturity variety. Its marketable curds are available in September in the northern plains and for this nursery, sowing is to be done in around middle of May. The curds are creamish white and non ricey. The yield potential of this variety is about 150q/ha. It is recommended for northern plains.
Pant Shubhra	This is a mid-season variety. This variety is characterized by semi-erect outer leaves. Inner leaves partially cover the curds which are compact, slightly conical, non-ricey and creamish white in colour which retains as such for a long time. The variety produces marketable curds by December–January. It takes about 120 days to produce marketable curds from nursery sowing. The yield potential of this variety is 250 qt/ha. It is recommended for H.P., J&K, hills of U.P., Punjab, Tarai region of U.P., Bihar and Jharkhand.
Varieties developed at TANU, Coimbatore	
Ooty.1	Compact uniform curds with attractive creamy white colour. Curd is free from ricyness and have good keeping quality. Average yield 47 t/ha. It is recommended for TN.

Name of variety/hybrid	Salient features
Cabbage	
Varieties developed at IARI, New Delhi	
Golden Acre	Plants with short stalk and small frame; heads round, very compact; cover-leaves dark green, cup shaped with prominent veins; ready in 60-75 days after transplanting. Yield 250 q/ha. It is recommended for all over India.
KGMR-1	Round, compact, green and covered head with outer leaf crimping; Medium waxy, green and slightly serrated leaves; Early in heading (60 days) Better staying capacity in the field after maturity (20 days); Average yield (350- 400 q/ha). It is recommended for Punjab, UP, Bihar, H.P. and Uttarakhand.
Pusa Ageti	Tropical type, produces seeds under sub-tropical conditions; forms marketable head at a temperature range of 15 - 30°C but day temperature should not be above 35°C, recommended for March and August - October planting in northern plains; head weight 600- 1200g; ready for harvest 75-90 days after transplanting. Yield 11 0-330q/ha depending upon the planting time. It is recommended for all over India.
Pusa Drum Head	Heads large (3-4 kg), flat; cover-leaves light green; field resistant to black leg *(Phoma lingam)* disease; ready for harvest in 80-90 days after transplanting. Yield 300 q/ha. It is recommended for throughout India
Pusa Mukta	Plants with short stalk; heads flattish round, medium size; leaves light green with slightly wavy margins; resistant to black rot *(Xanthomonas campestris);* ready for harvest 70-75 days after transplanting. Yield 300 q/ha. It is recommended for Punjab, UP, Bihar, W.B. and Assam.
Knol khol	
Varieties developed at IARI, New Delhi	
Purple Vienna	Early, dwarf; knobs globular, purplish, greenish white flesh; maturity 50-60 days after transplanting. It is recommended for all over India.
White Vienna	Early, dwarf, stems and leaves medium green; knobs globular to flattened globe, light green, creamy white flesh with delicate flavour. Yield 220 q/ha. It is recommended for all over India.
Sprouting Broccoli	
Varieties developed at IARI, New Delhi	
Pusa Broccoli Kt	Head compact, light green with small buds weighing 250- 400g; first picking 85-95 days after transplanting. Yield 125 q/ha. It is recommended for all over India.
Varieties developed at MPKV, Rahuri	
Ganesh Broccoli	Curds dark green, compact, dome shaped, big size and attractive with negligible stalk hollowness. Tolerant to downy mildew and bacterial blight. Suitable for rabi season, average yield 65 q/ha. It is recommended for Maharashtra.

Name of variety/hybrid	Salient features
Brussel's Sprouts	
Varieties developed at IARI, New Delhi	
Hills Ideal	Plant height 55-60 cm; bears 45-50 sprouts; sprouts compact with good flavour; first picking 80 days after transplanting. Yield 160 q/ha. It is recommended for north Indian hills.
Pea	
Varieties developed at IARI, New Delhi	
Arkel	Plants are dwarf (35-45 cm), bear double pods at lower nodes and solitary at upper nodes. Pods are attractive, dark green, 8-9 cm long, curved towards suture at distal end (sickle shaped), well filled with 7-8 seeds/pods with 40% shelling; suitable for fresh market and dehydration. Plants are highly susceptible to collar rot if sown under high temperature. Av. yield 75-80 q/ha in 50-55 days of crop duration. It is recommended for all over India.
Bonneville	Plants are medium-tall, flower in 55-60 days after seed sowing with two pods per peduncle. Pods are light green, straight, 8-9 cm long, having 6-7 grains per pod with 45% of shelling. Seeds are green and wrinkled. Av. yield 65-70 q/ha. It is recommended for throughout India
Pusa Pragati	Pods long (10 cm), green with 9 seeds per pod; first picking 60-65 days; resistant to powdery mildew. Yield 70 q/ha. It is recommended for all over India.
Varieties developed at IIHR, Bangalore	
Arka Ajit	Pods 8-9 cm long, Seeds bold, green and sweet. Resistant to both powdery mildew and rust, Pod yield 10 t/ha. It is recommended for Punjab, Tarai region of U.P., Rajasthan, Gujarat, Haryana, Delhi, Karnataka, Tamil Nadu and Kerala.
Varieties developed at IIVR, Varanasi	
Kashi Nandini	This is the earliest variety in the seed chain of the country. Average plant height is 50 cm and 50% plants bear flowers at 31 days after sowing. Plants are erect dark green having 7-8 pods/plant. Pods are 8-9 cm long, attractive, well filled having 8-9 seeds with shelling percentage 49-50. First harvesting can be done after 55 days of sowing. Average yield is 75-80 q/ha. It is recommended for J&K, H.P., Uttrakhand, Punjab, Tarai region of U.P., Bihar, Jharkhand, Karnataka, Tamil Nadu and Kerala.
Kashi Uday	Plant height is 60-65 cm and 50% plants flowering takes 34-35 days after sowing. Plants have dark green foliage and short internode with 8-10 pods/plant. Pods are 9-10 cm long, attractive with 8-9 bold seeds/pod, with shelling percentage 50-51. Harvesting start from 58-60 days after sowing and total yield of 85-90 q/ha can be obtained in 3-4 picking. It is recommended for Uttar Pradesh.

Name of variety/hybrid	Salient features
Varieties developed at GBUA & T, Pant Nagar	
Pant Matar-2	This is an early maturing variety. The first green pod picking is possible after 60 days of sowing. This variety is equivalent to Arkel in maturity duration and other features and gives about 10 percent higher yield than Arkel. The yield potential is about 100q/ha. It is recommended for J&K, H.P., Hills of UP, Punjab, Tarai region of U.P., Bihar and Jharkhand.
Pant Sabji Matar-3	Pods are long curved and well filled. Seeds are wrinkled with green seed coat colour. The first picking of green pods can be had in 60-65 day and thus Pant Sabji Matar-3 is of Arkel duration and belongs to early maturity group of pea varieties. Days to seed maturity vary from 110-120 days. Pant Sabji Matar-3 has longer pods than those in Arkel (about 9 cm against 8.5 cm), and contains on an average 8-9 seeds per pod. Its yield potential is about 90 q/ha. It is recommended for J&K, H.P., Hills of UP, Punjab, Tarai region of U.P., Bihar and Jharkhand.
Pant Sabji Matar-4	It is an early variety (70 days to green pod picking) and resistant to powdery mildew. It is leafless type. Green pod yield is 90 q/ ha. It is recommended for J&K, H.P., Hills of UP, Punjab, Tarai region of U.P., Bihar and Jharkhand .
Pant Sabji Matar-5	It is an early maturing variety, gives first picking in 60-65 days after sowing. Plants are dwarf (65cm) with green foliage. Pods are long and curved toward the tip. Seeds are wrinkled and green at maturity stage. Average green pod yield is 90-100q/ha. This variety is resistant to powdery mildew. It is recommended for J&K, H.P., Hills of UP, Punjab, Tarai region of U.P., Bihar and Jharkhand.
Pant Uphar	This variety is of intermediate duration between early cultivar Arkel and late maturing cultivar Bonneville. It is a vegetable variety which might fill the gap in time duration between the two dominant cultivars, Arkel and Bonneville. The first green pod picking in this variety is available in 75-80 days after sowing. The yield potential is about 100q/ ha. It is recommended for U.P. and Uttarakhand.
Varieties developed at MPKV, Rahuri	
Phule Priya	Suitable for rabi season, pods straight, green, tender, sweet, 8-10 greens per pod and tolerant to powdery mildew disease, suitable for rabi season, average yield 100-105 q/ha. It is recommended for Maharashtra.
Varieties developed at TANU, Coimbatore	
Ooty 1	Pureline selection. Plants dwarf. Pods long and dark green. Average yield 12 t/ha. It is recommended for T.N.
Pole type Ooty 1	Pureline selection. Plants grow to 2m tall. Pods are fleshy and stringless. Average yield 33 t/ha. It is recommended for T.N.

Name of variety/hybrid	Salient features
Varieties developed at CSAUAT, Kanpur	
Azad P-2	Late, light green, smooth, resistant to powdery mildew. Average yield 80-90 q/ha It is recommended for Punjab, Tarai region of U.P., Uttarakhand, Bihar and Jharkhand.
Azad P-3	Early, medium, bold seeded curved at distal end. Average yield 65-70 q/ha. It is recommended for UP.
Azad P-4	Medium, small pods, resistant to powdery mildew. Average yield 80-90 q/ha It is recommended for Punjab, Tarai region of UP, Uttarakhand, Bihar and Jharkhand.
Azad P-5	Pod medium, extended bearing upto March, resistant to P.M. Average yield 90-100 q/ha It is recommended for Punjab, Tarai region of UP, Uttarakhand, Bihar and Jharkhand.
Cowpea	
Varieties developed at IARI, New Delhi	
Pusa Komal	Plants arc bushy and flower in 45 days after seed sowing. Pods are light green, tender, 25-30 cm long and suitable for both spring summer and rainy season cultivation. Plants are resistant to bacterial blight. Av. yield 100-115 q/ha. It is recommended for Uttar Pradesh, Bihar, Punjab and Haryana.
Pusa Sukomal	Resistant to golden yellow mosaic virus and lcaf spot diseases. Av. yield 60 and 65 q/ha during summer and *kharif* seasons, respectively. Maturity 55-60 days. It is recommended for NCR.
Varieties developed at IIHR, Bangalore	
Arka Garima	Plants tall, photo insensitive. Pods light green, long, thick, round, fleshy and stringless. Suitable for vegetable purpose. Tolerant to heat and low moisture stress. Pod Yield 18 t/ha. It is recommended for M.P., Maharashtra, Karnataka, Tamil Nadu and Kerala.
Arka Samrudhi	Plants erect, bushy, photo insensitive. Pods green, medium thick, medium long round, tender, fleshy without parchment. Good cooking qualities. Pod Yield 19 t/ha. It is recommended for Rajasthan, Gujarat, Haryana, Delhi, M.P. and Maharashtra.
Varieties developed at IIVR, Varanasi	
Kashi Gauri	It is a dwarf and bush type (70-75 cm height), photoperiod insensitive and early variety suitable for sowing in both spring-summer and rainy seasons. It flowers in 35- 40 days and pods get ready for harvest in 45-50 days. It produces 35-40 pods per plant of about 25 cm long. The pods are fleshy and less fibrous, light green and free from parchment. The variety is resistant to cowpea golden mosaic virus and *Pseudocercospora cruenta* and produces about 100-125 q green pods per ha. It is recommended for Uttar Pradesh and notified for cultivation in Uttar Pradesh.

Name of variety/hybrid	Salient features
Kashi Kanchan	It is bush type and dwarf (50-60 cm height), photoperiod insensitive and early variety suitable for sowing in both spring-summer and rainy seasons. It flowers in 40-45 days and pods get ready for harvest in 50-55 days. It produces 40-45 pods per plant of 30-35 cm long. The pods are dark green, pulpy with less fibrous and free from parchment. The variety is resistant to cowpea golden mosaic virus and *Pseudocercospora cruenta* and produces about 150-200 q/ha green pods. It is recommended for Punjab, U.P., Bihar, Jharkhand, M.P., Chhattisgarh, Orissa, A.P. and Maharashtra states.
Kashi Unnati	It is a photoperiod-insensitive, dwarf and bush type (40- 45 cm height) early variety suitable for sowing in both spring-summer and rainy seasons. It flowers in 30-35 days and pods get ready for harvest in 40-45 days. It produces 40-45 pods per plant of 30-35 cm long. The pods are fleshy with less fibrous, light green and free from parchment. The average pod yield is about 125-150 q/ha. It is recommended for Uttar Pradesh and notify for cultivation in Uttar Pradesh.
Varieties developed at AAU, Anand, Maharashtra	
AVCP-1	Plant habit of the variety is bushy and compact with dark green foliage and broad leaves. The pods are medium size, tender and green in colour. Average yield is 73 q/ha. It is recommended for north east and Andaman & Nicobar Island.
Varieties developed at TANU, Coimbatore	
PKM-1	Pureline selection, Semi traiting with very long pod (40- 45cm green pod yield. Average yield 25 t/ha. It is recommended for TN.
Lab Lab Bean	
Varieties developed at IARI, New Delhi	
Pusa Sem-2	Pods dark green, very tender, stringless, semi-round, 16 to 17 cm long with 5-6 seeds per pod, 11-12 pods/spike; excellent organoleptic quality. Highly tolerant to viruses, anthracnose, aphid, jassid, pod borer and frost. First picking in 120-125 days. Yield 150 q/ha. It is recommended for NCR.
Pusa Sem-3	Pods green, meaty, very tender, string less, flat, 15 cm long with 5-6 seeds/pod, 9-10 pods/cluster; excellent organoleptic quality. Tolerant to viruses, anthracnose, aphid, jassid and pod borer. First picking in 115-120 days. Yield 170 q/ha. It is recommended for NCR.
Varieties developed at IIHR, Bangalore	
Arka Jay	Plants dwarf, bushy, erect and photo insensitive. Pods long, light green slightly curved, without parchment. Vegetable type with excellent cooking qualities. Tolerant to low moisture stress. Pod Yield 12 t/ha. It is recommended for Karnataka, Tamil Nadu and Kerala.

Name of variety/hybrid	Salient features
Arka Vijay	Plants dwarf, bushy, erect and photo insensitive. Pods short dark green. Seeds bold. Pods with characteristic aroma, without parchment. Vegetable type with excellent cooking qualities. Tolerant to low moisture stress. Pod Yield 12 t/ha. It is recommended for Karnataka, Tamil Nadu and Kerala.
Varieties developed at HAU, Hisar	
Hisar Kirti	Early variety, pods flat, dark green in colour with better shelf life. Average yield is 200-220 q/ha. It is recommended for Haryana.
Varieties developed at J.A.U., Junagadh	
GJIB-2	The pods of this variety are flat and medium-long in shape with green colour. Suitable for *late kharif* (15th Aug to 15th Sept). Average yield is 112 q/ha. It is recommended for Gujarat.
Varieties developed at CSAUAT, Kanpur	
Azad Sem-1	Flowers white 10-14 pods/punch, resistance to YMV. Average yield 125-130 q/ha. It is recommended for UP.
Kalyanpur type-2	Light green coloured pods, leaves broad in size with white flowers. Average yield 250-300 q/ha. It is recommended for UP.
Rajni	Stem colour purple in early stage, flowers purple, pods dark green. Average yield 150-175 q/ha. It is recommended for U.P.
Broad bean	
Varieties developed at IARI, New Delhi	
Pusa Sumeet	Attractive dark-green pods in clusters; first picking in 65- 70 days. Yield 180q/ha. It is recommended for NCR.
French bean	
Varieties developed at IARI, New Delhi	
Contender	Early, bush type; flowers pink; pods green, large with curved tips, stringless, meaty. Suitable for sowing in hills from April to June and in plains from last week of September to the beginning of October. First picking in 50 days. Yield 200 q/ha. It is recommended for throughout India
Pusa Parvati	Early, bush type; pods green, long with straight tips. Resistant to mosaic and powdery mildew. Suitable for sowing in hills from April to June and in plains from last week of September to first week of October and last week of January to first week of February; first picking in 50 days. Yield 225 q/ha. It is recommended for all over India.
Varieties developed at IIHR, Bangalore	
Arka Anoop	Plants bushy, photo-insensitive. Resistance to both rust and bacterial blight. Pods long (17 to 18 cm), flat and straight. Pod yield 20 t/ha. It is recommended for all over India.

Name of variety/hybrid	Salient features
Arka Komal	Plants erect and bushy, Photo insensitive, Bright green stem and foliage. Flat, green straight pods. Good transportation and cooking qualities. Pod yield 20 t/ha. Seed yield 1500 kg/ha. H.P., J&K, hills of U.P., M.P., Maharashtra, Karnataka, Tamil Nadu and Kerala.
Arka Suvidha	Plants bushy and photo-insensitive. Pods straight, oval, light green, fleshy, stringless and crisp. Pod yield 19 t/ha. It is recommended for H.P., J&K and hills of U.P.
Varieties developed at IIVR, Varanasi	
Kashi Param	This variety has been developed through pure line selection. Plants are determinate (70 cm long) and leaves are dark green. Pod is fleshy, length 14.7cm, round, dark green, pod yield 120-140 q/ha. It is recommended for J&K, H.P., Uttaranchal and M.P.
Varieties developed at G.B.U.A.&T, Pant Nagar	
Pant Anupma	The plant is bushy, dwarf and upright with green foliage. The pods are tender, smooth, round, non-stringy and translucent. The podding habit is such that the pods are fully covered by the leaf canopy and thus get protected against sunlight. The first picking can be had in 55-65 days after sowing. It is moderately resistant to bean common mosaic virus and angular leaf spot. The yield potential is 90 q/ha green pods. It is recommended for H.P., J&K, Hills of U.P., Punjab, Tarai region of U.P., Bihar and Jharkhand.
Pant Bean-2	The green pods are flat round, straight and non-stringy in the early stage. The seed are dark brown. Pant Bean-2 takes about 60 days for first green pod picking and about 130 days for seed maturity. It is moderately resistant to bean common mosaic virus. The green pod yield potential of this variety is 100 q/ha. It is recommended for U.P. and Uttarakhand.
Varieties developed at MPKV, Rahuri	
Phule Suyash	Attractive tender, straight to slightly curved smooth green pods. Tolerance to major pest and disease, suitable for kharif and rabi seasons, average yield 168 q/ha. It is recommended for Maharashtra.
Varieties developed at TANU, Coimbatore	
TKD.1	Pods are less fibrous. Seeds turn white on maturity. Average yield 6 t/ha (Green pods) and 3 t/ha (Dry seeds). It is recommended for TN.
YCD.1	Plants are semi-dwarf and bushy. Pods are slightly flat, long, broad and robust. Average yield 10 t/ha. (Green) and 6 t/ha (Dry seeds). It is recommended for T.N.
Varieties developed at CSAUAT, Kanpur	
Azad Rjmash	Highly attractive, smooth pods, stringh less. Average yield 65-70 q/ha. It is recommended for UP.

Name of variety/hybrid	Salient features
Carrot	
Varieties developed at IARI, New Delhi	
Nantes	Temperate type; small tops with green leaves; roots well shaped, small, orange, perfectly cylindrical, abruptly ending in small thin tail, sweet and self-coloured core with orange flesh. Suitable for sowing from mid-October to early December; maturity 90-100 days. Yield 120 q/ha. It is recommended for all over India.
Pusa Asita	Gives more than 30% higher yield over check Pusa Kesar besides bring self-black coloured and late bolter. Maturity 95-100 days. Av. Yield 300 q/ha. It is recommended for NCR.
Pusa Kesar	Short tops; roots tapering, deep red; core small and self- coloured; rich in carotene; maturity 80-90 days. Yield 300 q/ha. It is recommended for all over India.
Pusa Meghali	It has short tops with smooth roots, orange flesh, self coloured core and stumpy to slightly tapering roots. It is suitable for sowing as early (August-September) as well as late (October-November). It is preferred for early sowing and can be marketed in 110-120 days. Av. Yield 250-280 q/ha. It is recommended for Madhya Pradesh and Maharashtra.
Pusa Nayanjyoti	It is first orange coloured temperate carrot hybrid from Public sector in India. It is rich in beta-carotene (7.55mg/100g). Sowing time: Apr-Aug (Hills); Nov-Dec (North Indian plain). Roots are self-coloured with no core. Average yield 35-45t/ha. It is recommended for NCR & H.P.
Pusa Rudhira	Gives more than 50% higher yield over check. Pusa Kesar, besides, being self-core red coloured with delayed bolting. Maturity 85-90 days. Av. Yield 330 q/ha. It is recommended for NCR.
Pusa Vrishti	It is a new heat tolerant tropical carrot variety. It is suitable for early sowing beginning in July under North Indian plains. Maturity 85-90 days Av. Yield 250 q/ha. It is recommended for NCR.
Pusa Yamdagni	Temperate type; takes 86-130 days from sowing, earlier than Nantes by a week. Roots orange with self-coloured core, slightly tapering or semi-stumpy with medium tops; rich in carotene; maturity 90-100 days. Yield 120 q/ha. It is recommended for all over India.
Varieties developed at HAU, Hisar	
Hisar Gairic	Self-cored, high carotene, high yield, roots longer than Pusa Kesar. Average yield is 300 q/ha. It is recommended for Haryana.
Radish	
Varieties developed at IARI, New Delhi	
Japanese White	Roots pure white, 20-25 cm long, thick, stumpy, cylindrical, smooth, mildly pungent; tops medium with deeply cut leaves; maturity 55-60 days. Yield 375 q/ha. It is recommended for all over India.

Name of variety/hybrid	Salient features
Pusa Chetki	Its roots are medium large and stumpy, pure white, tender and smooth and mildly pungent. It becomes ready for harvesting in 40-50 days after sowing. Being tolerate to high temperatures it is suitable for sowing from the mid of March to the mid of August. It shows profuse seed set in the plains. It can be grown as a summer and a monsoon crop from April-September and has yield potential of 200-250q/ha. It is recommended for throughout India except temperate region.
Pusa Desi	Roots white with green shoulders, 30-35 cm long, medium thick, tapering, pungent; tops medium; leaves dark green. Suitable for sowing from mid-August to September in the northern plains; maturity 50-55 days. Yield 300 q/ha. It is recommended for all over India.
Pusa Himani	Roots white, 30-35 cm long, medium thick, tapering, with light green shoulders; flesh pure white, crisp, mildly pungent; tops short with cut leaves. Suitable for sowing from December to February in the northern plains and throughout the year except winter months in hilly regions; maturity 50-55 days. Yield 325 q/ha. It is recommended for all over India.
Pusa Mridula	Roots are globular with bright red skin, mildly pungent. Av. Yield 130 q/ha, Maturity 25 days. NCR.
Pusa Reshmi	Roots 30-35 cm long, medium thick, white with green shoulder, uniformly smooth, mildly pungent; tops medium, leaves light green. Suitable for mid-September to October sowing in the northern plains; maturity 55-60 days. Yield 350 q/ha. It is recommended for all over India.
Varieties developed at IIHR, Bangalore	
Arka Nishant	Roots long, marble white in colour with crisp texture and mild pungency. Resistant to pithiness, premature bolting, root branching and forking. Yield 35-40 t/ha. It is recommended for all over India.
Varieties developed at IIVR, Varanasi	
Kashi Hans	This has been developed through selection. It is suitable for September to February planting and harvesting can be done after 40-45 days of sowing. It can stand in the field for up to 10-15 days after commercial maturity. Leaves are soft and smooth like spinach. Roots are straight, tapering, 30-35 cm long, 3.5-4.2 cm diameter; yield of 430-450 q/ha. It is recommended for U.P., Punjab, and Jharkhand.
Kashi Sweta	This has been developed through selection from the Chetki materials. It is suitable for early harvesting (30-35 days after sowing). Roots are 25-30 cm long, 3.3-4.0 cm in diameter, straight, tapering with pointed tip; gives yield of 450-470 q/ha. It is recommended for U.P., Punjab, Bihar and Jharkhand.
Varieties developed at HAU, Hisar	
Hisar Sel-1	High yield, roots medium long, smooth white, tender and mildly pungent with white shoulder and short tops. Its root: shoot ratio (1.70: 1) is higher than the check varieties (1.3: 1). Average yield is 300-325 q/ha. It is recommended for Haryana.

Name of variety/hybrid	Salient features
Varieties developed at TANU, Coimbatore	
CO 1	Roots milky white, less fibrous; suitable for growing throughout the year. Average yield 22 t/ha. It is recommended for TN.
Turnip	
Varieties developed at IARI, New Delhi	
Purple Top White Globe	Temperate type; tops dark green, erect with cut leaves; roots large, round, smooth, bright purplish-red in upper half, lower portion creamy white. Flesh white, firm and crisp; suitable for sowing from October to December in plains; maturity 60-65 days. Yield 300 q/ha. It is recommended for all over India.
Pusa Chandrima	Temperate type; early maturing; roots medium to large, nearly globular, medium top, smooth, pure white skin with fine grains, sweet and tender flesh; suitable for October to December sowing. Yield 350 q/ha. It is recommended for all over India.
Pusa Kanchan	Tops light green; roots light red, medium large, round; flesh creamish yellow with good flavour and taste; suitable for sowing from September to mid-October in plains; maturity 55-60 days. Yield 250 q/ha. It is recommended for all over India.
Pusa Swarnima	Temperate type; roots flattish round with golden yellow skin; flesh medium amber colour, finely grained and tender; maturity 60-65 days. Yield 350 q/ha. It is recommended for all over India.
Pusa Sweti	Tropical and sub-tropical type; roots pure white, medium in size, round to flattish round, tender, flesh white; leaves finely lobed and light green; suitable for sowing from August to September in plains; maturity 40-45 days. Yield 250 q/ha. It is recommended for all over India.
Beetroot	
Varieties developed at IARI, New Delhi	
Crimson Globe	Tops medium; leaves bright green with maroon shade; roots globular to flattish globe, medium red skin, flesh medium dark red with indistinct zones, non-corrosive in taste. Yield 180 q/ha. It is recommended for all over India.
Detroit Dark Red	Tops small; leaves glossy, dark green and tinged with maroon veins; roots perfectly round with deep red skin; flesh deep blood red with light red zoning, tender and fine-grained. "Earthy" in taste when taken raw; maturity 80-100 days. Yield 180 q/ha. It is recommended for all over India.
Varieties developed at TANU, Coimbatore	
Ooty.1	Roots are blood red in colour with thin skin and good quality. Average yield 31 t/ha. It is recommended for T.N.

Name of variety/hybrid	Salient features
Bottle gourd	
Varieties developed at IARI, New Delhi	
Pusa Hybrid 3	Fruits green, slightly club shaped without a neck; suitable for easy packing and long distance transportation; first picking in 50-55 days. Yield 425 q/ha in summer and 470 q/ha in *kharif* season. It is recommended for all over India.
Pusa Naveen	It is free from crook-necked fruits. Its fruits are cylindrical, straight and each weighing 550 g. It is suitable for cultivation in spring-summer and rainy seasons. Av. Yield 350-400 q/ha. It is recommended for Maharashtra and M.P.
Pusa Samridhi	Fruits are long without neck. In spring-summer and 308 q/ha in *kharif*, an increase of 19% and 14% respectively over the check Pusa Naveen. Superior in nutritional qualities. Maturity 50-55 days. Av. Yield 272 q/ ha. It is recommended for NCR.
Pusa Sandesh	Vines medium long; fruits attractive green, round, deep oblate, medium sized weighing 600g; first picking in 55- 60 days in *kharif* and 60-65 days in summer. Suitable for both summer and *kharif* seasons. Yield 320 q/ha. It is recommended for NCR.
Pusa Santushti	Pear-shaped fruits. In spring-summer and 290 q/ha in *kharif*, an increase of 22% and 18% over best check Pusa Sandesh. Sets fruits under low temperature (10-12oC) as well as high temp. (35-40oC). Maturity 55-60 days. Av. Yield 260 q/ha. It is recommended for Delhi, Punjab, Uttarkhand, Bihar, M.P. and Maharashtra.
Pusa Summer Prolific Long	Its vines are vigorous in growth and bears long cylindrical fruits. It has yield potential of 250-300 q/ha. It is recommended for throughout India
Varieties developed at IIHR, Bangalore	
Arka Bahar	Fruits medium long and straight without crookneck. Light green shining fruit skin when tender (1 kg). Tolerant to blossom end rot. Yield 40-45 t/ha. It is recommended for M.P. and Maharashtra.
Varieties developed at IIVR, Varanasi	
Kashi Bahar	This is a long fruited hybrid with green vine and vigorous growth, fruit straight, light green, length 30-32 cm and average weight 780-850 g and yield 500-550 q/ha. It is recommended for U.P., Punjab, Bihar and Jharkhand.
Kashi Ganga	This is an early variety derived from the cross IC-92465 x DVBG-151. Fruits are light green, length 30 cm, diameter 7 cm, fruit weight 800-900 g and yield 480-550 q/ha. It is recommended for U.P., Punjab, and Jharkhand.
Varieties developed at AAU, Anand, Gujarat	
Anand Bottle Gourd -1	The fruits are long, tender, attractive, light green in colour, slightly rounded at stem end as well as blossom end with fine cluster. The fruit skin is smooth, non-hairy and has long shelf life. Average yield is 233.97 q/ha. It is recommended for middle Gujarat.

Name of variety/hybrid	Salient features
Varieties developed at GBUA & T, Pant Nagar	
Pant Sankar Lauki-1	The fruits are intermediate sized, long and somewhat cylindrical (about 35 cm long). The fruits are green. Vine length is about 5.5 m. The first picking is possible in about 60 days. It is suitable for planting in the plains as well as in the hills. The yield potential is 400 q/ha. It is recommended for U.P. and Uttarakhand.
Pant Sankar Lauki-2	The fruits are about 40 cm long, club shaped with smooth green colour. The first green fruit harvest is in 65 days. This is suitable for plains and hills both. It can be sown from March to July in plains and April to May in the hills. The seed rated is 6 kg / ha. The yield potential is 400 q/ha. It is recommended for U.P. and Uttarakhand.
Pant Lauki-4	This is the medium duration and high yielding variety of bottle gourd. It has long fruits (40 cm) of light green colour with light strips having hairs. Yield potential is about 300 q/ha. It is recommended for the plains and hill areas of North India.
Varieties developed at NDUA&T, Faizabad	
Bottle Gourd (NDBG-619)	Fruits are attractive with cylindrical shape, early maturity (55 days). Average yield 300 q/ha. It is recommended for U.P., M.P. and Maharashtra.
N. Bottle Gourd Hybrid-4	Fruits are cylindrical, uniform and attractive. Early maturity (50-55 days) and prolific bearer with yield potential of 700q/ha in summer, 1000q/ha in rainy and early winter seasons. It is recommended for U.P., Bihar, Punjab, Rajasthan and Haryana.
N. Dharidar (NDBG-208-1)	Fruits are long bottle shaped with striped green suitable for summer seasons. Average fruit yield 350 q/ha. It is recommended for UP.
N. Jyoti (NDBG-104)	Fruits are long slender and attractive. Early maturity (60 days). Average fruit yield during summer is 400 q/ha and 600 q/ha during rainy seasons. It is recommended for U.P.
N. Madhuri (NDBG-505)	Fruits are attractive with round shape. Highly palatable when cooked. Average yield is 1000 q/ha. It is recommended for U.P.
N. Rashmi (NDBG-1)	Fruits are long, bottle shaped, suitable for cultivation in summer and rainy seasons. Early maturity (60 days), average fruit yield 300-400 q/ha. It is recommended for U.P.
N. Shishir (NDBG-202)	It is a winter type, round, fruited variety and has a peculiar multifid/podate leaf shape. Resistance against anthracnose, downy mildew, powdery mildew and viral disease complex. Average yield is 850 q/ha. It is recommended for U.P.
N. Shivani (NDBG- 403)	It is a winter-type prolific bearer, very long slender fruited is suitable for kitchen garden purpose. Average yield is 1300 q/ha. It is recommended for U.P.

Name of variety/hybrid	Salient features
N.Bottle Gourd (NDBG-132)	Fruits are long slender and shaped, it is an early variety requiring 60 days for first picking in spring/summer crop. Average yield 300 q/ha. It is recommended for U.P., Haryana, Rajasthan and Gujarat.
Varieties developed at MPKY, Rahuri	
Samarat	Fruits are 30-40 cm long, green in colour, cylindrical in shape. Good for box packing, suitable for Kharif and summer seasons. It is recommended for Maharashtra.
Varieties developed at TANU, Coimbatore	
CO 1	Fruits round at the base and bottleneck at the top, light green, each weighs 2 kg. Average yield 36t/ha. It is recommended for TN.
TNAU Bottle gourd Hybrid CO 1	F1 hybrid developed by crossing NDBG 121 x Arka Bahar. Fruits are cylindrical, without crook neck, medium-sized (0.95-1.00 kg) and suitable for nuclear family. Suitable for bower system of cultivation. Average yield 80 t/ha. It is recommended for TN.
Varieties developed at CSAUAT, Kanpur	
Azad Nutan	Long, green. Average yield 350-400 q/ha. It is recommended for Punjab, Tarai region of U.P., Uttarakhand, Bihar and Jharkhand.
Bitter gourd	
Varieties developed at IARI, New Delhi	
Pusa Do Mousami	The fruits reach edible stage in about 55 days from sowing; fruits dark green, long, medium thick, club shaped with 7-8 continuous ridges, 18 cm long at edible stage, 8-10 fruits weigh one kg. It is recommended for Delhi, Punjab, Haryana, U.P. and Bihar.
Pusa Hybrid-1	Fruits medium long, medium thick, glossy-green; suitable for pickling and dehydration. Higher yielder than Pusa Vishesh and Pusa Do Mausami; first picking in 55-60 days. Suitable for growing in spring-summer season. Yield 200 q/ ha. It is recommended for NCR.
Pusa Hybrid-2	Its fruits are medium long and thick, glossy-green, suitable for pickling and dehydration. First picking starts at 55-60 days after seed, sowing. Av. Yield 190-200 q/ha. It is recommended for Punjab, U.P., Bihar, Jharkhand, Chhattisgarh, Orissa , A.P., Rajasthan, Gujarat, Haryana and Delhi.
Pusa Vishesh	It is a dwarf vine variety suitable for planting in high plant density. Fruits are glossy green, medium long and thick, suitable as vegetable for pickling and dehydration. It is recommended for Delhi, Punjab, Haryana, U.P. and Bihar.
Varieties developed at IIHR, Bangalore	
Arka Harit	Fruits spindle shaped, glossy green skin colour without tubercles. Medium-sized fruits, Thick flesh, moderate bitterness, with less seeds. Yield 12 t/ha. It is recommended for Karnataka, Tamil Nadu and Kerala.

Name of variety/hybrid	Salient features
Varieties developed at IIVR, Varanasi	
Kashi Urvasi	This variety has been derived from the cross IC-85650B x IC-44435A, having dark green and long fruits, mild projection, length 16-18 cm, fruit weight 90-110 g and yield 200-220 q/ha. It is recommended for U.P., Punjab, and Jharkhand.
Varieties developed at GBUA & T, Pant Nagar	
Pant Karela-1	The vine length is about 2 m. Fruits are thick, about 15 cm long with tapering ends. It takes about 55 days to first harvest it. It is suitable for planting in the hills. The yield potential is 150 q/ha. It is recommended for U.P. and Uttarakhand.
Varieties developed at MPKY, Rahuri	
Phle Green Gold	Fruits are dark green, 25-30 cm long with prickles, tolerant to downy mildew, suitable for kharif and summer season, average yield 230 q/ha. It is recommended for Maharashtra.
Varieties developed at TANU, Coimbatore	
CO 1	Fruits green, 30 - 35 cm long, 100 - 200 g each. Average yield 14 t/ha. It is recommended for T.N.
COBgoH 1	Fruits creamy white and weighs 300 each. Average yield 44 t/ha. It is recommended for T.N.
MDU.1	Fruits green and each weighs 410g. Average yield 32 t/ha. It is recommended for T.N.
Sponge gourd	
Varieties developed at IARI, New Delhi	
Pusa Chikni	It is suitable for cultivation in rainy and spring summer seasons. Fruits are smooth, dark green, and cylindrical. Single plant bears 15-17 fruits. Flowering starts in 60 days after sowing. Av. yield 200-225 q/ha. It is recommended for all over India.
Pusa Supriya	Its vines are 2.5-2.8 cm long with 4-5 branches. Leaves are dark green, smaller in size with six lobed petiole, tip pointed and simple. Fruits are pale green, smooth, 15-20 cm long, straight and slightly curved at the stem end, pointed distal end, non-hairy, flesh tender, suitable for spring summer and Kharif season. Fruits become ready for picking at 50-55 days after sowing in spring-summer and 44-48 days after sowing in kharif season. Av. Yield 130-140 q/ha. It is recommended for Punjab, U.P., Bihar, Jharkhand, Rajasthan, Gujarat, Haryana and Delhi.
Pusa Sneha	Fruits dark green. Av. yield 120 q/ha, Suitable for cultivation in both spring summer and kharif seasons. Maturity 50-55 days. It is recommended for NCR.
Varieties developed at CSAUAT, Kanpur	
Azad Taroi -1	Early, fruit green, smooth. Average yield 125-150 q/ha. Recommended for U.P.

Name of variety/hybrid	Salient features
Azad Taroi -2	Profuse fruiting, suitable for summer and rainy season. Average yield 140-160 q/ha. It is recommended for Punjab, Tarai region of U.P., Uttarakhand, Bihar and Jharkhand.
Ridge gourd	
Varieties developed at IARI, New Delhi	
Pusa Nasdar	Mid-season; flowering in 50 days; 15-20 fruits per vine, fruits ridged, light green. More suited for rainy season. It is recommended for all over India.
Pusa Nutan	Fruits long (25-30 cm), straight, attractive green, av. fruit wt. 105 g, flesh tender, suitable for spring summer and kharif season. Average yield 160 q/ha, Maturity 45-50 days. It is recommended for NCR.
Varieties developed at HAU, Hisar, Haryana	
Hisar Kalitori	Early long thin and straight fruited variety, tolerant to powdery mildew, suitable for rain fed-areas. Average yield is 80-100 q/ha. It is recommended for Haryana.
Varieties developed at J.A.U., Junagadh, Gujarat	
GJRGH-1	The fruits of GJRGH-1 are long in size with green colour. Suitable for kharif season. Average yield is 113.30 q/ha. It is recommended for middle Gujarat.
Varieties developed at AAU, Anand, Gujarat	
Gujarat Anand Ridge Gourd-1	This variety are medium-sized and elliptical shape with green fruit skin colour. The variety has less mosaic and downy mildew disease reaction. Average yield is 150.70 q/ha. It is recommended for middle Gujarat.
Varieties developed at GBUA & T, Pant Nagar	
Pant Torai-1	The main shoot is 5 m long. Fruits are 15-20 cm long and club shaped. It takes about 65 days to first harvest. It is specifically suitable for rainy season. The yield potential is about 100 qt/ha. It is recommended for U.P. and Uttarakhand.
Varieties developed at TANU, Coimbatore	
CO 1	Fruits long (60-75 cm) light green and weighs 300 g each. Average yield 14 t/ha. It is recommended for T.N.
CO 2	Fruits are very long (1 m) green and fleshy. Average yield 25 t/ha. It is recommended for T.N.
PKM.1	Fruits long, dark green with shallow grooves. Average yield 28 t/ha. It is recommended for T.N.
Cucumber	
Varieties developed at IARI, New Delhi	
Pusa Sanyog	Fruits 28-30 cm long, cylindrical, attractive, dark green with yellow spines, crisp flesh. First picking in 50 days. Yields 200 q/ha. It is recommended for North Indian hills.

Name of variety/hybrid	Salient features
Pusa Uday	The fruits are 28% higher over check (Poinsette). Suitable for cultivation both in spring-summer and rainy season. Maturity 50-55 days. Average yield is 155 q/ha. It is recommended for NCR.
Varieties developed at GBUA & T, Pant Nagar	
Pant Parthenocarpic khira-2	A parthenocarpic cucumber variety. It is suitable for poly house cultivation. No. of female flowers per vine is 551. Average fruit wt. is 630 g and yield potential is 2107 qt/ha. It is recommended for U.P. and Uttarakhand.
Pant Khira-1	The fruits are long (20cm) and cylindrical with light white stripes. The fruits attain first picking stage in 50-60 days. Recommended seed rate is 4 kg/ha. The yield potential is 150 q/ha. It is recommended for U.P. and Uttarakhand.
Pant Sankar Khira-1	The fruits are long, about 20 cm, cylindrical and green with light stripes. Vine length is about 120 cm. It takes 50 days to first picking the yield potential is 200 qt/ha. It is recommended for U.P. and Uttarakhand.
Parthenocarpic khira-3	It is a parthenocarpic cucumber variety. It is suitable for poly house cultivation. No. of female flowers per vine is about 465. Average fruit weight is 415 g and yield potential is 1992 qt/ha. It is recommended for U.P. and Uttarakhand.
Varieties developed at MPKV, Rahuri	
Himangni	Fruits are white in colour and resistant to bronzing. Suitable for kharif season, average yield 180 q/ha. It is recommended for Maharashtra.
Phule Shubangi	Fruits are green in colour. Tolerant to powdery mildew, suitable for kharif season, average yield is 180 q/ha. It is recommended for Maharashtra.
Varieties developed at TANU, Coimbatore	
CO 1	Ripe fruits are bright yellow with greenish yellow intermittent stripes. Average yield 25-28 t/ha. It is recommended for TN.
Long melon	
Varieties developed at IIHR, Bangalore	
Arka Sheetal	Fruits medium-long (length 22cm. and girth 7.8 cm) with light green skin colour covered with soft hair. Each fruit weighing 90-100g at marketable stage. Free from bitter principles with crisp texture. Seeds light tan colour. Yield 35 t/ha. It is recommended for Karnataka, Tamil Nadu and Kerala.
Varieties developed at HAU, Hisar	
Karnal Selection	A prolific bearer, fruits are tender light green in colour, long thin, flesh crisp with good flavour. Average yield is 90-100 q/ha. It is recommended for all over India.

Name of variety/hybrid	Salient features
Varieties developed at GBUA & T, Pant Nagar	
Pant Kakri-1	The vines are vigorous with long light green straight fruits. It is free from common diseases and insects. The first green immature fruit picking is possible in 50 days after sowing. Seed to fruit stage is about 90 days. Yield potential is 300q/ha. It is recommended for U.P. and Uttarakhand.
Muskmelon	
Varieties developed at IARI, New Delhi	
Pusa Madhuras	Fruits are roundish-flat, each weighing about 1 kg, sparsely netted with pale green strips, flesh is salmon-orange, juicy with 12-14% TSS. It has poor keeping quality. First picking starts 90-95 days after sowing. Average yield is 150-170 q/ha. It is recommended for Punjab, U.P., Bihar, Rajasthan, Gujarat, Haryana, Delhi, M.P., Maharashtra, Karnataka, Tamil Nadu and Kerala.
Pusa Sharbati	Fruits round with green stripes on moderately netted skin, weighs 0.9 to 1 kg, good keeping quality; flesh thick, bright orange with narrow seed cavity, cantaloupe flavour, sweet (T.S.S. 10-12%); maturity 80- 85 days. Yield is 160 q/ha. It is recommended for riverbeds and upland conditions of Northern Gangetic plains.
Varieties developed at IIHR, Bangalore	
Arka Jeet	Round but flat at both the ends. Skin colour dark green at early stage of fruit development, which turns into golden orange at full slip stage. Flesh white, medium texture, excellent flavour with high TSS (13-16%). Yield is 14-15 t/ha. It is recommended for Rajasthan, Gujarat, Haryana, Delhi, Karnataka, Tamil Nadu and Kerala.
Arka Rajahans	Fruits medium large, round to slightly oval, creamy white skin, netted with shallow vein tracks at full slip stage. Fruit skin white with green vein tracts when immature. Flesh white with pink seed cavity, crisp texture and mild flavour when ripe. Average fruit weight is 1 kg with 2-3 fruits per plant. Yield is 25 t/ha. It is recommended for Karnataka, Tamil Nadu and Kerala.
Varieties developed at IIVR, Varanasi	
Kashi Madhu	Plants have medium vine and leaves sparsely lobed and dark green, fruits are round, with open prominent green sutures, weight 650-725 g, half-slip in nature, thin rind, smooth and pale yellow at maturity, flesh salmon orange (mango colour), thick, very juicy, T.S.S. 13-14% and seeds are loosely packed in the seed cavity. Post harvest life is long with good transportability, tolerant to powdery and downy mildew, medium maturity and yield 200-270 q/ha. It is recommended for U.P., Punjab and Jharkhand.

Name of variety/hybrid	Salient features
Varieties developed at AAU, Anand, Gujarat	
Gujarat Muskmelon-3	The fruits are oblong shaped and it has soft, attracting green flesh textured with sweet taste and good aroma having 12 % TSS. Average yield is 127.05 q/ha. It is recommended for Punjab, Tarai region of U.P., Bihar, Jharkhand, M.P. and Maharashtra.
Watermelon	
Varieties developed at IARI, New Delhi	
Sugar Baby	It has medium long vines. Fruits are round with a black rind, deep pink flesh with 11-13% TSS, brown small seeds and each weighing 2-5 kg. Average yield is 150-170 q/ha in 85 days of crop duration. It is recommended for Chhattisgarh, Orissa, A.P., M.P., Maharashtra, Karnataka, Tamil Nadu and Kerala.
Asahi Yamato	Mid-season; vines medium with deeply cut, lobed leaves; fruits medium sized, 6-7 kg, round, pale green, smooth, non-striped skin; flesh deep red, sweet, TSS 9-11%; maturity 95-100 days. Yield 200 q/ha. It is recommended for North Indian hills.
Varieties developed at IIHR, Bangalore	
Arka Jyothi	F1 hybrid, round to oval fruits. Light green rind with green to blue-green angular stripes. Crimson flesh with TSS 12-14%. Average fruit weight 5 kg with 2-3 fruits per vine. Yield 60-80 t/ha. It is recommended for Karnataka, Tamil Nadu and Kerala.
Arka Manik	Fruits oval with light green rind colour and green to dark green stripes. Deep crimson flesh with granular texture, pleasant aroma, very sweet TSS 12-15%). Average fruit weight 6 kg. Seeds small with dark brown specks on the testa. Triple resistant to powdery mildew, Downey mildew and anthracnose. Yield 60 t/ha. It is recommended for Punjab, Tarai region of U.P., Bihar, Jharkhand, M.P., Maharashtra, Karnataka, Tamil Nadu and Kerala.
Varieties developed at TANU, Coimbatore	
	PKM.1 Suitable for arid, semi-arid, irrigated conditions; pulp pink in colour; high. Average yield 37 t/ha. It is recommended for TN.
Roundmelon	
Varieties developed at IIHR, Bangalore	
Arka Tinda	Fruits round with lush green shining skin covered with soft hair when young and tender (60 g), Yield 10 t/ha. It is recommended for Karnataka, Tamil Nadu and Kerala.
Varieties developed at HAU, Hisar	
Hisar Tinda (HT-10)	Early and high yielding, fruits round-medium sized, tender an tolerant to downy mildew and root rot wilt. Average yield 75-100 q/ha. It is recommended for Haryana.

Name of variety/hybrid	Salient features
Snapmelon	
Varieties developed at IARI, New Delhi	
Pusa Shandar	Fruits oblong, medium in size with creamy white skin, thick and light pink flesh. Average yield 380 q/ha, an increase of 32% over best local check. Maturity 45-50 days. It is recommended for NCR.
Pumpkin	
Varieties developed at IARI, New Delhi	
Pusa Hybrid 1	Fruits flattish round, medium sized, weighing 4.75 kg; flesh golden yellow. Suitable for both seasons; 25% higher yield than Pusa Vishwas. Yield 520 q/ha. It is recommended for NCR.
Pusa Vikas	Vines semi-dwarf to dwarf (2.0-2.5m long); leaves soft with light green or yellow spots; fruits small weighing 2 kg, flattish round; flesh yellow, rich in vitamin-A. Yield 300 q/ha. It is recommended for NCR.
Pusa Vishwas	Vigorous vegetative growth, dark green leaves with white spots including veins. Fruits are brown, spherical, each weighing 4-5 kg with thick and golden yellow flesh. Average yield 400-425 q/ha in 120 days of crop duration. It is recommended for J&K, H.P., Uttaranchal, Punjab, U.P., Bihar, Chhattisgarh, Orissa, A.P., M.P. and Maharashtra.
Varieties developed at IIHR, Bangalore	
Arka Chandan	Fruits round with pressed blossom end. Rind colour green with white patches when immature which turns to light brown colour with self-coloured patches on maturity. Thick orange flesh, solid cavity. Fruit weight 2- 3 kg. Rich in carotene (3333 IU). Yield 33 t/ha. It is recommended for Karnataka, Tamil Nadu and Kerala.
Arka Suryamukhi	Fruits round with pressed ends. Rind colour orange red with white stripes when mature. Fruit stalk round with flared ends. Average fruit weight 1 kg. Resistant to fruit fly. Yield 34 t/ha. It is recommended for Karnataka, Tamil Nadu and Kerala.
Varieties developed at IIVR, Varanasi	
Kashi Harit	Short vines and dark green leaves with white spots. Fruits are green, spherical, weight 2.5-3.0 kg at green stage. Yield potential of 300-350 q/ha in 65 days of crop duration. It is recommended for U.P., Bihar, Chhattisgarh and Uttarakhand.
Varieties developed at AAU, Anand, Gujarat	
Anand Pumpkin-1	The fruits are medium sized, globular in shape with deep yellow flesh colour. The fruits of this variety possess higher total soluble solids, carotene, total soluble sugars and protein. Average yield is 24.4 t/ha. It is recommended for Gujarat.
Varieties developed at NDUA&T, Faizabad	
N. Abhooshan	Fruits are in round shape with dark green striped and highly attractive. Average fruit yield 700 q/ha. It is recommended for UP.

Name of variety/hybrid	Salient features
N. Agrim	Fruits are small, round and stripeless dark green. Suitable for summer crop. Early maturity (55 days), average yield 300-400 q/ha. It is recommended for U.P., Delhi, Punjab, Haryana and Gujarat.
N. Amrit	Fruits are round in shape and light green mottled in colour. Suitable for summer crop. Mature fruit weight 5-6 kg. It is recommended for U.P.
Varieties developed at TANU, Coimbatore	
CO1	Fruits are globular, large-sized (8-10 kg); light pink skin and yellow flesh. Average yield 30 t/ha. It is recommended for T.N.
CO2	Fruits small (1.5 kg), flattish round, light brown skin and yellow flesh. Average yield 23 t/ha. It is recommended for T.N
Varieties developed at CSAUAT, Kanpur	
Azad Pumpkin-1	Fruit green medium broken white pattern, spherical. Average yield 450-500 q/ha. It is recommended for Punjab, Tarai region of U.P., Uttarakhand, Bihar and Jharkhand.
Ash gourd	
Varieties developed at IARI, New Delhi	
Pusa Ujjwal	Fruits are oblong, ellipsoid, rind greenish white while flesh is white with average fruit weight of -7.0 kg. Its fruits are ideal for long distance transportation. Average yield 480-500 q/ha (kharif season) and 410-425 q/ha (summer season). It is recommended for Karnataka, Tamil Nadu and Kerala.
Varieties developed at IIVR, Varanasi	
Kashi Ujawal	Fruits are globular, each weighing 10-12 kg and less seeded. Yield potential of 600-610 q/ha in 130-140 days of crop duration. It is recommended for Punjab, U.P., Bihar, Jharkhand, Karnataka, Tamil Nadu and Kerala.
Varieties developed at TANU, Coimbatore	
CO 1	Fruits globular, green, large, oblong - oval in shape, ash coated, weighing 6-8 kg. Average yield 25 t/ha It is recommended for T.N.
CO2	Selection. Small fruited, weighing 2-3 kg. Fruits are oblong or cylindrical and compressed on both sides. Average yield 20 t/ha. It is recommended for T.N.
TNAU Ash gourd Hybrid CO 1	F1 hybrid between PAG 3 x CO 2. Fruits are oblong and medium-sized. Average yield 90 t/ha. It is recommended for T.N.
Summer squash	
Varieties developed at IARI, New Delhi	
Australian Green	Very early, bush type; fruits dark green with longitudinal stripes of greenish white colour all over, 25-30 cm long, very tender and delicious. Suitable for late January- February sowings in the northern plains. It is recommended for all over India.

Name of variety/hybrid	Salient features
Pusa Alankar	Very early and high yielding F1 hybrid; fruits dark green with light coloured stripes, 20-25 cm long, slightly tapering towards the stem end, tender with delicious flesh. It is recommended for all over India.
Varieties developed at IIHR, Bangalore	
Patty Pan	An introduction from USA. White, disc shaped fruits with notches along the margin. Fruit weight 350-400g. Duration 75-80 days. Yield 55 t/ha. It is recommended for Karnataka, Tamil Nadu and Kerala.
Onion	
Varieties developed at IARI, New Delhi	
Pusa Madhvi	Plants are tall, bulbs are light red, flattish round with TSS 11-13%. Bulbs mature in 130-145 days. Average yield 300-325 q/ha in 135 days of crop duration. It is recommended for Madhya Pradesh and Maharashtra.
Pusa Ratnar	Plants are 30 cm tall. Dark green leaves with waxy bloom. Bulbs are bronze deep red, more exposed (above the ground at maturity), obviate to flat-globular, less pungent with 12% TSS and drooping neck with good storage quality. Av. Yield 325-350 q/ha in 150 days of crop duration. It is recommended for all over India.
Pusa Red	Plants are 55-65 cm tall with 6-9 leaves per plant. Bulbs are bronze red in colour, flat globular in shape with 12- 13% TSS content, weight 70-90 g, measure 5-6 cm across, less pungent and quite firm with good keeping quality. Average yield 250-300 q/ha in 125-140 days of crop duration. It is recommended for Punjab, U.P. and Bihar, Rajasthan, Gujarat, Haryana, Delhi, Karnataka, Tamil Nadu and Kerala.
Pusa White Round	Bulbs medium in size, roundish with attractive white colour, TSS 12-14%; very good keeping quality; maturity 125-130 days after transplanting. It is recommended for all over India.
Pusa White Flat	Bulbs are medium-large, flattish round with attractive white colour, 12-14% TSS and good keeping quality. This variety has yield potential of 300-325 q/ha 120-130 days of crop duration. It is recommended for all over India.
Sel. 126	Compact, flat globe, and brownish coloured bulbs. Average equatorial diameter of bulbs ranges from 4.5 – 6.0 cm, polar diameter ranges from 4.8-6.5 cm and single bulb weight ranges from 70.0 – 100.0 g. The variety is with high TSS, average around 17° ± 2 Brix. The variety is suitable for storage, drying, processing and export. Average yield is 25 t/ha. It is recommended for Delhi, UP, Haryana, Bihar, Punjab, Rajasthan, Gujarat, MP, Chhattisgarh and Orissa.

Name of variety/hybrid	Salient features
Varieties developed at IIHR, Bangalore	
Arka Kalyan	Flat Globe shaped bulbs with medium large size. Deep red coloured outer scales and fleshy succulent internal scales. Average bulb weight 130-180g. Pungent with TSS 11-13%. Moderately resistant to purple blotch caused by *Alternaria porii*. Yield 45 t/ha. It is recommended for Punjab, Tarai region of UP, Bihar, Jharkhand, Chhattisgarh, Orissa, A.P., Rajasthan, Gujarat, Haryana, Delhi, Karnataka, Tamil Nadu and Kerala.
Arka Niketan	Globe shaped bulbs with thin neck. Medium large size bulbs of attractive red coloured outer scales, inner scales arranged in tight concentric rings. Average bulbs weight 100-180 g. Bulbs very firm, highly pungent with TSS 12-14%. Excellent keeping quality (3-4 months). Average yield 37 t/ha. It is recommended for Punjab, Tarai region of U.P., Bihar, Jharkhand, Rajasthan, Gujarat, Haryana, Delhi, Karnataka, Tamil Nadu and Kerala.
Varieties developed at NHRDF, Nasik, Maharashtra	
Agrifound Dark Red	Kharif onion grown at Nasik. Bulbs are dark red in colour, globular in shape with tight skin with moderately pungent. Bulb diameter is 4-6 cm. Total soluble solid 12-13 % and Pyruvic acid 10.07 micro mole/g. It is also suitable for bulblets production for early Kharif crop. Mature in 95-110 days after transplanting. Average yield 300 q/ha. It is recommended for all over India.
Agrifound Light Red	Rabi onion grown in Dindori area of Nasik. Bulbs are globular in shape with tight skin, light red colour and 4 - 6 cm in diameter. Total soluble solid 13-14% and Pyruvic acid 12.20 micro mole/g. Mature in 110-120 days after transplanting and keeping quality good. Average yield 300 325 q/ha. It is recommended for all over India.
Agrifound Rose	Bulbs are globular round in shape, deep scarlet red in colour and 2.5-3.5 cm in size. Mature in 95-110 days from seed sowing. Total soluble solid is 16-18% and Pyruvic acid 10.27 micro mol/g. Average yield 190-200 q/ha. It is recommended for Andhra Pradesh and Karnataka.
Agrifound White	White onion grown during Rabi season in Nimar area of Madhya Pradesh. Bulbs are globular in shape, tight skin, silvery attractive white colour and 4-6 cm in diameter. Total soluble solid more than 15% and pyruvic acid 12 micro mol/g. Mature in 110-130 days after transplanting and keeping quality good. Average yield 250-300 q/ha and suitable for late Kharif and Rabi season. It is recommended for Chhattisgarh, Orissa, A.P., Rajasthan, Gujarat, Haryana, Delhi, M.P. and Maharashtra.

Name of variety/hybrid	Salient features
NHRDF Red-2 (L-355)	Plant straight, erect leaves, light red bulbs colour, globular round shape, thin neck and 4.0-60. cm in diameter. Total soluble solid 13-14% and pyruvic acid 12.0 micro mole/g. Mature in 100-120 days after transplanting and keeping quality good. Moderately resistant to foliar diseases like stemphylium blight. Average yield 300-350 q/ha. It is recommended for North East and Andaman & Nicobar Islands. Punjab, Tarai region of U.P., Bihar, Jharkhand, Rajasthan, Gujarat, Haryana and Delhi.
NHRDF-Red (L-28)	Bulbs are attractive very dark red in colour and 4.5-6 cm diameter. Total soluble solid is 13-14%. Variety is very popular among farmers because of its higher yield, attractive dark red colour of bulbs and better storage performance. Mature in 110-120 days after transplanting. Average yield 250-300 q/ha. It is recommended for Northern, Central and Western India.
Agrifound Red	Average size of cluster is 7.15 cm with weight of 65 – 67 g. Average numbers of bulblets per cluster 5.79. Size of bulblets is 3.64 cm, and average weight of single bulblets is 8.85 g. Colour of bulblets is light red. Total soluble solid is 17-19% and Pyruvic acid 10.13 micro mole/g. Mature in 65-67 days after planting. Average yield is 180 - 200 q/ha. It is recommended for Karnataka, Tamil Nadu and Kerala.
Varieties developed at PAU, Ludhiana, Punjab	
Punjab Red Round	Bulbs are shining red, medium to large in size and round in shape. Early maturing and high yielding variety. Keeping quality medium. Total soluble solid 12-14%. Average yield is 200-250 q/ha. Suitable for rabi season. It is recommended for Punjab, Tarai region of U.P., Bihar and Jharkhand.
Punjab Selection	Bulbs are red, globular in shape, 5-6 cm in diameter and quite firm. Bulbs mature 105-120 days after transplanting. Total soluble solid 14%. Keeping quality good. Average yield 200 q/ha. Suitable for rabi season. It is recommended for Punjab, Tarai region of U.P., Bihar, Jharkhand, M.P., Maharashtra, Karnataka, Tamil Nadu and Kerala.
Punjab-48	Bulbs are white and flattish round. Bulbs matures 120-130 days. Keeping quality medium. Total soluble solid 11-12%. Average yield 300 q/ha. Suitable for rabi season. It is recommended for Punjab, Tarai region of U.P., Bihar and Jharkhand.
Varieties developed at CSA, Kanpur	
Kalyanpur Red Round	Bulbs are bronze red, globular in shape. Bulbs mature 105-120 days after transplanting. Keeping quality good. Total soluble solid 13-14%. Average yield 250-300 q/ha. Suitable for rabi season. It is recommended for Punjab, Tarai region of U.P., Bihar and Jharkhand.

Name of variety/hybrid	Salient features
Varieties developed at RAU, Udaipur	
Udaipur-102	Bulbs are white, round to flat in shape and 4.5-6.5 cm in size. Bulbs mature 120 days after transplanting. Keeping quality is medium. Total soluble solid 12%. Average yield 300-500 q/ha. Suitable for rabi season. It is recommended for Rajasthan, Gujarat, Haryana and Delhi.
Udaipur-103	Bulbs are red but lighter than Udaipur-101, oblate globular shape. Bulbs mature 100-110 days after transplanting. Keeping quality is medium. Total soluble solid 12-14%. Average yield is 200-300 q/ha. Suitable for rabi season. It is recommended for Rajasthan, Gujarat, Haryana and Delhi.
Varieties developed at Department of Agriculture, Maharashtra	
N-53	Bulbs are red, flattish round in shape, medium to large and mildly pungent. Bulbs mature 90-100 days after transplanting. Keeping quality is poor. Total soluble solid 11-12 %. Average yield 250 q/ha. Suitable for rabi seasons. It is recommended for all over India
Varieties developed at MPKV, Rahuri Maharashtra	
Baswant-780	Bulbs are crimson red, globular in shape, mildly pungent, less bolting and twins. It is recommended for Rajasthan, Gujarat, Haryana and Delhi.
Phule Safed	Bulbs are white and globular in shape. Bulbs mature 120-130 days after transplanting. Total soluble solid 13%. Suitable for dehydration. Average yield 250-300 q/ha. Suitable for rabi season. It is recommended for M.P. and Maharashtra.
Phule Smrat	Bulbs dark red, globular is shape, thin neck. Bulbs mature 100-110 days after transplanting. Keeping quality is medium. Average yield 200-250 q/ha. Suitable for kharif and early rabi. It is recommended for M.P. and Maharashtra.
Phule Survarna	Bulbs are yellow, medium to big, less pungent. Keeping quality is good. Total soluble solid 11.5%. Average yield 240 q/ha. Suitable for late kharif and rabi season. It is recommended for M.P. and Maharashtra.
Varieties developed at TNAU, Coimbatore, Tamil Nadu (Multiplier)	
Co-1	Red colour medium-sized bulblets, 7-8 per plant. Average weight of bulblets in 55-60 g/clump. Bulbs mature 90 days after sowing. Fairly pungent with medium total soluble solid. Average yield 100 q/ha. Suitable for kharif and rabi seasons. It is recommended for Karnataka, Tamil Nadu and Kerala.
Co-2	Bulblets are crimson colour, moderately big in size, 7-9 bulblets/plant. Average weight of bulblets is 60-65g per clump. Bulblets mature 65 days after sowing. Pungent with high total soluble solid. Keeping quality is good. Average yield 120 q/ha. Moderately resistant to thrips and Alternaria blight. Suitable for kharif and rabi seasons. It is recommended for Karnataka, Tamil Nadu and Kerala.

Name of variety/hybrid	Salient features
Co-3	Bulblets are pink in colour, bigger than Co-2 and 8-10 bulblets/plant. Average weight is 75 g/clump. Bulblets matures is 65 days after sowing. Keeping quality is good. Total soluble solid 13%. Average yield is 160 q/ha. Moderately resistant to thrips. Suitable for kharif and rabi seasons. It is recommended for Karnataka, Tamil Nadu and Kerala.
Co-4	Bulbs are light brown, 8-10 bulblets/plant. Average weight is 90 g/clump. Bulbs mature 60-65 days after sowing. Average yield 180 q/ha. Moderately resistant to thrips. Suitable for kharif and rabi seasons. It is recommended for Karnataka, Tamil Nadu and Kerala.
Co-5	Bulbs are attractive pink high yielding with bold bulblets. It is free flowering type with seed setting ability and can be propagated through seeds as well as bulblets. Average yield 189 q/ha. Suitable for kharif and rabi seasons. It is recommended for Karnataka, Tamil Nadu and Kerala.
MDU-1	Bulbs are bright red, uniform, big size and 10-11 bulblets/plant. Average weight of bulblets is 75 g/clump. Bulblets mature 60-75 days after sowing. Keeping quality is good. Average yield is 150 q/ha. Tolerant to lodging due to thick erect leaves. Suitable for kharif and rabi seasons. It is recommended for Karnataka, Tamil Nadu and Kerala.
Varieties developed at HAU, Hisar	
Hisar-2	Bulbs bronze-red in colour, flattish globular in shape, more pungent, better shelf life. Average yield is 200- 250 q/ha. It is recommended for all over India.
Hisar Onion-3	Bulbs globular shaped, tight skinned with thin neck, high yield, less bolting, good storage life. Average yield is 320-350 q/ha. It is recommended for Haryana.
Varieties developed at DOG Research, Rajguru Nagar, Pune	
Bhima Kiran	Bulb attains immediate light red colour after harvest. Bulbs are oval to round in shape, have very less number of bolters and doubles in rabi (less than 5%), thin neck, TSS 12%, good keeping quality suitable for storage up to 5-6 months. Matures in 130 days after transplanting. The average marketable yield potential is up to 41.5 t/ha during rabi season with national average yields of 30.3 t/ha. It is recommended for Delhi, UP, Haryana, Bihar, Punjab ,Maharashtra, Karnataka and AP.
Bhima Raj	Dark red and oval shaped bulbs with single centre and thin neck. The TSS range is from 10.0 to 11.0%. Fruit matures within 120-125 days after transplanting with absolutely no bolters with average yield ranges from 25-30 t/ha with high percentage of marketable bulbs during rabi. It can be stored up to 3 months during rabi season. The highest yield potential is 40-45 t/ha. It is recommended for Rajasthan, Gujarat, Haryana and Delhi.

Name of variety/hybrid	Salient features
Bhima Red	Bulbs are attractive red in colour with round shape. TSS ranges from 10.0 to 11.0%. Bolters are less than 5% during late kharif and almost no bolters during rabi. It can be stored up to 4 months during late kharif season. This variety can be grown during rabi season also for immediate marketing as it can be stored up to 3 months during rabi. It matures in about 115-120 days after transplanting during late kharif and rabi season. Average yield in late kharif season range from 48.0- 52.0 t/ha. It is recommended for Maharashtra, M.P, Karnataka, and Gujarat.
Bhima Shakti	Suitable for late kharif as well as rabi season. Bulbs attain immediate attractive red colour after harvest. Shape is round. Very less number of bolters 2.15% & 0.36% and doubles 5.69% & 1.38% during late kharif and rabi season, respectively. TSS 11.8%. Small to medium neck thickness, uniform neck fall during rabi and more than 70% neck fall during late kharif. Bulbs have very good storage life up to 5-6 months. Bulbs mature in 130 days after transplanting during late kharif and rabi season. It is tolerant to thrips. Marketable yield during late kharif is 45.9 t/ha and during rabi 42.7 t/ha in Maharashtra with national average yield of 29.18t/ha during rabi season. It is recommended for Delhi, UP, Haryana, Bihar, Punjab, Rajasthan, Gujarat, MP, Chhattisgarh, Orissa, Maharashtra, Karnataka and Andhra Pradesh.
Bhima Shubra	Bulbs are attractive white in colour, oval to round in shape bulb, have less number of bolters up to 6% during late kharif and no bolters during kharif and doubles less than 3% in late kharif and kharif season, thin neck, TSS 10.4% in kharif and 11.7% in late kharif, matures in 112 day after transplanting during kharif and 125 days in late kharif. Bulbs can be stored for 2-3 months during late kharif. It has capacity to tolerate environmental fluctuation, hence can be cultivated in all the three seasons. Average marketable yield during kharif is 24.16 t/ha and during late kharif 38.9 t/ha can fill the gap for processing from October to February. It is recommended for Maharashtra.
Bhima Shweta	Bulbs are attractive white in colour, round in shape, very less number of bolters and doubles less than 3% in rabi season, thin neck, TSS 11.5%, matures in 110-115 day after transplanting during rabi, medium in keeping quality up to 3 months during rabi season. Average marketable yield during rabi season is 35.9 t/ha with national average yield of 28.2 t/ha. It is tolerant to thrips. It is recommended for Delhi, UP, Haryana, Bihar, Punjab, MP, Chhattisgarh, Orissa, Maharashtra, Karnataka and Andhra Pradesh.
Bhima Super	Bhima Super is suitable for kharif season and can be cultivated during late kharif. The variety possesses medium red coloured round bulbs with tapering towards neck. The TSS is from 10-11%. Bulbs attain maturity after 100-110 days after transplanting during kharif season. The variety produces quality bulbs with maximum number of single centered bulbs with an average yield of 26-28 t/ha in kharif and 40 to 45 t/ha in late kharif cultivation. It is recommended for Maharashtra, Karnataka, and Gujarat.

Name of variety/hybrid	Salient features
Varieties developed at MPKV, Rahuri	
Baswant 780	Bulbs are globular in shape with crimson red colour, suitable for kharif season, average yield 250-300 q/ha. It is recommended for Maharashtra.
N-2-4-1	Bulbs are globular in shape with brick red colour. Suitable for rabi season, average yield 350 q/ha. It is recommended for Maharashtra.
Phule Samarth	Early maturity, thin bulb neck, less twins, lusterious dark red in colour, globular bulbs. Suitable for kharif and late kharif season, average yield 280-400 q/ha. It is recommended for Maharashtra, T.N., Kerala, Rajasthan, Gujarat, Delhi, Haryana.
Varieties developed at TANU, Coimbatore	
CO 2	Bulbs bigger (10 - 15 g) and crimson in colour; photo-insensitive type. Average yield 12 t/ha. It is recommended for T.N.
CO 3	Bulbs light pink, bold (10-15 g) bulbs store well for 120 days. Average yield 16 t/ha. T.N.
CO 4	Bulbs light pink, bold (12-15 g) with better consumers' appeal; bulbs store well for 150 days. Average yield 19t/ha. It is recommended for T.N.
CO(On) 5(Seed propagated)	Bulbs reddish pink, bold, 80-90 g/cluster. Average yield 300 kg/ha. (Seed) and 19 t/ha. (Bulb) It is recommended for T.N.
Palak	
Varieties developed at IARI, New Delhi	
Pusa Bharati	Leaves smooth, tender, green, without any red pigmentation; higher vitamin B and C content than Pusa Jyoti; leaves ready for first picking 30-40 days; recommended for cultivation in winter and early-spring in north India. Yield 500 q/ha. It is recommended for all over India.
Pusa Harit	Plants erect; leaves upright, thick, dark green, slightly wavy margins, short petiole; quick rejuvenation after cuttings; bolting very late. Under hill conditions, if grown during summer, continues to give leaf cuttings throughout the year. Yield 260 q/ha. It is recommended for all over India.
Pusa Jyoti	Leaves thick, dark green, fleshy, puckered, giant size, wavy margins; late bolter; rich in vitamins and minerals. Suitable for September sowing. Yield 425 q/ha. It is recommended for all over India.
Varieties developed at TANU, Coimbatore	
Ooty.1	Higher carotene in content. Rich in vitamins. It is recommended for T.N.
Amaranthus	
Varieties developed at IARI, New Delhi	
Pusa Kiran	Sutable for sowing in rainy season. Leaves and stem glossy green; first harvest after 25-30 days. Yield 350 q/ha. It is recommended for all over India.

Name of variety/hybrid	Salient features
Pusa Kirti	Suitable for sowing in summer season. Leaves green, broad and ovate; stem green and tender; first harvest after 30-35 days. Yield 500 q/ha. It is recommended for all over India.
Pusa Lal chaulai	Upper surface of leaves deep red or magenta and lower surface purplish-red; stem deep red; first harvesting after 35 days in spring summer and 25 days in kharif seasons. Yield 450 q/ha in summer and 400 q/ha in kharif season. It is recommended for all over India.
Varieties developed at TANU, Coimbatore	
CO 1	Can be grown throughout the year (25 days - Mulaikeerai) (50-60 days for Thandukeerai). Average yield 7 t/ha. It is recommended for T.N.
CO 3	Plants dwarf; clipping type. Average yield 31 t/ha. It is recommended for T.N.
CO 4	A grain amaranthus type, seeds are rich in protein (19.55 %) and amino acid like lysine (7.5 mg per 100 g) can be substituted for minor millet like Ragi and Thenai. Average yield 3 t/ha. It is recommended for T.N.
CO 5	Single plant selection from A-166-1. Double coloured type with dual purpose. Suitable for container cultivation. Average yield 30 t/ha. It is recommended for T.N.
Fenugreek	
Varieties developed at IARI, New Delhi	
Pusa Early Bunching	Early, upright branching; leaves green. Suitable for sowing in October-November. It is recommended for all over India.
Pusa Kasuri	Late; leaves green rosette type; flowers yellow; pod sickle shaped. Suitable for October-November sowing. It is recommended for all over India.
Varieties developed at HAU, Hisar	
Hisar Madhavi	Plant erect, green stem, foliage dark green, resistant to powdery mildew and tolerant to downy mildew. Average yield is 20-25 q/ha. It is recommended for Haryana.
Hisar Manohar	Foliage dark green, seed green in colour resistant to downy and powdery mildew diseases. Average yield is 20-22 q/ha. It is recommended for Haryana.
Hisar Mukta	Early, light green foliage, resistant to downy mildew, seed bold brown green, resistant to downy mildew. Average yield is 23-26 q/ha. It is recommended for all over India.
Hisar Sonali	Quick growing, dual purpose and high yielding. Average yield is 20-25 q/ha. It is recommended for all over India.
HM-219	High yielding, resistance to powdery mildew. Average yield is 20-22 q/ha. It is recommended for all over India.

Name of variety/hybrid	Salient features
Vegetable mustard	
Varieties developed at IARI, New Delhi	
Pusa Sag-1	Leaves large, broad, glabrous, smooth and attractive green, higher in carotene and ascorbic acid. Maturity 35 days. Average yield 700 q/ha. It is recommended for NCR.
Chenopodium (Bathua)	
Varieties developed at IARI, New Delhi	
Pusa Bathua-1	Plants grow up to a height of 2.25 m, leaves and stems purplish green, leaves tender, first picking in 45 days after planting. Yield 300 q/ha. It is recommended for NCR.
Lettuce	
Varieties developed at IARI, New Delhi	
Great Lakes	Heading type, large firm heads; leaves dark green, blistered; resistant to tip burn. It is recommended for all over India.
Chinese Yellow	Early, loose leaf variety; leaves light green, crisp, tender. It is recommended for all over India.
Parsley	
Varieties developed at IARI, New Delhi	
Moss Curled	Plants dwarf, compact; leaves dark green, exceedingly fine-cut, serrated and deeply curled. Excellent for garnishing and decorative purposes. It is recommended for all over India.
Okra	
Varieties developed at IARI, New Delhi	
Perkins Long Green	Plants very tall; leaves thick, with rough surface; fruits bright green, along with eight edges. Suitable for hilly areas only. It is recommended for all over India.
Pusa A 4	Tolerant to yellow vein mosaic virus, tolerant to aphids and asides, fruits dark green, 12-15 cm long; first picking in 45 day. Yield 140 q/ha. It is recommended for all over India.
Pusa Sawani	Fruits are smooth with 5 ridges, dark green and 18-20 cm long when fully developed. It is distinguished by the presence of a purple patch at the base of the yellow petal on both the sides. It is free from the YVMV under field conditions and is suitable for plains of northern India in virus-free period (spring-summer season). In hills, it can be sown form April to May depending on the altitude. Average yield 150-175 q/ha. It is recommended for all over India.
Varieties developed at IIHR, Bangalore	
Arka Abhay	Fruits lush green, tender and long. Fruits free from spines having delicate aroma. Good keeping and cooking qualities. Field tolerant to yellow vein mosaic virus (YVMV). Yield 18 t/ha. West Bengal and Assam.

Name of variety/hybrid	Salient features
Arka Anamika	Fruits lush green, tender and long. Fruits free from spines having 5-6 ridges, delicate aroma. Good keeping and cooking qualities. Field tolerant to YVMV. Yield 20 t/ha. It is recommended for all regions.
Varieties developed at IIVR, Varanasi	
Kashi Kranti	Medium tall in growth having 2-3 branches. First flower appears after 39-41 days after sowing on 5th nodes during rainy season and 4th node during summer season. Average fruits/plant are 21-23 having dark green in colour and 10-11 cm in length. Yield potential is 140-150 q/ha. It is recommended for Uttar Pradesh, Bihar, Jharkhand and Punjab.
Kashi Pragati	Plants are tall, height 130-175 cm, with 1-2 effective branches. First flower appears after 36-38 days after sowing on 4th nodes during rainy season and 3rd node during summer season. Fruits are 8-10 cm in length at marketable stage, 23-25 per plants and yield 180-190 q/ha during rainy and 130-140 q/ha during summer season. It is recommended for Chhattisgarh, Orissa and A.P.
Kashi Vibhuti	Dwarf growth, 60-70 cm plant height during rainy and 45-50 cm during summer seasons. It bears 2-3 branches with short inter-nodal length. Flowering starts on 4th to 5th nodes after 38-40 days after sowing. A plant bears 18-22 fruits with 8-10 cm length at marketable stages; yield 170-180 q/ha. It is recommended for Rajasthan, Gujarat, Haryana and Delhi.
Varieties developed at HAU, Hisar	
HBH-142 (F1 hybrid)	High yielding hybrid, suitable for rainy season, fruiting starts on third/fourth node, resistant to yellow vein mosaic disease. Average yield is 135-150 q/ha. It is recommended for all over India.
Hisar Naveen	High yielding variety, tolerant to YVM virus disease and suitable for rainy season. Average yield is 110-120 q/ha. It is recommended for all over India.
Hisar Unnat	Suitable for rainy season, fruiting starts on third/fourth node, resistant to yellow vein mosaic disease. Average yield is 100-110 q/ha. It is recommended for all over India.
Varsha Uphar	Plant medium tall, suitable for rainy season, fruiting starts on third/fourth node, resistant to yellow vein mosaic disease. Average Yield is 90-100 q/ha. It is recommended for all over India.
Varieties developed at AAU, Anand, Gujarat	
Gujarat Anand Okra-5	The fruits of this variety are dark green colour with attractive shape and size. The variety has less Y V M V and low population of sucking pest. Average yield is 141.11 q/ha. It is recommended for middle Gujarat.
Varieties developed at MPKV, Rahuri	
Phule Utkarsha	Fruits are green, straight, slender and shiny. Suitable for kharif and summer seasons, average yield 231 q/ha. It is recommended for Maharashtra.

Name of variety/hybrid	Salient features
Varieties developed at TANU, Coimbatore	
CO 1	Plants medium tall; pods pink, long, tender and less fibrous; tolerant to yellow vein mosaic. Average yield 14 t/ha. It is recommended for T.N.
CO 2	Fruits fairly long with hairy surface. Average yield 17 t/ha. It is recommended for T.N.
CO 3	Fruits medium-sized and dark green. Moderately resistant to yellow vein mosaic disease. Average yield 18 t/ha. It is recommended for T.N.
COBhH 1	F1 hybrid developed by crossing Varsha Uphar x PA 4(T) Dark green long fruits, slender, less fibrous and sparsely pubescent. Resistant to YVMV disease. Average yield 23 t/ha. It is recommended for TN.
Varieties developed at CSAUAT, Kanpur	
Azad Bhindi -1	Resistant to YMV, Early fruiting. Average yield 100-125 q/ha. It is recommended for U.P.
Azad Bhindi-3	Red fruit suitable for both crop less susceptible to YMV. Average yield 125-150 q/ha. U.P.
Garlic	
Varieties developed at NHRDF, Nasik	
Agrifound Parvati-2 (G-408)	Bulbs are bigger in size 5.0 - 6.0 cm in diameter, creamy white colour, diameter of cloves 1.5 - 1.6 cm, 1 - 14 cloves/bulb. Total soluble solid, dry matter and pyruvic acid content is 36 - 37%, 37 - 38% and 22 – 23 micro moles/g respectively. Average yield is 175 – 200 q/ha. It is recommended for Himachal Pradesh, Uttarakhand, J&K and Sikkim.
Agrifound Parvati (G-313)	Bulbs are of bigger in size 5.0 - 6.5 cm in diameter, creamy white colour with pinkish tinge, diameter of cloves 1.5 - 1.8 cm, 10 - 16 cloves/bulb. Total soluble solid, dry matter and pyruvic acid content is 35 - 36%, 37 - 38% and 22 - 23 micro moles/g respectively. Medium in storage and average yield is 175 - 225 q/ha. It is recommended for Himachal Pradesh, Uttar Pradesh, Uttarakhand and Northern hills.
Agrifound white (G-41)	Bulbs are compact, silvery and white with creamy flesh. A bigger elongated clove with 20-25 in number and diameter of bulb is 3.5 to 4.5 cm. Variety is susceptible to purple blotch and stempthylium blight. Total soluble solid is 41%, dry matter 42.78% and pyruvic acid content is 27 micro mole/g. Good for storage and average yield is 130 q/ha. It is recommended for Maharashtra and Madhya Pradesh.
Yamuna Safed (G-1)	Bulbs are compact, silvery white skin with creamy flesh, diameter 4.0 - 4.5 cm. Sickle shaped cloves with 25 – 30 in number and diameter of cloves 0.8 to 1.00 cm. Variety is tolerant for purple blotch, stemphylium blight and thrips. Total soluble solid, dry matter and pyruvic acid content is 38% and 39.5 % and 29 micro mole/g respectively. Good for storage and average yield 150 - 175 q/ha. It is recommended for all over India.

Name of variety/hybrid	Salient features
Yamuna Safed-2 (G-50)	Bulbs are compact, attractive, white creamy flesh and diameter 3.5 - 4.0 cm. Number of cloves 35 - 40, diameter of cloves 0.75 - 1.4 cm and clove size index 1.75 - 2.5 cm2. Total soluble solid, dry matter and pyruvic acid content is 38 - 40%, 40 - 41% and 26 micro mole/g respectively. Average yield is 150 - 200 q/ha. It is recommended for northern India, Punjab, Haryana and Madhya Pradesh.
Yamuna Safed-3 (G-282)	Bulbs are creamy white and bigger sized and 4.5 - 6 cm in diameter. Diameter of clove 1.2 - 1.5 cm and cloves/bulb is 15 - 16. Total soluble solid, dry matter and pyruvic acid content is 38 - 42%, 39 - 43% and 25 micro mole/g respectively. Medium in storage and average yield is 175 200 q/ha/ Variety is suitable for export It is recommended for Madhya Pradesh, Maharashtra, Haryana, Gujarat, Punjab, Rajasthan, Uttar Pradesh and Chhattisgarh.
Yamuna Safed-4 (G-323)	Bulbs are silvery white and bigger sized 3.5 - 4.0 cm in diameter. Diameter of clove 1.2 - 1.25 cm, clove size index 2.50 - 3.00 cm^2 and number of cloves 20 - 25/ bulb. Total soluble solid and dry matter content is 40 - 42% and 41 - 42% respectively. Average yield is 175 - 200 q/ha. It is recommended for north India and Madhya Pradesh.
Yamuna Safed-5 (G-189)	Dark green straight erect leaves, sturdy plant growth, erect shape wide 8 - 10 leaves/plant. Bulbs compact, attractive creamy white colour, creamy flesh, big size bulbs (4.5 - 5.0 cm in diameter) and cloves, oblong in shape, number of cloves per bulbs 22 30. Variety is table purpose pungent with good aroma. Total soluble solids and dry matter content is 39% and 41 % respectively. Good storer and better adaptability and tolerant to stemphylium blight, purple blotch and thrips under field condition. Average yield is 150 – 160 q/ha. It is recommended for North-East and Andaman & Nicobar Island, Punjab, Tarai region of U.P., Bihar, Jharkhand, Rajasthan, Gujarat, Haryana and Delhi.
Varieties developed at DOGR, Rajguru Nagar, Pune	
Bhima Omkar	This variety has been recommended for cultivation in the states of Delhi, Gujarat, Haryana and Rajasthan. It matures in 120-135 days and average yield is 8-14 t/ha. It produces medium-sized compact white bulbs.
Bhima Purple	This variety with attractive purple skinned bulbs has been recommended for cultivation in the states of Andhra Pradesh, Bihar, Delhi, Haryana, Karnataka, Maharashtra, Punjab and Uttar Pradesh. It matures in 120-135 days and the average yield is 6-7 t/ha.
Varieties developed at GBUA & T, Pant Nagar	
Pant Lohit	It is moderately resistant to purple blotch. Bulbs are compact and bold, weighing 18-20g. There are 15-18 cloves per bulb having light purple colour. It is a high yielder with the bulb yield of 125-130 q/ha. It is recommended for U.P. and Uttarakhand.

Name of variety/hybrid	Salient features
Varieties developed at HAU, Hisar	
HG-17	High yielding, good storage life, bulb weight is better than other varieties and has least incidence of purple blotch disease. Average yield 120-125 q/ha. It is recommended for Haryana.
Varieties developed at TANU, Coimbatore	
Ooty.1	Bigger bulbs weighing 30 - 40 g each. Average yield 17 t/ha. It is recommended for T.N.
Potato	
Varieties developed at CPRI, Shimla	
Kufri Chandramukhi	An early maturing (90-100 days) variety producing medium-tall and semi-compact plants with red-violet flowers. Tubers are white cream, ovoid with shallow eyes. The variety is susceptible to major diseases, however it is suitable for suitable for processing also. Yield potential is 20-25 t/ha. It is recommended for North Indian plains and plateau region.
Kufri Arun	A medium maturing (90-110 days) variety tall and semi-compact plants with red-violet flowers. Tubers are red, ovoid with medium deep eyes. The variety is moderate resistant to late blight. Yield potential is 30-35 t/ha. It is recommended for North Indian plains.
Kufri Ashoka	An early maturing (70-80 days) variety producing short and semi-compact plants with red-violet flowers. Tubers are white cream, ovoid with medium-deep eyes. The variety is susceptible to major potato diseases. Yield potential is 25-30 t/ha. It is recommended for North Indian plains.
Kufri Badshah	A medium maturing (100-110 days) variety producing tall and semi-compact plants with white flowers. Tubers are white cream, ovoid with shallow eyes. The variety possesses moderate resistance to late blight and early blight and resistant to potato virus X. Yield potential is 30-35 t/ha. It is recommended for North Indian plains and plateau region.
Kufri Bahar	A medium maturing (100-110 days) variety producing medium-tall and semi-compact plants with white flowers. Tubers are white cream, ovoid with medium deep eyes. This early bulking variety is immune to wart, tolerant to Gemini viruses and has a slow rate of degeneration. Yield potential is 30-35 t/ha. It is recommended for North Indian plains.
Kufri Chipsona-1	A medium maturing (90-110 days) variety producing medium-tall and semi-compact plants with white flowers. Tubers are white cream, ovoid with shallow eyes. The variety possesses resistance to late blight and is suitable both for preparation of chips and French fries. Yield potential is 30-35 t/ha. It is recommended for North Indian plains.

Name of variety/hybrid	Salient features
Kufri Chipsona-3	A medium maturing (100-110 days) variety producing tall and semi-compact plants with white flowers. Tubers are white cream, ovoid with shallow eyes. The variety possesses resistance to late blight. The tubers have high dry matter, low reducing sugars and low phenols. Variety is highly suitable for making chips and French fries. Yield potential is 30-35 t/ha. It is recommended for North Indian plains.
Kufri Chipsona-4	An early maturing (90-100 days) variety producing tall, compact plants with white flowers. Tubers white cream, round with shallow eyes. The variety possesses resistance to late blight. The tubers have high dry matter, negligible glyco alkaloid content, suitable for processing. Yield potential is 30-35 t/ha. It is recommended for Madhya Pradesh, West Bengal and Karnataka.
Kufri Frysona	A medium maturing (100-110 days) variety producing tall and open plants with red-violet flowers. Tubers are white cream, long-oblong with shallow eyes. The variety is field resistant to late blight and immune to wart. Variety possess high dry matter, low reducing sugars and low phenols. First Indian potato variety suitable for making French fries. Yield potential is 30-35 t/ha. It is recommended for North Indian plains.
Kufri Girdhari	A medium maturing (110-135 days) variety producing medium-tall and open plants with white flowers. Tubers are white cream, ovoid with shallow eyes. The variety possesses high level of field resistance to late blight in foliage and tubers. Yield potential is 30-35 t/ha. It is recommended for Indian hills.
Kufri Himalini	A medium maturing (110-120 days) variety producing medium-tall and semi-compact plants with red-violet flowers. Tubers are white cream, ovoid with medium deep eyes. The variety possesses moderate resistance to late blight. Yield potential is 30-35 t/ha. It is recommended for North Indian hills.
Kufri Himsona	A medium maturing (110-135 days) variety producing tall and semi-compact plants with red-violet flowers. Tubers are white cream, round with shallow eyes. Variety is moderately resistant to late blight resistant and immune to wart. Suitable for processing. Yield potential is 15-20 t/ha. It is recommended for Indian hills.
Kufri Jyoti	It is a medium maturing (90-100 days) in plains and 110- 130 days in hills) producing medium-tall and semi-compact plants with white flowers. Tubers are white cream, ovoid with shallow eyes. It is suitable for processing also when grown in warmer areas of central plains. The variety carries moderate resistance to early and late blight. It is immune to wart pathogen. It has wide adaptability, early bulking and slow rate of degeneration. Yield potential is 20-25 t/ha in the hills and 30 t/ha in plains. It is recommended for North and South Indian hills, North Indian plains and plateau.

Name of variety/hybrid	Salient features
Kufri Kanchan	A medium maturing (110-135 in summers and 90-100 days in winter) variety producing medium tall, semi-compact plants with blue-violet flowers. Tubers are red, ovoid with medium deep eyes. The variety is field resistant to late blight, immune to wart and slow rate of degeneration. Yield potential is 25-30 t/ha. It is recommended for North Bengal hills and Sikkim.
Kufri Khyati	An early maturing (75-85 days) variety producing medium-tall and semi-compact plants with white flowers. Tubers are white cream, ovoid with medium deep eyes. Moderately resistant to early and late blight. Early bulking variety suitable for high cropping intensity. Yield potential is 25-30 t/ha. It is recommended for North Indian plains.
Kufri Lauvkar	An early maturing (75-80 days) variety producing medium tall, semi compact plants with white flowers. Tubers are round, white cream with medium deep eyes. The variety is suitable for processing when grown in warmer areas of central plains. Tolerant to heat and susceptible to major diseases. Yield potential is 20-25 t/ha. It is recommended for Plateau region.
Kufri Pukhraj	An early maturing (70-90 days) variety producing medium-tall and semi-compact plants with white flowers. Tubers are yellow, ovoid with medium-deep eyes and develop faint purple colour on exposure to sun. The variety is resistant to early blight, moderately resistant to late blight and immune to wart. Early bulker and suitable for low input ecosystem. Yield potential is 35-40 t/ha. It is recommended for North Indian plains and plateau region.
Kufri Pushkar	A medium maturing (90-110 days) variety producing medium-tall and compact plants with white flowers. Tubers are yellow, ovoid with medium deep eyes. The variety is moderately resistant to late blight, resistant to early blight and immune to wart. Yield potential is 30-35 t/ha. It is recommended for North Indian plains.
Kufri Sadabahar	A medium maturing (90-100 days) variety producing medium-tall and compact plants with white flowers. Tubers are white cream, ovoid with shallow eyes. The early bulking variety possesses moderate resistance to late blight. Yield potential is 30-35 t/ha. It is recommended for Uttar Pradesh and adjoining areas.
Kufri Shailja	A medium maturing (100-120 days) variety producing medium tall and semi-compact plants with red-violet flowers. Tubers are white cream, ovoid with medium deep eyes. The variety possesses moderate resistance to late blight. Yield potential is 30-35 t/ha. It is recommended for North Indian plains.

Name of variety/hybrid	Salient features
Kufri Sindhuri	It is a late maturing (110-120 days) variety producing medium tall and open plants with flowers of red-violet colour. Tubers are medium in size, round in shape, red in colour with deep eyes. It is moderately resistant to early blight, tolerant to PLRV and has a slow rate of degeneration. Suitable economic yields even under low fertility conditions. Yield potential is 30-35 t/ha. It is recommended for North Indian plains.
Kufri Surya	An early maturing (75-90 days) variety producing tall and semi-compact plants with red-violet flowers. Tubers are white cream, oblong with shallow eyes. This is a heat tolerant variety that can be grown in areas having night temperature above 20°C. Suitable for early planting in plains. This variety is immune to wart and tolerant to hopper burn. Yield potential is 20-30 t/ha. It is recommended for North Indian plains and plateau region.
Kufri Swarna	A medium maturing (100-110 days) variety producing medium-tall and semi-compact plants with white flowers. Tubers are white cream, round with shallow eyes. The variety is resistant to early blight, late blight, cyst nematodes and immune to wart. Yield potential is 30-35 t/ha. It is recommended for South Indian hills.
Ivy gourd (Kundru)	
Varieties developed bat IGKVV, Raipur, Chhattisgarh	
Indira kundru-05	Fruits are oblong size, green colour with white strips Suitable for table purpose. Average yield - 1019 q/ha. It is recommended for Chhattisgarh Plains.
Indira kundru-35	Long sized and light green fruits. Suitable for table purpose. Average yield 936.88 q/ha. It is recommended for chhattisgarh plains.

Root and Tuber Crops

Name of the variety	Pedigree	Avg. Yield (t-ha-¹)	Maturity (Months)	Cooking quality	Salient features and Recommended area
Cassava					
Sree Jaya (CI-649)	Selection from indigenous germplasm	26-30	6-7, early maturing	Excellent	Short duration suitable for low land and becoming popular in rainfed plains of Andhra Pradesh.
H-165 Hybrid	Chadayamangalam Vella X Kalikalan	33-38	8-9	Good	Short duration variety, popular in hill rainfed of Tamil Nadu and Andhra Pradesh.
H-226 Hybrid	M4 X Ethakaruppan	30-35	10	Good	Most popular variety in Tamil Nadu in rainfed plains and, some parts of Andhra Pradesh, very much suitable for industries. Comes up well in irrigated plains also.

Name of the variety	Pedigree	Avg. Yield (t-ha-[1])	Maturity (Months)	Cooking quality	Salient features and Recommended area
H-97 Hybrid	Manjavella-X Acc.No.300	25-35	10	Good	Starchy variety suitable for industries suited to upland rainfed and irrigated plains.
Sree Padmanabha	Exotic germplasm	38	9-10	Excellent	Irrigated plains of Tamil Nadu. CMD resistant variety.
Sree Harsha (2-14)	Triploid (OP-4(2x) x H-2304 (4x)	35-40	10	Good	Starchy variety for industrial areas, irrigated plains.
Sree Prabha (TCH-2)	Top cross hybrid	40-45	10	Excellent	Upland rainfed and irrigated plains of Tamil Nadu.
Sree Prakash (S-856)	Selection from Indigenous germplasm	30-35	7, early maturing	Good	Short duration suitable as sequential cropping in rice fallows.
Sree Rekha (TCH-1)	Top cross hybrid	45-48	10	Excellent	High yield, less fibre. Upland rainfed and irrigated plains of Tamil Nadu.
Sree Sahya (11-2304) hybrid	Multiple (Acc No.468, 174, 3024, 1310, 82,3939, 3588 and M4	35-40	10-11	Good	Long duration crops , Upland rainfed, hardy variety suited to marginal soils.
Sree Vijaya (CI-731)	Selection from indigenous germplasm	25-28	6-7, early maturing	Excellent	Hybrid Suitable for low land and becoming popular in irrigated plains of Tamil Nadu. Suitable as intercrop in coconut gardens.
Sree Visakham Hybrid	(Acc No.1501 X S-2312 (exotic)	35-38	10	Good	High in carotene content, Cooking quality good, suitable as sequential crop in rice fallows and inter crop in coconut gardens.
Sweet potato					
Gouri (85-16)	Hybrid (H-219 x H-42)	19.0	110-120	Fair	Orissa State, upland and low land conditions.
Goutam	Selection from Polycross	18.9	105-110	Very good	Uplands and hilly areas.
H-41 Hybrid	Norin x indigeneus cultivar	20 – 25	120	Cooks easily, Sweet, low Fibre	Kerala
H-42 Hybrid	Velladamph X Triumph	22 – 25	120	Cooks easily, Sweet less fibre	Kerala

Name of the variety	Pedigree	Avg. Yield (t-ha-^{1})	Maturity (Months)	Cooking quality	Salient features and Recommended area
Kalinga	Selection from OP seed	17.2	105-110	Excellent	High starch, good for food and fodder.
Kishan	Selection from polycross	17	110-120	Good, Sweet and mealy	High starch, sandy loam and black sandy soils.
Sankar (85-70)	Hybrid H-219 x S-73	13.73	120	Excellent	Orissa State upland and low land conditions.
Sourin	Selection from Polycross	16.2	105-110	Good	Rainfed and irrigated, withstands mid season drought.
Sree Vardhini (76 – OP 219)	Selection from OP Seed progeny of S-13	25	100 –105	Cooks well with attractive colour	Kerala State, carotene rich, dual purpose variety, upland and low land.
Sree Arun	Recurrent selection from polycross	20 - 28	90	Good	Upland and low land conditions.
Sree Varun	Seedling selection from CIP, Peru	20 - 28	90	Good	Upland and low land conditions.
Sree Kanaka	Hybrid (S.187 x H. 633)	10 - 15	75-85	Good	Very high carotene, upland
Sree Bhadra (S-1010)	Selection from seedling population from Nigeria	20 – 27	90	Excellent	Kerala State, popular in Bihar. Trap crop against root knot nematode.
Sree Nandini (76 – OP 217)	Selection from OP Seed progeny of S-32	20 – 25	100 –105	Excellent Cooks well, Sweet	Kerala State, paddy fallows
Sree Retna (X-108-2)	Hybrid (S-187x Sree Vardhini)	20 – 26	90 –105	Excellent	Kerala State
Varsha (H-268)	Double cross Hybrid (Acc. No. 39 x Acc No.3) X (Acc No 1871 x Acc No 1103)	25	120	Good	Maharashtra State
Yams					
Greater Yam					
Orissa elite	Clonal selection	25	6-7	Excellent	Excellent cooking quality.
Sree Karthika	Clonal selection	30	9	Good	Excellent cooking quality.
Sree Keerthi	Clonal selection	25-30	9-10	Good	Large sized tubers intercropping in banana and coconut.
Sree Roopa	Clonal selection	25-30	9-10	Good	Intercropping in coconut.

Name of the variety	Pedigree	Avg. Yield (t-ha-[1])	Maturity (Months)	Cooking quality	Salient features and Recommended area
Sree Shilpa	Hybrid (Da-140 X Sree Keerthi)	28	8	Good	First hybrid, oval medium sized tubers, early harvesting.
White Yam					
Sree Dhanya	Seedling selection	20	9	Good	First dwarf variety, no need for staking, suitable for closer Spacing.
Sree Priya	Seedling selection from Umidika	35	9-10	Excellent	Intercropping in banana and Coconut.
Sree Subhra	Seedling selection from Iwo	35	9-10	Excellent	Drought tolerant.
Lesser Yam					
Sree Kala	Exotic selection - Kombi	20	7 1/2	Excellent	Oval smooth tubers.
Sree Latha	Selection from germplasm	25	8	Good	Intercropping in banana and Coconut.
Elephant Foot Yam					
Gajendra		50 - 80	180 - 200	Excellent	It is very popular all over India.
Sree Athira	Hybrid selection from cross Am-15 x Am-45	40.5	9-10	Good	First genetically improved variety, high tuber yield.
Sree Padma	Selection from germplasm from Wayanad, Kerala	42	8-9	Very good	First variety from Kerala, fairly tolerant to mosaic disease and field tolerant to collar rot.
Taro					
Muktake shi	Clonal selection from Orissa	20	5-6	Excellent	Non acrid, leaf blight resistant, field tolerant to Dasheen mosaic virus, aphids, cut worm and scale insects, suitable for upland and lowland during summer and rainy season.
Pani Saru 1	Local landrace from Kantilo	15.7	6-7	Good	Field tolerant to leaf blight, suitable for water logged / submerged condition with long keeping quality of tubers.
Pani Saru 2	Local landrace from Begunia	13	6-7	Good	Field tolerant to leaf blight, suitable for water logged / swampy / submerged and upland condition of Orissa.

Name of the variety	Pedigree	Avg. Yield (t-ha-[1])	Maturity (Months)	Cooking quality	Salient features and Recommended area
Sree Kiran	Hybrid (C-303 x C-383)	17.5	6-7	Good	First hybrid taro in India, long keeping quality of tubers, moderately susceptible to leaf blight.
Sree Pallavi	Selection from germplasm from Meghalaya	16	7	Very good	Tall variety with large number of small-sized tubers. Field tolerant to leaf blight and Dasheen mosaic virus.
Sree Rashmi	Selection from germplasm	18	7	Very good	Acrid-free edible leaves, corms and cormels. Gives economical yield under low input levels.

ooo

CHAPTER 11

Production Technology of Vegetable Crops

S.K. Tyagi
Scientist (Horticulture)
Krishi Vigyan Kendra, Khargone (M.P.)

Amaranthus

Basic Information and Facts

Scientific name	*Amaranthus blitum* (chhoti chaulai), *Amaranthus tricolor* (Badi chaulai)
Family	Amaranthaceae
Origin	Central and South America, India, South –East Asia
Chromosome number	2n = 32
Inflorescence type	Spike
Sex form	Monoecious and Dioecious
Pollination	Cross pollinated (anemophilous) crop

Amaranth (***Amaranthus* sp.)** is a widely used leafy vegetable belonging to the family **Amaranthaceae.**

- They are relatively short-lived annuals that grow vigorously, are drought resistant and adapt readily to new environments.
- Inflorescence terminal and axillary, branched spikes, flower small, regular, mostly unisexual, **monoecious.**
- In general the cultivated species are **monoecious.**
- All the **monoecious species** of Amaranthus exhibit **protogyny,** and the inflorescence structure ensures adequate **cross pollination.**
- *Amaranth* is **a cross-pollinated crop.** Cross-pollination occurs mainly **by wind.**

Production Considerations

Climate	It is also grown successfully in hot summer season and humid conditions of kharif season.
Soil	Well drained loamy soils with slightly acidic nature and warm climate are suitable.
Improved varieties	Pusa Chhoti Chaulai, Pusa Badi Chaulai, Pusa Kirti (most suitable for summer), Pusa Kiran (for rainy season), Pusa Lal Chaulai (red pigmented variety), Arka Suguna, Arka Arunima.
Seed rate	2.0 kg/ha.
Sowing time	**Summer crop:** February- March and **Rainy season crop:** June- July.
Spacing	20 cm × 15 cm.
Manures and fertilizers	Apply FYM 25 t/ha, *Azospirillum* 2 kg and *Phosphobacteria* 2 kg/ha, N 50 kg, P 50kg/ha and K 20 kg per ha.
Irrigation	Irrigate before and after sowing and at weekly intervals after germination.
Yield	60-80 q/ha.

Beetroot

Basic Information and Facts

Scientific name	*Beta vulgaris* L. ssp. *vulgaris*
Family	Amaranthaceae
Origin	Mediterranean region
Chromosome number	2n= 18
Inflorescence type	Spike
Fruit type	Aggregate

- Beetroot (***Beta vulgaris* L.**) belongs to the family Amaranthaceae.
- Beets are believed to originate along the coasts of the Mediterranean region).
- The **inflorescence** is a large **spike,** which normally develops in the second year.
- **Fruits** are mostly **aggregate** formed by the cohesion of two or more fruits and held together by a swollen perianth (calyx) base and thus forming an irregular dry cork-like body known as seed ball or seed.
- Beets contain a pigment called ***betalains*** that is used for dyes.
- Purple colour of table beets is due to ***betacynin*** and yellow pigmentation is due to ***betaxanthin.***
- Beetroot is rich in protein, carbohydrates, Ca, P, Fe and vitamin C.

Production Considerations

Climate	It is generally considered as a cool weather crop. It grows best during winter in plains.
Soil	It comes up well in all types of friable soils. The suitable soil pH is 6.0-7.0. It is considered to be a cool weather crop.
Improved varieties	Detroit Dark Red, Crimson Globe
Seed rate	7-9kg/ha
Sowing time	**Hills:** March-July, **Northern Plains:** September-November and **Southern Plains:** July-November.
Spacing	45-60cm x 8-10cm
Manures and fertilizers	Apply FYM at 20 t/ha and 60:160:100 kg NPK/ha as basal and 60 kg N/ha after 30 days.
Irrigation	Irrigate the field copiously immediately after sowing and afterwards as and when necessary.
Yield	25 - 30 t/ha in 120 days.
Storage	Temperature: 0°C and RH: 90%

Spinach/Vilayati palak

Basic Information and Facts

Scientific name	*Spinacia oleracea* L.
Family	Amaranthaceae
Origin	Perssia (Iran)
Chromosome number	2n= 12
Sex form	Dioecious
Pollination	Cross pollinated crop

- Spinach (***Spinacia oleracea*** **L.**) is an edible flowering plant in the family **Amaranthaceae.**
- Spinach is thought to have originated in ancient **Persia** (modern **Iran** and neighbouring countries).
- It is a **wind-pollinated** crop with extremely fine pollen. It is a **dioecious** species.

Production Considerations

Climate	Spinach is a long-day crop and is grown mainly in the cool season.
Soil	Spinach grows in any type of well drained soil. However loamy soil is ideal. It is sensitive to acidity and should be grown on soil in the pH range of 6-5-7.5.
Improved Varieties	Virginia Savoy, Early Smooth Leaf
Seed rate	37 - 45 kg/ha
Season	September- October

Spacing	20 × 10 cm
Manures and fertilizers	FYM 25 t/ha, N 80 kg, P 60 kg and K 60 kg/ha
Irrigation	Winter spinach crop needs irrigation at 10-15 days intervals depending on soil type. January- planted spinach crops need irrigation at 6 to 7 days interval in summer months.
Yield	5-6 t/ha

Beet leaf/ Palak

Basic Information and Facts

Scientific name	*Beta vulgaris* var. *bengalensis* L.
Family	Amaranthaceae
Origin	Indo-China
Chromosome number	2n=2x= 18
Pollination	Cross-pollinated crop

- Spinach beet also known as Palak or Palong is closely related to beetroot and Swiss chard.
- Botanically it is known as ***Beta vulgaris* var. *bengalensis*** and belongs to the family Amaranthaceae.
- Plant of spinach is an herbaceous annual for the edible leaf production while it is a biennial for seed production. Leaves are long with entire margin and long petioles.
- Flowers are small, sessile, bisexual, perfect and are borne in the axil of leaf in a group of 2-3, with 5 stamens, flowers produce abundant, small and light pollen grains which are carried by wind. Hence the crop is highly **cross-pollinated** (a wind-pollinated crop).

Production Considerations

Climate	It is predominantly a cool season crop but can be grown throughout the year under mild temperature conditions.
Soil	It is highly tolerant to saline conditions and can be successfully grown in saline sodic soils.
Improved varieties	All Green, Pusa Bharati, Pusa Harit , Pusa Jyoti , Ooty.1 , Jobner Green, HS 23, Punjab Green
Seed rate	**Winter crop:** 10 - 15 kg/ha **Summer crop:** 25-30 kg/ha
Season	September-November and February-March
Spacing	20 cm × 5-10 cm
Manures and fertilizers	FYM 25 t/ha, N 87.5 kg, P 60 kg and K 60 kg/ha.
Irrigation	The spring summer crop need frequent irrigation at 6-7 days intervals whereas autumn winter crop requires irrigation at about 10-15 days interval.
Yield	150-200 q/ha

Onion

Basic Information and Facts

Scientific name	*Allium cepa* L.
Family	Amaryllidaceae (previously classified under Liliaceae)
Origin	Central Asia
Chromosome number	2n= 2x=16
Inflorescence type	Umbel
Pollination	Highly cross-pollinated crop due to its protandry
Pollinator	Honey bees

- The onion (***Allium cepa* L.)** belongs to the family **Amaryllidceae,** plant consists of shallow fibrous roots, a bulb, and green, tubular leaves
- Available data indicates that the genus name ***Allium*** comes from the Celtic word ***all*=pungent** and ***Cepa*** in Latin meaning onion.
- The genus ***Allium*** includes some of the most ancient cultivated crops, including *A. sativum* (garlic), *A. cepa* (onion), *A. schoenoprasum* (chives), *A. ampeloprasum* (great-headed or elephant garlic), *A. tuberosum* (Chinese or garlic chives), *A. fistulosum* (Japanese bunching onion) and *A. chinense* (rakkyo; when listing members of the genus, Allium is abbreviated as ***"A."***).
- Flowers are borne in simple **umbels** at the apex of a floral stem which is hollow and round in cross section and somewhat swollen at the middle or near the base.
- The **flower is protandrous** in nature and stigma becomes receptive when shedding of pollen is over.
- **Onion is cross-pollinated** in nature and bees, flies and other insects to achieve the full potential of onion seed.
- **Onion bulbs** range in shape from flat to globular to oblong, and usually are white, yellow, or red in colour.
- Pungency is due to **'allylpropyldisulphide'** and it varies with cultivar, growing & storage conditions, bulb maturity etc.
- **Red colour varieties** are more **pungent** followed by brown, yellow and whites.
- **Red colour** onion is due to **anthocyanin** pigment and **yellow colour** due to **querectin.**
- Onions when cut release **allicin**, a volatile sulfur gas that irritates the eyes.
- Hybrid-onion (*Allium cepa*) seed is produced using systems of **cytoplasmic-genic male sterility (CMS).**
- Male sterility in onion is controlled by the combination of a cytoplasmic factor 'S' together with a recessive nucleus gene in its homozygous form (ms).
- These male sterile lines are maintained by pollination with a maintainer line of construction 'N ms ms'.

Production Considerations

Climate	Mild season without extremes of heat and cold. It requires 13-21°C temperature for vegetative growth before bulb initiation and 16-25°C for bulb development and 25-30°C for bulb maturation.
Soil	Red loam to black soils with good drainage facilities. It grows well in pH range of 6-7.
Improved varieties	**Red coloured:** Agrifound Dark Red, Agrifound Light Red, Arka Niketan, Arka Kalyan, Pusa Madhavi, Pusa Ratnar, Pusa Red, Pusa Riddhi ,Udaipur 101, Udaipur 103, Bhima Raj, Bhima Red. **White skinned varieties:** Pusa White Flat, Pusa White Round, Punjab-48, Udaipur-102. **Yellow skinned varieties:** Brown Spanish (Long day variety, suitable for growing in hills), Early Grano (Good for salad, suitable for green onions). **Kharif Onion:** Agrifound Dark Red, Arka Kalyan, Arka Pragati, N-53, Arka Niketan. **Multiplier Onion:** Agrifound Red, CO-1, C-2 (resistant to purple blotch), CO-3 (resistant to thrips), CO-4 (moderately resistant to thrips), MDU-1. **Small Onion:** Agrifound Rose (pickling type, suitable for export), Arka Bindu.
Seed rate	8-10kg/ha.
Season	**Kharif season:** May-June in South, Maharashtra, and part of Gujarat as well as north and north-east-India. **Rabi or late Kharif:** August-September in Maharashtra. **Rabi:** September-October in south, November-December in Maharashtra and part of Gujarat, October-November in north and north-eastern part.
Spacing	15 × 10 cm
Manures and fertilizers	Apply well rotten farmyard manure @ 200-300 q/ha, nitrogen @ 60-150 kg, phosphorus @ 35-150 kg and potassium @ 25-120kg per hectare depending upon the soil test, cultivar and growing season.
Irrigation	Irrigate at the time of planting of seedlings and third day and later at weekly intervals. Withhold irrigation 10 days before harvest.
Weed management	Application of Oxyflurofen @ 23.5% EC (1.5 -2.0 ml/L)/ Pendimethalin @ 30% EC (3.5-4ml/L) before transplanting or at the time of transplanting followed by one hand weeding at 40-60 days after transplanting is recommended for efficient weed management.
Growth regulator	**Borax (1000-1500 ppm)** or **iron sulphate (1000-1500 ppm)** about 15 days before harvesting is recommended for checking the sprouting of bulbs during storage.
Yield	**Rabi crop:** 250-300 q/ha, **Kharif crop:** 200-250 q/ha
Curing	Onion bulbs should be adequately cured because curing or drying of bulbs is an important process to remove the excess moisture from the outer skin and neck of onion bulbs.
Storage	For effective long storage of onion the parameters essential to be looked after are the bulb size, choice of cultivars, cultivation practices, time of harvest, field curing, removal of tops, drying, grading, packing, storage conditions (optimum storage range of **relative humidity 65% to 70%** with the **temperature ranging between 25C and 30°C**).

Coriander

Basic Information and Facts

Scientific name	*Coriandrum sativum* L.
Family	Apiaceae (previously classified under Umbelliferae)
Origin	Southern Europe and Asia minor
Chromosome number	2n = 22
Inflorescence type	Compound umbel
Pollination	Cross pollinated crop
Fruit type	Schizocarp
Chemical constituents	D-linalool (68%), gerniol (10%), pinene (10.5%)

- Coriander (***Coriandrum sativum*** **Linn.**) also called cilantro, are dried fruits of plant *coriandrum sativum* L., belonging to family **Apiceae.**
- The genus ***Coriandrum*** includes the cultivated plant ***C. sativum*** and the wild species ***C. tordylium.***
- **Coriander** is an annual herb and, according to the climatic conditions, is cultivated as a winter annual crop.
- All the parts of coriander such as leaf, seed and fruits have culinary as well as medicinal use.
- The germination is **epigaeal** and the plant has a **tap root.**
- The **stem** is more or less erect and **sympodial, monochasial-branched**, sometimes with several side branches at the basal node.
- The **inflorescence** is a **compound umbel**.
- The **flowers** are **protandrous.** The central flowers of the umbellets are staminiferous or sometimes sterile. The inner flowers of the umbellets are staminate.
- **Coriander** has an inferior ovary and the five calyx teeth surrounding the **stylopodium** are still visible in the ripe fruit.
- The **protandry** and the observations made suggest that **geitonogamy** is common and **xenogamy** is possible.
- The fruits are **schizocarp**, globular or ovate, with a diameter of up to 6 mm.
- The chromosome number of coriander is **2n=22**, as the cytologic investigation of eight genotypes by Das and Mallick (1989) demonstrated.
- The most important constituents are the **essential oil** and the **fatty oil**.
- The essential oil content of the weight of ripe and dried fruits of coriander varies between **0.03 and 2.6%,** and the content of fatty oil varies between **9.9 and 27.7%**.
- The extracted essential oil is used in the flavouring of a number of food products and in soap manufacture.
- **Endosperm** consists of fixed oil globules whereas vittae consist of volatile oil.
- Coriander consists of **volatile oil (0.3 to 1%).** The volatile of coriander consist of **D-linalool (68%)** ,gerniol (10%) , pinene (10.5%). It also consist of coriandryl acetate, L-borneol, fixed oil(13%).

- **Coriander** consist of Water (7.3g), Food energy (279kcal), Protein (21.83g), Fat (4.76g), Carbohydrates (52.10g), Ash (14.02g), Calcium (1.246mg), Phosphorus (481.00mg), Sodium (211.00mg), Potassium (4.466mg), Iron (42.46mg) Vit –A (566.7mg).
- **Coriander is a facultative cross-pollinator** and after selfing the studied plants did not show obvious inbreeding depression.

Production Considerations

Climate	Cool and comparatively dry, frost free climate. Coriander performs well at a temperature range of 20 – 25 °C
Soil	Well drained silt or loamy soils are suited for cultivation. For rainfed cultivation soil should be clay in nature and the pH should be 6 – 8.
Seed rate	10-12kg/ha for irrigated crop and 20-25kg/ha for rainfed crop.
Season	**Kharif crop:** June- July, **Rabi crop:** October- November.
Spacing	20 x 15 cm.
Manures and fertilizers	**NPK**- 60:60:30 kg /ha.
Irrigation	First irrigation should be given immediately after sowing and the second on the third day and subsequent irrigations at 7-10 days interval.
Weed management	Pre-emergence spray of herbicide Fluchloralin 700 ml in 500 lit/ha.
Growth regulator	Spray CCC @ 250 ppm one month after sowing for inducing drought tolerance in rainfed crops.
Yield	**Rainfed Grain yield:** 300-400 kg/ha. **Irrigated :** 500-600 kg/ha, **Leaf yield :** 6-7 t/ha.

Celery

Basic Information and Facts

Scientific name	***Apium graveolens* L.**
Family	Apiaceae (previously classified under Umbelliferae)
Origin	Mediterranean region
Chromosome number	2n=22
Inflorescence type	Compound umbels
Fruit type	Schizocarp

- The celery plant (***Apium graveolens* L.**) belonging to family **Apiaceae** is a hardy biennial, occasionally annual, widely cultivated for its **fleshy leafstalk** used as a vegetable and seeds which yield essential oil.
- The seeds contains **2-3% essential oil** and **17-18% fatty oil**.
- The essential oil has d-selenene, sedlanolide and sedanoic acid anhydride contributing to its flavour and 60% of d-limonene.The oil is pale yellow in colour,

contains d-limonene (60%), ß-selinene (10-12 %), sedanoic acid anhydride (0.5 %) and sedanolide (2.5-3.0 %). The later two components contribute to its characteristic odour.

- The leaves are a rich source of minerals like Ca, P, Fe, vitamin A and vitamin C for which it is consumed in salad.

Production Considerations

Climate	It is a winter (Rabi) crop, needs warm days and cool nights, low humidity and plenty of sun-shine. Generally, cool temperature at sowing and warm dry weather at seed ripening is ideal.
Soil	The crop thrives well on silt loam to loamy soil (pH 6.6– 7.5) with moderate organic matter. It is not suited to clayey soil and cannot withstand water-logging.
Varieties	Standard Beared, Wright Grove Grant, Fork Hook Emperor, Florida Golden
Seed rate	10-125 g/ha
Sowing time	August- September
Spacing	60 x 20 cm
Manures and fertilizers	FYM 20 to 30 t/ha at land preparation and 100-150 kg of N/ha (in 3 splits), P_2O_5 and K_2O. One-third of N, together with P_2O_5 and K_2O at 40 to 60 kg/ha.
Irrigation	The crop is given irrigation at fortnightly intervals in early part and weekly nearing seed maturity. Irrigation at CPE (cumulative pan evaporation) 25/50 is optimum for high seed yield (13q/ ha).
Weed management	Use of basalin (2 – 2.5 l./ha) as preplanting application.
Yield	33,0-56.0 tonnes/ha
Storage	At 0° C and 95-98% RH.

Carrot

Basic Information and Facts

Scientific name	***Daucus carota* L.**
Family	Apiaceae (previously classified under Umbelliferae)
Origin	South Western Asia especially **Afghanistan**
Chromosome number	2n=2x=18
Inflorescence type	Compound umbel
Pollination	Cross pollinated crop
Male sterility	CMS (Cytoplasmic Male Sterility)
Fruit type	Schizocarp

- Carrots are a root vegetable also known as ***Daucus carota* L.,** a member of the **Apiaceae** family
- It is one of the most important root vegetable plants in the world.
- Although cultivated for over 2000 years, and originally used only as a medicinal plant, the domestic carrot (***Daucus carota* var. *sativus, Apiaceae***) remains an important world crop with production expanding rapidly in Asia.
- Originally wild in many parts of Europe and Asia, it was first domesticated in **Afghanistan,** considered to be the primary center of diversity, and from there spread to Europe, the Mediterranean, and Asia, with Turkey recognised as a second center of diversity.
- Carrots are a **biennial plat**, which means that the life cycle of the plant takes two years to complete.
- The inflorescence is a **compound umbel.** Carrot flower is **protandrous.**
- Flowers are arranged in umbellets and the umbellets into an umbel.
- Carrot flowers are **epigynous** with five small sepals, five petals, five stamens, and two carpels.
- The carrot embryo sac is **monosporic** (developing from the chalazal macrospore) and 8-nucleate.
- The carrot fruit is a **bilocular schizocarp**, which dries and splits upon maturity to yield two mericarps (achenes) with one seed each.
- The carrot has **nine pairs of chromosomes**, and little variation in length exists between each chromosome.
- In carrot, the stamens mature first and are thus **protandrous in nature.** Due to this **cross-pollination** is a rule.
- **Carrot hybrids** descent from two or three parents, are therefore called **2-way and 3-way hybrids**.
- The **motherline** is paternally **sterile.** This sterility is a result of a change in the cytoplasm.
- This **cytoplasmatic sterility (CMS)** is a natural characteristic in carrot.
- Making controlled hybrid crosses in carrot is done by using **CMS (cytoplasmic male sterility)** which was a revolutionary discovery first made in onions (Henry Jones, 1930s) and later in carrot (1946 Welch and Grimball, brown anther type; 1953 Henry Munger, petaloid type).
- Hansche and Gabelman reported **digenic control of male sterility** with either of two genes, **dominant *(Ms-4)*** or a **recessive *(ms-5)***.
- **The yellow/orange** colour of western carrots is caused by the plastid-bound pigment **carotenoids, carotene, and xanthophyll.**
- **White carrots** contain only traces of pigment, mainly carotene and **xanthophyll** (Ladizinsky, 1998).
- **Purple carrots** contain **anthocyanins**, a powerful antioxidant, whilst red carrots contain **lycopene,** good for eye health, and also found in tomato (Rubatzky et al., 1999).

Production Considerations

Climate	It is predominantly a cool season crop. A temperature range of 7.2 to 23.9°C is suitable for seed germination and 18.3 to 23.9°C for better root growth. The optimum temperature for better colour development of roots is 15.6-21.1°C.
Soil	The carrot crop needs deep loose loamy soil. It requires a soil pH range of 6.0 to 7.0.
Varieties/ hybrids	**Asiatic or tropical type:** Pusa Kesar, Pusa Meghali (orange coloured), Pusa Vristi, Pusa Rudhira, Pusa Ashita (black coloured), Hisar Gairic **European type or temperate type:** Pusa Yamdagini, Jeno, Imperator, Chantaney, Danvers, Early Nantes, Nantes, Nantes Half Long, Ooty l, Pusa Nayanjyoti (hybrid)
Seed rate	5-6 kg/ha
Season	August to October
Spacing	30 cm – 8-10 cm
Manures and fertilizers	Farmyard manure 100 q/ha, 50-90 N : 40-80 P_2O_5 : 40-80 K_2O/ha
Irrigation	A pre-sowing irrigation is to be given to ensure better seed germination. Carrots require abundant and well-distributed water supply.
Weed management	For effective weed control, a pre-emergence application of **Propazine** @ 1.12 kg/ha has to be done.
Yield	**Asiatic types:** 250-300 q/ha. **European types:** 100-150 q/ha
Storage	At **temperature 0-4.4** °C with **93-98% RH** roots can be stored for 3-4 months.

Colocasia/ Taro

Basic Information and Facts

Scientific name	***Colocasia esculenta* (L.) Schott.**
Family	Araceae
Origin	Indo– Malayan region
Chromosome number	2n= 28
Inflorescence type	Spadix
Pollination	Cross pollinated crop
Fruit type	Berry

- Taro (***Colocasia esculenta* L.)** belongs to the genus ***Colocasia*,** within the sub-family Colocasioideae of the **monocotyledonous** family **Araceae.**
- Taro originates from the **Indo– Malayan region.**
- It is an herbaceous plant, 1–1.5 m in height.

- **Subterranean stem (corm)** globular to cylindrical, variable in size; epidermis smooth or scaly; flesh more or less fibrous, white, pink, yellowish, mauve or violet.
- Leaves **peltate** and **cordate**, light green to dark purple, 30–80 x 20–50 cm; petiole thick, variable in size and colour, uniform, speckled or striped.
- ***C. esculenta,*** commonly known in its crop form as taro, is mainly cultivated for its **starchy stem tubers and corms.**
- It is an **allogamous, protogynous** species, for which the main pollinators are insects.
- Chromosome numbers reported for taro include 2n = 22, 26, 28, 38, and 42. The disparity in numbers may be due to the fact that taro chromosomes are liable to unpredictable behaviour during cell division. The most commonly reported results are **2n = 28 or 42.**
- It is a perennial herb that has heart-shaped leaves and grows 1-2 metres high. It is one of the most widespread root crops of the humid tropics.
- Leaf blades 60 cm (24 in) long and 50 cm (20 in) wide, arrowhead shaped, with upper surface dark green and velvety; leaves peltate (stalked from back of blade); petioles large, succulent, often purplish near the top.
- **Inflorescence** on a fleshy stalk shorter than leaf petioles; part of fleshy stalk enveloped by a long yellow bract (spathe).
- Each inflorescence consists of a **spadix** covered by a spathe.
- The **spadix** is divided into a female part (lower part), a sterile region, a male part, and a sterile appendix. Female (pistillate) flowers are sessile and green, with well developed ovaries and stigmas.
- **Flowers** tiny, densely crowded on upper part of fleshy stalk, with **female flowers below and male flowers above.**
- **Fruit:** a small **berry**, in clusters on the fleshy stalk.

Production Considerations

Climate	Colocasia requires warm and moist climate with a mean temperature of 21-27° C.
Soil	It comes up well in loamy soils with a pH range of 5.5-7.0.
Improved varieties	Co 1, Panchamukhi and Satamukhi (Kovvur), Sree Pallavi, Sree Rashmi, Sree Kiran
Planting material and seed rate	**Cormels** weighing about 20-25 g form good planting material. Seed rate of 800 kg/ha is required.
Season	Under rainfed condition, planting during April to June is ideal. June – July and February – March. If grown as irrigated crop, it can be raised throughout the year.
Spacing	Plough the field to a fine tilth and form ridges and furrows at a spacing of 45 cm. In sandy loam soil, pit method is followed. Plant at a spacing of 45 cm in furrows. The cormels may be planted to a depth of 2.5 to 7.5 cm.

Mulching	Planted seed tubers take 30 to 45 days for sprouting. Mulching helps to hasten sprouting and control weed growth.
Manures and fertilizers	Apply 25 tonnes of FYM, 20 kg N, 30 kg P and 60 kg K/ha as basal and 20 kg N, 30 kg P and 60 kg K/ha 45 days after planting.
Irrigation	Once in a week.
After cultivation	Weeding and earthing up should be done 45 to 60 days after planting. Deep cultivation should be avoided. Small inefficient suckers from the mother plants have to be removed along with the second weeding. It requires profuse irrigation and shade.
Yield	8 – 10 t/ha in 180 days.

Elephant foot yam/ Suran/ Jimmikand

Basic Information and Facts

Scientific name	***Amorphophallus paeoniifolius*** **(Dennst.) Nicolson.; Syn.** ***Amorphophallus campanulatus*** **Blume**
Family	Araceae
Origin	South east Asia
Chromosome number	2n=28
Inflorescence type	Spadix
Pollination	Cross pollinated crop

- ***Amorphophallus paeoniifolius*** **(Densst)** **(Aracea)**, commonly known as elephant foot yam, is a highly potential tropical tuber crop.
- Single leaf (sometimes double), erect, composite, deeply lobed, with three main divisions each of which is then divided into multiple, elongate segments, pale green; long (50 cm) and thick petiole, rugose and often spiny, greenish-grey or purplish, dotted with little whitish markings.
- When the leaf falls **inflorescence** appears. This is a **spadix** 20–25 cm long comprising in order from bottom to top cylindrical female flowers, conical male flowers, and then a sterile appendage that is spongy, phallic, crumpled and adorned with purplish grooves.
- The stink of the inflorescence is reminiscent of rotting carrion, and it attracts flies.
- As it ages the **spathe** folds back to form a collar with wavy edges, yellow at the base and purplish at the apex.
- Numerous fruits stuck to one another, the size of small peas, red, 1 cm in diameter; peduncle erect.
- Seeds **polyembryonic**. Corm in the shape of a thick disc with a central depression, reddish brown, irritating to the skin when it is young, 20–25 cm in diameter and can weigh up to 15 kg.

Production Considerations

Climate	It requires humid and warm weather during the vegetative phase and cool and dry weather during the corm development period. It thrives well with a mean annual temperature of 30-35°C.
Soil	Rich red-loamy soil with a pH range of 5.5-7.0 is preferred. It is a tropical and subtropical crop.
Propagation	Through **tubers.**
Improved varieties	Sree Padma, Sree Athira Hybrid, Gajendra, Kusum
Planting season	April – May is the planting season.
Planting	The tuber is cut into 400-500g small bits in such a way that each bit has at least a small portion of the ring around each bud. Cut pieces are planted in beds at 45 cm x 90 cm spacing or pit of 60 x 60 x 45 cm size is dug and planted.
Manures and fertilizers	Apply 25 tonnes of FYM/ha during last ploughing. The recommended dose of NPK/ha is 80:60:100 kg. Apply 40:60:50 kg NPK/ha at 45 days after planting along with weeding and intercultural operations. Top dress with 40:50 N and K one month later along with shallow intercultural operations.
Irrigation	It is mostly raised as a rainfed crop. However, irrigation is required when monsoon fails where it is grown on a large scale.
Yield	30 – 35 t/ha in 240 days.

Asparagus

Basic Information and Facts

Scientific name	***Asparagus officinalis* L.**
Family	Asparagaceae
Origin	Temperate Europe and Asia
Chromosome number	2n=20
Sex form	Dioecious
Pollination	Cross-pollinated crop

- Asparagus (***Asparagus officinalis*** **L.**) is a herbaceous, **dioecious,** perennial plant that is 1 to 3 m in height, with either staminate or pistillate flowers.
- The **plant** is composed of **ferns**, a **crown**, and the **root system**.
- The fern is a **photosynthetically active modified stem.**
- The crown is a **series of rhizomes (underground root-like stems)** attached to the base of the main plant.
- New crown buds, from which spears (immature ferns) arise, are formed the previous year.
- As a **dioecious** crop, asparagus is inevitably **cross-pollinating**. Male and female flowers are borne on different plants.

- The **immature shoots or spears** are eaten. They are thick and juicy.
- The tender shoots contain a white crystalline substance, **asparagine,** which is used in medicine as diuretic in cardiac dropsy and chronic gout.

Production Considerations

Climate	Asparagus favours temperate climates. Optimal root and fern (foliage) growth occurs from 15.5° to 23.8°C.
Soil	Well drained soils are a must for successful production, and very sandy soils are preferred.
Improved varieties	Marth Washington, Mary Washington
Propagation	Through seeds or crowns
Seed rate	4kg/ha
Season	**Hills:** March- May, **Plains:** July- November
Spacing	15-20 cm x 5-7cm
Manures and fertilizers	A basal dose of FYM 20-25 t/ha and N 20-30 kg, P 80-100 kg and K 60-80 kg/ha.
Irrigation	The first irrigation is given at the time of planting and thereafter at 10-15 days intervals.
Harvesting	The aerial shoots called spears arise the buds on the rhizome six months after transplanting of crowns or seedlings. The first cutting should be done two year of planting.
Yield	1.2 – 3.0 tonnes/ha
Storage	At 0°C and 90% RH for a period of 3-4 weeks.

Globe Artichoke

Basic Information and Facts

Scientific name	***Cynara scolymus* L.**
Family	Asteraceae (previously known as Compositae)
Origin	Mediterranean region
Chromosome number	2n= 34
Inflorescence type	Capitulum
Pollination	Cross-pollinated crop
Fruit type	Achenes

- Globe artichoke **(*Cynara scolymus* L.)** is an herbaceous perennial that is grown for its **tender, edible, immature flower buds.**
- *Cynara cardunculus* L. is a **diploid (2n = 34),** mostly **cross-pollinated**, species that belongs to the family **Asteraceae.**
- Cross-pollination is largely assured by **protandry,** and the principal pollination vector is the **honeybee**.

- The fruits are **achenes** of elliptical shape, dark coloured, with a prominent pappus that contributes to wind dispersal.
- The edible parts of the globe artichoke are the **immature composite inflorescences (heads or buds, more formally referred to as capitula),** and these are consumed worldwide as both a fresh and a canned delicacy.
- It is an **allogamous** and **entomophilous** plant, native to the **Mediterranean region.**
- **Cynarin,** the main constituent in globe artichokes, has a powerful effect on the production of bile and fat digesting enzymes, stimulating liver function and lowering cholesterol levels.

Production Considerations

Climate	It is a cool season crop that can be grown successfully at a temperature of 12-18°C.
Soil	Glove artichoke grows on many types of soil but deep, sandy, well-drained soil with adequate organic matter is ideal. The optimum soil pH is 6-7.5.
Exotic varieties	Emerald, Green Glove, Purple Glove, Romanesco, Grand Beurre, Purple de Jesi
Propagation	Through seeds or suckers or off-shoots.
Time of planting	February- March
Spacing	2.4m x 1.8m
Manures and fertilizers	A basal dose of 25 t/ha of FYM and N 60-100 kg, P 50-60 kg and K 150-200 kg/ha.
Irrigation	Artichokes require adequate soil moisture during the vegetal and reproductive growth phases. A soaking irrigation is given at the time of planting and in early stage frequent irrigation is necessary for good establishment of crop.
Weed management	Application of Trifluralin 1.2 kg+ Diuron 1kg/ha before planting gives good control of broad leaved grass weeds.
Harvesting	Artichokes should be harvested when the buds have achieved maximum size. In plants raised from suckers, the first harvest commences 6-7 months after planting. Artichokes are cut by hand with 7,5 to 10 cm of stem remaining with the bud.
Yield	10-12 tonnes/ha
Storage	Artichokes will store well for three to four weeks under proper cold storage conditions. Ideal storage temperature and humidity conditions are **32° to 34°F and 90% to 95% humidity.**

Lettuce

Basic Information and Facts

Scientific name	*Lactuca sativa* L.
Family	Asteraceae (previously known as Compositae)
Origin	Middle East (Egypt and Iran)
Chromosome number	2n= 18
Inflorescence type	Capitulum
Pollination	Self-pollinated crops
Fruit type	Aachene

- Lettuce (***Lactuca sativa***) is the most important crop in the group of leafy vegetables.
- The genus ***Lactuca*** L. belongs to the family **Asteraceae,** the largest of the dicotyledonous families (Judd et al. 1999; Funk et al. 2005).
- Recent evidence indicates that the origin of cultivated lettuce is **polyphyletic** (DeVries 1997).
- The region of the **Middle East (Egypt and Iran)** is considered a centre of lettuce origin.
- ***Lactuca sativa*** is an annual glabrous herb with a thin tap root and an erect stem 30–100 cm tall, branched in the upper part.
- **Leaves** are spirally arranged, forming a dense rosette or a head before bolting.
- The **inflorescence** of lettuce **(capitulum),** contains approximately 24 florets. They are highly developed for **self-pollination** and the crop is therefore largely **self-fertilizing.**
- **Anthocyanin** can be distributed on the cotyledons and true leaves, stems and ligules.
- The **involucre** is 10–15 mm long, cylindrical; involucral bracts are broadly to narrow lanceolate, light green, with white margins, erect at the stage of fruit maturity.
- The **fruit (achene)** has 5 to 7 setose ribs on each side, a beak and a white pappus. Its length (including beak) is 6–8 mm, and its colour is white, cream, gray, brown or black.
- It is a **diploid** with a basic chromosome number of ***n* = 9**.

Production Considerations

Climate	Lettuce is a cool season crop that grows best within a temperature range of 13 to 16°C.
Soil	The plant grows well on a wide variety of soils ranging from light sand to heavy clay, whoever, best results are obtained on fertile loams that are rich in organic matter.
Improved varieties	Great Lakes, Chinese Yellow, Slow Bolt
Seed rate	400-500 g/ha
Season	October- November
Spacing	45 × 45 cm
Manures and fertilizers	FYM 10-15 t/ha, N 50 kg/ha, P 90 kg/ha and K 25 kg/ha.
Irrigation	Pre-sowing irrigation is required in nursery/seed-sown field. Similarly it requires a good irrigation after transplanting. Subsequently, weekly irrigation is sufficient.
Yield	10-15 t/ha

Ceylon spinach/ Malabar spinach/ Indian spinach

Basic Information and Facts

Scientific name	*Basella alba* L.
Family	Basellaceae
Origin	South Asia
Chromosome number	2n= 48
Pollination	Self-pollinated species

- ***Basella alba* L.** (Synonym: *Basella rubra* Roxb.) is an extremely heat tolerant, fast growing perennial vine which belongs to family **Basellaceae**.
- It is commonly known as **Malabar spinach, Indian spinach, Ceylon spinach, vine spinach**, **climbing spinach**, **East-Indian spinach, Chinese spinach** and **cyclone spinach**.
- Basella is native to tropical **Southern Asia**, probably originated from **India or Indonesia.**
- It appears to be a **self-pollinated species** judging from the characteristics of its flowers and performance at AVRDC.
- The plant is a climbing vine with thick fleshy stem and leaves.
- The succulent leaves with petiole and tender stems are cooked and eaten as a vegetable.
- Flowers are pink or white, small sessile in clusters on elongated thickened peduncles
- *Basella alba* is a good source of vitamin A, vitamin C, vitamin B_9 (folic acid), calcium, magnesium, iron and several vital antioxidants in the plant.
- *Basella alba* contains **basellasaponins** and peptide, phenolic compounds.

Production Considerations

Climate	Basella grows very well in warm moist climate. It can be successfully grown under partial shade.
Soil	A wide range of soil from sandy loam to clayey soil with sufficient organic matter will be the best suited.
Propagation	Through seeds as well as stem cuttings
Seed rate	12-15kg/ha
Spacing	60 x 20 cm
Manures and fertilizers	20-30 tonnes FYM and 60:60:40 kg NPK/ha
Yield	15-20 tonnes green matter per ha.

Cole Crops

- Cole crops are one of the most important groups of cool season vegetables. Initially, these crops were confined to Europe and other temperate regions.
- But after the Second World War, their cultivation has spread to tropics and subtropics as well.
- All the members of this group have originated primarily from their common ancestor wild cabbage (***Brassica oleracea* var. *oleracea***) and are fully cross-compatible with each other.
- In general, these are **highly cross-pollinated crops** and have shown preponderance of non-additive gene action for most of the economic traits thereby suggesting the relevance of heterosis breeding.
- In the absence of appropriate genetic mechanism for the production of hybrid seed with ease at relatively lower cost, heterosis in a crop is of no practical value.
- Fortunately, the genetic phenomena of **sporophytic self-incompatibility** and **male sterility** (particularly **cytoplasmic male sterility**) exist in cole crops and have proved instrumental in commercialization of hybrid seed production.

Self-incompatibility

- The term "self-incompatibility" refers to the partial or complete incapacity of a plant, producing functional gametes, to set viable seed on self-pollination. Alternatively, it is the failure of the genetically identical gametes to unite in fertilization.
- In cole crops, **sporophytic self-incompatibility (SSI)** exists wherein the incompatibility reaction is governed by the genotype of the pollen producing plant and it is **controlled by a series of alleles at a single locus 'S'.**
- SSI system has been confirmed in kale (Thompson, 1957), sprouting broccoli (Sampson, 1957), cabbage (Adamson, 1965) and cauliflower (Hoser-Krauze, 1979).
- However, its use in cabbage hybrid seed production had been demonstrated much earlier (Odland and Noll, 1950).

Sprouting Broccoli

Basic Information and Facts

Scientific name	***Brassica oleracea* var. *italica* Plenck**
Family	Brassicaceae (previously known as Cruciferae)
Origin	Eastern Mediterranean region and Italy
Chromosome number	2n= 18
Inflorescence type	Cyme
Pollination	Highly cross-pollinated crop
Pollinating agents	Honey bees and flies
Self-incompatibility system	Sporophytic self-incompatibility (SSI)
Male sterility	Genic male sterility

- Broccoli is the most nutritious member of the **Cole group** belonging to family **Brassicaceae.**
- Its marketable plant part is the **"head" composed of immature flowers or florets.**
- Basic chromosome number is 9 with **2n=18.**
- Broccoli has **sporophytic self-incompatibility (SSI)** system **controlled by a single gene with multiple alleles known as 'S' alleles.**
- **Genic male sterility** has been reported in calabrese and purple broccoli by Cole (1959) and Brocher (1996).
- Genic male sterility is **governed by the recessive male sterility gene (ms)** mutated from the male fertile gene (ms).
- In broccoli, several male sterile genes ms1, ms2, ms4, ms5 and ms6 have been identified which are not allelic.
- Cytoplasmic male sterility has not been found apparently in *B. oleracea*. It was transferred into *B. oleracea* from ***B. nigra*** *by* **Pearson** through backcrosses and embryo rescue technique.
- The inflorescence in broccoli is a **cyme.**
- It is a **highly cross-pollinated crop** in nature with cross-pollination upto 95%.
- Main pollinating agents are honeybees and flies.
- Broccoli is highly nutritive and is a rich source of vitamins, minerals, protein etc.
- It contains about 130 times more vitamin A contents than cauliflower and 22 times more than cabbage.
- It is also a rich source of **sulphoraphane,** a compound associated with reducing the risk of cancer.

Production Considerations

Climate	It is mainly grown as a winter crop in most parts of the country, however in high hills it may be grown as a spring summer crop. An optimum temperature of 12-18°C is suitable for proper head development.
Soil	Broccoli needs fertile soil with good moisture holding capacity for its cultivation.
Improved varieties	Palam Samridhi, Palam Haritika, Palam Kanchan, Palam Vichitra, Pusa Broccoli Kt Selection-I and Punjab Broccoli-I.
Seed rate	300-400g/ha
Season	Mid-August to mid-September in north Indian plains.
Spacing	40 x 30 cm, 45 x 45 cm, 50 x 45 cm, 50 x 50 cm.
Manures and fertilizers	15-20 tonnes FYM and 100-125:60-80:25-50 kg NPK/ha
Irrigation	Provides ample soil moisture as the heads develop, especially during dry periods.

Weed management	Application of **Alachlor (Lasso)** @ 2kg a.i./ha or **Trifluralin**@ 0.5 kg/ha or **Fluchloralin** @ 0.5 kg/ha before transplanting is beneficial for controlling annual and broad leaved weeds. **Pendimethalin** (Stomp) @1.2 kg a.i. /ha or **Oxyflurofen** (Goal) @ 600 ml/ha) can also be used before transplanting if there is problem of annual weeds only.
Harvesting	The compact clusters of **unopened flower buds and their attached sections of stems** are the edible portions of broccoli. Its green buds develop in one large central head surrounded by several smaller side shoots. Cut the central head with 5 or 6 inches of stem after the head is well developed but before it begins to loosen, separate or the flowers start to open. Removing the central head stimulates the side shoots, which grow from the axis of the lower leaves, to develop for later harvesting. Continue to harvest broccoli for several weeks
Yield	15-20t/ha.

Brussels Sprouts

Basic Information and Facts

Scientific name	***Brassica oleracea* var. *gemmifera* DC.**
Family	Brassicaceae (previously known as cruciferae)
Origin	Mediterranean region
Chromosome number	2n= 18
Inflorescence type	Racemose
Pollination	Highly cross-pollinated crop
Pollinating agents	Honeybees and flies
Self incompatibility system	Sporophytic self-incompatibility (SSI)
Male sterility	Genic male sterility

- The Brussels sprout (***Brassica oleracea* var. *gemmifera***) is a **cool-season cole crop** that is related to broccoli, cabbage and cauliflower.
- Basic chromosome number is 9 with **2n=18.**
- The sprouts are **buds or small heads** that are produced in the leaf axils (the space between the base of the leaf and the stem above it).
- The edible part of Brussels sprouts is **swollen axillary buds**, known as 'sprouts' or 'buttons' or 'mini-cabbages'.
- Inflorescence: **Brussels sprouts** is a **raceme.**
- It is a **highly cross-pollinated crop** in nature and cross pollinates upto 72%.
- Main pollinating agents are honeybees and flies.
- Brussels sprouts has **sporophytic self-incompatibility (SSI)** system **controlled by a single gene with multiple alleles known as 'S' alleles.**
- **Genic male sterility** has been reported in Brussels sprouts by **Johnson (1958).**
- Sprouts mature starting at the base of the stem and working upward.

Production Considerations

Climate	Mild winter and cool summer are ideal conditions for growing of this crop. In India it is mainly grown as a winter crop. Plants take optimum growth when temperature is 16°-20° C.
Soil	The sandy loam soils are preferred for early and clay loam for main season crops. The pH should be between 6.0 and 6.5 for optimum growth.
Improved varieties	Jade Cross, Hills Ideal
Seed rate	300-400g/ha
Season	**Plains:** August- September
Spacing	60 x 45 cm, 60 x 60 cm
Manures and fertilizers	15-20 tonnes FYM and 100-150:50-150:100 200 kg NPK/ha
Irrigation	The crop is irrigated immediately after transplanting of seedlings and subsequent irrigation depends upon the climate and soil conditions.
Weed management	Application of **Alachlor (Lasso)** @ 2kg a.i./ha or **Trifluralin**@ 0.5 kg/ha or **Fluchloralin** @ 0.5 kg/ha before transplanting is beneficial for controlling annual and broad leaved weeds. **Pendimethalin** (Stomp) @1.2 kg a.i. /ha or **Oxyflurofen** (Goal) @ 600 ml/ha) can also be used before transplanting if there is problem of annual weeds only.
Harvesting	The small sprouts or buds form heads 1 to 2 inches in diameter. They may be picked (or cut) off the stem when they are firm and about 1 inch in size. The lower sprouts mature first.
Yield	3 - 5 t/ha.
Storage	Store in a cooler or cold cellar at 36°F/2°C with 95– 98% relative humidity. They will store for 4–6 weeks under these conditions.

Cabbage

Basic Information and Facts

Scientific name	***Brassica oleracea* var. *capitata* L.**
Family	Brassicaceae (previously known as cruciferae)
Origin	Mediterranean region
Chromosome number	2n= 18
Inflorescence type	Raceme
Pollination	Cross-pollinated crop
Self incompatibility system	Sporophytic self-incompatibility
Fruit type	Silique

- Cabbage (*B. oleracea var. capitata*) is a member of the genus *Brassica and the mustard family,* Brassicaceae.
- Basic chromosome number is 9 with **2n=18.**

- The economic part of cabbage which we eat is termed as **"head".**
- Cabbage is a temperate crop, **thermoperiod** is the most important factor for its flower induction.
- The inflorescence is an unbranched and indeterminate terminal **raceme** measuring 50–100 cm (20–40 in) tall, with flowers that are yellow or white.
- Each flower has four petals set in a perpendicular pattern, as well as four sepals, six stamens, and a superior ovary that is two-celled and contains a single stigmaand style. Two of the six stamens have shorter filaments.
- The fruit is a **silique** that opens at maturity through dehiscence to reveal brown or black seeds that are small and round in shape.
- The **glycoproteins** present in the stigma hinder the pollen germination and penetration.
- **Protogyny** is the phenomenon where the stigma matures earlier than the anthers.
- Cabbage is a **cross-pollinated crop**, pollination occurs mainly through **bees.** Bumble bees are also some times observed.

Genetic Mechanisms

- Cabbage has **sporophytic self-incompatibility (SSI)** system **controlled by a single gene with multiple alleles known as 'S' alleles.**
- Although sporophytic self-incompatibility system was confirmed in cabbage by Adamson (1965), but the methods of seed production by utilizing self-incompatible parental lines had been suggested by Odland and Noll (1950) much earlier.
- Recessive **nuclear male sterility** has been reported by **Rundfeldt (1960)** and **Nieuwhof (1961)** but not commercialized due to the problems encountered in its maintenance and the absence of effective seedling markers.
- **Pearson (1972)** reported male sterility controlled by *Brassica nigra* nuclear-cytoplasm interaction but this could not be used on commercial scale due to partial opening of petals and less developed nectar glands leading to poor bee activity.
- **Ogura male sterility (1968)** was reported in Japanese radish. No restorer nuclear genes could be found in Japanese genotypes but the same could be obtained in European radish.
- **Bannerot *et al.* (1974)** could introduce cabbage nucleus into the cytoplasm of Ogura male sterile radish (Rcytoplasm) by repeated back crossing.
- This CMS in cabbage had the problem of chlorosis at low temperature (below 12°C). This drawback was overcome by replacing radish chloroplasts with *Brassica oleracea* chloroplasts through protoplast fusion by Robertson *et al.* (1987).
- **Sigareva and Earle (1997)** standardized the protoplast fusion procedure to transfer desirable male sterility cytoplasm from broccoli into cabbage much more rapidly than conventional back cross method.
- No fertility restorer genes have been found for Rcytoplasm (Ogura) induced male sterility in cole crops (*Brassica oleracea*).
- This implies that all *Brassica* spp. including cole crops act as maintainers of R-cytoplasm (Ogura) induced male sterility.

- Fang *et al.* (1997) have also reported the presence of a dominant male sterile gene in cabbage. Its usage may not be preferred over R-cytoplasm (Ogura)-induced male sterility because of the possible problems associated in maintaining the male sterile lines in homozygous dominant condition.

Commercialization of Hybrid Seed Production

- Initially, it was only the sporophytic self-incompatible system which was commercialized for hybrid seed production.
- In this system, two heterotic but cross-compatible S-allele lines are grown in alternate rows in a panmictic plot to allow for free cross-pollination and the hybrid seed is harvested from all the plants of both the parental S-allele lines.
- IARI, Regional Station, Katrain, has developed cabbage hybrid **KGMR-1** by using SSI and the same has been identified through AICRP on vegetable crops.
- With the successful transfer of R-cytoplasm (Ogura)-induced CMS system in cole crops (including cabbage), there has been a lot of interest in its use for hybrid development during for over a decade now.
- Cabbage has anti-cancerous properties, it protect against bowel cancer due to the presence of **indole-3-cabinol.**

Production Considerations

Climate	It thrives best in a relatively cool moist climate. It can withstand extreme cold and frost relatively better than cauliflower. The optimum temperature for growth and head formation is 15-20°C whereas the growth is checked above 25°C.
Soil	It is grown in varied types of soils ranging from sandy loam to clayey. It requires a pH range of 5.5 to 6.5. Most cabbages are somewhat **tolerant to salt.**
Seed rate	375-500g/ha
Improved varieties	Golden Acre , KGMR-1, Pusa Ageti , Pusa Drum Head, Pusa Mukta
Season	**Hills:** January – February, July – August and September – October. **Plains:** August – November.
Spacing	**Early varieties :** 45 cm × 30 cm or 30 cm × 30 cm (round & smaller heads) **Late varieties :** 60 cm × 45 cm or 60 cm × 60 cm
Manures and fertilizers	Mix 200-250 q/ha farmyard manure thoroughly at the time of field preparation. Application of 120-180 kg nitrogen, 75-80 kg phosphorus and 60-75 kg potassium per hectare is required to raise a healthy crop of cabbage.
Irrigation	Irrigation may be applied at 10-15 days intervals according to the season and soil but optimum soil moisture should be maintained regularly.

Weed management	Application of **Alachlor (Lasso)** @ 2kg a.i./ha or **Trifluralin**@ 0.5 kg/ha or **Fluchloralin** @ 1.5 kg a.i./ha before transplanting is beneficial for controlling annual and broad leaved weeds. **Pendimethalin** (Stomp) @1.2 kg a.i. /ha or **Oxyflurofen** (Goal) @ 600 ml/ha) can also be used before transplanting if there is problem of annual weeds only.
Growth regulator	Foliar spray of **Cytozyme (0.2%)** after 15 and 30 days of transplanting of seedlings has improved curd, head and knobs yield significantly.
Yield	**Early varieties:** 25-30t/ha, **Late season varieties:** 40-50t/ha.

Cauliflower

Basic Information and Facts

Scientific name	***Brassica oleracea* var. *botrytis* L.**
Family	Brassicaceae (previously known as cruciferae)
Origin	Mediterranean region
Chromosome numbers	2n= 18
Type of inflorescence	Racemose
Pollination	Cross-pollinated crop
Self incompatibility system	Sporophytic self-incompatibility
Fruit type	Siliqua

- Cauliflower is one of several vegetables in the species *Brassica oleracea*, in the family **Brassicaceae.**
- Basic chromosome number is 9 with **2n=18.**
- It is an annual plant that reproduces by seed.
- Typically, only the head (the *white curd*) is eaten. The cauliflower head is composed of white **inflorescence meristem.**
- Cole crops (*Brassica oleracea*) are characterized by diverse morphological forms that likely resulted from selection for various edible parts. Examples include, **leaves that form a head** [cabbage (*B. oleracea* var. *capitata* L.)], **non-heading leafy types** {kale and collard greens (*B. oleracea* var. *acephala* DC), chinese kale [*B. oleracea* var. *alboglabra* (L.H. Bailey) Musil]}, an **enlarged stem** [kohlrabi (*B. oleracea* var. *gongylodes* L.)], **axillary buds** {brussels sprouts [*B. oleracea* var. *gemmifera* (DC) Schultz]}, and **immature inflorescences** [broccoli (*B. oleracea* var. *italica*), cauliflower (*B. oleracea* var. *botrytis*)].
- In *Brassica* species, self-incompatibility is controlled **sporophytically** by a single Mendelian genetic locus, the **S-locus.** .
- It is a **cross-pollinated crop**. There are different groups based on their characteristics.
- Multiple pollination mechanisms, e.g., self-incompatibility and male sterility, not only encourages cross-pollination but is also found useful in the commercial hybrid seed production of the crop.

Production Considerations

Climate	It requires cool moist climate. It can grow at an average temperature of 5°-8°C to 25°-28°C. The optimum temperature for growth of young plants is around 23°C and 17°-20°C in later stages.
Soil	Cauliflower can be grown in all types of soil with good fertility and good water holding capacity. It prefers a soil reaction ranging from pH **6 to 6.5.**
Varieties/ hybrids	**Early varieties -** Early Kunwari, Kashi Kunwari, Pusa Katki, Pusa Deepali, Pusa Early Synthetic, Improved Japanese, Pant Gobhi-3 **Mid season varieties -** Pant Shubhra, Pusa Synthetic, Pusa Shubhra, Gobhi-4, Hisar-1 **Late varieties -** Pusa Snowball-1, Snowball-16, Pusa Snowball K-1, Pusa Himjyoti
Seed rate	Early varieties: 600-750g Mid-early season varieties: 500g Mid-late varieties: 400 g Late varieties: 300g
Season	Mid-May to Mid-October.
Spacing	**Early varieties:** 45 x 30cm **Mid and late season varieties:** 60 x 45cm
Manures and fertilizers	FYM: 200-250 q/ha and nitrogen, phosphorus and potash @ 120-180: 75-80: 60-75 kg/ha.
Irrigation	Once in a week.
Weed management	Application of **Alachlor (Lasso)** @ 2kg a.i./ha before transplanting is beneficial for controlling annual and broad leaved weeds. **Pendimethalin (Stomp)** @1.2 kg a.i./ha or **Oxyflurofen (Goal)** @ 600 ml/ha) can also be used before transplanting if there is problem of annual weeds only.
Growth regulator	Foliar spray of **Cytozyme (0.2%)** after 15 and 30 days of transplanting of seedlings has improved curd, head and knobs yield significantly.
Yield	**Early varieties:** 100-150 q/ha. **Mid and late season varieties:** 150-300 q/ha. **Snowball group** may produce yield upto 400 q/ha.

Knol-khol

Basic Information and Facts

Scientific name	***Brassica oleracea* var. *gongylodes* L.**
Family	Brassicaceae (previously known as cruciferae)
Origin	Central countries of northern Europe
Chromosome numbers	2n= 18
Type of inflorescence	Racemose
Pollination	Cross pollinated cop
Self incompatibility system	Sporophytic self-incompatibility

- Knol-khol (***Brassica oleracea* var. *gongylodes* L.**) is a **cole crop** which belongs to family **Brassicaceae.**
- Basic chromosome number is 9 with **2n = 18.**
- The inflorescence in knol khol is a **raceme.**
- It is a **highly cross-pollinated crop** in nature and cross-pollination upto 91%.
- Main pollinating agents are honey-bees and flies.
- Knol khol has **sporophytic self-incompatibility (SSI)** system **controlled by a single gene with multiple alleles known as 'S' alleles.**

Production Considerations

Climate	Knol-khol thrives best in a relatively cool, moist climate. It grows well with a monthly average temperature of 15°-25°C.
Soil	A soil rich in manures and fertilizers produces excellent knobs. Sandy loam is ideal for an early crop and clay or silt loam for higher yield and late crop.
Improved varieties	Purple Vienna, White Vienna and Palam Tender Knob.
Seed rate	800-1000g/ha
Season	Planting time under Northeastern plains is September-October.
Spacing	30-40 × 20-25 cm
Manures and fertilizers	Application of 30 t/ha of FYM and nitrogen, phosphorus and pot-ash @ 75- 100: 60-80: 60-80 kg/ha, respectively is required to raise a healthy crop of knokhol.
Irrigation	Knol-khol requires a continuous supply of moisture for uniform growth and development of knobs. First irrigation is done immediately after transplanting and thereafter irrigation is done when needed.
Weed management	Application of **Trifluralin**@ 0.5 kg/ha and **Desmetrayne**@1kg/ha before transplanting control both monocot and diocot weeds.
Yield	12 - 20 t/ha.

Radish

Basic Information and Facts

Scientific name	***Raphanus sativus* L.**
Family	Brassicaceae (previously known as cruciferae)
Origin	Mediterranean region
Chromosome numbers	2n= 18
Type of inflorescence	Raceme
Pollination	Cross pollinated crop
Pollination mode	Entomophilous
Self incompatibility system	Sporophytic self incompatibility
Male sterility	Gene- cytoplasmic male sterility
Fruit type	Silique

- The radish *(Raphanus sativus* **L.)** is an edible root vegetable *of the* **Brassicaceae** *family.*
- It has 18 chromosomes *(2n=18).*
- The characteristic pungent flavour is due to the presence of **volatile isothiocynates.**
- The edible portion of radish **root** develops from both primary root and the hypocotyls Radish roots vary in size, shape and other external characters as well as the length depending on cultivars.
- The **flowers** are **hermaphrodite**, insect-pollinated and small.
- .The inflorescence is the typical terminal **raceme** of the crucifers, and the flowers are white, rose or lilac in colour.
- Radish is highly **cross-pollinated crop**, pollination occurs primarily by **honeybees.**
- The mode of pollination is **entomophilous.**
- Dehiscence generally takes place at warmer temperature condition.
- Radish has a **sporophytic system of self-incompatibility** determined by **S alleles** at one locus.

Male Sterility

- Ogura (1968) reported **gene- cytoplasmic male sterility** in radish which is due to interaction between recessive nuclear gene *ms ms* and sterile (*s*) cytoplasm.
- Later it has been reported that there are two pairs of *ms* genes in the nucleus (Kalloo & Bergh, 1993) hence the genotype of the male sterile line (A) could be $sms_1\ ms_1\ ms_2\ ms_2$, maintainer line (B) $Nms_1\ ms_1\ ms_2\ ms_2$; pollen parental line (C) $NMS1\ MS_1$ $MS_2\ MS_2$.
- Ogura (1968) described the first sterile cytoplasm in a Japanese population of radish.

Production Considerations

Climate	It is predominantly a cool season crop and best adapted to cool or moderate climate. Roots of best size, flavour and texture are developed at about 10-15°C.
Soil	Sandy loam soils with high organic matter. The optimum soil pH is 5.5 to 6.8.
Varieties/ hybrids	**Asiatic/Tropical types:** Kashi Shweta, Kashi Hans, Pusa Chetki, Pusa Reshmi, Japanese White, Kalyanpur No.1, CO-1, Chinese Pink, Arka Nishant **European/Temperate types:** White Icicle, Pusa Himani, Rapid Red White Tipped, Scarlet Globe
Seed rate	**Asiatic type:** 10 kg/ha **European type:** 12-14 kg/ha
Season	**European type:** September-March, **Asiatic type:** August-January
Spacing	**European type:** 30cm × 5-10cm, **Asiatic types:** 45 cm × 6-8 cm
Manures and fertilizers	Apply FYM at 25 t/ha and 50-90 kg N, 50-80 kg P and 40-80 kg K/ha.
Irrigation	Once in a week.

Weed management	Application of Pendimethalin 1.2 kg a.i./ha or Alachlor 1.5 kg a.i./ ha or Fluchloralin (Basalin)@ 0.9 kg a.i./ha or Isoproturan 1.0 kg a.i./ha or metalachlor @ 1.0 kg a.i./ha in 750 litres of water as pre-emergence is very useful for effective weed control.
Yield	**European type:** 50-80 q/ha, **Asiatic type** :200-500 q/ha

Turnip

Basic Information and Facts

Scientific name	***Brassica rapa*** **L.**
Family	Brassicaceae (previously known as cruciferae)
Origin	Mediterranean region
Chromosome number	2n= 2x=20
Inflorescence type	Terminal raceme
Pollination	Cross-pollinated crop
Self incompatibility system	Sporophytic self-incompatibility
Fruit type	Silique

- The turnip (***Brassica rapa*** subsp. ***rapa***) is a root vegetable commonly grown in temperate climates worldwide for its **bulbous tap root.**
- Turnip has a chromosome number of **2n = 2x = 20.**
- The fleshy thickened underground portion of turnip is actually the **hypocotyls.**
- The inflorescence is a **terminal raceme** on the main stem.

Production Considerations

Climate	Turnip is best adapted to a cool or moderate Climate. The roots develop the best flavour, texture and size at a temperature of 10-15°C.
Soil	The well-drained, sandy loam soil is ideal for its cultivation.
Varieties/ hybrids	**Asiatic/tropical/subtropical type:** Pusa Kanchan, Pusa Sweti, Punjab Safed 4. **European/temperate type:** Purple Top White Globe, Golden Ball, Snow Ball, Early Milan Red Top, Pusa Chandrima, Pusa Swarnima. In India, Purple Top White Globe (PTWG) is the most common variety.
Seed rate	3-4kg/ha
Season	**Hills:** July- October **Plains:** September–December
Spacing	30-45 x 7.5-15 cm
Manures and fertilizers	Application of nitrogen, phosphorus and potash @ 50-90: 50-80: 50-80 kg per hectare respectively is required to raise a healthy crop of turnip.
Irrigation	Pre-sowing irrigation is essential for better germination of seeds. The water requirement would be about 300mm. In general, the crop is irrigated at 7 days intervals till harvest.
Yield	20-25 t/ha

Sweet Potato

Basic Information and Facts

Scientific name	*Ipomoea batatas* (L.) Lam.
Family	Convolvulaceae
Origin	South America
Chromosome number	X=15, 2n= 90
Ploidy level	Hexaploid
Inflorescence type	Cyme
Pollination	Cross-pollinated crop
Fruit type	Capsule

- The sweet potato (***Ipomoea batatas***) is a **dicotyledonous** plant that belongs to the **family Convolvulaceae.**
- It is a perennial, although it is grown as an annual.
- The sweet potato is a plant that probably originated in or near northwestern **South America.**
- The number of chromosomes in the sweet potato plant is **2n = 6x = 90.** This indicates that it is a **hexaploid** plant with a basic **chromosome number x = 15.**
- The **inflorescence** is generally a **cyme** in which the peduncle is divided into two axillary peduncles; each one is further divided into two after the flower is produced **(biparous cyme).**
- The **flower is bisexual**. Besides the calyx and corolla, they contain the stamens that are the male organs or androeclum and the pistil that is the female organ or gynoeclum.
- The **fruit is a capsule,** more or less spherical with a terminal tip, and can be pubescent or glabrous.
- The **storage roots are the commercial part of the sweet potato plant.**
- It is a creeper whose stems have internodes from which roots arise. Just beneath the surface of the soil, thickening of the adventitious roots produces a dozen or so **tubers,** fusiform or globular, smooth or furrowed, white, yellow, orange, red or violet.
- The plant is a herbaceous perennial vine, bearing alternate heart-shaped or palmately lobed leaves and medium-sized sympetalous flowers.
- **Leaves** are simple, arranged in spirals, entire or deeply lobed, oval, green or with a violet hue, 15 x 15 cm; petiole 5–30 cm long.
- The **edible tuberous root** is long and tapered, with a smooth skin whose colour ranges between yellow, orange, red, brown, purple and beige.

Production Considerations

Climate	Sweet potatoes are moderately drought- tolerant. A warm humid climate with reliable rainfall gives best yields.
Soil	It can be grown in loamy soil with a pH range of 5.6-6.6.
Improved varieties	CO3, CO - CIP 1, Sree Nandini, Sree Vardhini, Kiran, Sree Bhadra, Sree Rethna, Gouri and Sankar.
Propagation	**Vine cuttings**
Seed rate	Plant the **terminal vine cuttings (83,000/ha)**. The cuttings should be 20-45 cm in length with 3-5 nodes and to be collected from matured vines aged 3 months and above.
Season	June – July and September.
Spacing	60 x 20 cm.
Manures and fertilizers	Apply 25 t FYM/ha and 40:40:60 kg NPK/ha.
Irrigation	Irrigate before planting, on 3rd day and thereafter once a week. Stop irrigation one week before harvest.
Growth regulator	Application of **ethephon** (250 ppm), **CCC** (50-150 ppm), **Kinetin** (50-100 ppm) and **GA_3** has resulted in higher tuber yield.
Yield	5-15 t/ha

Cucurbitaceous Vegetables

- The **Cucurbitaceous** or vine crop family includes many important vegetables collectively referred to as cucurbits.
- Cucurbitaceae is one of the leading vegetable crops family of tropical and subtropical zone of the globe.
- The crops are known as reservoir of several essential nutrients, minerals, vitamins, dietary fibres and a number of nutraceuticals and phytomedicinal compounds.
- Most of the plants are warm season annuals and trailing herbs.
- Cucurbits have diverse daily use such as in salad (cucumber), dessert (watermelon and muskmelon), candy (wax gourd), utensil (bottle gourd), musical instruments (bottle gourd) and above all are a means of fresh kitchen vegetables.
- From plant breeding point of view, the crops are **cross-pollinated** and **entomophilies.**
- The crops are having diverse sex forms ranging from monoecious, dioecious, hermaphrodite, androecious, gynoecious, to gynomonoecious etc.

Botany

- **Roots:** The root system of all the commercial cucurbits is extensive but shallow. Upon germination of the seed, the plants soon develop a strong tap-root which may penetrate the soil to a depth of 3 or 4 to 6 ft. Numerous horizontal laterals develop rapidly and spread widely in the soil, although branching of the tap-root is not extensive below the 2 ft level.

- **Stems:** All cucurbits are alike in their general stem morphology. The stems are branched (3 to *8),* prostrate, trailing, hirsute to scabrous and usually angled in cross section.
- **Fruits:** The fruits of the cultivated cucurbits vary largely in size, shape and colour. The fruit is often a kind of modified berry called a **pepo.** The fruit peduncle is 5 to 8 angular. The edible portion is the placentae in cucumber and watermelon, pericarp with very little mesocarp in pumpkin, while the whole fruit in bottle gourd.
- **Seeds:** The seeds of the cultivated cucurbits vary in size, shape, colour, the presence or absence of a margin and in the type of scar formed at the hilum. In general, each seed has a firm testa of several layers, a thin collapsed perisperm and endosperm and a large embryo. The embryo consists of two large, flat, leaf-like cotyledons and a small radical.

Taxonomy

- The Cucurbitaceae, also called cucurbits and the gourd family, are a plant family genera consisting of 965 species in around 95.
- The family Cucurbitaceae is divided into five sub-families: Fevilleae, Melothrieae, Cucurbitaceae, Sicyodeae, and Cyclanthereae.
- The most important cultivated genera are Cucurbita, Cucumis, Citrullus, Lagenaria and Luffa found in the subfamily Cucurbitaceae and Sechium found in the sub-family Sicyoiodeae.

Sex Expression and Sex Forms in Cucurbits

- Cucurbitaceous plants produce three types of flowers depending on the species and variety. One plant my contain more than one type of flowers, and more than one sex forms may be available in a species.
- According to the sex forms, cucurbitaceous plants can be grouped as follows :
  - **Hermaphrodite:** All flowers of a plant are bisexual (available in some varieties of *Luffa, Cucumis* and *Benincasa* spp.)
  - **Monoecious***:* Male and female sexes are in different flowers of the same plant (the most common sex form).
  - **Andromonoecious:** Male and hermaphrodite flowers are in the same plant.
  - **Gynomonoecious:** Female and hermaphrodite flowers are in the same plant.
  - **Trimonoecious or Androgynomonoecious:** Male, female and hermaphrodite flowers are in the same plant.
  - **Dioecious:** Male and female flowers are in different plants. Plants bearing male flowers are called androecious and those bearing female flowers are called gynoecious plants.
  - **Gynodioecious:** In dioecious species some plants of a variety bear only female flowers and others bear hermaphrodite flowers.
  - **Sub-gynoecious:** When gynoecious plants of dioecious species produce some male orhermaphrodite flowers.
- Monoecious sex form is the most common one in this family.

- Hermaphroditism is the original sex form from which monoecious and dioecious sex forms have been evolved.

Flowering

- Flowering in cucurbits normally starts in about 40-45 days after sowing depending upon the weather condition.
- The sequence of flowering follows a set of pattern, namely **(i) male phase:** First few nodes bear only the staminate flowers, **(ii) Mixed phase:** both pistillate and staminate flowers appear in few nodes in the main axis and secondary branches in cycles and **(iii) Female phase:** few nodes produce mostly the pistillate flowers.
- In a typical monoecious sex form of cucurbits, the ratio of staminate and pistillate flowers may range from 25 to 30:1 to 15:1, the later condition is advantageous and economical, because consequently it results in higher fruit set and yield.
- Generally high nitrogen, long days and high temperature promote greater number of staminate flowers.

Pollination and Fruit Set

- Most of the cucurbits are **cross-pollinated** because of their **monoecious** flowering behaviour.
- It should be pointed out that there are two popular cucurbitaceous vegetables—**Kakrol** and **pointed gourd**, which are of **dioecious** flowering habit leading to 100% cross-pollination.
- Hermaphrodite flowers also fail to set seed by self-pollination sometimes as the pollen grains are sticky.
- However, self-pollination between two sexes of the same plant is also common in cucurbits.
- Poor fruit setting in cucurbits is primarily assumed as the failure of pollination.
- However, this situation may be improved by hand pollination.
- Fruit setting is largely dependent on varietal characteristics, nutrition, disease status and environment.
- It also depends on the number of fruits already present in a plant. Therefore, harvesting of edible-mature fruits encourages new fruit setting.

Ash gourd

Basic Information and Facts

Scientific name	***Benincasa pruriens* f. *hispida* (Thunb.)**
Family	Cucurbitaceae
Origin	Java, Japan
Chromosome number	2n = 24
Sex form	Monoecious
Pollination	Cross-pollinated crop
Fruit type	Pepo

- *Benincasa* is a **monotypic genus** belonging to the family **Cucurbitaceae** with a single species, *B. hispida* (Thunb.) Cogn.
- Crop originated in Asia specifically in **Java and Japan.**
- Genus ***Benincas*a** is **monotypic** and the only species *is B. hispida.*
- Ash gourd is diploid with **2n=24**.
- This **monoecious** crop produces large male flowers with long pedicels and female flowers with densely haired ovary and short peduncle on same plant. Corolla is yellow in colour and large in size.
- Ash gourd has separate male and female flowers on the same plant. **Cross-pollination** is by insects
- Ratio of staminate to pistillate flowers is **34:1**.
- Anthesis takes place at 4.30 – 7.30 a.m. and anther dehiscence is at 3.00 – 5.00 a.m. Stigma is receptive from 8 hours before to 18 hours after anthesis.
- In India and Cuba, mature fruits are "candied", commonly known as **"petha".**

Production Considerations

Climate	It grows well in warm, humid tropical climate. The temperature of 22- 35°C is ideal.
Soil	A deep loamy soil with pH range of 6.5-7.5 is suitable.
Varieties/ hybrids	Pusa Ujjwal, Kashi Ujawal, CO 1, CO2, TNAU Ash gourd Hybrid CO1
Seed rate	5-7 kg/ha
Season	June-July and February-March
Spacing	1.5-2.0m × 0.60-1.2 m
Manures and fertilizers	FYM 15-20t/ha and 45-60:50-60:60-80kg NPK/ha.
Irrigation	Irrigate the basins before dibbling the seeds and thereafter once a week.
Yield	10-15 t/ha in 140 days.

Bitter gourd

Basic Information and Facts

Scientific name	***Momordica charantia* L.**
Family	Cucurbitaceae
Origin	Tropical Asia
Chromosome number	2n= 22
Sex form	Monoecious
Pollination	Cross-pollinated crop
Fruit type	Pepo

- The bitter gourd, ***Momordica charantia*** L. belong to the family **Cucurbitaceae**, subfamily Cucurbitoideae, tribe Joliffieae Schard and subtribe Thladianthinae Pax.
- The bitter gourd *M. charantia* L. is a diploid, **2n = 2x = 22.**
- The genus *Momordica* has only six valid species in India, which can be grouped under two headings: *M. charantia* L. and *M. balsamina* L. representing the **monoecious group** and *M. dioica* Roxb., *M. sahyadrica* Joseph & Antony, M. *cochinchinensis* (Lour.) Spreng., and *M. subangulata* Blume (ssp. renigem (G. Don) W.J.J. deWilde) representing the **dioecious group**.
- Plants are **monoecious** annuals with medium size climbing vines with thin stem and tendrils.
- Bitter gourd is a **cross-pollinated crop** and cross-pollination occurs through bees.
- The bitterness of bitter gourd is due to the cucurbitacin-like alkaloid **momordicine** and **triterpene glycosides**.
- Cytogenetic studies confirmed the diploid chromosome number **(2n = 22)** of *M. charantia*.
- Bitter gourd fruits are a good source of carbohydrate, protein, vitamins, and minerals and have the highest nutritive value among cucurbits.

Production Considerations

Climate	A long period of warm, dry weather with 30-35°C temperature is optimum.
Soil	Sandy loam rich in organic matter with good drainage and pH range of 6.5-7.5.
Varieties/ hybrids	Pusa Do Mousami, Pusa Hybrid-, Pusa Hybrid-1, Pusa Vishesh, Arka Harit, Kashi Urvasi, Pant Karela-1, Phule Green Gold, CO 1, COBgoH 1, MDU.1
Seed rate	4.5-5.0 kg/ha
Season	January- March and June-July
Spacing	1.5-2.0m × 1.0m
Manures and fertilizers	Apply FYM 20 t/ha and 100 g of NPK 6:12:12 mixture per pit as basal and 10 g of N per pit 30 days after sowing. Apply *Azospirillum* and *Phosphobacteria* 2 kg/ha and *Pseudomonas* @ 2.5 kg/ha along with FYM 50 kg and neem cake @ 100 kg before last ploughing.
Irrigation	Irrigate the basins before dibbling the seeds and thereafter once in a week.
Growth regulator	Application of **GA_3 (5-10 ppm) and BA (25 ppm) at** 2 and 4 leaf stages have been found effective to enhance number of female flowers, early harvest, more number of fruits/plant and high fruit yield.
Yield	**Varieties:** 15 t/ha in140 - 150 days. **Hybrids:** 40 t/ha in 180 days.

Bottle gourd

Basic Information and Facts

Scientific name	*Lagenaria siceraria* **(Molina) Standl.**
Family	Cucurbitaceae
Origin	South Africa and India
Chromosome number	2n= 22
Sex form	Monoecious
Pollination	Cross pollinated crop
Fruit type	Pepo

- Bottle gourd (***Lagenaria siceraria*** **(Mol.) Standl.**), of the family **Cucurbitaceae**, is a climbing perennial plant widely cultivated as a vegetable crop in tropical countries, such as India, Japan and Thailand.
- The bottle gourd flowers are borne singly on the axils of the leaves, the males on long peduncles and the females on short peduncles.
- The flowers are white and attractive, up to 4 inches in diameter, with spreading petals.
- The ovary is inferior and in the shape of the fruit. Otherwise, the male and female flowers are similar in appearance.
- The anthers are borne on short filaments grouped at the center of the flower.
- The stigmas are short, thickened, and branched.
- The brownish seeds are numerous in a whitish green pulp.
- Each seed is somewhat rectangular in shape with grooved notches near the attached end.
- Bottle gourd is a **cross-pollinated crop** and cross-pollination occurs through honey bees, moths and other night insects.

Production Considerations

Climate	Slightly wet to semi-dry ecological condition is suitable for this crop.
Soil	Sandy loams rich in organic matter with good drainage and a pH range from 6.5 to 7.5.
Varieties/ hybrids	Pusa Hybrid 3, Pusa Naveen, Pusa Samridhi, Pusa Sandesh, Pusa Santushti, Pusa Summer Prolific Long, Arka Bahar, Kashi Bahar, Kashi Ganga, Anand Bottle Gourd -1, Pant Lauki-4, Pant Sankar Lauki-1, Pant Sankar Lauki-2, Bottle Gourd (NDBG-619), N. Bottle Gourd (NDBG-132), N. Bottle Gourd Hybrid-4, N. Dharidar (NDBG-208-1), N. Jyoti (NDBG-104), N. Madhuri (NDBG-505), N. Rashmi (NDBG-1), N, Shishir (NDBG-202), N. Shivani (NDBG- 403), Samarat , CO 1, TNAU Bottle gourd Hybrid CO 1, Azad Nutan
Seed rate	3-6 kg/ha
Season	June-July and December-January
Spacing	Dig pits of 30 cm x 30 cm x 30 cm size at 2.5 x 2 m spacing.
Manures and fertilizers	Apply FYM 20 t/ha and 40-60:40-60:60-80 kg NPK/ha. Apply *Azospirillum* and *Phosphobacteria* 2 kg/ha and *Pseudomonas* 2.5 kg/ha along with FYM 50 kg and neem cake @ 100 kg before last ploughing.
Irrigation	Irrigate the field before dibbling the seeds and thereafter once a week.

Growth regulator	Application of **NAA (25 ppm), ethrel (250 PPM), boron (3 ppm) and calcium (5 ppm)** at 2 and 4 leaf stages have been found effective to enhance number of female flowers, early harvest, more number of fruits/plant and high fruit yield.
Yield	20 t/ha.

Cucumber

Basic Information and Facts

Scientific name	***Cucumis sativus* L.**
Family	Cucurbitaceae
Origin	India
Chromosome number	2n= 14
Sex form	Monoecious
Pollination	Cross pollinated crop
Fruit type	Pepo

- Cucumber (***Cucumis sativus*** **L.**) is a widely cultivated plant, belonging to the family **Cucurbitaceae** and having a chromosome number, **2n =14.**
- It is **monoecious** (separate male and female flowers on the same plant) and its axillary flowers are actinomorphic and rarely bisexual.
- Cucumber is a **cross-pollinated crop** with an extent of 0 – 5% of self–pollination.
- It is pollinated by bees that harvest the nectar produced by the flowers; **honeybees** are more effective than wild bees. The pollen is sticky, which makes wind pollination improbable.

Production Considerations

Climate	The optimum temperature favouring anthesis and dehiscence is 13-18°C.
Soil	Sandy loam rich in organic matter with good drainage and pH range of 6.5-7.5.
Varieties/ hybrids	Pusa Sanyog, Pusa Uday, Pant Khira-1, Pant Parthenocarpic khira-2, Pant Sankar Khira 1, Parthenocarpic khira-3, Himangni, Phule Shubangi, CO 1
Seed rate	2.5 kg/ha
Season	June-July and January-April
Spacing	1.5-2.5m x 0.6-0.9m
Manures and fertilizers	Apply FYM 40 t/ha and 200:125:125kgNPK/ha.
Irrigation	Irrigate the field before dibbling the seeds and thereafter once in a week
Weed management	
Growth regulator	Application of **GA_3 (5-10 ppm), ethereal (CEPA 2chloroethylphosphonic acid) @ 150-200 ppm and NAA (100 ppm)** at 2 and 4 leaf stages have been found effective to enhance number of female flowers, early harvest, more number of fruits/plant and high fruit yield.
Yield	15-20t/ha.

Ivy gourd

Basic Information and Facts

Scientific name	*Coccinia grandis* (L.) Voigt
Family	Cucurbitaceae
Origin	India
Chromosome number	2n= 24
Sex form	Dioecious
Pollination	Cross pollinated crop
Fruit type	Pepo

- ***Coccinia grandis***, the **ivy gourd**, also known as **scarlet gourd** and **Kowai**, is a tropical vine.
- This plant is a perennial climber with single tendrils and glabrous leaves.
- Ivy gourd is a **dioecious** (male flowers are produced on separate plants to female flowers) perennial herb with tuberous roots and thick stems, which can grow to several meters in length and up to 12 cm in diameter.
- Its leaves are arranged alternately along the stems; they vary from heart to pentagon shape and are up to 10 cm wide and long.
- Flowers are large, white and star-shaped. The calyx has five subulate, recurved lobes, each 2–5 mm long on the hypanthium; peduncle 1–5 cm long.
- The fruit is red (when ripe), ovoid to elliptical, 25–60 mm long, 15–35 mm in diameter, hairless on stalks 10–40 mm long.
- Seeds are tan-coloured and 6–7 mm long. The roots and stems are succulent and probably enable the plant to survive prolonged drought.

Production Considerations

Climate	It requires warm and humid climate with an ideal temperature of 20-30°C.
Soil	The ideal soil for ivy gourd is well drained and fertile.
Improved varieties	Indira kundru-05, Indira kundru-35
Propagation	Through **stem cuttings**
Time of planting	June- July and February- March
Spacing	2.0m × 2.0m
Manures and fertilizers	60:40:40 kg NPK/ha
Irrigation	Crop should be irrigated once in a week during hot months.
Yield	10-12t/ha.

Muskmelon

Basic Information and Facts

Scientific name	*Cucumis melo* L.
Family	Cucurbitaceae
Origin	Tropical Africa
Chromosome number	2n= 24
Sex form	Monoecious
Pollination	Cross pollinated crop
Fruit type	Pepo

- Muskmelon (***Cucumis melo*** **L.**), is a member of the **Cucurbitaceae** family, which also includes cucumbers, watermelons and honeydew, Persian, casaba, and crenshaw melons.
- Melons are divided into two groups: *Citrullus* (watermelons) and *Cucumis* (the muskmelon -cantaloupe group).
- Muskmelons are annual plants with a trailing vine growth.
- Muskmelons contain two blossom types: perfect (having both male and female parts) and male (staminate) flowers.
- Perfect flowers upon pollination and fertilization will develop into the familiar fruits.
- Muskmelons set fruit in cycles where several fruit are set per plant in each cycle.
- Muskmelons require bees for pollination. For greater yields and larger melons, place one or two strong hives of honeybees per acre adjacent to fields of muskmelons.
- Botanically, a melon is a kind of berry, specifically a **"pepo".**
- Muskmelons produce two types of flowers—perfect flowers, having both male and female parts, and staminate flowers, having only male parts.

Production Considerations

Climate	The optimum temperature for germination of seeds is 27°-30°C.
Soil	Sandy loam rich in organic matter with good drainage and pH range of 6.5-7.5.
Varieties/ hybrids	Pusa Madhuras, Pusa Sharbati, Arka Jeet, Arka Rajahans, Kashi Madhu, Gujarat Muskmelon-3
Seed rate	2.5-3.5 kg/ha
Season	January to February.
Spacing	1.5-2.5m × 0.6-1.20m
Mulches	Plastic mulch warms the soil, conserves water, helps to control weeds, allows earlier planting and maturity, and reduces ground rot of the fruit.
Manures and fertilizers	FYM 25-40t/ha and 125:62:62 kg NPK/ha.
Irrigation	Irrigate the field before dibbling the seeds and thereafter once in a week.
Growth regulator	Application of **Ethrel (250 ppm)** at 2 and 4 leaf stages have been found effective to enhance number of female flowers, early harvest, more number of fruits/plant and high fruit yield.
Yield	25 tonnes/ha.

Pointed gourd

Basic Information and Facts

Scientific name	*Trichosanthes dioica* Roxb.
Family	Cucurbitaceae
Origin	India
Chromosome number	2n= 22
Inflorescence type	Recimose
Sex form	Dioecious
Pollination	Cross pollinated crop
Fruit type	Pepo

- Pointed gourd (***Trichosanthes dioica* Roxb.**) is one of the most important **cucurbit** vegetables in tropical and subtropical regions of the world.
- Cucurbits are predominantly monoecious, but dioecism occurs in pointed gourd (*Trichosanthes dicica*), spine gourd (*Momordica dioica*), ivy gourd (*Coccinia indico*), and some feral forms.
- Pointed gourd is morphologically distinct from other cucurbitaceous species due to its well-established **dioecism**, perennial nature, and vegetative means of propagation.
- Pointed gourd has tuberous roots and a long taproot system.
- Inflorescence is **recimose**; flowers are sessile, solitary,bracteates with oblong-cylindrical calyx tube.
- The fruit (**pepo**) of *T. dioica* is globose, oblong, and smooth, where the edible portion is mainly the pericarp with a little mesocarp.
- As a **dioecious species**, *T. dioica* is a strictly **cross-pollinated** vegetable.
- The chromosome number of *T. dioica* is 2n-2x=22.

Production Considerations

Climate	It is a warm season crop. Hot and humid climate is best for its development.
Soil	It can be grown successfully on both sandy loam and loamy soils.
Improved varieties	Green oval, green long striped, pale-white oval
Propagation	Through **vine cuttings**
Seed rate	2000-2500 cuttings/ha
Season	August (upland) and November (riverbeds)
Spacing	2.0m × 2.0m
Manures and fertilizers	FYM 20-25t/ha and 90:60:60 kg NPK/ha.
Irrigation	Once in a week.
Yield	6-8t/ha in first year and 14-15t/ha in second year.

Pumpkin

Basic Information and Facts

Scientific name	***Cucurbita moschata*** **(Duchesne) Duchesne ex Poir.**
Family	Cucurbitaceae
Origin	North western part of South America
Chromosome number	2n = 40
Sex form	Monoecious
Pollination	Cross-pollinated crop
Fruit type	Pepo

- Pumpkin **(*Cucurbita moschata* (Duchesne) Duchesne ex Poir)** with its origin in the **north western part of South America** is the most important cultivated cucurbit in India.
- The term pumpkin designates many plant species of the family ***Cucurbitaceae*** which belong to the genus ***Cucurbita***.
- Pumpkins and squash are the edible fruits of four of the five cultivated species of the genus *Cucurbita*.
- Generally, the term "pumpkin" applies to round or nearly round (oblate, globose, ovate, obviate, etc.) fruits and the term squash applies to non round fruits of edible *Cucurbita*.
- Some botanists believe that two of the species of *Cucurbita, C. pepo* and *C. moschata*, are of **American origin.**
- Pumpkin breeding involves the hybridization of primarily three domesticated *Cucurbita* species: *C. pepo, C. maxima* and *C. moschata*.
- Sex expression in ***Cucurbita*** is relatively stable and all species including *Cucurbita moschata* are **monoecious.**
- Flowers are unisexual, solitary in the leaf axils.
- The staminate flowers produce pollen with large granules, which are pollinated by the **honey bee** (*Apis mellifera*), bumble bee (*Bombus* spp.) and, most commonly, squash bee (*Peponapis pruinosa)*.
- Pumpkin reproduces sexually via **cross-pollination**.
- Pumpkin is an excellent source of **vitamin A,** with its orange colour indicating high content of-carotene that is transformed by the organism into **vitamin A**.
- The principal antioxidant compounds of pumpkin are **carotenoids.**

Production Considerations

Climate	A temperature of 18-30°C is optimum. Short days, low temperatures (18-30°C), high relative humidity, bright sunshine are ideal for its cultivation.
Soil	Sandy loam rich in organic matter and with good drainage. The pH range from 6.5-7.5 is found ideal.

Varieties/ hybrids	Pusa Hybrid 1, Pusa Vikas, Pusa Vishwas, Arka Chandan, Arka Suryamukhi, Kashi Harit, Anand Pumpkin- 1, N. Abhooshan, N. Agrim, N. Amrit, CO1, CO2, Azad Pumpkin-1
Seed rate	3-3.5 kg /ha
Season	June- July and February- March
Spacing	2.5 m x 2 m. Pit size 30 cm x 30 cm x 30 cm
Manures and fertilizers	FYM 20-25t/ha and 80-100:50-80:40-50 kg NPK/ha.
Irrigation	During initial stages of growth, irrigation is given at an interval of 3-4 days. During flowering and fruiting stages, irrigation should be given on alternate days.
Weed management	Pre-plant incorporation of Besulide @4-6kh.ha or Alchlor 2.5kg/ha as pre-emergence can be used for weed control.
Growth regulator	Application of **Ethrel (250 ppm)** at 2 and 4 leaf stages have been found effective to enhance number of female flowers, early harvest, more number of fruits/plant and high fruit yield.
Yield	20-30 t/ha in 120-140 days.

Ridge gourd

Basic Information and Facts

Scientific name	***Luffa acutangula* (L.) Roxb.**
Family	Cucurbitaceae
Origin	Sub-tropical Asian regions including India
Chromosome number	2n= 26
Sex form	Monoecious
Pollination	Cross pollinated crop
Fruit type	Pepo

- Ridged gourd (***Luffa acutangula* L. (Roxb.)**] belonging to the family **Cucurbitaceae** is a well known vegetable crop in India.
- It is a **monoecious** and **cross-pollinated crop**.
- Fruit is demulcent, diuretic and nutritive and it can be grown throughout the year.

Production Considerations

Climate	It grows very well in a warm hot climate, the optimum temperature being 25-30°C.
Soil	Sandy loam rich in organic matter with good drainage and a pH range from 6.5-7.5.
Varieties/ hybrids	Pusa Nasdar, Pusa Nutan, Hisar Kalitori, GJRGH-1, Gujarat Anand, Ridge Gourd 1, Pant Torai-1

Seed rate	3.5-5.0 kg/ha
Season	February- March and June-July
Spacing	1.5-2.5 m x 0.60-1.20m
Manures and fertilizers	FYM 15-20t/ha and 40-60:30-40:30 kg NPK/ha
Irrigation	Irrigate the basins before dibbling the seeds and thereafter once a week.
Growth regulator	Application of **Ethrel (250 ppm)** at 2 and 4 leaf stages have been found effective to enhance number of female flowers, early harvest, more number of fruits/plant and high fruit yield.
Yield	8-12 t/ha in 125 days.

Round gourd/Tinda

Basic Information and Facts

Scientific name	***Praecitrullus fistulosus* (Stocks) Pangalo**
Family	Cucurbitaceae
Origin	Tropical Africa
Chromosome number	2n= 22
Sex form	Monoecious
Pollination	Cross pollinated crop
Fruit type	Pepo

- Round gourd (***Praecitrullus fistulosus***) is cultivated as a vegetable in India, Pakistan and Afghanistan.
- The origin is probably **northwestern India**, where wild types may still be found in the wild.
- Botanically, it is an annual, climbing or trailing herb, with robust, villous hairy stem, tendriller and slender.
- The plant is **monoecious,** flowers are usually solitary in leaf axils, comparatively small, regular, pentamerous, calyx campanulate, petals connate, yellow, hairy, male flowers with three stamens, female flowers with inferior, globose and hairy ovary.
- The fruit is a type of berry called a **pepo** by Gerald Carr. The fruit is approximately spherical, and 5–8 cm in diameter. Fruit about the size of a small turnip, depressed at each end, hispid when young afterwards glabrous.

Production Considerations

Climate	It grows well under warm and dry climate but can be grown both under hot as well as mild climates.
Soil	Sandy loam rich in organic matter with good drainage and pH range of 6.5-7.5.
Varieties/ hybrids	Arka Tinda, Hisar Tinda (HT-10)

Seed rate	3.5 -7.0kg/ha
Season	January-February and June- July
Spacing	1.5-2.5 m x 0.60-1.20 m
Manures and fertilizers	FYM 20-25t/ha and 50: 100: 50kg NPK/ha.
Irrigation	Irrigate the field before dibbling the seeds and thereafter once a week.
Weed management	Alchlor 2.5 litre/ha as pre-emergence can be used for weed control.
Yield	10 t/ha in 90 days.

Snake gourd

Basic Information and Facts

Scientific name	***Trichosanthes cucumerina***
Family	Cucurbitaceae
Origin	India
Chromosome number	2n= 24
Sex form	Monoecious
Pollination	Cross-pollinated crop
Fruit type	Pepo

- Snake gourd (***Tricosanthes cucumerina***) is a well known plant, the fruit of which is mainly consumed as a vegetable.
- It is a tropical or subtropical vine, its variety *T. cucumerina* var. *anguina* raised for its strikingly long fruit,
- It is an annual climber belonging to the family **Cucurbitaceae.**
- It is a **monoecious** annual herb climbing by 2–3-branched tendrils upto 5 to 6 meters high or less.
- The leaves are alternate, simple with no stipules.
- The staminate inflorescences are long-peduncled and axillary, with six to fifteen flowers.
- Flowers are unisexual, regular, and white in colour with green and hairy calyx.
- The seeds are half-ellipsoid, somewhat compressed, undulate, hard, rugose, nearly one centimeter long, greyish-brown, sculptured, margin undulate and imbedded in a soft foetid with red pulp.
- The chemical constituents present in *T. cucumerina* are cucurbitacin B, cucurbitacin E, isocucurbitacin B, 23,24- dihydroisocucurbitacin B, 23,24-dihydrocucurbitacin E, sterols 2 -sitosterol stigmasterol.

Production Considerations

Climate	It requires warm and humid climate for best growth.
Soil	Sandy loam rich in organic matter with good drainage and a pH range of 6.5-7.5.
Improved varieties/ hybrids	CO1, CO2, CO4, TA19, Konkan Sweta
Seed rate	5-6 kg/ha.
Season	June-July and December- January.
Spacing	1.5-2.5 m x 0.60-1.20 m.
Manures and fertilizers	FYM 10-15t/ha and 40-60: 30-40: 30-40kg NPK/ha.
Irrigation	Irrigate the basins before dibbling the seeds and thereafter once in a week.
Yield	8-10 t/ha

Spine gourd

Basic Information and Facts

Scientific name	*Momordica dioica* Roxb. ex Willd.
Family	Cucurbitaceae
Origin	Indo –Malayan region
Chromosome number	2n = 28
Sex form	Dioecious
Pollination	Cross pollination crop
Fruit type	Pepo

- Spine gourd (***Momordica dioica*** **Roxb. ex. wild**) is a perennial, rhizomatous, **dioecious** climber belonging to the **cucurbitaceae** family with **chromosome number 2n=28,** under the genus *Momordica* and is known by various names like Kakrol, Kartoli, Kankad, Teasel gourd or Bhat kerala.
- It originated in the **Indo-Malayan region.**
- It is **dioecious**, perennial in nature, having tuberous roots.
- Leaves 1.5-4 inches long, cordate, acute more or less 3-5 lobed; Flowers large, dioecious and yellow in colour.
- Fruit 1-3 inches long, shortly beaked, densely covered with soft spines.
- The green fruit is extensively used as a vegetable.
- Fruit of spine gourd are free from cholesterol and are and a great source of energy (45.74 Kcal.) with adequate amount of water (84%), protein (3.1 g/100g) and important minerals and vitamins.

Production Considerations

Climate	Spine gourd is a warm season crop and can be successfully cultivated in the plains and hills of sub-tropical and tropical regions. Plenty of sunshine and low humidity are ideal conditions for its growth.
Soil	A well-drained sandy loam soil with a pH between 5.5 and 6.5 is the most ideal soil condition for spine gourd production.
Variety	**Indira Kankoda I (RMF 37)** is developed by Ambikapur (Chhattisgarh) based Indira Gandhi Agricultural University, RMD College of Agriculture and Research Station.
Propagation	Root tubers are commercially used as planting material in spine gourd.
Planting time	The best time of planting tubers is February- March.
Seed rate	3000-5000 tubers/ha.
Spacing	2m × 1 m
Manures and fertilizers	Well rotten FYM @15-20 t/ha is applied at the time of land preparation. NPK @ 120:80:80 kg/ha is added with full dose of P and K before sowing and half dose of N at the time of vining and remaining half dose of N before flowering
Irrigation	Irrigation Just after sowing of seed or placement of cuttings irrigation is essential and it should continue for 4-5 days.
Harvesting	Spine gourd is harvested when their fruits are still young and tender. Fruits are ready for harvest within 75-80 days in first year and 35-40 days from second year. Fruits should be harvested at the mature green stage.
Yield	Under optimum crop management condition fruit yield is 75-100 q/ha.

Sponge gourd

Basic Information and Facts

Scientific name	***Luffa cylindrica* (L.) Roem.**
Family	Cucurbitaceae
Origin	Asian regions including India
Chromosome number	2n = 26
Sex form	Monoecious
Pollination	Cross pollinated crop
Fruit type	Pepo

- Luffa (***Luffa cylindrica*** **(L.)** Roem commonly called **sponge gourd**, vegetable sponge, bath sponge or dish cloth gourd, is a member of the **cucurbitaceous family**.
- The number of species in the genus *Luffa* varies from 5 to 7.
- Only two species *L. cylindrica* and ribbed or ridge gourd (*L. acutangula* (L.) Roxb) are domesticated.

- Two wild species are *L. graveolens* and *L. echinata.* These two cultivated and one wild species, *L. echinata* exist in Nepal.
- Luffa is diploid species with 26 chromosomes **(2n = 26)** and a **cross-pollinated** crop.

Production Considerations

Climate	It grows very well in warm hot climate, the optimum temperature being 25-30°C.
Soil	Sandy loam rich in organic matter with good drainage and a pH range from 6.5-7.5.
Varieties/ hybrids	Pusa Chikni, Pusa Sneha, Pusa Supriya, Azad Taroi -1, Azad Taroi -2, CO 1, CO 2, PKM.1
Seed rate	2.5-3.5 kg/ha
Season	February- March and June-July
Spacing	1.5-2.5 m x 0.60-1.20m
Manures and fertilizers	FYM 15-20t/ha and 40-60:30-40:30 kg NPK/ha.
Irrigation	Irrigate the basins before dibbling the seeds and thereafter once a week.
Growth regulator	Application of **Ethrel (250 ppm)** at 2 and 4 leaf stages have been found effective to enhance number of female flowers, early harvest, more number of fruits/plant and high fruit yield.
Yield	15-20 tonnes/ha in 125 days.

Summer squash

Basic Information and Facts

Scientific name	***Cucurbita pepo* L.**
Family	Cucurbitaceae
Origin	North America
Chromosome number	2n = 40
Sex form	Monoecious
Pollination	Cross pollinated crop
Fruit type	Pepo

- Squash *(Cucurbita pepo)* commonly called as summer squash is a well known vegetable belonging to the family **Cucurbitaceae.**
- Native to North America, it has been observed growing wild in the southern United States and northeastern Mexico.
- *C. pepo* is an annual herb with climbing, creeping, or in some varieties bushy, 5-angled stems up to 15 m long.
- The shallow root system is branched, growing from a well-developed taproot.
- The stems are scabrous and setose, branching, often rooting at the nodes.

- The petioles are setose, grooved, 6-24 cm long, and estipulate. The plant bears tendrils borne at 90 degrees to the leaf insertion, which are coiled, and 1-6-branched.
- On bushy plants, tendrils may be poorly developed. The thin leaves are simple, alternate, broadly ovate to deltoid, basally cordate, apically acute, palmately lobed with 5-7 lobes, marginally toothed, scabrous, palmately veined, 20-30 cm long, and 10-35 cm broad.
- It is monoecious (unisexual flowers, with male and female on the same plant) and bears solitary actinomorphic flowers (~ 10 cm across) that produce nectar.
- The calyx is campanulate with 5 free sepals; each sepal is linear and 0.9-3 cm long, smaller on pistillate flowers.
- The yellow corolla is campanulate, 5-parted (occasionally 6-parted) with erect to spreading petals that are apically acute, approximately 5-10 cm long, and ca. 3 cm broad on smaller flowers.
- Staminate flowers are borne on a 3-20 cm long peduncle, and have 3 stamens with free filaments (~ 1.5 cm long) and connivent twisted anthers (~ 1 cm long). Pistillate flowers are borne on shorter peduncles, only 2-5 cm long, and have an inferior 1-locular globose to cylindrical ovary, and a thick style with 3 bi-lobed stigmas.
- The peduncle of the developing fruit is hard, slightly bent, expands at the junction of the fruit, and ranges from 5.6 to 15.1 mm in diameter
- Squash is a naturally **cross-pollinated crop** and self-pollination occurs to the extent of 0 – 5%.

Production Considerations

Climate	It can be grown both in high and low temperature conditions. The plants can be grown successfully between a temperature range of 24-27°C.
Soil	A well drained sandy loam soil with high organic content is considered best.
Varieties/ hybrids	Pusa Alankar, Punjab Chappan Kaddu-1, Patty Pan
Seed rate	2-4kg/ha
Season	February- March and June –July
Spacing	1 m x 1 m
Manures and fertilizers	FYM 20-25t/ha.
Irrigation	Irrigation should be given at an interval of 3-4 days depending upon season and type of soil.
Growth regulator	To improve fruit set during early February and March foliar application of Ethrel (250 ppm) at 2 and 4 true leaf stage of plants have resulted in early harvest, increased number of fruits and total fruit yield.
Yield	20-30t/ha.

Watermelon

Basic Information and Facts

Scientific name	*Citrullus lanatus* (Thunb.) Matsumura & Nakai
Family	Cucurbitaceae
Origin	Southern Africa
Chromosome number	2n = 22
Sex form	Monoecious
Pollination	Cross pollinated crop

- Watermelon [***Citrullus lanatus* (Thunb.) Matsumura & Nakai**] belongs to family **Cucurbitaceae** is widely cultivated for its edible fruits.
- Cultivation of watermelon is now widespread in all tropical and sub-tropical regions of the world and is mostly grown for fresh consumption of the juicy and sweet flesh of mature fruit.
- Watermelon is thought to have originated in **southern Africa**, as growing wild throughout the region, and showed maximum diversity (Vander Vossen et al., 2004).
- Watermelon is a **diploid with 22 chromosomes**.
- Watermelon grows as a trailing vine; the stems are thin, hairy, angular, grooved and have tendrils at each node.
- The development of flowers and fruits are promoted by high light intensity and high temperature.
- Flowers are staminate (male), perfect (hermaphroditic), or pistillate (female).
- However, **monoecious t**ypes are most common, but there are **andromonoecious** (staminate and perfect) types also.
- Flowers are yellow coloured approximately 1-3 cm in diameter.
- Male flowers are in the axil of spoon-shaped, 8-10 x 3-4 mm bract; on an elongate, 1-5 cm long, villous peduncle.
- Calyx tube is broadly campanulate, villous; sepals as long as the calyx tube, narrowly lanceolate.
- Corolla is 2.5-3 cm in diam, usually greenish on outside, pale yellow inside, villous.
- Female flowers are borne at every seventh or eighth node with inferior ovary; peduncles 2-6 cm long.
- Ovary oblong, lanate; style slender, 4-5 mm long.
- Sex expression in watermelon is highly influenced by environmental factors (temperature, humidity, light) and nutrition besides being genetically controlled by a single pair of alleles.
- Locus 'a' determines sex producing monoecious (AA) vs. andromonoecious (aa) flowers (Guner and Wehner, 2004; Kumar et al., 2013).
- Monoecious (staminate and pistillate flowers on the same plant) sex expression promotes allogamy, while andromonoecious (staminate and perfect flowers on same plant) sex expression favours autogamy (Martin et al., 2009).

Production Considerations

Climate	Plant growth is optimum under 28°-30° C, while fruiting is better at 24°-27° C. higher temperatures are beneficial during ripening.
Soil	Sandy loam rich in organic matter with good drainage and pH range of 6.5-7.5.
Varieties/ hybrids	Asahi Yamato, Sugar Baby, Arka Jyothi, Arka Manik, PKM.1
Seed rate	3.5-5.0 kg/ha
Season	December-January
Spacing	2.5-3.5m x 0.9-1.20m
Manures and fertilizers	Apply FYM 20 t/ha, 100:60:60kgNPK/ha. Apply *Azospirillum* and *Phosphobacteria* @ 2 kg/ha and *Pseudomonoas* @2.5 kg/ha along with FYM 50 kg and neem cake 100 kg before last ploughing.
Irrigation	Irrigate the field before dibbling the seeds and thereafter once a week.
Growth regulator	Application of **Tri-iodo benzoic acid (TIBA) @ 25-50 ppm** at 2 and 4 leaf stages have been found effective to enhance number of female flowers, early harvest, more number of fruits/plant and high fruit yield.
Yield	30-40t/ha

Yams

Basic Information and Facts

Scientific name	***Dioscorea* spp.**
Common Names:	Greater yam, water yam, white yam *or* winged yam
Family	Dioscoreaceae
Origin	South-East Asia
Chromosome number	2n = 20, 40, 60, 80
Ploidy level	Diploid, tetraploids, hexaploids and octoploids
Sex form	Dioecious
Pollination	Cross-pollinated crop
Fruit type	Capsule

- Yams, belonging to the genus ***Dioscorea,*** family **Dioscoreaceae**, are an important food crop in tropical and subtropical regions.
- Yam plants are **dioecious monocotyledonous** vines, cultivated for their starchy tubers.
- Yam is a multi-species, **polyploidy**, and **vegetatively propagated crop**, exclusively propagated by means of "setts," small tubers or small pices of tuber.
- Out of the 10 cultivated yams, *Dioscorea rotundata, D. cayenensis, D. alata* and *D. trifida* are the major species, while the six others are often referred to as minor ones.
- The species *D. rotundata and D. cayenensis* (also referred to as *D. rotundata – cayenensis complex)* are the most commonly cultivated in the world and represent 95% of yam production worldwide.
- ***D. rotundata*** plants have **2*n* = 40 chromosomes**, whereas ***D. cayenensis*** includes accessions with **2*n* = 60 and 80 chromosomes.**

- ***D. alata***, or the greater yam, is the most widely cultivated yam species throughout the tropics. It is a polyploidy species that includes accessions with **2*n* = 40, 60 and 80 chromosomes.**
- ***D. trifida*** is the most important edible *Dioscorea* in tropical America. Its flesh is considered to be the most flavourful of all the yams. This species has **2*n* = 80 chromosomes.**
- **Fufu** is an important product made from yam.

Production Considerations

Climate	Temperature in the range of 20-30ºC promotes better growth. Day length greater than 12hrs during the initial stages and shorter day length during the later part of the growing season favour satisfactory tuber production.
Soil	Yams require a fertile, free draining soil and grow best in sandy loam soil having Ph 5-7.
Propagation	Through tubers
Improved varieties/ hybrids	**Greater Yam:** Sree Keerthi, Sree Roopa, Sree Shilpa Hybrid, Sree Karthika, Orissa elite **White Yam:** Sree Priya, Sree Subhra, Sree Dhanya **Lesser Yam:** Sree Latha, Sree Kala
Time of Planting	March- June
Spacing	**Lesser yam and white yam:** 75cm x 75cm **Greater Yam:** 90cm x 90cm
Manures and fertilizers	FYM 10-12tinnes/ha and 80:60:80kg NPK/ha.
Irrigation	For uniform sprouting, yams should be irrigated immediately after planting. The second irrigation can be given if the soil is dry before the onset of the monsoon.
Harvesting	Yams are harvested **7-9 months** after planting. In India, yams are harvested during **December-January.**
Yield	**Greater Yam:** 33-40 tonnes/ha, **White Yam:** 18-25 tonnes/ha, **Lesser Yam:** 25-30 tonnes/ha

Cassava/Tapioca

Basic Information and Facts

Scientific name	***Manihot esculenta* Crantz**
Family	Euphorbiaceae
Origin	Brazil
Chromosome number	2n= 20
Inflorescence type	Panicle
Sex form	Monoecious
Pollination	Cross pollinated crop

- Cassava (***Manihot esculenta* Crantz**) belonging to the family **Euphorbiaceae,** is a native of **Brazil.**
- It is **diploid**, and emerged from successive hybridizations between wild species.
- Cassava is a **monoecious** plant producing both male and female flowers on the same inflorescence and is **cross-pollinated.**
- **Inflorescence** terminal, with male flowers and broader female flowers, both with a calyx with five triangular lobes.
- **Fruit** globular, 1.5 cm long, bearing six fine wings, containing three seeds that are explosively dispersed three to six months after pollination.
- The edible portion of cassava is a starchy root. A mature cassava root may be anything from 15 to 100 cm in length and from 0.5 to 2.0 kg in weight, subject to variety and growing conditions.
- Cassava contains about 1% protein and some **30-35% of amyloses and amylopectins** on a dry weight basis; it is thus a predominantly starchy food.
- The main constituent of the tuber is **starch.** The **yellow colour** of the flesh is due to the presence of **carotene,** the precursor of **vitamin A.**

Production Considerations

Climate	Cassava is **not frost-tolerant,** and is grown mainly in lowland tropical areas. Thrives best in tropical, warm humid climate with well distributed rainfall of over 100 cm per annum. Cultivated upto an elevation of 1000 m.
Soil	Any well drained soil preferably red lateritic loam with a pH range of 5.5-7.0.
Improved varieties	CO2, CO3, CO (TP) 4, MVD 1, H 165, H 226, Sree Visakham (H.1687), Sree Sahya (H.2304), Sree Prakash (S. 856), Sree Vijaya, Sree Jaya, Sree Rekha and Sree Prabha, **CTCRI CO (Tp) 5 (Sree Padmanabha)** resistance to cassava mosaic disease.
Propagation	Through cuttings
Time of planting	Plant throughout the year under irrigation.
Spacing	**Irrigated:** 75 x 75 cm (17,777 setts) and 90 x 90 cm (12,345 setts). **Rainfed:** 60 x 60 cm (27,777 setts). **Under Kanyakumari conditions:** 90 x 90 cm (12, 345 setts).
Manures and fertilizers	Apply 25 t FYM/ha and 75:50:75 kg N, P_2O_5 and K_2O/ha.
Irrigation	First irrigation is given at the time of planting. Life irrigation is given on the 3rd day followed by once in 7 – 10 days upto 3rd month and once in 20 – 30 days upto 8th month.
Yield	**Irrigated:** 40 - 50 t/ha, **Rainfed:** 20 - 25 t/ha.

Broad bean

Basic Information and Facts

Scientific name	*Vicia faba* L.
Common names	Fava bean, faba bean, field bean, bell bean, English bean, horse bean and Windsor bean
Family	Fabaceae (formerly Leguminosae)
Origin	Northern Africa and south-west Asia
Chromosome number	2n = 12
Pollination	Cross- and self-pollinated crop
Fruit type	Legume

- Broad bean (***Vicia faba*** L.) belongs to the **Fabaceae** family. It is also referred to as faba bean, horse bean and field bean.
- It is native to **northern Africa and south-west Asia**, where it is extensively cultivated.
- Faba bean is a **diploid species, 2n=12** (six homologous pairs) and partly out-crossing; cross-pollination is reported to range from 8% to 84%, with an average of 35% (Bond and Poulsen 1983).
- The reproductive system of faba bean is a mixture of **cross- and self-pollination.**
- The plant is rigid and erect, growing to 0.5 to 1.7 m tall.
- The leaves are 10–25 cm long, pinnate with two to seven leaflets and have a distinct grey-green colour.
- The pods are green, broad and leathery.

Production Considerations

Climate	Broad bean is a hardy plant and is the only bean which is grown as an autumn and a winter crop.
Soil	Thrives in almost all soils with a pH range of 6.5-7.5. It tolerates salinity to some extent.
Improved varieties	Pusa Sumeet, SWS 1 (Suttan White Seeded), BR 1 (Bihar Black Seeded) and BR 2 (Bihar Yellow Seeded).
Seed rate	70-100 kg/ha.
Season	July – August and November - December
Spacing	45 × 15 cm
Manures and fertilizers	Apply 25 t of FYM and 20kg N 50 kg P and 40 kg K/ha as basal dose.
Irrigation	Immediately after sowing, the field has to be irrigated. Thereafter, Light irrigation should be given at a regular interval of 12-15 days.
Yield	Pod yield 7-10 tonnes/ha.

Cluster bean

Basic Information and Facts

Scientific name	*Cyamopsis tetragonoloba* (L.)Taub.
Family	Fabaceae (formerly Leguminosae)
Origin	India
Chromosome number	2n = 14
Inflorescence type	Raceme
Pollination	Self-pollinated crop
Fruit type	Legume

- Guar or cluster bean (***Cyamopsis tetragonoloba*** **(L.) Taub.**) belongs to family **Fabaceae.**
- It is native to the **Indian subcontinent.**
- The importance of legume has been highly appreciated as a source of **guar gum.**
- The seed of cluster bean contains about **30-33% gum in the endosperm.**
- The discovery of the **galactomannan gum** in the endosperm during 1948, led to this hitherto insignificant plant gaining importance as an industrial crop.
- The gum is utilized for many food items like ice creams, baked and dairy products etc.
- Moreover, its gum also used in many other industries like pharmaceuticals, cosmetics, mining, textile, paper, oil drilling, explosive industry etc.
- Traditionally, pods of the cluster bean are used for vegetable purpose.
- Its plant, seed and straw are good source of nutritive fodder and feed for livestock. Cluster bean is also raised as a green manure and cover crop.
- Being a leguminous crop, it enriches the soil fertility by fixing the atmospheric nitrogen.
- *Cyamopsis tetragonoloba* grows upright, reaching a maximum height of up to 2–3 m. It has a main single stem with either basal branching or fine branching along the stem.
- Flower is complete and hermaphrodite with papilionaceous corolla usually having ten stamens. Each stamen has two anther lobes which get burst at maturity to liberate pollen grains when stigma is receptive. Thus ensures the **self pollination.**
- Usually, mature seeds are white or gray, but in case of excess moisture they can turn black and lose germination capacity.
- The chromosome number of guar is **2n=14.**
- The mucillagenous seed flour is valued as a guar gum (**glactomannan**) and used in textile, paper, cosmetic and oil industries. It is a useful absorbent for explosives.
- Recently, the oil industry has started using guar gum in hydraulic fracturing.
- India alone contributes more than 80% of global guar production followed by 15% in Pakistan.

Production Considerations

Climate	Cluster bean is a tropical plant. It requires warm growing season. The crop requires 30 to 35°C temperatures at the sowing time for proper germination and 32 to 38° C temperature encourages good vegetative growth, but high temperature at flowering stage can result in pre-mature flower drop.
Soil	Cluster bean can be grown in a variety of soils. The crop grows best on well drained, upland sandy loam and loam soils. It does not thrive well on very heavy and water logged soils. It also does not thrive well in saline and alkaline soils. It can be raised successfully in the soils having pH 7 to 8.5.
Improved varieties	Pusa Mausami, Pusa Naubahar, Goma Manjari and Pusa Sadabahar
Seed rate	15 kg ha.
Season	June - July and February- March
Spacing	45-50 cm × 10 cm
Manures and fertilizers	FYM 15t/ha and 20:40:20 kg NPK/ha, respectively.
Irrigation	Irrigate the field immediately after sowing then at weekly intervals.
Weed management	**Pendimethalin @ 2.5 -1 to 3.30 L ha** can be applied by mixing with 500 L of water as pre-emergence application (within 2 days of sowing). After that one manual weeding at 30 DAS or post emergence -1 application of **imazythypr @ 400 gm ha** mixing with 500 litre water at 20-25 days after sowing can be applied for controlling weeds of cluster bean field.
Yield	A good crop under favourable climatic conditions yields about 300 quintals of green fodder or 15 quintals of dry seeds or 60 quintals of green pods per hectare.

Varieties of Guar for different Regions

S. No.	Region	Varieties
1.	Northern India	Pusa Navbahar, Pusa Mausami, Pusa Sadabahar, GH-10, HFG-119, Ageta Guara-111, Guar No.2, FS-277, Durgapur Safed, S-299-7, G-1, G-2, G-4, HG-12, D-111, D-128, G-429, HG-75, B-17, B-2796, IGFRI-5212
2.	Southern India	Pusa Navbahar, Pusa Sadabahar, GH-10, HFG-119, Saradbahar, IC 11704, P-28-1, CP-78, S-299-7, IC-11521, IC-11388

Cowpea

Basic Information and Facts

Scientific name	*Vigna unguiculata* (L.)Walp.
Family	Fabaceae (formerly Leguminosae)
Origin	Africa
Chromosome number	2n= 22
Inflorescence type	Raceme
Pollination	Self-pollinated crop
Fruit type	Legume

- Cowpea (***Vigna unguiculata*** (L.) Walp.) belongs to the family **Fabaceae.**
- Cowpea originated in **Africa** and is widely grown in Africa, Latin America, Southeast Asia and in the southern United States.
- It is commonly cultivated as a nutritious and highly palatable food source in India.
- The seed is reported to contain **24% crude protein**, 53% carbohydrates, and 2% fat (FAO, 2012).
- It is an annual herbaceous legume that can reach more than 80 cm in height.
- The plant has a deep taproot with numerous spreading lateral roots that help stabilise the soil.
- Flowers are self-pollinating and may be white, dirty yellow, pink, pale blue or purple in colour. They are arranged in **raceme or intermediate inflorescences** in alternate pairs
- Cowpea is a **self-pollinated crop** which is encouraged by the arrangements of the floral parts.

Production Considerations

Climate	A warm season crop, suitable to humid tropics and subtropical zones. The germination is better at 12-15°C temperature and the crop thrives best between 21 and 35°C temperature.
Soil	It grows in all types of well drained soils with pH range of 5.5 to 6.5.
Varieties/ hybrids	Pusa Komal, Pusa Sukomal, Arka Samridhi, Arka Garima, Arka Suman, Bidhan Barbati-1, Bidhan Barbati-2, Kashi Kanchan, Kashi Gauri, Kashi Unnati. **Photo insensitive varieties**: Kashi Kanchan, Kashi Gauri, Kashi Unnati, Arka Garima and Arka Samridhi.
Seed rate	25 kg/ha
Season	**Spring-summer crop:** February-March, Rainy season: June- July. **South India:** December-January for spring-summer crop.
Spacing	45-60 cm × 10-15 cm (bush type), 75 cm × 20-25 cm (indeterminate types).
Manures and fertilizers	Apply FYM 25 t/ha and, yield 50:80:80 kg NPK /ha.
Pinching	Before flowering, the tendrils should be pinched thrice for getting bushy plants.
Irrigation	Cowpea is a hardy crop which comes up well under rainfed condition. For higher yields the crop should be irrigated regularly at 5-7 days intervals.
Yield	50-80 q/ha green pods, the dry seed yield being 12-15q/ha.

French bean

Basic Information and Facts

Scientific name	*Phaseolus vulgaris* L.
Family	Fabaceae (formerly Leguminosae)
Origin	Southern Mexico and central America
Chromosome number	2n = 22
Pollination	Self-pollinated crop
Fruit type	Legume

- French bean (***Phaseolus vulgaris* L.**) is an herbaceous annual plant in the family **Fabaceae** which is grown as a pulse and green vegetable.
- It is valued for its **protein rich (23%)** seeds. Seeds are also rich in calcium, phosphorus and iron.
- The fresh pods and green leaves are used as vegetable.
- Vavilov (1951) reported **Mexico and Central America** as the primary and Peruvian-Ecquadrion-Bolvian region of South America as the secondary center of French bean.

Production Considerations

Climate	A mean air temperature of 15°-25°C is optimum for its growth and high pod yield.
Soil	Well drained loamy soils with pH range of 5.5-6.0.
Seed rate	**For bush varieties:** 65 kg/ha., **for pole varieties:** 25-30 kg/ha.
Varieties/ hybrids	**Pole types:** Kentucky Wonder, Pusa Himalata, SVM-1, VL-12, VL- 17 **Bush types:** Arka Komal, Arka Suvidha, Contender, Kashi Param, Azad Rajmah-1, Premier, VL Boni-1, VLRajmah-63, Pant Anupama
Season	**Hills:** March- May **Plains:** July- September and January- February
Spacing	45cm × 15cm (Bush type) and 90cm x 10-15cm (Pole type).
Manures and fertilizers	Apply FYM 25 t/ha at the last ploughing. N at 90 and P at 125 kg/ha should be applied on one side of the ridges. For rainfed conditions of Shevaroy hills, apply as a basal dose of 62.5 kg/ha of Phosphorus as superphosphate and with another half of 62.5 kg/ha Phosphorus as FYM enriched super phosphate.
Irrigation	Immediately after sowing, third day and thereafter once a week.
Weed management	Pre-emergence application of Alachlor 3litres/ha or Pendimethalin @ 4 litres/ha or Thiobencarb @ 4 litres/ha or Fluchloralin @ 2.5 litres/ha.
Growth regulator	Spray of PCPA @ 2ppm or NAA 5-25 ppm induces pod formation when pods do not set normally at prevailing temperatures.
Yield	8-10 tonnes/ha in **bush varieties:** 12-15 tonnes/ha in **pole varieties.**

Lablab Bean/ Indian bean

Basic Information and Facts

Scientific name	*Lablab purpureus* (L.)Sweet, *Dolichos lablab* L. (ambiguous synonym)
Family	Fabaceae (formerly Leguminosae)
Origin	India or South-East Asia
Chromosome number	2n = 22
Inflorescence type	Raceme
Pollination	Self pollinated crop
Fruit type	Legume

- ***Lablab purpureus* L. (Sweet)** usually known as *Dolichos* bean, Hyacinth bean or Field bean is one of the most ancient crops among cultivated plants.
- **Linneus (1754)** described the species under ***Dolichos L.***
- **Verdcourt (1980)** finally assigned it to monotypic genus Lablab and now widely known as ***Lablab purpureus*** L. (Sweet) although some, still, refer to it as *Dolichos Lablab.*
- It is a native **to India or South-East Asia**. It is probably of an Asian origin and has been under cultivation since ancient times. Its wild forms are found in India.
- It is a bushy, semi-erect, perennial herb, showing no tendency to climb.
- Young immature pods are cooked and eaten like green beans (older pods may need to be de-stringed).
- *Labalab* is predominantly **self-fertilizing.** Chromosome number **2n = 22.**
- Inflorescence is a stiff axillary **raceme** with many flowers.
- Flowers are white, pink, red or purple coloured, in clusters of 4-5, each with 2 large basal bracts, anthers uniform, upper 2 sepals connate purple to pink or white, 2-4 at each node in an elongating raceme upto 2.5 cm long.
- Pods flat or inflated, pubescent or smooth, papery, straight or somewhat curved, white, green or purplish in colour and can vary in length from approximately 5 - 20 cm and in width from 1 - 5 cm.
- Seeds variable in size and colour, normally upto 12.5 cm in length, colour ranging from white, red, brown, black or speckled, rounded or oval, hilum white, prominent, approximately 10 mm in length, average weight of 100 seeds is 25 - 40 g. **Germination is epigeal.**

Production Considerations

Climate	It is a cool season and drought tolerant crop. Lablab bean responds to photo periods, there being some short day and some long day types. However now **photo-insensitive varieties** like **Arka Jay** and **Arka Vijay** are available which can be grown throughout the year.
Soil	Well drained loamy soil with pH range of 6.5-8.5.

Improved varieties	**Bush types:** CO 6, CO 7, CO 8, CO 9, CO 10, CO 11, CO 12, CO 13, COGB 14, Arka Jay and Arka Vijay. **Pandal types:** CO 1, CO 2, CO 3, CO 4, CO 5 and Pusa Early Prolific.
Seed rate	25 kg/ha for Bush type and 5 kg/ha for pandal type.
Season	**Bush type** - Throughout year; **Pandal type** - July – August.
Spacing	Dibble single seed 30 cm apart on one side of the ridge formed at a spacing of 60 cm for bush type. For pandal type, 2 - 3 seeds/pit at 2 x 3 m spacing. Spacing for CO 1 Dolichos bean is 1 x 1 m.
Manures and fertilizers	Apply FYM 25 t/ha at the last ploughing. 20kg N, 60kg P_2O_5 and 60 kg K_2O/ha.
Irrigation	Immediately after sowing and on 3rd day, thereafter once a week.
Yield	**Pandal type** : 12 – 13 t/ha, **Bush type** : 8 – 10 t/ha.

Garden Pea

Basic Information and Facts

Scientific name	*Pisum sativum* L.
Common names	Matar (Hindi, Nepali) Pea; split pea, garden pea, seed pea, shelling pea, combining pea, field pea, dry pea, vining pea (English)
Family	Fabaceae (formerly Leguminosae)
Origin	Ethiopia and western Asia
Chromosome number	2n = 14
Inflorescence type	Raceme
Pollination	Self-pollinated crop
Fruit type	Legume

- ***Pisum sativum***, the common pea (also known as the garden or field pea), is an herbaceous annual in the **Fabaceae** (formerly Leguminosae) family.
- The origin and progenitors of *Pisum sativum* are not well known. The **Mediterranean region, western and central Asia and Ethiopia** have been indicated as centres of origin.
- Recently the Food and Agriculture Organisation (FAO) designated **Ethiopia and western Asia** as centres of diversity, with secondary centres in southern Asia and the Mediterranean region.
- The pea is a cool season annual vine that is smooth and has a bluish-green waxy appearance.
- Leaves are alternate, pinnately compound, and consist of two large leaflike stipules, one to several pairs of oval leaflets, and terminal tendrils.
- Inflorescences occur in the leaf axils, and consist of **racemes** with one to four flowers.
- Flowers have five green fused sepals and five white, purple or pink petals of different sizes.

- Pods swollen or compressed, straight or curved on short stalk.
- Ripe seeds are round, smooth or wrinkled, and can be green, yellow, beige, brown, red-orange, bluered, dark violet to almost black, or spotted.
- Pea is a **self-pollinated crop.**
- The plant is a **diploid (2n = 14).**

Production Considerations

Climate	Pea is a cool season crop and requires frost-free weather particularly at flowering and pod formation stage Temperature of 13-19°C is favourable for its cultivation.
Soil	Well drained loamy soil with optimum pH range of 6-7.5.
Improved varieties	**Early wrinkle seeded:** Arkel, Pusa Pragati, Matar Ageta 6, Azad P3, Pant Sabzi Matar 3 (PSM-3), VL Ageti Matar 7 (VL-7), Vivek Matar -10, Kashi Nandini (VRP-5) , Kashi Uday(VRP-6), Palam Triloki . **Main season wrinkle seeded varieties:** Bonneville, Lincoln, Azad P-1, Punjab-89, Palam Priya, Vivek Matar-6, Vivek Matar -8, Vivek Matar -9,Arka Ajit. **Edible poded peas:** Sylvia, Punjab Mithi Phali, Arka Sampoorna.
Seed rate	**Early crop:** 100-120kg/ha, **Mid-season and late varieties:** 75-100 kg/ha.
Season	**North India- Early varieties: September. Main season varieties:** October to mid-November.
Spacing	**Early varieties:** 30cm × 5cm, **Main season varieties:** 45 x 10 cm
Manures and fertilizers	Apply FYM @ 20 t/ha and 60 kg N, 80 kg P and 70 kg K/ha as basal and 60 kg N/ha 30 days after sowing.
Irrigation	Once in a week.
Weed management	Pre-emergence application of Alachlor @ 3litres/ ha or Pendimethalin @ 3litres/ha or Fluchloralin @ 2.5 litres/ha may take care of weeds in the initial growth stages.
Growth regulator	Application of GA_3 (50 ppm) twice at 30 days after sowing and again 15 days later has been found effective in increasing early and total yield.
Yield	**Early varieties:** 60-85q/ha., **Main season varieties:** 100-150q/ha

Fenugreek

Basic Information and Facts

Scientific name	***Trigonella foenum-graecum* L. (common methi)** ***Trigonella corniculata* (Kasuri methi)**
Family	Fabaceae
Origin	South-eastern Europe and West Asia
Chromosome number	2n= 16
Pollination	Self-pollinated crop
Fruit type	Legume

- Fenugreek (***Trigonella foenum-graecum***) is an annual plant in the family **Fabaceae**, with leaves consisting of three small obovate to oblong leaflets.
- It is an annual legume, **diploid (2*n* = 16)** plant.
- The plant grows to a height of about three feet, has three part leaves, the long slender stems bear tripartite, toothed, grey-green obovate leaves, 20-25 mm (3/4-1 in) long.
- *Trigonella foenum-graecum* has long stalked leaves up to 5 cm long stipules triangular, lanceolate, leaflets about 2.5 cm long, obovate to obanceolate.
- The root is a mass of fingery structures.
- The sissile axillary flowers are white or pale yellow.
- The thin, sword-shaped pods are 10-15 cm (4-6 in), with a curved beak-like tip, each carrying 10-20 seeds.
- The plant radiates a spicy odour which persists on the hands after touching. Wild and cultivated varieties exist. Flowers are 1-2, axillary, sessile, racemed, whitish or lemon yellow that bloom from June to July.
- Pod 5.7 cm long with a persistent beak, hairy with 10-20 seeds.
- Fenugreek seed contains 45-60% carbohydrates, mainly mucilaginous fiber **(galactomannans)**, **20-30% proteins** high in lysine and tryptophan, 5 - 10% fixed oils (lipids), pyridine alkaloids, mainly trigonelline (0.2 - 0.38%), choline (0.5%), gentianine and carpaine, the flavonoids apigenin, luteolin, orientin, quercetin, vitexin and isovitexin, free amino acids, such as 4-hydroxyisoleucine (0.09%), arginine, histidine and lysine, calcium and iron, saponins (0.6 - 1.7%), glycosides yielding steroidal sapogenins on hydrolysis (diosgenin, yamogenin, tigogenin, neotigogenin), cholesterol and sitosterol, vitamins A, B_1, C and nicotinic acid and 0.015% volatile oils (n-alkanes and sesquiterpenes).

Production Considerations

Climate	Being a cool season crop, it is fairly tolerant to frost and freezing weather. It can also be grown as a hot weather crop.
Soil	Clay loam soil is the ideal for its cultivation though can be grown in all types of soils.
Improved varieties	Lam Selection 1, Pusa Early Bunching, Kasuri Selection, CO-I Fenugreek, Rajendra Kranti, RMT-1, Methi No. 47, Methi No. 14, and Palam Soumya, Hisar Madhavi, Hisar Manohar , Hisar Mukta, Hisar Sonali, HM-219
Seed rate	Seed rate for common methi is 25 kg/ha and that for Kasuri type is 20 kg/ha
Season	September- December.
Spacing	20-30 cm × 5-7 cm
Manures and fertilizers	FYM 10-15t/ha and 20:60:30 kg NPK/ha.
Irrigation	Frequent irrigation is necessary to obtain quick growth of the crop. Irrigation during early vegetative and grain formation stages is more critical than later stages.
Yield	12-15q of seed/ha and 8-10 tonnes/ha leaves.

Chinese potato

Basic Information and Facts

Scientific name	*Plectranthus rotundifolius* (Poir.) Spreng. Syn. *Solenostemon rotundifolius* (Poir) J.K. Morton
Common name	Coleus potato, Hausa potato, Country potato
Family	Lamiaceae
Origin	Tropical Africa
Chromosome number	2n = 64,84

- ***Plectranthus rotundifolius***, commonly known as "Chinese potato" in India is a perennial herbaceous plant of the mint family **(Lamiaceae)** native to **tropical Africa.**
- *Plectranthus rotundifolius* (Poir.) Spreng. is also known by common names. In English, this includes Hausa potato, frafra potato, coleus potato, country potato, Sudan potato, Madagascar potato and Chinese potato.
- Chinese potato is an erect, semi-succulent annual herb. It is bushy from the base, up to 30 cm tall, prostrate or ascending, and has a succulent stem and somewhat thick leaves.
- The flowers are small and may be white, blue, pinkish or pale violet in colour, being borne on **distal inflorescences** with slender false spikes measuring up to 15 cm in length.
- These flowers are hermaphroditic and the fruits consist of four nutlets which rarely develop.
- Small tubers are produced in clusters at the base of the stem.
- **The tubers** are eaten as a main starchy staple or part of it in combination with legumes, vegetables or cereals.

Production Considerations

Climate	Comes up well in shade with warm humid climate. It requires very good rainfall for its growth and cannot withstand drought conditions.
Soil	The crop requires loamy and well drained fertile soil.
Improved Variety	Sree Dhara
Propagation	Through vine cuttings.
Seed rate	For the production of vine cuttings, tubers are first planted on flat beds. Nursery raised with one kilogram tuber is sufficient to plant 75-100m^2.
Spacing	60cm × 30 cm
Manures and fertilizers	FYM at 25 t/ha as basal and NPK at 30:60:150 kg/ha. Apply 30 kg N 30 days after planting at the time of earthing up along with 2 kg of *Azospirillum*. In case, the soil has eroded from the base of the plant, give one more earthing up at 30 days later to promote tuber formation.
Irrigation	At weekly intervals.
Harvesting	When the leaves and stems dry out and tubers have reached maturity, the plants are harvested with their tubers still attached by gently excavating them.
Yield	10 to 15 t/ha in 150 days.

Okra

Basic Information and Facts

Scientific name	*Abelmoschus esculentus* (L.) Moench
Family	Malvaceae
Origin	India and Africa
Chromosome number	2*n*= 130
Pollination	Often cross pollinated crop
Fruit type	Capsule

- Lady's finger or okra or bhendi ***(Abelmoschus esculentus* (L.) Moench**) is the sixth important vegetable crop of India belonging to the family **Malvaceae.**
- The most frequently observed somatic chromosome number, however, is **2*n*=130**, although Dutta and Naug (1968) suggest that the numbers 2n=72, 108, 120, 132 and 144 are in regular series of polyploids with n=12.
- Okra is an upright annual, herbaceous 3 to 6 feet tall plant with a hibiscus-like flower It is a tropical direct sown vegetable with a duration of 90-100 days.
- Leaves are alternate and usually palmately five lobed, whereas the flower is axillary and solitary.
- The flowers are borne vertically only on the orthotropic axis every two or three days. The flower is axillary and solitary, borne on a peduncle 2.0 – 2.5 cm long.
- The flowers are almost always bisexual and actinomorphic. The perianth consists of 5 valvate, distinct or basally connate sepals and 5 distinct petals that are usually basally adnate to the androecium.
- Okra has perfect flowers (male and female reproductive parts in the same flower) and is self-pollinating.
- Although insects are unnecessary for pollination and fertilization in case of okra, the flowers are very attractive to bees and the plants are cross-pollinated. The cross-pollination upto the extent of 4-19% (Purewal and Randhawa, 1947; Choudhury and Anothai, Choomsai, 1970; Shalaby, 1972) with maximum of 42.2% (Mitidieri and Vencovsky 1974) has been reported. The extent of cross-pollination in a particular place will depend upon the cultivar, competitive flora, insect population and season, etc.
- Okra is categorized as **often cross-pollinated** crop.
- The fruit is an elongated, conical **capsule,** comprising for the most part, five cavities contain ovules.

Production Considerations

Climate	Okra requires a long, warm and humid growing period. It can be successfully grown in hot humid areas. For normal growth and development a temperature between 24°C and 27°C is preferred.
Soil	It is adaptable to a wide range of soils from sandy loam to clayey loam. The most ideal pH range for its cultivation is 6.0 to 6.8.

Varieties/ hybrids	**Open-pollinated varieties:** Pusa Sawni, Parbhani Kranti, Punjab Padmani, Arka Anamika, Arka Abhay, Pusa A-4, Varsha Uphar, Hisar Unnat, Hisar Naveen, Azad Kranti, Azad Bhindi1, Kashi Pragati, Kashi Vibhuti, Kashi Kranti,Phule Utkarsh, Gujarat Anand Okra-5.
Seed rate	18-22 kg/ha for spring crop and 8-10 kg/ha for kharif crop.
Season	June - July and February-March
Spacing	60 x 30 cm for branching types and 45 x 30 cm for non- branching types.
Manures and fertilizers	FYM @200-250 quintals per ha should be applied at the time of field preparation. In addition, apply 60-75 kg N, 50-60 kg phosphorus (P_2O_5) and 50-60 kg potassium (K_2O) per hectare.
Irrigation	After germination, irrigate at weekly intervals.
Weed management	Spray **Oxyflourfen** @ 0.15 kg ai/ha **or Fluchloralin** @ 1.0 kg ai / ha or **Metolachlor** @ 0.75 kg a.i / ha as pre emergence application on third day after sowing + one hand weeding is effective to keep crop weed free.
Growth regulator	Seed treatment with cycocel (100 ppm) for 24 hours followed by foliar application on 20th and 40th days after sowing has resulted in highest fruit set and yield. Seed treatment with GA_3 (45 ppm) has increased the plant height, flowering and fruit yield. Soaking the seeds for 24 hours in GA_3 at 10 mg/litre or IAA at 100 mg/litre or NAA at 100 mg/litre of water has enhanced the seed germination.
Yield	8-10 tonnes/ha during spring-summer and 12.5-15 tonnes/ha during rainy season.

Arrowroot

Basic Information and Facts

Scientific name	***Maranta arundinacea* L.**
Family	Marantaceae
Origin	Andean region of South America
Chromosome number	2n = 18

- ***Maranta arundinacea*** L., commonly known as West Indian arrowroot is an important starchy medicinal plant which finds uses in traditional food and medicine from time immemorial.
- It is a widely cultivated root crop for its **starchy rhizomes** that form the source for extraction of easily digestible starch which finds use as infant and convalescent food.
- It belongs to the family **Marantaceae** having 31 genera and about 550 species distributed throughout the tropics.
- Arrowroot is a perennial plant growing to a height of between 0.3 m (1 ft) and 1.5 m (5 ft).
- Its leaves are lanceolate. The edible part of the plant is the **rhizome.**
- Twin clusters of small white flowers bloom about 90 days after planting.
- Rhizomes are fleshy, cylindrical, and grow from 20 cm (8 in) to 45 cm (18 in) long.

- The rhizomes are used for the production of a very fine, easily-digested starch, which appears on world markets as a dry white powder known as arrowroot starch. It is valued as a foodstuff, particularly for infants and invalids, and is used in biscuits, cakes and puddings.

Production Considerations

Climate	It prefers hot and humid climate. The temperature of 20-30°C favours its growth.
Soil	A slightly acidic, fertile, deep, sandy loam to loamy soil with better drainage facility is most-suited for its cultivation.
Propagation	Through rhizomes.
Seed rate	25q/ha
Time of planting	Early-June
Spacing	90cm × 90 cm
Manures and fertilizers	FYM 10tonnes/ha
Irrigation	The crop is predominantly grown as a rainfed crop. Optimum moisture supply throughout the growth periods gives better yield.
Yield	25-30 tonnes/ha

Drum stick

Basic Information and Facts

Scientific name	***Moringa oleifera* Lam.**
Family	Moringaceae
Origin	African tropics and North-West India
Chromosome number	2n = 28
Inflorescence type	Panicle
Self incompatibility system	Highly cross-pollinated crop

- ***Moringa oleifera*** is the most widely cultivated species of the genus ***Moringa*,** which is the only genus in the family Moringaceae.
- It is a fast-growing, evergreen, deciduous tree.
- It can reach a height of 10-12 m and the trunk can reach a diameter of 45 cm.
- The flowers are fragrant and bisexual, surrounded by five unequal thinly veined yellowish-white petals.
- The bisexual, oblique, stalked, axillary and heteromorphic flowers are **highly cross-pollinated** due to **heteromorphism.**
- The fruit is a hanging, three-sided brown capsule of 20–45 cm size which holds dark brown, globular seeds with a diameter of approximately 1 cm.
- The seeds have three whitish papery wings and are dispersed by wind and water.

Production Considerations

Climate	It is a tropical plant and grows well in plains. The optimum temperature for better growth is 25-35°C.
Soil	Sandy loam soils are most suitable for its cultivation with pH around 6.0-7.5 and good drainage. Water logging and heavy clay soils are not suitable.
Improved varieties	Perennial: Jaffna (yazhpanam), Chavakacheri murungai, Chemmurungai, Palmurungai and Puna murungai. Annual moringa: PKM-1, PKM-2, GKVK-1, GKVK-2, GKVK-3, Dhanaraj.
Propagation	Perennial moringa: though stem cuttings. Annual moringa: through seeds.
Seed rate	Annual moringa: 625 g/ha
Season	Perennial moringa: June to August Annual moringa: June-July
Spacing	Perennial moringa: Dig pits of size 60×60×60 cm at a spacing of 5x5 m Annual moringa: Dig pits of size 45 cm x 45 cm x 45 cm with a spacing 2.5 × 2.5 m High density planting at 1.5 X 1.0 m spacing with two plants/hill.
Manures and fertilizers	A fertilizer dose of 45:15:30 g of NPK/pit may be applied 3 months after sowing. Apply 45 g of N/pit after 6 months when the crop is in bearing.
Irrigation	Irrigate before sowing, on 3rd day after sowing and subsequently at 10 to 15 days interval according to soil type.
Ratooning	Cutting down the plant to a height of one meter from the ground level can be practiced after one year to allow ratooning of the crop.
Pollarding	Pollarding or pruning following harvesting is recommended to promote branching, increased pod production and easy harvesting. This is done during winters (November- December) when no fruit production is seen and start bearing four or five months after ratooning.
Yield	The **annual moringa** tree bears 250-400 fruit depending on the type. **Perennial types** raised through cuttings take nearly a year to bear fruit. In general, the yield during the first two years of fruit-bearing is low (80-90fruit/ year) and gradually increases to 500-600 fruit/tree/ year by fourth to fifth years.

Chekkurmanis

Basic Information and Facts

Scientific name	***Sauropus androgynus* (L.) Merr.**
Family	Phyllanthaceae
Origin	India and Burma region
Sex form	Monoecious
Pollination	Highly cross pollinated crop

- Chekkurmanis (***Sauropus androgynus*** **(L.) Merr.**) belongs to the family **Phyllanthaceae.**
- It is a popular leafy green perennial vegetable with high level of provitamin A, vitamin B, C, A and K, carotenoids, protein, fibre and minerals like potassium, calcium, phosphorus, magnesium and iron.

Production Considerations

Climate	A warm humid climate with good rainfall is best suited and it grows luxuriantly at lower elevations at 500m above MSL.
Soil	It comes up well in all type of soils. The growth and yield are high when they are grown in rich, well drained sandy loam or semi-laterite soils.
Propagation	Through seeds and stem cuttings.
Seed rate	One lakh cuttings/ha.
Time of planting	April-may
Spacing	45 × 60 cm
Manures and fertilizers	5 - 10 kg FYM per plant.
Yield	The annual productivity is 30-50 tonnes/ha with a per plant yield of 1-3 kg leaves per year.

Rhubarb

Basic Information and Facts

Scientific name	*Rheum rhabarbarum* L.
Family	Polygonaceae
Origin	Temperate Asia, probably Siberia
Chromosome number	2n = 44
Inflorescence type	Panicles

- Rhubarb (***Rheum rhabarbarum***) is a species of plant in the family **Polygonaceae**
- It is a winter-hardy herbaceous perennial grown for its edible **leaf stalks**.
- The genus ***Rheum*** (Polygonaceae), commonly known as rhubarb, is composed of 60 species, primarily distributed in mountainous and desert regions of northern and central Asia (Wang et al., 2005).
- Culinary rhubarb is generally considered a **tetraploid, 2n = 44**; however, rhubarb species are **2n = 22, 44, and 66** (Chin and Youngken, 1947; Englund, 1983).

Production Considerations

Climate	Rhubarb is a cold-resistant plant. Under a mild winter coupled with rainless summer, the plant exhibits the habit of dormancy during summer and grows well, during winter and spring season.
Soil	A well drained loamy soil with good organic matter would be the most ideal soil for rhubarb.
Improved varieties	Victoria, Linnaeus, McDonald, Ruby, Valentino, Sunrise, Strawberry and Cherry Red
Propagation	Through division of crowns
Time of planting	Spring and autumn
Spacing	1.2-1.8m × 0.6-1.2m
Manures and fertilizers	FYM 50 tonnes/ha and 100-150kfg of each NPK/ha
Harvesting	Rhubarb is harvested from the second year of planting. During harvesting the stalks are pulled (they are not cut) and then the leaf blades are cut off.

Curry leaf

Basic Information and Facts

Scientific name	*Murraya koenigii* (L.) Spreng.
Family	Rutaceae
Origin	India
Chromosome number	2n=18
Inflorescence type	Cyme

- ***Murraya Koenigii***, belongs to the family **Rutaceae**, commonly known as curry-leaf tree.
- The plant originated in the Tarai region of Uttar Pradesh, and at present it is cultivated in Burma, Ceylon, China, Australia and the Pacific Islands.
- It is an important leafy vegetable. Its leaves are widely used in Indian cookery for flavouring foodstuffs.
- Flowers, bisexual, white, funnel-shaped, sweetly scented, stalked, complete, ebracteate, regular, actinomorphic, pentamerous, hypogynous, the average diameter of a fully opened flower being 1.12 cm.
- Inflorescence, a **terminal cyme**, each bearing 60 to 90 flowers; calyx, 5-lobed, persistent, inferior, green; corolla, white, polypetalous, inferior, with 5 petals, lanceolate; length, 5 mm; androecium, polyandrous, inferior, with 10 stamens, dorsifixed, arranged into circles of five each; smaller stamens, 4 mm long whereas the longer ones, 5 to 6 mm; gynoecium, 5 to 6 mm long; stigma, bright, sticky; style, short; ovary, superior.
- The major constituent responsible for the aroma and flavour has been reported as pinene, sabinene, caryophyllene, cadinol and cadinene.

Production Considerations

Climate	Curry leaf is well grown in warm climate. The optimum temperature requirement is 26° to 37°C.
Soil	Red sandy loam with good drainage and high organic matter is best suited for its cultivation.
Improved varieties	Sen Kaambu, Dharwad-1, Dharwad-2
Propagation	Through seeds.
Season	July- August
Spacing	1.2 -1.5m x 1.2 -1.5m
Manures and fertilizers	After each harvest, 20 kg of FYM/plant is applied and mixed with soil.
Irrigation	Immediately after planting the pits are irrigated. On the third day the second irrigation is given, then the irrigation is given once a week.
Yield	**Vth year onwards:** 5000 kg/ha once in 3 months, which work out to 20,000 kg/ha/year.

Brinjal or Eggplant

Basic Information and Facts

Scientific name	*Solanum melongena* L.
Family	Solanaceae
Origin	Indo-Burma region
Chromosome number	2n = 2x = 24
Inflorescence type	Scorpioid cyme
Pollination	Often cross pollinated crop
Self-incompatibility system	Gametophytic self-incompatibility (GSI)
Male sterility	Genic male sterility
Fruit type	Fleshy berry

- Brinjal, eggplant or aubergine belongs to the family **Solanaceae** and is known under the botanical name ***Solanum melongena* L.**
- The family contains **75 genera** and **over 2000 species**, out of which, about **150-200 are tuber bearing** and belong to section Tuberarium. The majority of species (**about 1800**) are **non-tuber bearing**.
- The chromosome number of many species under non-tuberous group is fairly stable as **2n =2x = 24.**
- It is a herbaceous annual with erect or semi spreading habits.
- It is a **day-neutral plant.**
- In "Origin of cultivated plants" published in 1886 De Candolle, stated that the species *S. melongena* has been known in India from ancient times and regarded it

as a native of Asia. **Vavilov (1928)** was of the opinion that its centre of origin was in the **Indo-Burma region.**

- Cultivars of eggplant (*Solanum melo*ngena L.) bear **hermaphrodite flowers** in clusters as well as singly.
- The flowers are large, violet coloured and solitary or in clusters of two or more.
- Flower consists of calyx: sepals 5, united, persistent; corolla: petals 5, united, usually cup shaped; Androecium : stamens 5, alternate with corolla; Gynoecium: carpels are united, ovary superior.
- Four types of flowers have been reported depending on the length of styles, viz. (i) long-styled with large ovary, (ii) medium-styled with medium size ovary, (iii) Pseudoshort-styled with rudimentary ovary and (iv) true short-styled with very rudimentary ovary.
- The fundamental **inflorescence type** in eggplant is a **scorpioid cyme.**
- The fruit is pendent and is **fleshy berry** borne signally or in clusters.
- Brinjal is a **self-pollinating crop** but cross-pollination also occurs in it. This is because it has pronounced **heterostyly** which favours cross-pollination.
- The anthesis and dehiscence in brinjal are mainly influenced by the daylight, temperature and humidity**).**
- Usually anthesis starts from **7-30 am** and continues up to **11 am** Peak time for anthesis is **8-30** to **10-30 am.** The pollen dehiscence starts from **9-30** to **10 am.**
- As per Indian minimum seed certification standards, an **isolation distance of 300 meters** is required for production of foundation seeds of varieties/hybrids in case of brinjal.
- It has been reported that on an average, the oblong-fruited eggplant cultivars are rich in total soluble sugars, whereas the long-fruited cultivars contain a higher content of free reducing sugars, anthocyanin, phenols, glycoalkaloids (such as solasodine), dry matter, and amide proteins (Bajaj et al., 1979).
- Bitterness in eggplant is due to the presence of **glycoalkaloids** which are of wide occurrence in plants of Solanaceae family.
- The **glycoalkaloid contents** in the Indian commercial cultivars vary from **0.37 mg/100 g** fresh weight to **4.83 mg**) (Bajaj et al., 1981).
- Generally, the high content of glycoalkaloids (20mg/100 g fresh weight) produce a bitter taste and off flavour. The discoloration in eggplant fruit is attributed to high polyphenol oxidase activity.
- The cultivars which are least susceptible to discoloration are considered suitable for processing purposes.
- As brinjal is susceptible to several pests and diseases such as wilt, Phomopsis , little leaf and root-knot nematodes and to insects such as shoot and fruit borer, jassids, epilachna beetle, etc. the development of pest resistant varieties is a major challenges.
- Among the related wild species, ***S. sisymbriifolium*** and ***S. torvumare*** particularly interesting on account of their resistance to three of the most serious eggplant diseases (bacterial wilt, Verticillium wilt and nematodes).

Production Considerations

Climate	A long and warm growing season with a mean temperature of 21°-27°C is most favourable for its successful production.
Soil	Well drained and fertile sandy loam soils rich in organic matter with a pH of 5.5-6.6 are best suited for its cultivation.
Important Varieties	**Long type:** Kashi Taru, Pusa Purple Long, Pusa Purple Cluster, Pant Samrat, Azad Kranti, Pusa Anupam, Punjab Sadabahar, Punjab Barsati, Arka Nidhi, Arka Sheel, Arka Keshav, Utkal Keshari, Swetha **Round type:** Pusa Purple Round, Pusa Kranti, Pusa Uttam, Pusa Upkar, Pant Rituraj, Kalyanpur T-3, Azad B-1, Azad B-2, Hisar Shyamal, Jamuni Gola, Punjab Bahar, Aruna, MDU-1, Swarna Mani, Utkal Tarini **Green brinjal:** Green Brinjal Long-1, CO-1, Phule Harit, Arka Kusumakar, Arka Shirish, Utkal Madhuri
Seed rate	200-300 g/ ha
Season	December – January and May – June
Spacing	60 × 45cm, 75 × 60cm, 75 × 75cm
Manures and fertilizers	**Basal dose:** FYM 25 t/ha, NPK 50:50:30 kg/ ha. **Top dressing:** 50 kg N/ha on 30th day of planting or during earthing up.
Mulching	Mulch with black LDPE sheets of **25 micron thickness** and bury both the ends into the soil to a depth of 10 cm.
Irrigation	After establishment of seedlings, irrigate at weekly intervals.
Weed management	Apply Pendimethalin 1.0 kg a.i. / ha or Fluchloralin 1.0 kg a.i / ha as pre-emergence herbicide, followed by hand weeding once at 30 days after planting.
Growth regulator	The beneficial effects of growth substances on fruit set in brinjal has been observed by using **PCPA (20 ppm), 2,4-D (2.5 ppm) and NAA (60 ppm)** at intervals of one week over a period of 60-70 days from commencement of flowering.
Yield	20-30 tonnes/ha.

Capsicum (Sweet pepper / Bell pepper)

Basic Information and Facts

Scientific name	***Capsicum annuum* L.**
Family	Solanaceae
Origin	Central and South America
Chromosome number	2n=24
Inflorescence type	Cyme
Pollination	Often cross-pollinated crop
Fruit type	Berry

- Sweet peppers (***Capsicum annuum*** **L.**) originated from **central and South America** where numerous species were used centuries before Columbus landed on the continent.
- The ***Capsicum*** **genus** belongs to the **Solanaceae** family and includes 27 recognized species.
- The taxonomic structure of the genus was established from a multidisciplinary approach using numerical taxonomy, cross fertility and cytogenetics, biochemical, geographical and ethnobotanical providing evidence for 5 distinct cultivated species: *C. annuum L., C. frutescens L., C. chinense Jacq., C. baccatum L.* and *C. pubescens* Ruiz et Pav., which originated from distinct domestication events and primary diversification centers, where related wild species still coexist.
- Bell pepper plants are herbaceous and either annual or perennial in their native habitat.
- Their shoot architecture is determined by their sympodial growth habit.
- Plant height varies from 50 to 90 cm under field conditions, but under greenhouse conditions, plant height may vary from 0.7 to 2.5 m.
- The pepper fruit is a **berry** composed of three or four carpels.
- The main parts are the peduncle, calyx, pericarp, placenta, and seeds.
- More than 20 viruses are known to infect chilli and sweet peppers in India. Among these, *Chilli veinal mottle virus* (ChiVMV) of the *Potyvirus* genus, *Cucumber mosaic virus* (CMV) of the *Cucumovirus* genus, chilli infecting *begomovirus* causing *leaf curl diseases* (referred here as *Chili leaf curl virus*, ChiLCV) and *Groundnut bud necrosis virus* (GBNV) or *Peanut bud necrosis virus* (PBNV) of the *Tospovirus* genus are currently considered to be the most important.
- ChiVMV is transmitted by **aphids** in a nonpersistent manner and is the most prevalent *potyvirus* in India.
- ChiLCV transmitted by **whitefly** has emerged as a serious problem throughout India.
- In recent years, *tospoviruses* (PBNV/GBNV) transmitted by **thrips** have emerged as a major threat for chilli cultivation in India. Tospovirus, (PBNV) was widely recorded on tomato and chilli peppers across 14 Indian states representing southern, northwestern, northeastern and central regions (Kunkalikar et al., 2011).

Production Considerations

Climate	It cannot withstand heavy rains during flowering or fruit set. Ideal temperature for flowering is 26°-28°C day and 16°-18°C night.
Soil	Although a wide range of soils are suitable, red or sandy loam soil with a pH of 5.5-6.8 is ideal
Varieties/ hybrids	California Wonder, Pusa Deepti, Arka Basant, Arka Gaurav, Arka Mohini
Seed rate	400-500g/ha
Season	June-July and October-November

Spacing	60 x 30cm, 45 x 45cm
Manures and fertilizers	Apply FYM 25 t/ha and NPK 50:60:60 kg /ha as basal and 40 kg N/ha each on 30, 60 and 90 days of planting as top dressing.
Irrigation	Irrigation at weekly or 10 days interval.
Growth regulator	Two foliar sprays of **NAA (10 ppm)** at flower initiation followed by another spray after 15 days reduced the flower drop and improved fruit set and fruit size.
Yield	15 tonnes/ha in 150-160 days.
Storage	**Temperature:** 7-10°C, **RH:** 90-95%.

Chilli (Hot pepper)

Basic Information and Facts

Scientific name	*Capsicum annuum* L.
Family	Solanaceae
Origin	South America
Chromosome number	2n = 2x = 24
Inflorescence type	Cyme
Pollination	Often cross-pollinated crop
Fruit type	Berry

- Hot pepper (***Capsicum annuum*** **L.**) belongs to the **Solanaceae** family, native to **South America**.
- **Genus *Capsicum*** includes 22 wild species and three varieties as well as five domesticated species and their wild relatives.
- The domesticated ***Capsicum*** spp. Can be grouped into two groups depending upon the cross-compatibility.
- The **white flowered** which includes *C, annum, C chinense C. fruitescens* (*C. annum* group) and *C. baccatum var. baccatum* and *C. baccatum var. pendulum* (*baccatum* group).
- The **purple flowered** comprising *C. pubescens, C. eximinum* and *C. cardenasii.*
- All the species in the genus have n=12 except *C. ciliatum* and *C. scolnikianum* which have n=13.
- The species of the genus are classified into two groups, according to the number of chromosomes: one with 2n=2x=24 chromosomes and another with 2n=2x=26 chromosomes.
- Domesticated species belong to the first group (Moscone et al. 2007, Pozzobon et al. 2006 and Pozzobon and Schifino-Wittmann 2006).
- Chilli is considered as one of the most important commercial spice crops and is widely used universal spice, named as ***wonder spice***.
- Capsicum is derived from the **Greek word "Kapsimo"** meaning "to bite."

- Genus *Capsicum* is divided into three sections by Hunziker - Monotypic Tubocapsicum , Pseudoacnistus and Capsicum.
- Chilli was introduced into India from **Brazil** in the 16th century by the **Portuguese**
- Chillies are referred to as chillies, chile, hot pepper, bell pepper, red pepper, pod pepper, cayenne pepper, paprika, pimento, and capsicum in different parts of the world.
- The original Mexican term, ***chilli*** (now *chile* in Mexico) came from the Nahuatl word *chilli* or *xilli*, referring to a larger *Capsicum* variety cultivated since 3000 BC, as evidenced by remains found in pottery from Puebla and Oaxaca.
- The **world's hottest chilli "Naga Jolokia"** is cultivated in the hilly terrain of Assam in a small town Tezpur, India.
- **India** is the largest producer and exporter of chillies.
- India has immense potential to export different types of chillies required by various markets around the world. It is the leader in export, with **25% share in world trade**, followed by China with 24% share in total global export.
- .***Capsicum annuum*** is an annual sub-shrub, the flowers of which are borne singly and fruits usually pendent, which provide red peppers, cayenne, paprika and chillies and sweet pepper (bell pepper) a mild form with large inflated fruits.
- ***Capsicum frutescence*** is a perennial chilly with small sized pods which are highly pungent. It is commonly known as **'bird chilly'** and **'Tabasco'**.
- Capsicum plants are herbaceous or semi-woody annuals or perennials.
- The leaves are ovate, tapering to a sharp point, measuring up to 15 cm, dark green on the upper surface and pale green on the lower surface.
- The glabrous stem is angular and the inflorescence terminal, flowers are axile in placentation borne at nodes. The corolla is whitish and calyx greenish.
- The flowers are small, white and borne singly or in clusters of 2 or 3 in the axils of the leaves.
- Capsicum and chilli are generally **self-pollinating**, although **cross-pollination is also common**.
- Chilli flower is normally solitary but occasionally borne in small **cymes** of leaf axils.
- The calyx is five lobed and corolla is five-parted and white, but occasionally purple in colour. The five stamens attached to the base of corolla are separated. The bluish anthers dehisce by splitting longitudinally. The single style is usually longer than the stamens and stigma is club-shaped. The ovary generally has three locules. Pepper tend to blossom and set fruit earlier under short day conditions.
- Capsicum flowers produce nectar at the base of the flower, with the quantity of nectar produced varying between varieties.
- Peppers are generally **self-pollinated**, but some **cross-pollination** can occur between and within the cultivars of the two species (*C. annuum* and *C. frutescens*).
- The fruit is a **berry** and may be green, yellow, or red when ripe. The fruits are of diverse shapes and sizes depending upon the variety.

- A few wild species (*C. Eximium, C. Cardenasii*) are self-incompatible with a **gametophytic system** as are other wild solanaceae species.
- Dry chilli is extensively used as spice. It is also used as an ingredient in curry powder and in seasonings.
- Bird chilli is used in making hot sauces as pepper sauce and Tabasco sauce.
- Paprika, Bydagi chilly, Warangal chapatta and similar high colour less pungent varieties are widely used for colour extraction.
- The pungency for the red peppers and the colour value for the paprikas are the most important parameters.
- Pungency in chilli is due to **capsaicin ($C_{18}H_{27}NO_3$)**, a colourless, odourless oil-like compound found in the fruit of a plant. It is a tasteless, odourless white crystalline substance.
- The **Scoville organoleptic test** is a refined, systematic approach to measure the pungency level.
- It was the first laboratory test reported and used to measure the pungency in chillies .
- One **'Scoville Heat Unit'** corresponds to about fifteen parts per million (ppm) of the capsaicin.
- The pure capsaicin has the maximum possible Scoville rating of **15 million units** (Scoville, 1912).
- **Capsaicin** and **dihydrocapsaicin** together make up 80- 90 per cent of the capsaicinoids found in hot peppers.
- In ***Capsicum annuum*** species, the total capsaicinoid content ranges from **0. 1 to 1.0 percent** and the capsaicin to dihydrocapsaicin ratio is about **1: 1.**
- In ***Capsicum frutescens,*** the total content of these compounds ranges from **0.4 to 1.0%** with the ratio around 2: 1.
- The whole chilli consists of **40% pericarp** containing an inner sheath known as dissepiment, **56% seeds** and **4% stalks**.
- The pericarp contains almost all of the pungency, whereas the chilli seeds contain only traces of pungency with a **capsaicin** content of **0.005%.**
- The capsaicin content of chilli oleoresin ranged from **2.5 to 3%.**
- The pungency of the pericarp is mostly concentrated in the dissepiment.
- The colour value is the principal criterion for assessing the quality of paprika.
- The pigent content ranged from 0.1 to 0.8 percent. The extractable colour value is measured by a spectrophotometric process, usually expressed in terms of **ASTA** colour value (American Spice Trade Association).
- Common extractable colour values present in the industry are 85, 100, 120 and 150 (Anu and Peter, 2000).
- It is reported that approximately 20 carotenoids contribute to the colour of *Capsicum* powder.
- The major red colour develops from **capsanthin** and **capsorubin,** while the yellow orange colour is from **beta-carotene** and **violaxanthin.**

- **Capsanthin,** the major carotenoid of ripe fruits, contribute up to **60%** of the total carotenoids.
- Other carotenoids present are capsorubin, zeaxanthin, lutein, kryptoxanthin and alpha and beta-carotene.
- This colour is highly popular among food and beverage processors for its use as a colourant, since this being a 'natural plant colour'.
- Capsicum has a tonic and carminative action. Taken inordinately it may cause gastro-enteritis.
- **Oleoresin capsicum** is used in pain balms and vaporubs.
- Dehydrated green chilly is a good source of vitamin 'c'.

Production Considerations

Climate	During the growing period which extends over 4 months, maximum temperature ranging from 20-30°C and a minimum temperature not below 10°C is ideal.
Soil	Well drained loamy soil rich in organic matter with pH range 6.5-7.5.
Seed rate	400-500g/ha
Season	June-July and October-November
Spacing	60 x 30cm and 45 x 45cm
Manures and fertilizers	Apply FYM 25 t/ha and NPK 50:60:60 kg /ha as basal and 40 kg N/ha each on 30, 60 and 90 days of planting as top dressing. **Apply 250g of Mycorrhiza** in two splits at transplanting and 30 DAT along with 100% N and K and 50% P for higher yield and saving of Phosphorous. **Foliar spray** of Zn SO4 @ 0.5 per cent thrice at 10 days interval from 40 days after planting. Spray 19:19:19 + Mn @ 1 % at 60 days after planting.
Irrigation	Irrigate at weekly intervals.
Weed management	Apply **Pendimethalin** 1.0 kg a.i. / ha or **Fluchloralin** 1.0 kg a.i. / ha as pre emergece herbicide followed by hand weeding once 30 days after planting.
Growth regulator	Spray of **Planofix (-NAA formulation)** @ 100 ml/ 450 litres of water per hectare at flowering stage followed by second spray three weeks after the first spray reduces flower drop and thus, improve fruit set and yield.
Yield	**Varieties:** 2 - 3 t/ha of dry pods or 7 - 16 t/ha of green chillies. **Hybrids:** 25t/ ha of green chillies.
Storage	**Temperature:** 7-10°C, **RH:** 90-95%.

Improved cultivars/ hybrids

List of chilli cultivars recommended for national release in India (1975-2014)

Cultivar type (developed by)	Name (recommended zone/s)
Open pollinated (all 24 by public sector)	G-4 or Bhagyalakshmi, G-5 or Andra Jyothi (I, IV, V, VI, VII, VIII), K-2 (IV, VI), J-218 (I, IV, V, VI, VII), Bhaskar or X-235 (IV, V, VI, VIII), Muslawadi (V), Arka Lohit/Sel-1 (V, VII, VIII), LCA206-B (V, VI, VII, VIII), Phule C-5(VII), AKC-86-39 (VII), BC-14-2 (V, VI), RHRC-Cluster Erect (VII), PMR57/88-K (VII), LCA-334 (III, IV, V, VII), ASC-2000-02 (VII), KA-2 or Kashi Anmol (IV), LCA-353 (IV, V, VII),BC-25 (V, VI, VII), PC-7 (V), HSHP-154 (VIII), IVPBC-535 or Kashi Sinduri-Paprika (VIII), PC-56 (I, IV, V, VII, VIII), VR- 338 (II), ACS-06-2 (VIII)
Hybrid (10 by private and 5 by public sector)	HOE-888 (IV, VIII), ARCH-236 (IV), Sungrow-86-235 (IV, VIII), ARCH-228 (IV, V, VI), Arka Meghana or MSH-172 (IV, V, VI & VIII), Arka Sweta or MSH-149 (IV, VI, VIII), CCH-2 or Kashi Surkh (II, IV, V, VI), KCH-3 (IV), CCH-3 or Kashi Early (IV, V, VIII), BSS-453 (II), NCH-587 (IV, VII), VNR-332 or Rani (IV, VIII), HH-41786 (VII), BSS-378 (VII), VNR-Vidya (IV)

Zones: I = Humid western Himalayan region, II =Humid Bengal-Assam basin, III = Humid eastern Himalyan and bay islands, IV = Sub-humid Satlej Ganga Alluvial plains, V = Sub-humid eastern and south eastern plains, VI = Arid western plains, VII = Semiarid plateau and central highlands, VIII = Humid, semi-arid Western Ghats and Karnataka plateau.

Potato

Basic Information and Facts

Scientific name	*Solanum tuberosum* L.
Family	Solanaceae
Origin	South America
Chromosome number	2n = 4x = 48
Inflorescence type	Cyme
Pollination	Self-pollinated crop
Self incompatibility system	Gametophytic self-incompatibility
Fruit type	Berry
Edible Part	Underground stem

- The potato (***Solanum tuberosum***) belongs to the **Solanaceae family.**
- Worldwide, the potato comes forth in terms of production after wheat, maize, and rice.
- It originated in the high Andean hills of **South America**, from where it was first introduced into Europe towards the end of 16th century through Spanish conquerors.

- The potato was first domesticated in the region of modern-day southern **Peru** ***and extreme northwestern*** **Bolivia** between 8000 and 5000 BC.
- ***It has since spread around the world and become a*** **staple crop** in many countries.
- The potato is a **dicot** plant belonging family Solanceae and the genus ***Solanum.***
- About 72% of the species are diploid (2n=24) and nearly 12% tetraploid (2n=48). The rest are triploid (2n=36), pentaploid (2n=60) and hexaploid (2n=72).
- The cultivated potato, *Solanum tuberosum* L., is an **autotetraploid (2n=4x=48)** species that displays tetrasomic inheritance.
- The major species grown worldwide is ***Solanum tuberosum (a*** tetraploid ***with 48*** chromosomes), and modern varieties of this species are the most widely cultivated.
- Tetraploid *S. tuberosum* ***is self-compatible, although most of the related diploid species are self-incompatible.***
- Pollen sterility occurs frequently in *S. tuberosum*, and ovule sterility occasionally; many varieties do not produce any botanical seed.
- Self-incompatibility occurs commonly in tuber-bearing *Solanum* ***species, especially at the diploid level. The locus S with multiple alleles controls*** **gametophytic self-incompatibility.**
- The **potato tuber** is an enlarged portion of an **underground stem.**
- Anatomically, the potato tuber is a typical stem consisting of the following principal regions: the periderm, the cortex, the vascular cylinder, the outer medulla representing the internal phloem and the inner medulla or pith with only a narrow parenchyma zone in the interior and without phloem elements.
- The internal phloem is the principal storage tissue of the tuber. The inner medulla forms the central axis of the tuber and is not as rich in starch.
- Potato plants may develop from seed or from tubers.
- The potato stem system consists of stems, stolons and tubers.
- The terminal **inflorescences** are **cymes** that are 5-11 cm long and generally found in the distal half of the plant.
- Potato flowers are bisexual. They possess all four essential parts of a flower; calyx, corolla, male elements and female elements.
- Potato is predominantly **self-pollinated**, although some cross-pollination is often accomplished by bumblebees (Caligari, 1992; Acquaah, 2007).
- **In 1845,** a plant disease known as late blight, caused by the fungus *Phytophthora infestans*, spreads rapidly through the poorer communities of **western Ireland,** resulting in the crop failures that led to the **Great Irish Famine.**
- The year **2008** was declared the **International Year of the Potato** ***by the*** **United Nations,** noting that the potato ***is a staple food in the diet of the world's population, and affirming the need to focus world attention on the role that the potato can play in providing*** food security ***and eradicating*** poverty.
- Potato (*Solanum tuberosum*) can be propagated both **vegetatively** (clonally) and **sexually.**
- Sexual or botanical seed of potato commonly called **"True Potato Seed" (TPS)** is a radical alternative to seed tubers for raising commercial potato crops.

- **Seed plot technique:** This technique was developed in 1970s to enable healthy seed potato production in the sub-tropical Indian plains under low aphid period.
- **Late blight** caused by ***Phytophthora infestans*** (Mont.) is the most devastating disease of potato world over.
- ***S. demissum*** is the most potential wild species that has been used for imparting **late blight resistance** in Solanum tuberosum worldwide due to resistance is controlled by dominant, major genes and provided complete control of the disease.
- **Wart disease of potato** is caused by fungus, ***Synchytrium endobioticum* (Schilb).** This disease was first reported in **1953 in North Bengal Hills.**
- The resistance genes are available in a number of varieties of *S. tuberosum*. Besides, a number of wild species like *S. boliviense, S acaule, S microdontum, S. demissum, S. sparsipilum, S. polytrichon, S. simplicifolium, S. chacoence* f.sp. *boergerii, S. vernei* and *S. spegazzinii* are known to have resistance to the disease.
- **Potato cyst nematodes** (*Globodera* spp.) is one of the major potato pest throughout the world. Quarantine or regulatory actions are imposed against them in most countries.
- Viruses have systemic distribution in the host and leads to degeneration of seed stocks. At least twelve viral diseases are known to infect potato crop. Among them, viruses **PVX, PVY, PVS, PVA** and **leaf roll (PLRV)** are important.
- **Resistance sources:**
    - **PVX:** *S. tuberosum* ssp. *andigena* and *tuberosum*, *S. acuale, S. sucrense, S. chacoense* and *S. curtilobum.*
    - **PVY:** *S. stoloniferum, S. demissum, S. chacoense, S. candiophyllum, S. microdontum, S. polyadenium* and *S. multidissectum.*
    - **PVS:** *S. tuberosum* ssp. *tuberosum* varieties like Great Scot, Kerries Pink, Red Skin, Eclipse, British Queen, Up-to-date, Duke of York and Foxlyna etc.
    - **PLRV:** *S. acuale, S. berthaultii, S. etuberosum, S. brevidense* and *S. tuberosum* ssp. *andigena.*
- Potato is affected by six bacterial diseases, viz. bacterial wilt or brown rot (*Ralstonia solanacearum* (Smith), soft rot of stem and tuber (*Erwinia carotovora, Bacillus spp., Pseudomonas spp.*), ring rot (*Clavibacter michiganensis* ssp. *sepedonicus* (Spieck and Koth), common scab (*Streptomyces spp.*), pink eye (*Pseudomonas spp*) and leaf spot (*Xanthomonas vesicatoria*).
- In India, the potato cyst nematode was first detected in **1961 by Jones at Ootacamund in Tamil Nadu.**
- The species possessing resistance against potato cyst nematode are *S. ehrenberjii, S. vernei, S. chacoense, S. phureja, S. demissum, S. gourlayi, S. microdontum, S. sucrense, S. tarijense, S. acaule, S. fendleri, S. multidissectum, S. oplocense, S. sparsipilum* and some accessions of *S. tuberosum* ssp. *andigena.*
- Potato is a highly nutritious, easily digestible, wholesome food containing carbohydrates, proteins, minerals, vitamins and high quality dietary fibre. A potato tuber contains 80% water and 20% dry matter consisting of 14% starch, 2% sugar, 2% protein, 1% minerals, 0.6% fibre, 0.1% fat, and vitamins B and C in adequate amount.

- The **Central Potato Research Institute (CPRI)** established in the year **1949** is the premier institute working on research on potato and is situated at **Shimla in Himachal Pradesh.** The varieties released from CPRI have **'Kufri'** as their first name.
- So far 50 potato varieties have been bred for different agro-climatic regions of the country with 28 varieties alone for north Indian plains.
- Varieties viz. Kufri Chipsona-1, Kufri Chipsona-2, Kufri Chipsona-3, Kufri Frysona have been developed for processing purposes.
- Cultivars **Kufri Sheetman** and **Kufri Dewa** possess resistance to frost.

Production Considerations

Climate	It is a cool season crop and can tolerate moderate frost. It requires 20°C soil temperature for better germination. Young plant growth is good at 24°C but later growth is favoured by a temperature of 18°C. **No tuberization takes place when the night temperature exceeds 23°C.** Maximum tuberization occurs at 20°C. Tuber formation stops completely at about 29-30°C.
Soil	It can grow in almost all types of soil. The well drained clay loam soil is considered as ideal for its cultivation. On sandy loam soil, crop can be successfully grown provided Manures and fertilizers is done heavily and the crop is irrigated properly and timely. **It produces best when soil pH is 6.0-6.5.**
Improved varieties	**Early varieties:** These varieties are ready for harvest in 70-80 days such as Kufri Ashoka, Kufri Chandermukhi, Kufri Jawahar, and Kufri Lauvkar. **Main season varieties:** They are ready for harvest in 90-95 days. Among the white coloured varieties, Kufri Jyoti, Kufri Sutlej, Kufri Pukhraj, Kufri Megha, Kufri Badshah, Kufri Anand, Kufri Bahar, Kufri Sadabahar, Kufri Deva, Kufri Sherpa, Kufri Swarna, Kufri Shailza, Kufri Surya, Kufri Himalini, Kufri Girdhari and Kufri Khyati are important. **Late varieties:** Kufri Jeevan, Kufri Neelamani, Kufri Khasigaro, Kufri Naveen **Varieties for processing:** Kufri Chipsona-1, Kufri Chipsona-2, Kufri Chipsona-3, Kufri Frysona **Cultivars resistance to frost:** Kufri Sheetman and Kufri Dewa
Seed rate	**Tubers:** 20-25q/ha for 15g seed, 25-30q/ha for 30g seed and 30-35q/ha for 45g seed size. **TPS:** 45-50g/ha
Dormancy breaking	**Thiourea (Sodium Potassium thiocynate)** @ **1-2%** solution which is used as a treatment to cut tubers for one to one and half hours and about 1 kg of thiourea is sufficient for 10 quintals of seed tuber. Tubers are kept in **5ppm solution of GA_3 for 10 seconds.** The tuber are then treated with aqueous solution of thiourea for one hour followed by dipping in 2 ppm solution of GA for 10 seconds. The tubers from cold storage are warmed up at **16°C for 10-14 days** before sowing which sprout quickly and give good germination stand.

Season	Plains: October – November
Spacing	**35-45 mm sized tubers:** 60 x 20cm, **45-55 mm sized tubers:** 60 x 25cm
Manures and fertilizers	FYM-25t/ha, NPK: 120: 80:60kg/ha.
Irrigation	The total water requirement varies between **350-550mm** depending upon soil type, climate and crop duration. Pre-planting irrigation is advantageous for uniform germination. Second irrigation is given after about a week and subsequent as and when required.
Weed management	Pre-emergence application of **fluchloralin**@ 1 kg a.i. per ha or **alachlor** @ 1 kg a.i. per ha or **pendimethalin** @ 1.8 kg a.i. per ha or **atrazine** @ 1.0 kg a.i. per ha can effectively control weeds. Application of **Tok-e-25** @ 2.5kg a.i. per ha as post-emergence application at about 2-3 leaf stage is also helpful in managing the weeds.
Growth regulator	The freshly harvested tubers have a rest period for 8 weeks and as a result fail to sprout if sown immediately. This dormancy can be broken by dipping tubers in 5ppm solution of **GA_3** for 10 seconds or in **thiourea (sodium potassium thiocynate) @ 1%** solution for one hour. In other studies, spray of ascorbic acid (100 ppm), borax (25 ppm) and cytozyme crop plus (0.2%) on the standing crop have also improved tuber yield significantly.
Yield	**Early varieties:** 200 q/ha **Main season / late varieties:** 300 q/ha
Storage	It is crucial that the storage area is dark, well ventilated and for long-term storage maintained at temperatures near 4 °C (39 °F).

Tomato

Basic Information and Facts

Scientific name	***Solanum lycopersicum* L. (Syn. *Lycopersicon esculentum* Miller)**
Family	Solanaceae
Origin	Central and South America
Chromosome number	2n=2x=24
Inflorescence type	Cyme
Pollination	Self-pollinated crop
Self incompatibility system	Gametophytic self-incompatibility (GSI) [wild, not cultivated]
Male sterility system	GMS, or genic male sterlity
Fruit type	Berry

- Tomato (***Solanum lycopersicum* L.**) belongs to the family **Solanaceae**.
- The cultivated tomato, ***Solanum lycopersicum*** L., is the second most consumed vegetable crop after potato and unquestionably the most popular garden crop in the world.

- In 1753, Linnaeus placed the tomato in the genus *Solanum* (alongside the potato) as *Solanum lycopersicum*.
- In 1768, Philip Miller moved it to its own genus, naming it *Lycopersicon esculentum*.
- Genetic evidence has now shown that Linnaeus was correct to put the tomato in the genus *Solanum*, making *Solanum lycopersicum* the correct name. Both names, however, will probably be found in the literature for some time.
- The species originated in **Central** and **South America.**
- Tomato is a model species for classical genetic and genomic studies. It has a chromosome number of **2n=2x=24** and a modest genome size (950 Mb/haploid genome).
- Tomato has perfect flowers grouped in compound inflorescences known as a **cyme**.
- Each inflorescence usually includes 4 to 8 flowers, and each plant may produce as many as 20 or more successive inflorescences during its life cycle.
- Tomato flower has a 5 to 10 parted calyx which persists until fruit matures.
- Tomato fruit is classified as a **berry**. As a true fruit, it develops from the ovary of the plant after fertilization, its flesh comprising the pericarp walls.
- Tomato is a predominantly **self-pollinated crop** with some amount of natural cross-pollination.
- The **self-incompatibility system in** tomato is **gametophytic** and regulated by a single, multiallelic *S* locus.
- The first commercially available **genetically modified food** was a variety of tomato named the **Flavr Savr,** which was engineered to have a longer shelf life.
- Tomato is characterized by two types of plant : (i) **Determinate type**-inflorescence occurs more frequently in almost every internode until terminate ones are formed and elongation ceases at this point (ii) **Indeterminate type**-inflorescence cluster occurs at every third internode and the main axis continues growing indefinitely.
- The tomato fruit is ripening by **climacteric** mechanism (as opposed to **non-climacteric**). Climacteric ripening is accompanied by a peak in respiration and a concomitant burst of **ethylene** (important phytohormone).
- The characteristic pigmentation of red tomato fruit is due to the deposition of **lycopene**, the predominant carotenoid found in tomato fruit, and beta-carotene, which are associated with the change from green to red as chloroplasts are transformed to chromoplasts.

Production Considerations

Climate	Tomato, a warm season vegetable, is grown extensively in cool season also. The optimum temperature required for its cultivation is 15°-27°C.
Soil	Well drained loamy soil rich in organic matter with a pH range of 6.5 - 7.5.
Seed rate	**Open pollinated:** 400-500g/ha **Hybrid:** 125-175g /ha
Season	June-July and October- November
Spacing	60 × 45cm, 75 × 60cm, 75 × 75cm

Manures and fertilizers	**Basal dose:** FYM 25 t/ha, NPK 75:100:50 kg / ha. Borax 10 kg and Zinc sulphate 25kg/ ha. **Top dressing:** 75 kg N/ha on 30th day of planting or during earthing up.
Micronutrient spray	Foliar spray of ZnSO4 @ 0.5 per cent thrice at 10 days interval from 40 days after planting. Spray 19:19:19 + Mn @ 1 % at 60 days after planting.
Mulching	Mulch with black LDPE sheets of **25 micron thickness** and bury both the ends into the soil to a depth of 10 cm.
Irrigation	After establishment of seedlings, irrigate at weekly intervals.
Weed management	Apply Pendimethalin 1.0 kg a.i./ha or Fluchloralin 1.0 kg a.i / ha as pre-emergence herbicide, followed by hand weeding once at 30 days after planting.
Growth regulator	Application of **Parafen (sodium salt of p-chlorophenoxy acetic acid 4-PCPA)** at 30 ppm in winter and early spring under poor natural light conditions, improved fruit formation, ripening and fruit yield. During summer season, foliar application of **PCPA (50 ppm) or Ethrel (250 ppm)** at turning stage of fruits accelerated ripening, produced early fruit yield, and also improved the fruit quality (vitamin C, T.S.S. and lycopene content).
Staking	Stake the plants 30 days after planting with 1 - 1.5 m tall stakes. Remove the side branches up to 20 cm from ground level.
Yield	20-25 tonnes/ha.

Important Varieties

1. **Determinate types:** Kashi Vishesh, Kashi Anupam, Kashi Amrit, Pusa Early Dwarf, Pusa Gaurav, Azad T-6, Azad T-2, Punjab Kesri, Punjab Chhuhara, S-12, Hisar Arun, Hisar Lalit, HS-101, HS-102, Mani Leima, Narendra Tomato-2, CO-2, CO- 3, Utkal Pallavi.
2. **Indeterminate types:** Pusa Ruby, Sel-120, Narendra Tomato- 4, Narendra Tomato-5, Pant T-3, Pant Bahar, Marglobe, Arka Meghali, Arka Saurabh, Arka Vikas, Solan Vajr, Utkal Kumari.
3. **Varieties for fresh market:** Pusa Early, Dwarf, Pusa Ruby, Pusa 120, Pant T-3, Arka Vikas, Arka Saurabh, CO-3, Punjab Kesari, Pant Bahar.
4. **Varieties for distant market:** Pusa Gaurav, Roma, Punjab Chhuhara, Pusa Uphar.
5. **Varieties for processing:** Pusa Gaurav, Rousa, Punjab Chhuhara, Pusa Uphar, Arka Saurabh.
6. **Varieties resistant to abiotic stresses:** Pusa Sheetal (low temperature), Pusa hybrid 1 (High temperature), Pusa Sadabahar (high and low temperature regime).
7. **Varieties resistant to leaf curl:** Sankranthi, Nandhi and Vybhav.
8. **Functional male sterile mutants :** Katrain Hybrid 1 and 2.
9. **Hybrids:** Kt-4, Pusa Hybrid 1, Pusa Hybrid 2, Pusa Hybrid 4, Arka Vardan, Arka Meghali, Arka Vishal, Arka Samrat, Kashi Hybrid 1, Kashi Hybrid 2.

OOO

CHAPTER 12

Seed Production of Vegetable Crops

N.K. Gupta[1] and [2]Nityamanjari Mishra
[1]*Professor (Horticulture)*
College of Agriculture, Indore (M.P.)
[2]*Jr. Scientist (Horticulture)*
RRTTS, (OUAT) Keonjhar, Odisha

The availability of quality seed is of utmost importance for increasing the vegetable production. Vegetable growers recognize quality seed of improved varieties as the most strategic resource for higher and better vegetable yields. Although India ranks second in vegetable production, the quality vegetable seed production in the country has been insufficient.

The growth of plant and the quality of seed production are strongly influenced not only by genetic factors but also by the environmental condition, in which production is undertaken. Emphasis should always be laid on those factors which contribute to and affect seed quality like selection of crop and variety, seed source, rouging, harvesting and post harvest operations etc.

The packages of seed production technology vary from location to location and from crop to crop. But a broad general recommendation can be adopted which could be suitably modified on the basis of individual vegetables and the growing area.

Seed

Seed is a mature ovule, consisting of an embryonic plant together with a store of food all surrounded by a protective coat.

Table 12.1: *Category of seed*

S. No.	Category of seed	Description	Colour of tag
1.	Nucleus seed	The seed or the research material produced by the breeder by various selection procedures in a pure line variety or clone is known as nucleus seed.	------
2.	Breeder seed	Breeder seed is produced from nucleus seed under the supervision of a qualified plant breeder in a research institute of Agricultural University.	Yellow

3.	Foundation seed	Foundation seed is the progeny of breeder seed and is produced by State Farm Corporation of India, National Seed Corporation, State seed Corporation under technical control of qualified plant breeders or technical officers.	White
4.	Certified seed	Certified seed is the progeny of foundation seed and its production is supervised and approved by certification agency.	Azar blue
5.	Labeled Seed	There is another category of seed known as Labeled Seed in which case there is no certification but labeling is compulsory.	Opel Green

Seed Certification Standards

- In India seed certification standards have been prescribed for foundation and certified seed.
- There are two types of standards; **field standards,** which apply to standing crop and **seed standards** which are applicable at seed level.
- Field standards include isolation requirement, maximum permissible level of off types, inseparable other crop plants, objectionable weed plants, pollen shedders (in male-sterile or A lines), plants infected by seed borne diseases etc.
- Seed standards relate to genetic purity, physical purity, germination, other crop seeds, weed seeds, moisture content etc.

Isolation Distance

- Maintaining proper isolation distance is a must for obtaining pure seeds of any particular variety. The isolation distance is decided based on the nature of pollination in a particular crop and the class of seed being raised.
- The recommended isolation distances for different vegetables are provided in the following table:

Table 12.2: *Minimum isolation requirement for seed production of vegetable crops*

S. No.	Crop	Botanical Name	Minimum Isolation Distance (Meters)		Minimum Isolation Distance (Meters) for hybrids	
			Foundation Seed	Certified Seed	Foundation Seed	Certified Seed
	Potato	*Solanum tuberosum* L.	5	5	-	-
	Tomato	*Lycopersicum esculentum* Mill.	50	25	200	100
	Brinjal	*Solanum melongena* L.	300	150	200	200
	Capsicum and chilli	*Capsicum annuum* L. *Capsicum frutescens* L.	500	250	500	250

S. No.	Crop	Botanical Name	Minimum Isolation Distance (Meters)		Minimum Isolation Distance (Meters) for hybrids	
			Foundation Seed	Certified Seed	Foundation Seed	Certified Seed
	Okra	*Abelmoschus esculentus(L.) Moench*	500	250	-	-
	Onion	*Allium cepa* L.	1000	500	1200	600
	Watermelon	*Citrullus lanatus* (Thunb.) Matsumura et. Nakai	1000	500	1500	1000
	Cabbage	*Brassica oleracea* var. *capitata* L.	1600	1000	-	-
	Cauliflower	*Brassica oleracea* (L.) var. *botrytis* L.	1600	1000	-	-

Source: Website of Central Seed Certification Board Department of Agriculture & Co-operation Ministry of Agriculture Government of India, New Delhi

Seed Standards of Genetic Purity

All certified seed lots shall conform to the following Minimum Standards for genetic purity unless otherwise prescribed:

Table 12.3: *Seed Standards of Genetic Purity*

Class	Standards for Minimum Genetic Purity (%)
Foundation	99.00
Certified: (i) Varieties, composites, synthetics & multilines	98.00
(ii) Hybrids	95.00
(iii) Hybrids of cotton, TPS, muskmelon, brinjal &tomato	90.00
(vi) Hybrid castor	

Physiological Quality of Seeds

- **Germination:** High germination is essential to maintain optimumplant population with recommended seed rate.
- **Seed Vigour:** Vigorous seeds are a pre-requisite to establish optimum plant population even under adverse conditions.
- **Seed Health:** Seeds free from seed-borne pathogens and insects are essential to maintain plant population and soil health.

Seed Processing

Objective

- Removal of excess moisture.
- Enhancement of physical purity and freedom from pest.

- Removal of impurities, immature seeds, plant materials (dockages), mechanically damaged seed.
- Enhance storability.

Table 12.4: *Suitable machineries of seed processing*

Physical difference	Suitable machineries
Seed Size- small to bold	Air screen cleaner-cum-grader
Density- ill filled, immature, light to dense seed	Specific gravity separator
Shape- Round to oval	Spiral separator
Surface texture- Smooth to wrinkled and rough	Roll mill/ dodder mill
Colour of the seed- Light to dark	Electronic colour separator
Seed vigour- Size of the plumule, internal status of the seed (Hollowness, insect infestation, impact damage or seed health	Soft X- ray sorter

Seed Testing, Labelling and Storage

- After processing of a given lot, the representative sample is sent to a notified seed testing laboratory for analysis. Seed testing is done as per **ISTA** Rules.
- If the seed test report is satisfactory, i.e. the seed meets the prescribed seed standards, then a given seed lot is approved and tags and certificates are issued by the certification agency to the seed producer.
- The validity period is nine months from the date of test at the time of initial certification.
- The validity period can be further extended for six months provided on re-testing the seed conforms to the prescribed standards w.r.t. physical purity, germination and insect damage for all seeds except vegetatively propagating material, which lot shall be examined for seed standards specified for the respective crop.
- Seed being a living entity is highly sensitive to ambient weather conditions, viz. high relative humidity and temperature, which deteriorates its viability and vigour.
- Improper handling of seeds also causes mechanical injury and lower down its germination and storability. Therefore, during post-harvest processing and storage seeds must be handled properly and protected from high relative humidity and temperature, insect pests and rodents.

Seed germination

Germination is the growth of an embryonic plant contained within a seed; it results in the formation of the seedling.

Table 12.5: *Types of seed germination*

S. No.	Type of germination	Description
1.	Epigeal germination	In this, the cotyledons are raised out of the soil and generally become green and photosynthetic. In dicots, they are pushed up by rapid extension of the hypocotyl before the growth of the epicotyl.

2.	Hypogeal germination	In this type of germination, the cotyledons remain underground. The hypocotyl growth is restricted. The epicotyl grows to raise the first leaves out of the soil.
3.	Viviparous germination	This is a special type of germination occurring in mangrove plants. These plants generally grow in salty lakes, sea coasts and deltas. Here, the seed germinates while still attached to the parent plant.

Improved Varieties and Hybrids of Vegetable Crops

Table 12.6: *Crop-wise popular varieties and hybrids of vegetable crops released from ICAR/ SAUs/other organizations*

S. No.	Name of crop	Name of Institutes/ Universities	Varieties/ Hybrids
1.	Amaranthus	IARI, New Delhi	Pusa Kiran, Pusa Kirti, Pusa Lal chaulai
		TANU, Coimbatore	CO-1, CO-3, CO-4, CO-5
2.	Beetroot	IARI, New Delhi	Crimson Globe, Detroit Dark Red
		TANU, Coimbatore	Ooty.1
3.	Beet leaf/ Palak	IARI, New Delhi	Pusa Bharati, Pusa Harit , Pusa Jyoti
		TANU, Coimbatore	Ooty.1
4.	Onion	IARI, New Delhi	Pusa Madhvi, Pusa Ratnar, Pusa Red, Pusa White Flat, Pusa White Round, Sel. 126
		IIHR, Bangalore	Arka Kalyan
		NBPGR, New Delhi	Arka Niketan, Pusa Ratnagar
		NHRDF, Nasik, Maharashtra	Agrifound, Agrifound Dark Red, Agrifound Light Red, Agrifound Red, Agrifound Rose, NHRDF Red-2 (L-355), NHRDF-Red (L-28)
		PAU, Ludhiana, Punjab	Punjab Red Round, Punjab Selection, Punjab-48
		CSA, Kanpur	Kalyanpur Red Round
		RAU, Udaipur	Udaipur-102, Udaipur-103
		NIPHAD	N-53
		MPKV, Rahuri Maharashtra	Baswant-780, Phule Samarth, Phule Survarna
		TNAU, Coimbatore, TN	CO-1, CO-2, CO-4, CO-5, MDU-1
		HAU, Hisar	Hisar-2, Hisar Onion-3
		DOG Research, Rajguru Nagar, Pune	Bhima Kiran, Bhima Raj, Bhima Red, Bhima Shakti, Bhima Shubra, Bhima Shweta, Bhima Super

S. No.	Name of crop	Name of Institutes/ Universities	Varieties/ Hybrids
5.	Carrot	IARI, New Delhi	Nantes, Pusa Asita, Pusa Kesar, Pusa Meghali, Pusa Nayanjyoti, Pusa Rudhira, Pusa Vrishti, Pusa Yamdagni,
		HAU, Hisar	Hisar Gairic
6.	Colocasia/ Taro	CTCRI, Thiruvananthapuram	Sree Rashmi, Sree Pallavi, Muktakeshi, Sree Kiran, Pani Saru 1, Pani Saru 2
7.	Elephant foot	CTCRI, Thiruvananthapuram	Sree Padma, Sree Athira Hybrid , Gajendra
8.	Lettuce	IARI, New Delhi	Great Lakes, Chinese Yellow
9.	Sprouting broccoli	IARI, New Delhi	Pusa Broccoli Kt 1
		MPKV, Rahuri	Ganesh Broccoli
10.	Brussels sprouts	IARI, New Delhi	Hills Ideal
11.	Cabbage	IARI, New Delhi	Golden Acre , KGMR-1, Pusa Ageti , Pusa Drum Head, Pusa Mukta
12.	Cauliflower	IARI, New Delhi	Improved Japanese, Pusa Deepali, Pusa Early Synthetic, Pusa Himjyoti, Pusa Hybrid-2, Pusa Kartik Sankar, Pusa Meghna, Pusa Paushja, Pusa Sharad, Pusa Shubra, Pusa Shukti, Pusa Snowball K-1, Pusa Snowball Kt-25
		IIVR, Varanasi	Kashi Kunwari
		HAU, Hisar	Hisar-1
		G.B.U.A.&T, Pant Nagar	Pant Gobhi-2, Pant Gobhi-3, Pant Shubhra
		TANU, Coimbatore	Ooty.1
13.	Knol khol	IARI, New Delhi	Purple Vienna, White Vienna
14.	Radish	IARI, New Delhi	Japanese White, Pusa Chetki, Pusa Desi, Pusa Himani, Pusa Mridula
		IIHR, Bangalore	Arka Nishant
		IIVR, Varanasi	Kashi Hans, Kashi Sweta
		HAU, Hisar	Hisar Sel-1
		TANU, Coimbatore	CO 1
15.	Turnip	IARI, New Delhi	Purple Top White Globe, Pusa Chandrima, Pusa Kanchan, Pusa Swarnima, Pusa Swati
16.	Sweet potato	CTCRI, Thiruvananthapuram	H-41,H-42, Varsha (H-268), Sree Nandini (76 – OP 217), Sree Vardhini (76 – OP 219), Sree Retna, Sree Bhadra (S-1010), Gouri (85-16), Hybrid Sankar (85-70), Sree Arun, Sree Varun, Kalinga, Sree Kanaka, Goutam, Sourin, Kishan

S. No.	Name of crop	Name of Institutes/ Universities	Varieties/ Hybrids
17.	Ash gourd	IARI, New Delhi	Pusa Ujjwal
		IIVR, Varanasi	Kashi Ujawal
		TANU, Coimbatore	CO1, CO2, TNAU Ash gourd Hybrid CO 1
18.	Bitter gourd	IARI, New Delhi	Pusa Do Mousami, Pusa Hybrid-1, Pusa Vishesh
		IIHR, Bangalore	Arka Harit
		IIVR, Varanasi	Kashi Urvasi
		G.B.U.A.&T, Pant Nagar	Pant Karela-1
		MPKY, Rahuri	Phule Green Gold
		TANU, Coimbatore	CO 1, COBgoH 1, MDU.1
19.	Bottle gourd	IARI, New Delhi	Pusa Hybrid 3, Pusa Naveen, Pusa Samridhi, Pusa Sandesh Pusa Santushti, Pusa Summer Prolific Long
		IIHR, Bangalore	Arka Bahar
		IIVR, Varanasi	Kashi Bahar, Kashi Ganga
		AAU, Anand, Gujarat	Anand Bottle Gourd -1
		G.B.U.A.&T, Pant Nagar	Pant Lauki-4, Pant Sankar Lauki-1, Pant Sankar Lauki-2
		NDUA&T, Faizabad	Bottle Gourd (NDBG-619), N. Bottle Gourd (NDBG-132), N. Bottle Gourd Hybrid-4, N. Dharidar (NDBG-208-1), N. Jyoti (NDBG-104), N. Madhuri (NDBG-505), N. Rashmi (NDBG-1), N. Shishir (NDBG-202), N. Shivani (NDBG- 403)
		MPKY, Rahuri	Samarat
		TANU, Coimbatore	CO 1, TNAU Bottle gourd Hybrid CO 1
		CSAUAT, Kanpur	Azad Nutan
20.	Cucumber	IARI, New Delhi	Pusa Sanyog, Pusa Uday
		G.B.U.A.&T, Pant Nagar	Pant Khira-1, Pant Parthenocarpic khira-2, Pant Sankar Khira-1, Parthenocarpic khira-3
		MPKV, Rahuri	Himangni, Phule Shubangi
		TANU, Coimbatore	CO 1
21.	Ivy gourd	IGKVV, Raipur, Chhattisgarh	Indira kundru-05, Indira kundru-35

S. No.	Name of crop	Name of Institutes/ Universities	Varieties/ Hybrids
22.	Longmelon	IIHR, Bangalore	Arka Sheetal
		HAU, Hisar	Karnal Selection
		G.B.U.A.&T, Pant Nagar	Pant Kakri-1
23.	Muskmelon	IARI, New Delhi	Pusa Madhuras, Pusa Sharbati
		IIHR, Bangalore	Arka Jeet, Arka Rajahans
		IIVR, Varanasi	Kashi Madhu
		AAU, Anand, Gujarat	Gujarat Muskmelon-3
24.	Pumpkin	IARI, New Delhi	Pusa Hybrid 1, Pusa Vikas, Pusa Vishwas
		IIHR, Bangalore	Arka Chandan, Arka Suryamukhi
		IIVR, Varanasi	Kashi Harit
		AAU, Anand, Gujarat	Anand Pumpkin- 1
		NDUA&T, Faizabad	N. Abhooshan, N. Agrim, N. Amrit
		TANU, Coimbatore	CO1, CO2
25.	Ridge gourd	CSAUAT, Kanpur	Azad Pumpkin-1
		HAU, Hisar, Haryana	Hisar Kalitori
		J.A.U., Junagadh, Gujarat	GJRGH-1
		AAU, Anand, Gujarat	Gujarat Anand, Ridge Gourd-1
		G.B.U.A.&T, Pant Nagar	Pant Torai-1
		TANU, Coimbatore	CO 1, CO 2, PKM.1
26.	Round gourd	IIHR, Bangalore	Arka Tinda
		HAU, Hisar	Hisar Tinda (HT-10)
27.	Snapmelon	IARI, New Delhi	Pusa Shandar
28.	Sponge gourd	IARI, New Delhi	Pusa Chikni, Pusa Sneha, Pusa Supriya
		CSAUAT, Kanpur	Azad Taroi -1, Azad Taroi -2
29.	Summer squash	IARI, New Delhi	Pusa Alankar
		IIHR, Bangalore	Patty Pan
30.	Watermelon	IARI, New Delhi	Asahi Yamato, Sugar Baby
		IIHR, Bangalore	Arka Jyothi, Arka Manik
		TANU, Coimbatore	PKM.1
31.	Greater Yam	CTCRI, Thiruvananthapuram	Sree Keerthi, Sree Roopa, Sree Shilpa Hybrid, Sree Karthika, Orissa elite
32.	White Yam	CTCRI, Thiruvananthapuram	Sree Priya, Sree Subhra, Sree Dhanya

S. No.	Name of crop	Name of Institutes/ Universities	Varieties/ Hybrids
33.	Lesser yam	CTCRI, Thiruvananthapuram	Sree Latha, Sree Kala
34.	Cassava (Tapioca)	CTCRI, Thiruvananthapuram	H-97 H-165, H-226, Sree Visakham, Sree Sahya 11-2304), Sree Prakash (S-856), Sree Harsha (2-14), Sree Jaya (CI-649), Sree Vijaya (CI-731), Sree Rekha (TCH-1), Sree Prabha (TCH-2), Sree Padmanabha
35.	Broad bean	IARI, New Delhi	Pusa Sumeet
36.	Cowpea	IARI, New Delhi	Pusa Komal, Pusa Sukomal
	Cowpea	IIHR, Bangalore	Arka Garima, Arka Samrudhi
	Cowpea	IIVR, Varanasi	Kashi Kanchan, Kashi Unnati
	Cowpea	AAU, Anand, Maharashtra	AVCP-1
	Cowpea	TANU, Coimbatore	PKM-1
37.	French bean	IARI, New Delhi	Contender
		IIHR, Bangalore	Arka Anoop, Arka Komal, Arka Suvidha
		IIVR, Varanasi	Kashi Param
		G.B.U.A.&T, Pant Nagar	Pant Anupma, Pant Bean -2
		MPKV, Rahuri	Phule Suyash
		TANU, Coimbatore	TKD.1, YCD.1
		CSAUAT, Kanpur	Azad Rjmash
38.	Lablab bean/ Indian bean	IARI, New Delhi	Pusa Sem-2, Pusa Sem-3
		IIHR, Bangalore	Arka Jay, Arka Vijay
		HAU, Hisar	Hisar Kirti
		J.A.U., Junagadh	GJIB-2
		CSAUAT, Kanpur	Azad Sem-1 , Kalyanpur type 2, Rajni
30.	Garden pea	IARI, New Delhi	Arkel , Bonneville, Pusa Pragati
		IIHR, Bangalore	Arka Ajit
		IIVR, Varanasi	Kashi Nandini , Kashi Uday
		G.B.U.A.&T., Pant Nagar	Pant Matar-2, Pant Sabji Matar-3, Pant Sabji Matar-4, Pant Sabji Matar-5, Pant Uphar
		MPKV, Rahuri	Phule Priya
		TANU, Coimbatore	Ooty 1, Pole type Ooty 1
		CSAUAT, Kanpur	Azad P-2, Azad P-3 , Azad P-4, Azad P-5
40.	Fenugreek	IARI, New Delhi	Pusa Early Bunching, Pusa Kasuri
		HAU, Hisar	Hisar Madhavi, Hisar Manohar, Hisar Mukta, Hisar Sonali, HM-219

S. No.	Name of crop	Name of Institutes/ Universities	Varieties/ Hybrids
41.	Okra	IARI, New Delhi	Perkins Long Green, Pusa A 4, Pusa Sawani
		IIHR, Bangalore	Arka Abhay, Arka Anamika
		IIVR, Varanasi	Kashi Kranti, Kashi Pragati, Kashi Vibhuti
		HAU, Hisar	HBH-142 (F1 hybrid), Hisar Unnat, Varsha Uphar
		AAU, Anand, Gujarat	Gujarat Anand Okra-5
		MPKV, Rahuri	Phule Utkarsha
		TANU, Coimbatore	CO 1, CO 2, COBhH 1
		CSAUAT, Kanpur	Azad Bhindi -1, Azad Bhindi-3
42.	Brinjal	IARI, New Delhi	DBHL-20, DBL-02, Pusa Ankur, Pusa Anupama, Pusa Bindu, Pusa Hybrid-5, Pusa Hybrid-6, Pusa Hybrid-9, Pusa Kranti, Pusa Purple Cluster, Pusa Purple Long, Pusa Shyamla, Pusa Upkar, Pusa Uttam
		IIHR, Bangalore	Arka Anand, Arka Keshav, Arka Navneeth, Arka Neelkanth, Arka Nidhi, Arka Sheel, Arka Shirish, Arka Kusumakar
		IIVR, Varanasi	Kashi Komal, Kashi Prakash, Kashi Sandesh, Kashi Taru
		HAU, Hisar	BR-112, Hisar Jamuni, Hisar Pragati, Hisar Shyamal HLB-12, PH-4
		JAU, Junagadh	GJB-3
		AAU, Jorhat, Assam	JC-1, JC-2
		AAU, Anand, Gujarat	Anand Brinjal Small Round-2, GBH-1 Oblong Brinjal- 1
		G.B.U.A.&T, Pant Nagar	Pant Brinjal Hybrid, Pant Rituraj, Pant Samrat
		MPKY, Rahuri	Krishna, Phule, Phule Arjun
		TANU, Coimbatore	CO-1, CO-2, COBH, COBH-2, KKM1, MDU1, PKM1 PLR (BR) 2, PPI 1, TNAU Brinjal VRM 1
		CSAUAT, Kanpur	Azad B-2, Azad B-3, Azad B-4, Azad Hybrid
43.	Capsicum	IARI, New Delhi	California Wonder, Pusa Deepti
		IIHR, Bangalore	Arka Basant, Arka Gaurav, Arka Mohini

S. No.	Name of crop	Name of Institutes/ Universities	Varieties/ Hybrids
44.	Chilli	IARI, New Delhi	Pusa Jwala, Pusa Sadabahar
		IIHR, Bangalore	Arka Harita, Arka Lohit, Arka Meghana, Arka Suphal, Arka Sweta
		IIVR, Varanasi	Kashi Anmol, Kashi Early, Kashi Surkh
		HAU, Hisar	Hisar Shakti
		AAU, Anand, Gujarat	Chilli 101, Chilli 111, Chilli 112, Chilli 131, Chilli hybrid-1
		GBUA & T, Pant Nagar	Pant C-1
		MPKV, Rahuri	Phule Jyoti, Phule Mukta
		TANU, Coimbatore	CO 1, CO 2, CO 3, CO 4, K1, K2, MDU.1, PKM 1, PLR1, PMK.1,TNAU Chilli Hybrid Co1
45.	Potato	CPRI, Shimla	Kufri Arun, Kufri Ashoka, Kufri Badshah, Kufri Bahar, Kufri Chandramukhi, Kufri Chipsona-1, Kufri Chipsona-3, Kufri Chipsona-4, Kufri Frysona, Kufri Girdhari, Kufri Himalini, Kufri Himsona, Kufri Jyoti, Kufri Kanchan, Kufri Khyati, Kufri Lauvkar, Kufri Pukhraj, Kufri Pushkar, Kufri Sadabahar, Kufri Shailja, Kufri Sindhuri, Kufri Surya, Kufri Swarna
46.	Tomato	AAU, Anand, Gujarat	Anand Tomato- 3, Gujarat Tomato-2
		CSAUAT, Kanpur	Azad T-5, Azad T-6, KTH-1 (Hybrid), KTH-2 (Hybrid)
		GBUA & T, Pant Nagar	Pant Bahar, Pant poly house tomato hybrid-1, Pant poly house tomato-1, Pant T-3
		HAU, Hisar	Hisar Arun, Hisar Lalima, Hisar Lalit, HS-101, HS-102, HS-110
		IARI, New Delhi	Pusa Gaurav, Pusa Hybrid-2, Pusa Hybrid-4, Pusa Hybrid-8, Pusa Rohini, Pusa Ruby, Pusa Sadabahar, Pusa Sheetal, Pusa Uphar, Pusa-120
		IIHR, Bangalore	Arka Abha, Arka Abhijit, Arka Ananya, Arka Saurabh Arka Vardan, Arka Vikas, Arka Vishal
		IIVR, Varanasi	Kashi Amrit, Kashi Anupam, Kashi Hemant, Kashi Sharad, Kashi Vishesh
		MPKV, Rahuri	Dhanashree, Phule Raja
		TNAU, Coimbatore	CO-1, CO-2, CO-3, COTH-2, COTH1, Paiyur1, PKM1

Cultivation Practices

Table 12.7: *Time of nursery sowing and transplanting in transplanted seed crops inNorthern Plains*

S. No.	Name of the vegetable	Time of nursery sowing	Time of Transplanting
1.	Tomato	September	October
2.	Brinjal	July	August
3.	Chilli	July	August
4.	Cauliflower		
	Early	June –July	July-August
	Mid	August- September	September-October
5.	Onion		
	Kharif	June	August (Bulbs are harvested and planted in November)
	Rabi	November	January (Bulbs are harvested in April-May, stored for about 5-6 months and planted in November)

Table 12.8: *Time of sowing in direct sown seed crops in Plains*

S. No.	Name of the vegetable	Time of sowing
1.	Okra	Feb.-March for summer June-July for kharif
2.	Vegetable Peas	October last week to November1st fortnight
3.	French bean	October last week to November 1st week
4.	Cowpea	August 1st fortnight
5.	Radish [Asiatic type]	Sept. last to October IInd week
6.	Carrot [Asiatic type]	Sept. last to October Ist week
7.	Spinach beet	October- November
8.	Methi	October-November
9.	Cucurbits	Feb.-March for summer June-July for kharif

Table 12.9: *Time of sowing in direct sown seed crops in temperate areas*

S. No.	Name of the vegetable	Time of sowing	
		1200 m and above	800-1200 m
1 .	Okra	June	May-June
2 .	Pea	October	October-November
3 .	French bean	June	June
4 .	Radish	Sept.- Oct	Oct.
5 .	Carrot	July-Aug.	July-Sept.
6 .	Spinach	July August	Aug-Sept
7 .	Cucumber	Mar-Apı	Feb-March
8 .	Turnip	Aug	Aug-Sept
9 .	Beet	July August	Aug-Sept

Table 12.10: *Seed Rate, Planting Distance and fertilizer application in different vegetable crops*

Crop	Seed rate/ha	Sowing/ Transplanting Distance (cm)	Fertilizers ($N:P_2O_5:K_2O$) kg/ha	Others
Tomato	400g	60×30(Det.) 100×60 (Indet.)	120:60:60	20kg/ha borax & 0.5% Zinc /ha
Brinjal	500g	100×90	120:60:60	
Chilli	650g	45×30	120:80:60	
Capsicum	700g	60×30	120:80:80	20kg/ha borax & 20kg/ha calcium carbonate
Vegetable peas	100-160kg	23×5-7	60:60:60	
French bean	50kg (Bush type) 40kg (Pole type)	45×15 (Bush type) 60×30 (Pole type)	120:60:60	
Cowpea	20kg (Bush type) 15kg (Pole type)	45x15 (Bush type) 60j×30 (Pole type)	40:60.60	
Okra	18kg (Summer) 10kg (Rainy)	60×15 60×30	120:60:60	
Cauliflower	600g (Early) 400g (Late)	40×30cm (Early) 60×45cm (Late)	120:60:60	25kg/ha borax & 1.5kg/ha ammonium molybdate
Cabbage	600g (Early) 400g (Late)	40×30cm (Early) 60×45cm (Late)	150:60:60	
Radish	8-10kg	Sowing 45×8 cm	100:50:50 100:80:60	10kg boron/ha

Carrot	8-10kg	Sowing 50×10 cm	80:60:80 60:60:80	
Onion	10kg	45x30 cm (Bulb spacing)	100:60:60	
	4 kg	5×1m	80:60:60	
Bitter gourd	5 kg	2×0.5m	80:60:60	
Sponge gourd	4 kg	5×1m	80:60:60	
Cucumber	3 kg	2×0.5m	80:60:60	
Muskmelon	3.5 kg	2×0.3m	80:80:60	
Watermelon	4 kg	2×0.5m	80:80:60	

Table 12.11: *Important roguing stages & permitted off types for production of different vegetable seeds*

Vegetable Crops		Minimum number of field inspections and roguing stages F	Off type (%) Maximum permitted	
			C	
1.	Cowpea	2: From flowering to fruiting	0.10*	0.20
2.	Garden pea	3: First before flowering, second at flowering and third at edible pod stage	0.10*	0.50*
3.	Chilli & Capsicum	3: First before flowering, second at flowering and third at the mature fruit stage	0.10*	0.20*
4.	Cauliflower, Cabbage & Knol khol	4: First before the marketable stage, second start of curd head formation, third when most plants have formed curd and fourth at flowering stage	0.10*	0.20**
5.	Lettuce	3: First before full grown stages in non-heading types, second full grown stage in non-heading type and third at flowering	0.10*	0.20*
6.	Carrot	4: First 20-30 days after sowing, second at lifting and replanting, third at flowering and fourth at maturity to verify the true nature of umbels	0.10*	0.20**
7.	Onion	4: (When seed crop is raised by the transplanting method) First early (20- 30 days after sowing), second when bulbs are lifted, third when bulbs are replanted and fourth at flowering 3: (When seed crop is raised by seed to seed method) First 20-30 days after sowing, second when bulbs are formed and third at flowering	0.10**	0.20**

8.	Radish and Turnip	3: First 20-30 days after sowing, second when lifted and replanted and third at flowering	0.10**	0.20**
9.	Okra	3: First before flowering, second at full flowering and fruiting and third at mature fruit stage.	0.10**	0.20**
10.	Tomato	3: First before flowering, second during flowering and the immature fruit stage and third at mature fruit stage	0.10*	0.20*
11.	Brinjal	3: First before flowering, second during flowering and the immature fruit stage and third at mature fruit stage	0.10*	0.20*
12.	Cucurbits	3: First before flowering, second during flowering and the immature fruit stage and third during mature fruit stage	0.10**	0.20**
13.	Spinach beet (Palak)	3: First at cotyledon leaf stage, second during vegetative growth, third at bolting and flowering	0.10**	0.20**

*Maximum permitted at the final inspection; ** Maximum permitted at and after flowering; F = Foundation seed; C = Certified seed.

Table12.12: *Seed standards for different vegetables*

Crop	Class of seed	Germination (%)	Pure seed (%)	Inert natter (%)	Other crop seed Max.	Weed seed Max.	Moisture (%)	
		(Min.)	(Min.)	(Max.)	(no/kg)	(no/kg)	Ordinary pack	Vapour proof pack
Tomato	FS CS	70 70	98 98	2 2	5 10	None None	8 8	6 6
Brinjal	FS CS	70 70	98 98	2 2	None None	None None	8 8	6 6
Chilli	FS	60	98	2	5	5	8	6
Capsicum	CS	60	98	2	10	10	8	6
	FS CS	65 65	99 99	1 1	None 5	None None	10 10	8 8
Cabbage	FS CS	70 70	98 98	2 2	5 10	5 10	7 7	5 5
Cauliflower Knol Kkol	FS CS	65 65	98 98	2 2	5 10	5 10	7 7	5 5
Radish	FS CS	70 70	98 98	2 2	5 10	10 20	6 6	5 5

Carrot	FS CS	60 60	95 95	5 5	5 10	5 10	8 8	7 7
Beet	FS CS	60 60	96 96	4 4	5 10	5 10	9 9	8 8
Spinach	FS CS	60 60	96 96	4 4	5 10	5 10	9 9	8 8
Amaranthus	FS CS	70 70	95 95	5 5	5 10	10 20	8 8	6 6
Lettuce	FS CS	70 70	98 98	2 2	None None	5 10	8 8	7 7
Onion	FS CS	70 70	98 98	2 2	5 10	5 10	8 8	6 6
Garden pea	FS CS	75 75	98 98	2 2	None 5	None None	9 9	8 8
Cowpea	FS CS	75 75	98 98	2 2	5 10	5 10	9 9	8 8
French bean	FS CS	75 75	98 98	2 2	None None	None 10	9 9	7 7
Cucurbits	FS CS	60 60	98 98	2 2	- -	- -	7 7	6 6

ꝎO

CHAPTER 13

Pesticides and its Uses

Ashish Shrivastava
Assistant Professor (Plant Pathology)
JNKVV, College of Agriculture, Ganj Basoda (M.P.)

Definition

Any substance or mixture of substances intended for preventing, destroying or controlling any pest, including vectors of human or animal disease, unwanted species of plants or animals causing harm during or otherwise interfering with the production, processing, storage, transport or marketing of food, agricultural commodities, wood and wood products or animal feedstuffs, or substances which may be administered to animals for the control of insects, arachnids or other pests in or on their bodies. The term includes substances intended for use as a plant growth regulator, defoliant, desiccant or agent for thinning fruit or preventing the premature fall of fruit. Also used as substances applied to crops either before or after harvest to protect the commodity from deterioration during storage and transport. (FAO)

Pesticide consumption in India as compared to other countries

- In developed countries like the USA, Europe, Japan, China, etc., pesticide use is 20 times more than in India.
- Per hectare consumption of pesticide in India is 381 g a.i./ha which is lower than the world average of 500 g a.i/ha.
- The present consumption of technical grade pesticides in the country during year 2010-11 was around 55,542 tonnes.
- India's consumption of pesticides is only 2% of the total world consumption.
- Among different classes of pesticides used in India the percent share of insecticides is (60%) followed by fungicides (19%), herbicides (16%), bio pesticides (3%) and others (3%).

Classification of Pesticides

Table 13.1: *Classification of pesticides based on target organisms*

S. No.	Pesticide Class	Primary Target/ Action	Example(S)
1.	Acaricide	Mites	Aldicarb, Bifenazate
2.	Algaecide	Algae	Copper sulphate
3.	Attractant	Attracts wide range of pests	Pheromones
4.	Avicide	Birds	Avitrol (aminopyridine)
5.	Bactericide	Bacteria	Copper complexes, streptomycin
6.	Bait	Wide range of organisms	Anticoagulants
7.	Biopesticide	Wide range of organisms	*Bacillus thuringiensis*
8.	Defoliant	Removes plant foliage	Tribufos
9.	Desiccant	Removes water	Boric acid
10.	Fumigant	Wide range of organisms	Aluminum phosphide
11.	Fungicide	Fungi	Azoxystrobin, chlorothalonil
12.	Herbicide	Weeds	Atrazine, glyphosate, 2,4-D
13.	Insect growth regulator	Insects	Diflubenzuron
14.	Insecticide	Insects	Aldicarb, Carbaryl, imidacloprid
15.	Molluscicides	Snails, slug	Metaldehyde
16.	Nematicide	Nematodes	Aldicarb, fenamiphos
17.	Piscicide	Fish	Rotenone
18.	Plant growth regulator	Regulates plant growth	Gibberellic acid, 2,4-D
19.	Predacide	Mammal predators	Strychnine
20.	Repellent	Vertebrates and invertebrates	DEET, methiocarb
21.	Rodenticide	Rodents	Warfarin
22.	Silvicide	Trees	Tebuthiuron
23.	Termiticide	Kills termites	Fipronil

Pesticide Dose Calculations

The success of a pesticide spraying in small areas or large fields depends upon accurate control of the application rate. After the equipment is accurately calibrated to apply the desired volume of spray fluid, one must determine how much chemical is required to put into the tank to apply the correct dosage recommended.

The calculation and formulae for various formulations of pesticides are given below:

For Emulsifiable Concentrate (EC) and Wettable Powder (WP) Formulations

$C_1 V_1 = C_2 V_2$

C_1 = Concentration of given formulation (%)

V_1 = Volume/amount of formulation required (ml or g)

C_2 = Concentration of spray fluid required (%)

V_2 = Volume/amount of spry fluid required (ml or g)

Example 1. How much spray fluid of 0.1% concentration can be prepared from 125g of carbaryl 50 WP?

C_1	=	50%
C_2	=	0.1%
V_1	=	125 g
V_2	=	?
V_2	=	$(C_1V_1)/C_2$
	=	62500 ml
	=	62.5 liters

Example 2. Calculate the amount of carbaryl 50 WP required to prepare 10 liters of 0.1% sprays material.

C_1 V_1	= =	50% ?
C_2	=	0.1%

V_2 V_1	= =	10 liters = 10000 ml $(C_2 V_2)/ C_1$
	= =	(0.1 × 10000)/50 20 g.

Example 3. How much quantity of cypermethrin 20 EC is required for spraying 0.025% spray fluid at the rate of 250 liter/ha for controlling the diamondback month on cauliflower over an area of one-twenty-fifth of a hectare?

C_1 V_1	= =	20% ?
C_2	=	0.025%

Rate of application = 250 liter/ha

For one-twenty-fifth of a hectare = 250 × liter/ha

V_2 C_1V_1 V_1	= = =	10 liters = 10000 ml C_2V_2 $(C_2V_2)/C_1$
	= =	(0.025 × 10000)/20 12.5 ml

Example 4. Chlorpyrifos 20% EC measuring 300 ml is added in 150 liters of water. Determine the percentage concentration of the pesticide in the spray fluid.

C_1 = 20 %

V_1 = 300 ml

C_2 = 150 liters =150,000 ml $C_1V_1 = C_2V_2$

C_2 = (20 × 300)/150000

= 0.04%

For Granules and Dust formulations

C_1V_1 = 100 RA

C_1 = Concentration of formulation available (%)

V_1	=	Amount of formulation required (g or kg)
R	=	Recommended rate of pesticide application [g or kg ingredient (a.i.)/ha]
A	=	Area to be treated (ha)

Example 1. Calculate the amount of 5% carbofuran granules applied @ 0.25 kg a.i. to one ha area.

C1 = 5%

V1 = ?

R = 0.25 kg

A = 1ha C1V1 = 100 RA

V1 = (100 RA)/C

= (100 × 0.25 × 1)/5

= 5 kg.

Example 2. How many kg of commercial formulation is required to treat 3000 m^2 when the recommended rate is 1 kg a.i./ha and per cent a.i. in the commercial granular formulation is 10%?

C1 = 10%

V1 = ?

R = 1 kg a.i./ha

A = 3000 m^2 = 0.3 ha

C1V1 = 100 RA

V1 = (100 RA)/C_1

$= (100 \times 1 \times 0.3)/10$

$= 3$ kg.

Example 3. How many kg of the commercial formulation is required to treat the 1000 m^2 area at the recommended rate of 0.75 kg/ha? The percentage a.i. in the commercial formulation is 5%.

C1 = 5%

V1 = ?

R = 0.75 kg

A = 1000 M^2 = 0.1 ha

C1V1 = 100 RA

V1 = (100 RA)/C_1

$= (100 \times 0.75 \times 0.1)/5$

= 1.5 kg

Calculation of Formulated Product Requirement

The leaflet supplied with the product will provide details on per cent active ingredient contents and dose of active ingredient required per hectare. Using this information we can work out the quantity of formulated product required for spraying for one hectare of land by using the following formula:

$$\text{Commercial product (g/ha)} = \frac{\text{Does in g.a.i./ha} \times 100}{\text{a.i. in the formulation}}$$

Examples

1. **Solid Formulations like WP, Dust:** To calculate quantity of Diafenthiuron 50 % WP required to cover 1 hectare of land @ 300 g active ingredient per hectare;

   = 300 g.a. /ha × 100

   50% a.i. in 1 kg

   = 600 g

2. **Liquid Formulations like EC, SC, SL:** To calculate quantity of Flubendamide 480 SC required to cover 1 hectare of land @ 48 g active ingredient per hectare;

   = 48 g.a.i./ ha × 100

   48% a.i. in 1 kg

   = 100 ml

3. **Granule Formulation:** To calculate quantity of Fipronil 0.3 GR required to cover 1 hectare of land @ 50 g active ingredient per hectare;

   = 50 g.a.i. × 100

   0.3 % a.i. in 1 kg

   = 16667g (16.67 kg)

List of pesticides which are banned, refused registration and restricted in use:

(As on 20th October 2015)

I. Pesticides / Formulations banned in India

A.	Pesticides Banned for manufacture, import and use .	
	1.	Aldicarb (vide S.O. 682 (E) dated 17th July 2001)
	2.	Aldrin
	3.	Benzene Hexachloride
	4.	Calcium Cyanide
	5.	Chlorbenzilate (vide S.O. 682 (E) dated 17th July 2001)
	6.	Chlordane
	7.	Chlorofenvinphos
	8.	Copper Acetoarsenite
	9.	Dibromochloropropane (DBCP) (vide S.O. 569 (E) dated 25th July 1989)
	10.	Dieldrin (vide S.O. 682 (E) dated 17th July 2001)
	11.	Endrin
	12.	Ethyl Mercury Chloride
	13.	Ethyl Parathion
	14.	Ethylene Dibromide (EDB) (vide S.O. 682 (E) dated 17th July 2001)
	15.	Heptachlor
	16.	Lindane (Gamma-HCH)
	17.	Maleic Hydrazide (vide S.O. 682 (E) dated 17th July 2001)
	18.	Menazon
	19.	Metoxuron
	20.	Nitrofen
	21.	Paraquat Dimethyl Sulphate
	22.	Pentachloro Nitrobenzene (PCNB) (vide S.O. 569 (E) dated 25th July 1989)
	23.	Pentachlorophenol
	24.	Phenyl Mercury Acetate
	25.	Sodium Methane Arsonate
	26.	Tetradifon
	27.	Toxaphene(Camphechlor) (vide S.O. 569 (E) dated 25th July 1989)
	28.	Trichloro acetic acid (TCA) (vide S.O. 682 (E) dated 17th July 2001)
B.	Pesticide formulations banned for import, manufacture and use	
	1.	Carbofuron 50% SP (vide S.O. 678 (E) dated 17th July 2001)
	2.	Methomyl 12.5% L
	3.	Methomyl 24% formulation
	4.	Phosphamidon 85% SL

C.	**Pesticide / Pesticide formulations banned for use but continued to manufacture for export**	
	1.	Captafol 80% Powder (vide S.O. 679 (E) dated 17th July 2001)
	2.	Nicotin Sulfate
D.	**Pesticides Withdrawn (Withdrawal may become inoperative as soon as required complete data as per the guidelines is generated and submitted by the Pesticides Industry to the Government and accepted by the Registration Committee. (S.O 915(E) dated 15th Jun,2006)**	
	1.	Dalapon
	2.	Ferbam
	3.	Formothion
	4.	Nickel Chloride
	5.	Paradichlorobenzene (PDCB)
	6.	Simazine
	7.	Sirmate (S.O. 2485 (E) dated 24th September 2014)
	8.	Warfarin (vide S.O. 915 (E) dated 15th June 2006)

II. Pesticides Refused Registration

S.No.	Name of Pesticides
1.	2,4, 5-T
2.	Ammonium Sulphamate
3.	Azinphos Ethyl
4.	Azinphos Methyl
5.	Binapacryl
6.	Calcium Arsenate
7.	Carbophenothion
8.	Chinomethionate (Morestan)
9.	Dicrotophos
10.	EPN
11.	Fentin Acetate
12.	Fentin Hydroxide
13.	Lead Arsenate
14.	Leptophos (Phosvel)
15.	Mephosfolan
16.	Mevinphos (Phosdrin)
17.	Thiodemeton / Disulfoton
18.	Vamidothion

III. Pesticides Restricted for use in the Country

S.No.	Name of Pesticides	Details of Restrictions
1.	Aluminium Phosphide	The Pest Control Operations with Aluminium Phosphide may be undertaken only by Govt./Govt. undertakings / Govt. Organizations / pest control operators under the strict supervision of Govt. Experts or experts whose expertise is approved by the Plant Protection Advisor to Govt. of India except [1]Aluminium Phosphide 15 % 12 g tablet and [2]Aluminum Phosphide 6 % tablet. *[RC decision circular F No. 14-11(2)-CIR-II (Vol. II) dated 21-09-1984 and G.S.R. 371(E) dated 20th may 1999]. [1]Decision of 282nd RC held on 02-11-2007 and, [2]Decision of 326th RC held on 15-02-2012.* The production, marketing and use of Aluminium Phosphide tube packs with a capacity of 10 and 20 tablets of 3 g each of Aluminium Phosphide are banned completely. (S.O.677 (E) dated 17th July, 2001)
2.	Captafol	The use of Captafol as foliar spray is banned. Captafol shall be used only as seed dresser. (S.O.569 (E) dated 25th July, 1989) The manufacture of Captafol 80 % powder for dry seed treatment (DS) is banned for use in the country except manufacture for export. (S.O.679 (E) dated 17th July, 2001)
3.	Cypermethrin	Cypermethrin 3 % Smoke Generator, is to be used only through Pest Control Operators and not allowed to be used by the General Public. [Order of Hon,ble High Court of Delhi in WP(C) 10052 of 2009 dated 14-07-2009 and LPA-429/2009 dated 08-09-2009]
4.	Dazomet	The use of Dazomet is not permitted on Tea. (S.O.3006 (E) dated 31st Dec, 2008)
5.	Diazinon	Diazinon is banned for use in agriculture except for household use. (S.O.45 (E) dated 08th Jan, 2008)
6.	Dichloro Diphenyl Trichloroethane (DDT)	The use of DDT for the domestic Public Health Programme is restricted up to 10,000 Metric Tonnes per annum, except in case of any major outbreak of epidemic. M/s Hindustan Insecticides Ltd., the sole manufacturer of DDT in the country may manufacture DDT for export to other countries for use in vector control for public health purpose. The export of DDT to Parties and State non-Parties shall be strictly in accordance with the paragraph 2(b) article 3 of the Stockholm Convention on Persistent Organic Pollutants (POPs). (S.O.295 (E) dated 8th March, 2006) Use of DDT in Agriculture is withdrawn. In very special circumstances warranting the use of DDT for plant protection work, the state or central Govt. may purchase it directly from M/s Hindustan Insecticides Ltd. to be used under expert Governmental supervision. (S.O.378 (E) dated 26th May, 1989)

7.	Fenitrothion	The use of Fenitrothion is banned in Agriculture except for locust control in scheduled desert area and public health. (S.O.706 (E) dated 03rdMay, 2007)
8.	Fenthion	The use of Fenthion is banned in Agriculture except for locust control, household and public health. (S.O.46 (E) dated 08th Jan, 2008)
9.	Methoxy Ethyl Mercuric Chloride (MEMC)	The use of MEMC is banned completely except for seed treatment of potato and sugarcane. (S.O.681 (E) dated 17thJuly, 2001)
10.	Methyl Bromide	Methyl Bromide may be used only by Govt./Govt. undertakings/ Govt. Organizations / Pest control operators under the strict supervision of Govt. Experts or Experts whose expertise is approved by the Plant Protection Advisor to Govt. of India [G.S.R.371 (E) dated 20thMay, 1999 and earlier RC decision]
11.	Methyl Parathion	Methyl Parathion 50 % EC and 2% DP formulations are banned for use on fruits and vegetables. (S.O.680 (E) dated 17thJuly, 2001) The use of Methyl Parathion is permitted only on those crops approved by the Registration Committee where honeybees are not acting as a pollinators. (S.O.658 (E) dated 04th Sep., 1992.)
12.	Monocrotophos	Monocrotophos is banned for use on vegetables. (S.O.1482 (E) dated 10thOct, 2005)
13.	Sodium Cyanide	The use of Sodium Cyanide shall be restricted for Fumigation of Cotton bales under expert supervision approved by the Plant Protection Advisor to Govt. of India. (S.O.569(E) dated 25thJuly, 1989)

Endosulfan*:- Endosulfan has been banned by the Supreme Court of India w.e.f. 13-05-2011 for production, use & sale, all over India, till further orders vide ad-Interim order in the Writ Petition (Civil) No. 213 of 2011.

Source: website of Central Insecticides Board & Registration Committee, Faridabad available at http://www.cibrc.nic.in/

CHAPTER 14

Major Diseases of Vegetable Crops

Y.K. Jain
Senior Scientist (Plant Pathology)
Zonal Agriculture Research Station, Khargone (M.P.)

Table 14.1: *Common and scientific names and pathogen of diseases of vegetables crops*

S. No	Crop	Common name	Scientific name	Pathogen	Remarks
1.	Leafy vegetables	Leaf spots	*Cercospora* sp. and *Aternaria* sp.	Fungus	
		White rot	*Sclerotinia sclerotiroum*	Fungus	
2.	Onion and garlic	Purple blotch	*Alternaria porri*	Fungus	
		Stemphylium leaf blight	*Stemphylium vesicarium* (Wallr.) Simmons	Fungus	
		Damping off	*Phythium species, Rhizoctonia solani, F. oxysporum and F. oxysporum* f.sp.*cepae*	Fungus	
		Blight	*Stemphylium vesicarium*	Fungus	
		Onion smut	*Urocystic cepulae*	Fungus	
		Downy mildew	*Peronospora destructor* (Berk). Casp.)	Fungus	
		Black mould	*Aspergillus niger* Van Tieghem.	Fungus	
		Fusarium basal rot / Basal plate rot	*Fusarium oxysporum f.sp cepaae*	Fungus	
		Bacterial soft rot	*Eriwinia carotovora*	Bacteria	
		Onion yellow dwarf virus		Virus	
3.	Carrot	Bacterial soft rot	*Eriwinia carotovora*	Bacteria	
		White rot	*Sclerotinia sclerotiorum*	Fungus	
		Collar rot	*Sclerotium rolfsii*	Fungus	

S. No	Crop	Common name	Scientific name	Pathogen	Remarks
4.	Colocasia	Leaf blight	*Phytophthora colocasiae*	Fungus	
		Mosaic	*Dasheen mosaic virus*	Virus	
5.	Cole crops	Damping -off	*Rhizoctonia solan, Phytophthora* spp. *And Pythium* spp.	Fungus	
		Curd rot	*Sclerotinia sclerotiorum, Erwinia carotovora*	Fungus	
		Alternaria leaf spot	*Alternaria brassicae* and *A. brassicicola*	Fungus	
		Stalk rot	*Sclerotinia sclerotiorum*	Fungus	
		Downy mildew	*Peronospora parasitica*	Fungus	
		White rust or white blisters	*Albugo candida*	Fungus	
		Black leg	*Phoma lingam*	Fungus	
		Club root	*Plasmodiophora brassicae*	Fungus	
		Bacterial black rot	*Xanthomonas campestris pv. Campestris*	Bacteria	
6.	Radish	White rust	Albugo candida	Fungus	
		Phyllody		MLO	Transmitted by leafhoppers
		Alternaria blight	*Alternaria raphani*	Fungus	
7.	Sweet Potato	Chlorotic leaf distortion	*Fusarium lateritium*	Fungus	
		Feathery mottle disease	Sweet potato feathery mottle potyvirus	Virus	
8.	Cucurbits	Powdery mildew	*Erysiphe cichoracearum and Sphaerotheca fuliginea*	Fungus	
		Anthracnose	*Colletotrichum orbiculare & C. lagenarium*	Fungus	
		Downy mildew	*Pseudoperonospora cubensis*	Fungus	
		Fruit rots	*Phytophthora cinnamomi, Pythium, Rhizoctonia, Phomopsis cucurbitae*	Fungus	
		Gummy stem blight	*Didymella bryoniae-* teleomorph *and Phoma cucurbitacearum* anamorph	Fungus	

S. No	Crop	Common name	Scientific name	Pathogen	Remarks
		Leaf spots	*Cercospora citrullina, Alternaria cucumerina and Corynespora melonis, Didymella bryoniae* (teleomorph) and *Phoma cucurbitacearum* anamorph	Fungus	
		Mosaic		Virus	Transmitted by aphids
		Bud necrosis		Virus	Transmitted by thrips
9.	Yams	Anthracnose/ leaf blight	*Colletotrichum gloeosporiodes*	Fungus	
		Leaf spot	*Cercospora carbonaceae*	Fungus	
		Mosaic	Yam mosaic virus	Virus	
10.	Cassava	Cassava mosaic disease	Indian cassava mosaic geminivirus	Virus	
		Brown leaf spot	*Cercospora henningsii*	Fungus	
		Anthracnose or die-back	*Colletotrichum manihotis*	Fungus	
		Tuber rot	*Phytophthora drechsleri*	Fungus	
11.	Beans	Anthracnose	*Colletotrichum lindemuthianum*	Fungus	
		Rhizoctonia web blight	*Rhizoctonia solani*	Fungus	
		Angular leaf spot	*Isariopsis griseola*	Fungus	
		Floury leaf spot	*Ramularia phaseoli*	Fungus	
		Fuscous blight/ Bacterial blight	*Xanthomonas campestris* pv. *fuscans*	Bacteria	
		Yellow mosaic		Virus	Transmitted by whitefly
		Southern bean mosaic		Virus	
		Lima bean mosaic		Virus	Transmitted by aphids
		Spotted wilt		Virus	Transmitted by thrips
12.	Pea	Powdery mildew	*Erysiphe polygoni*	Fungus	
		Ascochyta foot rot and blight	*Ascochyta pisi, A. pinodella and A. pinodes*	Fungus	
		Wilt	*Fusarium oxysporum f. sp. pisi*	Fungus	
		Rust	*Uromyces pisi and U. fabae*	Fungus	
		Bacterial blight	*Pseudomonas syringae pv. pisi*	Bacteria	

S. No	Crop	Common name	Scientific name	Pathogen	Remarks
13.	Okra	Yellow-vein mosaic	YVMV- Gemini virus	Virus	Transmitted by whitefly
		Enation Leaf Curl (ELCV)		Virus	
		Powdery mildew	*Erysiphe cichoracearum*	Fungus	
		Leaf Blight	*Cercospora abelmoschi and C. hibisina*	Fungus	
14.	Brinjal	Damping off	*Phytophthora Spp. and Pythium Spp*	Fungus	
		Phomopsis blight	*Phomopsis vexans*	Fungus	
		Pseudocercospora Leaf Blight	*Pseudocercospora sp.*	Fungus	
		Wilt	*Fusarium solani*	Fungus	
		Rhizoctonia root rot	*Rhizoctonia solani*	Fungus	
		Sclerotinia rot	*Sclerotinia sclerotiorum*	Fungus	
		Collar rot	*Sclerotium rolfsii*	Fungus	
		Bacterial wilt	*Ralstonia solanacearum*	Bacteria	
		Little leaf		Mycoplasma	Transmitted by leafhoppers
15.	Capsicum and chilli	Damping -off	*Phytophthora spp. And Pythium aphanidermatum*	Fungus	
		Leaf blight	*Alternaria alternata* and *Cercospora capsici*	Fungus	
		Dieback and Anthracnose	*Choanephora capsici, Colletotrichum capsici*	Fungus	
		Wilt	*Fusarium oxysporum sub spp. Licopersicae*	Fungus	
		Powdery mildew	*Erysiphe cichoracearum*	Fungus	
		Leaf curl		Virus	Transmitted by whitefly
		Spotted wilt		Virus	Transmitted by thrips
		Bacterial wilt	*Pseudomonas solanacearum*	Bacteria	
16.	Potato	Early blight	*Alternaria solanai*	Fungus	
		Late blight	*Phytophthora infestans*	Fungus	
		Black scurf	*Rhizoctonia solanai*	Fungus	
		Wart	*Synchytrium endobioticum*	Fungus	
		Dry-rot	*Fusarium* spp.	Fungus	
		Bacterial wilt	*Ralstonia solanacearum*	Bacteria	
		Latent or faint mosaic	PVX and PVS	Virus	

S. No	Crop	Common name	Scientific name	Pathogen	Remarks
		Severe mosaic	PVY	Virus	
		Rugose mosaic	PVX+PVY	Virus	
		Crinkle mosaic	PVX+PVA	Virus	
		Leaf roll	PLRV	Virus	
17.	Tomato	Damping-off	*Phytophthora spp. And Pythium spp.*	Fungus	
		Buckeye rot	*Phytophthora nicotiana var. parasitica*	Fungus	
		Alternaria blight	*Alternaria solani, A. alternata f.sp. lycopersici*	Fungus	
		Late blight	*Phytophthora infestans*	Fungus	
		Pseudocercospora leaf blight	*Pseudocercospora fuligena*	Fungus	
		Fruit rot	*Phytophthora lycopersici*	Fungus	
		Collar rot	*Sclerotium rolfsii*	Fungus	
		White rot	*Sclerotinia sclerotiorum*	Fungus	
		Grey leaf spot	*Stemphillium solani*	Fungus	
		Fusarium wilt	*Fusarium oxysporum sub spp.Lycopersici*	Fungus	
		Septoria leaf blight	*Septoria lycopersici*	Fungus	
		Bacterial wilt	*Ralstonia (Pseudomonas) solanacearum*	Bacteria	
		Bacterial canker	*Clavibacter michiganense sub* sp. *Michiganense*	Bacteria	
		Bacterial spot	*Xanthomonas campestris pv. vesicatoria*	Bacteria	
		Tomato mosaic virus		Virus	Transmitted by contact and seed
		Leaf curl		Virus	Transmitted by whitefly
		Spotted wilt		Virus	Transmitted by thrips
		Fern leaf		Virus	Transmitted by aphids

❍❍❍

CHAPTER 15

Major Insect Pests of Vegetables Crops

M.L. Sharma
Principal Scientist (Entomology) and Associate Director Research
Zonal Agriculture Research Station, Khargone (M.P.)

Table 15.1: *Common and scientific names of insect pests of vegetables crops*

S. No.	Crop	Common name of insect pests	Scientific name	Remarks
1.	Amaranth and Spinach	Aphids	*Aphis croccivora, Lipaphis erysimi*	
		Leaf-eating caterpillar	*Hymenia recurvalis* (Fb.)	
		Leaf-webber	Lamprosema indicata	
		Scale	*Coccus hesperidum*	
		Stem weevil	*Hypolixus truncatules*	
		Blue-beetle	*Altica caerulesces*	
2.	Beetroot	Cut-worm	*Agrotis ipsilon*	
		Leaf-eating caterpillar	*Spodoptera litura*	
3.	Onion and Garlic	Thrips	*Thrips tabaci* Lindeman	
		Maggot	*Hylemia antiqua*	
		Eriophyid mite	*Aceria tulipae* Keifer	
		Red spider mite	*Tetranychus cinnabarinus*	
		Cut-worm	*Agrotis ipsilon*	
4.	Carrot	Leaf hopper	*Amrasca punjabensis*	
		Cut-worm	*Agrotis ipsilon*	
5.	Cole crops	Diamond back moth (DBM)	*Plutella xylostella* L.	One of the most notorious pest throughout the world on cole crops. Loss may be 80-90% during hot periods of year in plains.

S. No.	Crop	Common name of insect pests	Scientific name	Remarks
		Leaf-webber	*Crocidolomia binotalis* Zeller	
		Mustard saw-fly	*Athalia lugens proxima* Kulg	
		Cabbage aphid	*Brevicoryne brassicae* Linn	
		Stem borer	*Hellula undalis* Zeller	
		Gram-caterpillar	*Helicoverpa armigera* Hübner	
		Tobaco caterpiller	*Spodoptera litura* Fabricius	
		Striped-flea beetle	*Phyllotereta striolata*	
		Painted bug	*Bagrada cruciferarum* Kirkaldy	
6.	Radish	Mustard saw-fly	Athalia lugens proxima	
		Aphids	*Myzus persicae, Lipaphis erysimi*	
		Striped-flea beetle	*Phyllotereta striolata*	
		Painted bug	*Bagrada cruciferum*	
7.	Sweet Potato	Sweet potato weevil	*Cylas formicarius*	
		Vine- borer	*Omphisa anastamosalis*	
8.	Cucurbitaceous Vegetables	Red pumpkin beetle	*Aulacophora foveicollis* Lucas & A. *lewisii* Baly	
		Fruit fly	*Bactocera cucurbitae* Coquillett	
		Aphids	*Aphis gossypii, Aphis malvae*	
		Serpentine leaf miner	*Liriomyza trifolii* Burgess	
		Blister beetle	*Mylabris pustulata*	
		Stink bug	*Coridius brunneus, C. Janus, C. observus*	
		Red spider mite	*Tetranychus neocaledonicus* Andre	
		Thrips	*Thrips palmi* Karny	
		Plume moth	*Sphenarches caffer*	
		Leaf-eating caterpillar	*Dipahania (=Margaronia) indica* Saund	
		Stem gall fly	*Lasioptera falcata*	

S. No.	Crop	Common name of insect pests	Scientific name	Remarks
9.	Yams	Aphids	*Aphis gossypii*	
		Thrips	*Calliothrips indicus*	
		Leaf eating caterpillars	Spodoptera litura	
10.	Cassava	Scale insect	*Aonidomytilus albus*	
		Spider mites	*Tetranychus cinnabarinus, T. neocaledonicus*	
11.	Cluster bean	Aphid	*Aphis craccivora*	
		Pod-borer	*Adisura atkinsoni*	
		Bugs	*Megacopta (Captosoma) cribraria*	
12.	Cowpea	Stem fly	*Ophiomyia phaseoli*	
		Pod-borer	*Muruca testulalis*	
		Bruchid	*Callasobruchus chinensis*	
		Aphid	*Aphis cracivora*	
		Bugs	*Nezara viridula, Anoplocnemis phasiana, Megacopta cribraria*	
		Leaf-eating caterpillar	*Spilasoma obliqua*	
13.	French bean	Stem fly	*Ophiomyia phaseoli*	
		Leaf hopper	*Empoasca herri*	
		Serpentine leaf-miner	*Liriomyza trifolii*	
		Thrips	*Calliothrips indicus, Megaleus distalis*	
		Red spider mite	*Tetranychus cinnaberinum*	
14.	Pea	Aphids	*Macrosiphym pisi*	
		Pod borer	Helicoverpa armigera Hübner	

S. No.	Crop	Common name of insect pests	Scientific name	Remarks
15.	Okra	Leafhopper	*Amrasca biguttula biguttula*	
		Shoot and fruit borer	*Earias vittella* (Fabricius)	
		Red spider mite	*Tetranychus cinnabarinus* (Boisduval)	
		White fly	*Bemisia tabaci*	Vector of yellow vein mosaic virus
		Aphids	*Aphis gossypii*	
		Jassids	Amrasca devastans	
		Red cotton bug	Dysdercus cingulatus	
16.	Moringa	Bud worm	*Noorda moringae*	
		Pod fly	*Gitona distigma*	
		Leaf caterpillar	*Noorda blitealis*	
		Bark caterpillar	*Indarbela tetraonis*	
17.	Brinjal	Shoot and fruit borer	*Leucinode orbonalis* Guen.	This is the most serious pest of brinjal throughout the country. Losses may be up to 60% in severe cases
		Red spider mites	*Tetranychus urticae Koch*	
		Aphids	*Aphis gossypii* Glover and *Myzus persicae* (Sulzer)	
		Epilachna beetle	*Epilachna vigintioctopunctata*, Fab.	
		Spotted leaf beetle (or Hadda beetle)	*Henosepilachna vigintioctopunctata* (Fabr.)	
		Grey weevil	*Myllocerus subfasciatus* G. or *M. maculosus*	
		Stem borer	*Euzophera perticella*	
		Thrips	*Thrips palmi, Thrips tabaci*	
		Mealy bug	*Coccidohystrix insolitus*	
		Leafhopper	*Amrasca biguttula biguttula* (= *Empoasca devastans* Dist.)	
		Ash weevil	*Myllocerus subfasciatus* Guerin	
		Brown leaf hopper	Cestius phycitis	

S. No.	Crop	Common name of insect pests	Scientific name	Remarks
18.	Capsicum and chilli	Chilli thrips	*Scirtothrips dorsalis* Hood	
		Aphids	*Aphis gossypi* and *Myzus persicae*	
		Yellow mite	Polyphagotarsonemus latus	Losses may be up to 80% in severe cases
		Whitefly	*Bemisia tabaci*	Vector of leaf curl disease
		Tobacco caterpillar	*Spodoptera litura*	
		Fruit borer	*Helicoverpa armigera* Hübner	
		White grub	*Anomala bengalensis, Holotricha consanguinea, Holotricha reynaudi*	
19.	Potato	Aphids	*Aphis gossypii, Myzus persicae*	
		White grubs	Holotrichia sp.	
		Whitefly	*Bemisia tabaci*	
		Tobacco Caterpillar	Spodoptera litura	
		Cutworm	*Agrotis ipsilon* (Hufn.)	
		Potato tuber moth	*Phthorimaea operculella* Z.	
		Mites	*Hemitarsonemus latus*	
20.	Tomato	Serpentine leaf-miner	*Liromyza trifolii* Burgess	
		Fruit borer	*Helicoverpa armigera* Hubner	
		Red spider mites	*Tetranychus urticae*	
		Mealy bug	*Ferrisia virgata* (Cockerell)	
		Leaf eating caterpillars	*Spodoptera litura*	
		Thrips	*Thrips tabaci, Calliothrips indicus*	Vector of spotted wilt
		White fly	*Bemisia tabaci*	Vector of leaf curl
		Tobacco caterpillar	*Spodoptera litura* Fabricius	

❍❍❍

CHAPTER 16

Physiological and Nutritional Disorders in Vegetable Crops

[1]Dr. Rashmi Vajpayee and [2]Dr. Pragnyashree Mishra

[1]*Scientist (Horticulture)*

RVSKVV, College of Agriculture, Gwalior (M.P.)

[2]*Assistant Professor (Horticulture)*

OUAT, Odisha

The productivity as well as the quality of **vegetable crops** is affected to a greater extent due to the physiological and nutritional disorders. Disturbance in the plant metabolic activities resulting from an excess or deficit of environmental variables like temperature, light, aeration and nutritional imbalances result in crop disorders.

Table 16.1: *Name of crop, physiological/ nutritional disorders, causes and their control measures*

S. No.	Crop	Physiological/ nutritional disorder	Causes	Control measures
1.	Beetroot	Internal black spot	Boron deficiency	Apply 10-15 kg of Borax/ ha.
2.	Sugarbeet	Speckled yellows' of sugarbeet	Manganese deficiency	
3.	Onion	Bolting	Transplanting of aged seedlings. Early sowing of seeds in the nursery beds, which result in the formation of small sets. Late transplanting of seedlings. Low temperature (10-12°C) for prolonged period.	Time of planting should be adjusted in such a way that the crop may expose to moderate temperature at bulbing. Sow nursery at proper time.
		Sprouting	Excessive moisture at maturity and supply of nitrogen.	Adjust time of planting in such a way that harvesting can be done in dry period. Stop irrigation as soon as bulbs reach maturity. Spray iron sulphate or borax @ 500-1000 ppm 2-3 weeks prior to harvesting.

S. No.	Crop	Physiological/ nutritional disorder	Causes	Control measures
4.	Carrot	Cavity spot	Calcium deficiency	
		Pigeon chest	Calcium deficiency	
		Splitting of roots	Sudden increase in soil moisture after prolonged drought.	Keeping the soil moisture at optimum condition.
		Forking	Excess moisture during the root development. It occurs on heavy soils due to soil compactness.	Keeping the soil moisture at optimum condition.
5.	Colocasia	Metsubre	Calcium deficiency	
6.	Cauliflower	Browning and hollow stem	Boron deficiency	Application of borax @20kg/ha
		Blindness	Poor fertility of the soil or damage to the terminal portion during handling at the time of planting or by insects, diseases etc.	Healthy and vigorous seedlings with terminal portion intact should be planted.
		Whiptail	Molybdenum deficiency	Application of molybdenum @ 1kg/ha.
		Buttoning	Planting of over-aged seedlings which do not get sufficient time to initiate growth before transformation to curding or selection of wrong cultivars means planting early variety late or root injury by insects or diseases result in buttoning.	
		Riceyness	Temperature higher or lower than the optimum required for curding or high application of nitrogen result in riceyness.	Proper management of soil moisture and fertility during the development of head or curd.
7.	Radish	Pore extent or pithiness	Delay in harvesting is the main reason for this disorder.	Therefore, harvesting should be done at an appropriate time.
8.	Turnip	Brown heart	Boron deficiency	
9.	Cucumber	Pillow	Calcium deficiency	

S. No.	Crop	Physiological/ nutritional disorder	Causes	Control measures
10.	French bean	Hypocotyl cracking	Calcium deficiency	
11.	Potato	Greening	Exposure of tubers to sunlight	Proper earthing up of tubers as the tuberization takes place.
		Hollow Heart	Rapid growth of tubers	Maintain soil moisture conditions to the optimum level. Avoid over fertilization particularly nitrogen.
		Black Heart	Sub-oxidation conditions under potato tuber storage as there is no aeration in the centre of the piles.	Provide proper ventilation. Keep potato tubers in layers. Do not store tubers in a heap.
		Knobbiness	It occurs due to uneven growth of tuber cells/ tissues.	Ensure frequent and optimum irrigation.
		Cracking	It is due to boron deficiency or uneven water supply.	Application of Borax @ 20kg/ha. Ensure frequent and optimum irrigation.
		Sun scalding	It occurs, generally, in the autumn crop when both the temperature and sunshine are high.	Water should be passed through the furrows to lower the soil temperature.
		Black spot	It occurs in vascular tissues within 3 days of mechanical injury.	Genetic makeup of the varieties. Provide proper storage and growing conditions.
		Freezing injury	It occurs due to the exposure of tubers to freezing temperatures during or after harvest. It takes place at -1.5°C or below temperature.	Avoid exposure of tubers to freezing temperature during storage or harvest.
		Sprouting	It is often a serious problem in storage.	It can be inhibited by spraying borax or iron sulphate @ 1000 1500 ppm about 2-3 weeks before harvesting. Chemicals like Chloro IPC (N-tetra chloro isopropyl carbonate) @ 0.5% also help in inhibiting sprouting.

S. No.	Crop	Physiological/ nutritional disorder	Causes	Control measures
12.	Tomato	Cat face	Abnormal growing condition	
		Fruit cracking	Boron deficiency	
		Blossom end rot	Calcium deficiency	
		Sun-scald	Excessive exposure to sunrays	
		Blotchy ripening	Potassium deficiency	
		Puffiness	Due to poor pollination	

CHAPTER 17

Post-Harvest Technology of Vegetable Crops

S.D. Patil
Associate Professor (Horticulture)
College of Agriculture, Dhule, Maharashtra

- All vegetables and root crops are living biological organisms, having a respiratory system, similar to that of humans.
- They continue their living processes after harvest. Respiration is the process by which plants take in oxygen and give out carbon dioxide.
- In accordance with the respiration rate most horticultural commodities can be classified as follows:
  - **Low respiration rate:** Garlic, onion and sweet potato.
  - **Moderate respiration rate:** Cabbage, carrot, lettuce, pepper and tomato.
  - **High respiration rate:** Cauliflower and green onion.
  - **Extremely high respiration rate:** Broccoli, peas, spinach and sweet corn.
- **Ethylene is** a natural product of plant metabolism and is produced by all tissues of higher plants.
- It is considered the natural aging and ripening hormone and is active even at small traces.
- Vegetables and root crops contain **65 to 95% water** and their post-harvest life depends on the rate at which they use up their stored food reserves and their rate of water losses.
- When food and water reserves are exhausted the produce dies and decays.

Post-harvest technology procedures

- Temperature and humidity are used to control shelf-life of commodities in refrigerated cold stores.

Temperature management practices

- Temperature management is the most important tool that we have to extend shelf-life of fresh horticultural commodities after harvesting the produce.

- Temperature management begins with a rapid removal of the field heat by using one of the following cooling methods:
  - Hydro cooling
  - In package ice
  - Top icing
  - Evaporative cooling
  - Room cooling
  - Forced air cooling
  - Serpentine forced air cooling
  - Vacuum cooling
  - Hydro-vacuum cooling.

Cold storage facilities should be well constructed and adequately equipped. They should have:

- Good construction, and insulation and vapour barrier
- Strong floor
- Adequate doors for loading and unloading
- Effective distribution of refrigerated air
- Properly located controls
- Enough refrigerated coil surface
- Capacity adequate to expected needs
- Appropriate stacking of the produce.

Control of relative humidity

Appropriate relative humidity is important to control the following:

- Water losses
- Decay development
- Incidence of some physiological disorders
- Uniformity of ripening.

- Proper relative humidity should be 85-95% for the majority of the fruits, 95-98% for vegetables (except dry onions and pumpkins at 70-75%) and 95-100% for some root vegetables.
- Relative humidity can be controlled by the following methods:
  - Addition of moisture to air by humidifiers
  - Regulation of air movement in relation to produce
  - Maintaining coil temperature to 1°C difference to air temperature
  - Wetting the floor in the storage room
  - Addition of crushed ice.

Controlled Atmosphere

- Controlled atmosphere means the addition or removal of gases resulting in an atmospheric composition surrounding the commodity that is different from that of the air 79% of nitrogen, 21% of oxygen and traces of carbon dioxide).
- Usually this involves reduction of oxygen and elevation of carbon dioxide, in a perfectly sealed room.
- The use of controlled atmosphere can be considered only as a supplement to the proper temperature and humidity procedures.
- Controlled atmosphere is used for a certain number of crops to extend shelf-life, reduce disorders such as chilling injuries, reduction of pathogens and some insect control.

Supplemental Procedures

The following treatments may be applied to horticultural commodities:

- Curing of certain roots, tubers and bulbs vegetables (see details on specific chapter);
- Sorting for defect elimination
- Waxing and other surface coatings
- Hot water treatment
- Treatment with post-harvest fungicides or bactericides
- Use of sprout inhibitors
- Special post harvest chemical treatment
- Fumigation for insect control
- Films wrapping
- Ethylene treatment for de-greening and ripening certain fruits, such as citrus, bananas and mangos.

Ripening Fruits with Ethylene Gas

- In the post-harvest physiology of most horticultural crops, ethylene plays an important role, sometime beneficial (improving quality of the produce by faster and more uniform ripening prior to retail distribution) and often deleterious (increasing the rate of senescence and reducing shelf-life).

Systems for Ethylene Treatments

- Handlers can equip existing rooms for use as ripening chambers or they can install specially built ones.
- Both need automatic control of temperature (for heating and cooling), humidity and ventilation.
- The room should be as tight as possible, to prevent leakage of gas, but not essentially hermetically sealed.

Amount of Gas Needed

- 100 ppm of gas.
- Higher concentration will not speed up the ripening process.
- Too much gas may result an explosive air-gas mixture.
- Safety precautions have to be followed.

Temperature and Humidity

- Optimum temperature varies from **18 to 25°C**
- At lower temperature ripening is slowed, from 25 to 30°C ripening may be inhibited and decay accelerated.
- Relative humidity should be as high as possible.

Other Technologies for using Ethylene

Fruit ripening could also be induced with the following methods:

- Explosion-proof ethylene mixed with an inert gas
- Ethylene generators, widely used in developed countries
- Ethephon
- Fruits already ripe (included in the high ethylene producing category) can be used in very small commercial operations, or at home, to ripen other fruits.

Harvesting and Field Handling

- Farm management is generally aimed at maximizing the yield of a crop from the area of land under cultivation while at the same time maximizing the return.
- The time, labour and capital expended in bringing the crop to maturity is rewarded by the financial return obtained by the grower during marketing.

Harvesting Factors

- The quality and conditions of produce sent to market and their consequent selling price are directly affected by the care taken during harvesting and field handling.
- Whatever the scale of operations or the resources of labour and equipment available, the planning and carrying out of harvesting operations must observe basic principles.

The objective of the grower should be:

- To harvest a good quality crop in good condition.
- To keep the harvested produce in good condition (protected from rain, sun or animals) until it is consumed or sold.
- To dispose of the crop to a buyer or through a market as soon as possible after harvest.
- To meet these objectives, success for the harvesting, field handling and marketing must depend on planning from the earliest stages of production, with particular regard to:

- crop selection and timing to meet expected market requirements;
- contact with buyers so that the crop can be sold at a good price when ready for harvest;
- planning harvest operations in good time, arranging for labour, equipment, material, cover space to protect the inputs and the harvested crop and transport;
- it is economically sound in terms of returns on investment to improve grading, packing and handling of the produce before it leaves the farm;
- considering the above the grower must ensure that all those working on the farm are properly trained;
- Provide full supervision at all stages of harvesting and field handling.

Maturity at Harvest

- Fruits and vegetables are considered to be commercially mature when at the stage of physiological development that consumers consider to be the most desirable.
- However the stage, at which a commodity reaches commercial maturity, varies greatly.
- Many leafy vegetables are commercially mature at an early stage of plant development while fruits are often ready for harvest at a fully developed stage.
- Maturity standards have been determined for many fruits and vegetables.
- Harvesting crops at the proper maturity allows handlers to begin their work with the best possible quality produce.
- Produce harvested too early may lack flavour and may not ripen properly, while produce harvested too late may be fibrous or overripe and have a shorter shelf-life.
- Vegetables are harvested over a wide range of maturities, depending upon the part of the plant used as food. Table 17.2 provides some examples of maturity indices of vegetable crops.

Table 17.1: *Maturity indexes for vegetables and root crops*

Crop	Index
Root, bulb and tuber crops	
Radish and carrot	Large enough and crispy (over-mature if pithy)
Potato, onion, and garlic	Tops beginning to dry out and topple down
Yam, bean and ginger	Large enough (over-mature if tough and fibrous)
Green onion	Leaves at their broadest and longest
Fruit vegetables	
Cowpea, yard-long bean, snap bean, sweet pea, and winged bean	Well-filled pods that snap readily
Lima bean and pigeon pea	Well-filled pods that are beginning to lose their greenness
Okra	Desirable size reached and the tips of which can be snapped readily

Snake gourd, and dishrag gourd	Desirable size reached and thumbnail can still penetrate flesh readily (over-mature if thumbnail cannot penetrate flesh readily)
Eggplant, bitter gourd, chistophine or slicing cucumber	Desirable size reached but still tender (over mature if colour dulls or changes and seeds are tough)
Tomato	Seeds slipping when fruit is cut, or green colour turning pink
Sweet pepper	Deep green colour turning dull or red
Muskmelon	Easily separated from vine with a slight twist leaving clean cavity
Watermelon	Colour of lower part turning creamy yellow, dull hollow sound when thumped
Flower vegetables	
Cauliflower	Curd compact (overmature if flower cluster elongates and become loose)
Broccoli	Bud cluster compact (overmature if loose)
Leafy vegetables	
Lettuce	Big enough before flowering
Cabbage	Head compact (over mature if head cracks)
Celery	Big enough before it becomes pithy

Source: Bautista, O.K. and Mabesa, R.C. (Eds). 1977. *Vegetable production*. University of the Philippines at Los Banos.

Packaging

The principal benefit of packaging is to provide protection against physical damages through inadequacies in handling and transport of a fresh produce. There are additional benefits, which can be derived from packaging:

- Barrier protection, to prevent contamination of produce by undesirable environmental agents such as dust and micro-organisms;
- No utilization of produce, to create relatively small units of produce, which are easier and quicker to handle than unpacked goods. It also gives better utilization of space during storage and transport.
- Trading aid to provide a standard size unit for market trading and thereby eliminating the need to weigh or count all items of produce being traded.
- Marketing aid, through the display of basic market information and attractive designs on the package.

Storage

- The term "storage" as now applied to fresh produce is almost automatically assumed to mean the holding of fresh fruit and vegetables under controlled conditions.

- Although this includes the large-scale storage of some major crops, such as carrots and potatoes, to meet a regular continuous demand and provide a degree of price stabilization it also meets the demands of populations of developed countries and of the richer consumers of developing countries, providing year round availability of various local and exotic fruits and vegetables of acceptable quality.
- In many developing countries, however, where seasonally produced plant foods are held back from sale and released gradually, storage in a controlled environment is not possible because of the cost and the lack of infrastructure development and maintenance and managerial skill.
- Even in developed countries, however, there are still many peoples who, for their own consumption, preserve and store fresh produce by traditional methods.

Recommended Storage Temperatures

Table 17.2: *Recommended temperature and relevant humidity and storage life for vegetables and root crops*

Product	Temperature		Relative humidity (percent)	Approximate storage life
	C	F		
Bean sprouts	0	32	95-100	7-9 days
Beans dry	4-10	40-50	40-50	6-10 months
Beans, green or snap	4-7	40-45	95	7-10 days
Beans, Lima, in pods	5-6	41-43	95	5 days
Beans, bunched	0	32	98-100	10-14 days
Beets, topped	0	32	98-100	4-6 months
Bitter melon	12-13	53-55	85-90	2-3 weeks
Broccoli	13-15	55-60	85-90	2-6 weeks
Cabbage, early	0	32	95-100	10-14 days
Cabbage, late	0	32	98-100	3-6 weeks
Carrots, bunched	0	32	95-100	2 weeks
Carrots, mature	0	32	98-100	7-9 months
Carrots, immature	0	32	98-100	4-6 weeks
Cashew apple	0-2	32-36	85-90	5 weeks
Cauliflower	0	32	95-98	34 weeks
Celery	0	32	98-100	2-3 months
Chayote squash	7	45	85-90	4-6 weeks
Cucumbers	10-13	50-55	95	10-14 days
Eggplants	12	54	90-95	1 week
Garlic	0	32	65-70	6-7 months
Ginger root	13	55	65	6 months

Greens, leafy	0	32	95-100	10-14 days
Leeks	0	32	95-100	2-3 months
Lettuce	0	32	98-100	2-3 weeks
Okra	7-10	45-50	90-95	7-10 weeks
Onion, green	0	32	95-100	34 weeks
Onion, dry	0	32	65-70	1-8 months
Onion set	0	32	65-70	6-8 months
Peppers, Chill (dry)	0-10	32-50	60-70	6 months
Peppers, Sweet	7-13	45-55	90-95	2-3 weeks
Pumpkins	10-13	50-55	50-70	2-3 months
Snow peas	0-1	32-34	90-95	1-2 weeks
Spinach	0	32	95-100	10-14 days
Squashes, Summer	5-10	41-50	95	1-2 weeks
Squashes, winter	10	50	50-70	2-3 months
Sweet potatoes	13-15	55-60	85-90	4-7 months
Taro root	7-10	45-50	85-90	4-5 months
Tomatoes, mature-green	18-22	65-72	90-95	1-3 weeks
Tomatoes, firm-ripe	13-15	55-60	90-95	4-7 days
Watermelons	10-15	50-60	90	2-3 weeks
Yams	16	61	70-80	6-7 months

Adapted from: McGeorge, B.M. 1989. Tropical Products Transport Handbook. USDA Office of Transportation, Agriculture Handbook 668.

Post-Harvest Losses

- The vegetable perished very rapidly due to its succulent in nature. Nearly, 30% losses have been observed during the post harvest handling of the vegetables, which reflect as loss of appearance, senescence, texture and nutritional value and quantitative losses due to lack of adequate facilities/ technologies i.e., packaging material, long term storage facilities in general and cool chain system in particular.
- Vegetable base processing industries not located in the production areas and even that existing industries are located in towns and cities which are very far away from production centre.
- Low cost technologies for post-harvest handling and packaging for domestic and export market have not been developed adequately. Transport facilities specially suited to vegetables are lacking.

Table 17.3: *Post harvest losses in vegetable crops*

Vegetables	Post harvest losses (%)	Remarks
Onion	6.01-40.0	Storage & transportation
Potato	30.0-40.0	Harvesting & Storage
Garlic	0.9-2.7	Storage
Tomato	6.7-33.5	Transportation & marketing
Brinjal	3.7-13.4	Marketing
Cabbage	3.0-13.4	Marketing
Cauliflower	10.0-13.2	Transportation & marketing
Beans	7.45	Transportation & marketing
Smooth Gourd	8.0-15.0	Transportation & marketing
Bottle gourd	7.0-10.0	Transportation & marketing
Okra	5.0-10.0	Marketing
Chilli	4.0-35.0	Farm and storage
Beetroot	10.0-15.0	Marketing
Radish	3.0-5.0	Marketing
Carrot	3.0-5.0	Transportation & marketing
Pointed gourd	19.81	Marketing storage
Garlic	0.9-2.7	Marketing storage

Various post-harvest processes can significantly reduced the losses and maintain the quality of vegetables for longer duration. Proper grading, washing, precooling, proper packaging, storage and transportation can reduce losses to larger extent.

Vegetable Preservation

Production and processing scenario of fruits and vegetables in India and the world

- The Green Revolution and subsequent efforts through the application of science and technology for increasing food production in India have brought self-reliance in food.
- The impetus given by the Government, State Agricultural Universities, State Departments of Agriculture and other organizations through the evolution and introduction of numerous hybrid varieties of cereals, legumes, fruits and vegetables and improved management practices have resulted in increased food production.
- However, the nation still faces the problem of the use of improper methods for the storage of food stuffs, leading to great wastage of the food produced. Such losses in the food front aggravate the existing syndromes of undernutrition and malnutrition.
- Fruits and vegetables, which are among the perishable commodities, important ingredients in the human dietaries. Due to their high nutritive value they make

significant nutritional contribution to human well-being. They are the cheaper and better source, the protective foods. If they can be supplied in fresh or preserved form throughout the year for human consumption, the national picture will improve greatly.

- The perishable fruits and vegetables are available as seasonal surpluses during certain parts of the year in different regions and are wasted in large quantities due to absence of facilities and know-how for proper handling, distribution, marketing and storage.
- Furthermore, massive amounts of the perishable fruits and vegetables produced during a particular season result in a glut in the market and become scarce during other seasons. Neither can they all be consumed in fresh condition nor sold at economically viable prices.
- In developing countries agriculture is the mainstay of the economy. As such, it should be no surprise that agricultural industries and related activities can account for a considerable proportion of their output. Of the various types of activities that can be termed as agriculturally based, fruit and vegetable processing are among the most important.
- Therefore, fruit and vegetable processing has been engaging the attention of planners and policy makers as it can contribute to the economic development of rural population. The utilization of resources both material and human is one of the ways of improving the economic status of family.
- In the post green revolution era, even though food grains have been taken care of, fruits and vegetables for want of simple technologies of processing, preservation and transport to various places of need, have suffered post-harvest losses, estimated to be more than 25% and only about 1% of the total fruits and vegetables produced are processed.
- All forms of preserved fruits are in the reach of only. the urban elite and the rural masses who produce more than 90% of these fruits and vegetables are usually deprived of their usage.
- Today, India is the second largest producer of fruits (44 million tonnes) and vegetables (87.5 million tonnes) as mentioned in Indian Horticulture Database-2000 published by National Horticulture Board. Our share in the world production is about 10.1% in fruits and 14.4% in vegetables.

Fruit and vegetable processing industries in India

- Fruits and vegetables and their products have gained considerable importance by contributing significantly to the economy of many countries in the world. In developing countries agriculture is the mainstay of economy.
- Among the various activities, which are termed agriculturally based, fruits and vegetables processing are the most important.
- India has arable land of 184 million hectares producing various agro stocks including fruits and vegetables (151.5 MT) and is considered as a goldmine as far as raw material base for food processing industries is concerned. India is the

second largest producers of fruits in the world producing 50 million tonnes and accounting for 10% of the world's fruit production.

- It is also the second largest producer of vegetables, producing 101 million tonnes, accounting for 15% of the world's production.
- Apart from fruits and vegetables, India also produces about 20 million tonnes of root and tuber crops. Considering the wide ranging and large raw material base that the country offers, along with the consumer base of over one billion people, the processing industry holds tremendous opportunities for large investment.
- Depending on factors like availability, socio-economic conditions, tradition, taste and culture, some of these fruits and vegetables are regularly consumed in the fresh and processed forms.
- Even though India is the second largest producer of fruits and vegetables in the world after China, the present quantity of fruit and vegetable processing is very meager (around 2.2%) as compared to 80% in USA, 70% France, 80% Malaysia and 30% Thailand.
- The fruit and vegetable processing in India is highly decentralized, small-scale industries accounting for 33%, organized 25%, unorganized 42% and large number of units in cottage/household and small scale sector, having capacities of up to 250 tonnes/annum.
- In the organized sector of India, there are about 5000 units and several thousands in the unorganized sector of fruit and vegetable processing are in the job.
- The installed capacity of fruit and vegetable processing industry has increased from two million tonnes in year 2000 to only 2.1 million tonnes in 2006. Since 2000, the industry has seen significant growth in ready to serve beverages, fruit juices/pulps, dehydrated and frozen fruit and vegetable products, pickles, processed mushrooms and curried vegetables and units engaged in these segments are export oriented.
- India has exported total of 0.325 million tonnes of processed products valued at Rs. 4.34 billion during 2004 - 05, which was significantly increased to 0.55 million tonnes in the year 2006 - 07 valued at Rs. 17.31 billion.

History

- Food processing techniques dates back to the prehistoric ages when crude processing techniques incorporated slaughtering, fermenting, sun drying, preserving with salt, and various types of cooking (such as roasting, smoking, steaming, and oven baking).
- Modern food processing technology in the 19th and 20th century was largely developed to serve military needs.
- **In 1809 Nicolas Appert** invented a vacuum bottling technique that would supply food for French troops, and this contributed to the development of tinning and then canning by Peter Durand in 1810.
- Pasteurization, discovered by **Louis Pasteur in 1862**, was a significant advance in ensuring the micro-biological safety of food.

Principles of preservation—Drying /dehydration-process-types pre-treatment's required-factors affecting rate of dehydration-Reconstitution- coefficient of rehydration

- Food preservation can be defined as the science which deals with the methods of prevention of decay or spoilage of food, thus allowing it to be stored in a fit condition for future use.
- In the preservation of foods by various methods, the following principles are involved:

1. Prevention or delay of microbial decomposition

(a) by keeping out microorganisms (asepsis);

(b) by removal of microorganisms, e.g., by filtration;

(c) by hindering the growth and activity of microorganisms, e.g., by low temperature, drying, anaerobic conditions, chemicals or antibiotics; and

(d) by killing the microorganisms, e.g., by heat or radiation.

2. Prevention or delay of self-decomposition of the food

(a) by destruction or inactivation of enzymes, e.g., by blanching;

(b) by prevention or delay of chemical reactions, e.g., prevention of oxidation by means of an antioxidant.

3. Prevention of damage by insects, animals, mechanical causes, etc.

To retain the natural taste and aroma of a product, it is necessary to preserve it soon after preparation, without allowing it to stand for any length of time. Various methods of preservation are employed and each has its own merits. The methods generally used are as under:

1. Asepsis (Absence of infection)

- Asepsis means preventing the entry of microorganisms. Maintaining of general cleanliness while picking, grading, packing and transporting of fruits and vegetables increases their keeping quality and the products prepared from them will be of superior quality.
- Washing or wiping of the fruits and vegetables before processing should be strictly followed as dust particles adhering to the raw material contain microorganisms and by doing so the number of organisms can be reduced considerably.

2. Preservation by High Temperature

- Coagulation of proteins and inactivation of their metabolic enzymes by the application of heat leads to the destruction of microorganisms present in foods.
- Further, heating can also inactivate the enzymes present in the food. Heating food to high temperatures can, therefore, help to preserve it. The specific treatment varies with:

Table 17.4: *Thermal Death Time (TDTl) of bacteria*

Bacteria	Time (min)	Temperature (°C)
Salmonella typhosa	4.3	60
Staphylococcus aureus	18.8	60
E. coli	20-30	57.3
Streptococcus thermophilus	15	70.77
L. bulgaricus	30	71

The spare-farming bacteria found in tomato juice, however, require heating at a higher temperature of 88°C for a much longer duration. Heat resistance of bacterial spores are also given in the Table.

Table 17.5: *Heat resistance of bacterial spores*

Spore	Time (min) to kill at 100°C
B. anthracis	1.7
B. subtilis	15-30
C. batulinum	100-300
Flat sourbacteria	aver 1000

Enzymes also require air (oxygen) at normal temperature far their action and can, therefore, be destroyed at a moderate temperature by removing air from the juice.

- Pectic enzymes which cause changes in flavour and also. bring about the clotting of particles in the juice can be destroyed by heating the juice for about 4 minutes at 85°C, or for one minute at 88°C.
- Usually juices, R.T.S. and nectar are pasteurized at about 85°C for 25 to 30 minutes according to the nature of the juice and the size of the container.
- Acidic fruit juices require a lower temperature and less time for pasteurization than the less acidic ones. Juice can be pasteurized in two ways: (i) by heating it at a low temperature for a long period, or (ii) by heating it at a high temperature for a short time only (HTST method). There are three methods of pasteurization:

(a) **Bottle or 'Holding' pasteurization:** This method is commonly used for the preservation of fruit juices at home. The extracted juice is strained or clarified as the case may be, and filled in bottles, leaving sufficient head space for the expansion of the juice during heating. The bottles are then sealed airtight and pasteurized.

(b) **Overflow method:** Juice is heated to a temperature about 2.5°C higher than the pasteurization temperature, and then filled in sterilized bottles up to the brim, taking care that during filling and sealing the temperature of juice does not fall below the pasteurization temperature. The sealed bottles arc pasteurized at a temperature 2.5°C lower than the filling and sealing temperature and then cooled. This method is very suitable for grape juice because it minimizes the adverse effect of air and the quality of the juice.

(c) **Flash pasteurization:** The juice is heated rapidly to a temperature of about 5.5°C higher than the pasteurization temperature and kept at this temperature for about a minute. The method has been developed specially for the canning of natural orange juice but can also be used for grape and apple juices. It has the following advantages:

- Loss of flavour is minimum
- Vitamins are not destroyed
- Effects economy of time and space.
- Keeps the juice uniformly cloudy, and
- Juice is heated uniformly and thus its cooked taste is minimum.

Sterilization

- Sterilization by definition means the destruction of all viable microorganisms.
- Heat sterilization is the most effective process of food preservation.
- It has severe effect on heat liable nutrients, particularly vitamins and mainly through Maillard reaction, the nutritional quality of proteins is reduced. By this method all microorganisms are completely destroyed due to high temperature.
- The time and temperature necessary for sterilization vary with the type of food.
- Fruit and tomato products should be heated at 100°C for 30 minutes so that the spore forming bacteria which are sensitive to high acidity may be completely killed. Vegetables like green peas, okra, beans. etc., being non-acidic and containing more starch than sugar, require higher temperature to kill the spore-forming organisms.
- Continuous heating for 30 to 90 minutes at 116°C is essential for their sterilization. Before using the empty cans and bottles should also be sterilized for about 30 minutes by placing them in boiling water.
- Temperatures above 100°C can only be obtained by using steam pressure sterilizers such as pressure cookers and autoclaves.

Table 17.6: *The major differences between pasteurization and sterilization are as under:*

S. No.	Pasteurization	Sterilization
1.	Partial destruction of microorganisms	Complete destruction of microorganisms
2.	Temperature below 100°C	Temperature 100°C and above
3.	Normally used for fruits	Normally used for vegetables

- A quick technique of "Aseptic Canning", using high temperature, has also been developed which not only reduces the sterilization time but also improves the quality of the product markedly.
- Another methods is "Hot Pack or Hot Fill", generally used in homes for preparation of jams and other products. Both are discussed here in brief.

3. Preservation by Low Temperature

- Microbial growth and enzyme reactions are retarded in foods stored at low temperatures. The lower the temperature, the greater the retardation. Low temperatures can be produced by (i) cellar storage (about 15°C), (ii) refrigeration or chilling (0 to 5°C), and (ii) freezing (-18 to -40°C).

I. **Cellar storage (about 15°C):** The temperature in cellars (underground rooms) where surplus food is stored in many villages is usually not much below that of the outside air and is seldom lower than 15°C. It is not low enough to prevent the action of many spoilage organisms or of plant enzymes. Decomposition is, however, slowed down considerably. Root crops, potatoes, onions, apples and. similar foods can be stored for limited periods in cellars during the winter months.

II. **Refrigeration or chilling (0 to 5°C):** Chilling temperatures are obtained and maintained by means of ice or mechanical refrigeration. Fruits, vegetables and their products can be preserved for a few days to many weeks when kept at this temperature. The best storage temperature for many foods is slightly above O°C but his varies with the product and is fairly specific to it. Besides temperature, the relative humidity and the composition of the air can affect the preservation of the food. Commercial cold storages with proper ventilation and automatic control of temperature are now used throughout the country (mostly in cities) for the storage of semi-perishable foods such as potatoes and apples. This has made such foods available throughout the year and has also stabilized their prices.

III. **Freezing (-18 to -40°C):** Freezing method is the most harmless method of food preservation. Microbial growth is inhibited and the rate of chemical reactions is slowed down at low temperatures. In commercial frozen storage the activity of meat enzymes is stopped while plant foods have to be blanched before freezing to avoid undesirable quality changes. At temperatures below the freezing point of water (-18 to - 40°C) growth of microorganisms and enzyme activity are reduced to a minimum. Most perishable foods can be preserved for several months if the temperature is brought down quickly (quick freezing) and the food kept at these temperatures. Foods can be quick frozen in about 90 minutes or less by: (i) placing them in contact with the coil through which the refrigerant flows, (ii) blast freezing in which cold air is blown across the food, and (iii) dipping in liquid nitrogen.

- Quick frozen foods maintain their quality and freshness when they are thawed (brought to room temperature) because only very small ice crystals are formed when foods are frozen in this manner. Many microorganisms can survive this treatment and become active and spoil the food if it is kept at higher temperatures. Frozen foods should, therefore, always be kept at temperatures, below -5°C. Enzymes in certain vegetables can continue to act even after being quick frozen and so such vegetables have to be given a mild heat treatment called blanching (above 80°C) before they are frozen to prevent development of off-flavours.
- Thc best way of preserving pure fruit juice is by freezing. Properly frozen juice retains its freshness, colour and aroma for a long time. This method is particularly useful in the case of juices whose flavour is adversely affected by heating. The juice is first desecrated and the vacuum filled with nitrogen

gas. It is then transferred into containers which are hermetically sealed and frozen. Moulds are sometimes not affected by this technique. Juice can be kept in good condition for a long time in frozen form at -12 to -17°C by excluding air. It is defrosted before consumption.

4. Preservation by Chemicals

- Microbial spoilage of food products is also controlled by using chemical preservatives which do not include salt, sugar, acetic acid, oils, alcohols, etc., but only microbial antagonists. The inhibitory action of preservatives is due to their interfering with the mechanism of cell division, permeability of cell membrane and activity of enzymes.
- Pasteurized squashes, cordials and crushes have a cooked flavour. After the container is opened, they ferment and spoil within a short period, particularly in a tropical climate. To avoid this, it is necessary to use chemical preservatives. Chemically preserved squashes and crushes can be kept for a fairly long time even after opening the seal of the bottle. It is, however, essential that the use of chemicals is properly controlled, as their indiscriminate use is likely to be harmful. The preservative used should not be injurious to health and should be non-irritant. It should be easy to detect and estimate.
- According to the British Food and Drug Act of 1928, a "preservative" is any substance which is capable of inhibiting, retarding or arresting the process of fermentation, acidification or other decomposition of food, but does not include common salt (sodium chloride), saltpetre (sodium or potassium nitrate), sugar, acetic acid or vinegar, alcohol or potable spirits, spices, essential oil or any other substance added to the food by the process of curing known as smoking.

The two important chemical preservatives permitted in many countries are:

(i) Sulphur dioxide (including sulphites), and

(ii) Benzoic acid (include benzoates)

These two are also allowed in India according to the **Fruit Product Order (F.P.O.) of 1955.**

I. Sulphur dioxide

- It is widely used throughout the world in the preservation of juice, pulp, nectar, squash, crush, cordial and other products. It has good preserving action against bacteria and moulds and inhibits enzymes, etc. In addition, it acts as an antioxidant and bleaching agent. These properties help in the retention of ascorbic acid, carotene and other oxidizable compounds. It also retards the development of nonenzymatic browning or discolouration (after killing the enzyme) of the product. It is generally used in the form of its salts such as sulphite, bisulphite and metabisulphite.
- **Potassium metabisulphite (K20.2S02 or K2S20S)** is commonly used as a stable source of sulphur dioxide. Being a solid, it is easier to use than liquid or gaseous sulphur dioxide. It is fairly stable in neutral or alkaline media but decomposed by weak acids like carbonic, citric, tartaric and malic acids. When added to fruit juice

or squash it reacts with the acid in the juice forming the potassium salt and sulphur dioxide, which is liberated and forms sulphurous acid with the water of the juice.

- Sulphur dioxide has a better preservative action than sodium benzoate against which are bacteria and moulds. It also retards the development of yeasts in juice, but cannot arrest their multiplication, once their number has reached a high value. It is well known that fruit juices with highly acidic do not undergo fermentation readily.
- The preservative action of the fruit acid is due to its hydrogen ion concentration. The pH for the growth of moulds ranges from 1.5 to 8.5, that of yeasts from 2.5 to 8.0 and of bacteria from 4.0 to 7.5. As fruit beverage like citrus squashes and cordials have generally a pH of 2.5 to 3.5, the growth of moulds and yeasts in them cannot be prevented by acidity alone. Bacteria, how-ever, cannot grow.
- The pH is, therefore, of great importance in the preservation of food product and by regulating it, one or more kinds of microorganism in the beverage can be eliminated .

Table 17.7: *The concentration of sulphur dioxide required to prevent the growth of microorganism at different pH levels*

pH	Organisms and sulphur dioxide concentration (ppm)			
	Saccharomyces ellipsoideus (Yeast)	Mucor (Mould)	Penicillium (Mould)	Mixed bacteria
2.5	200	200	300	100
3.5	800	600	600	300
7.0	above 5000	above 5000	above" 5000	above

- The toxicity of sulphur dioxide increases at high temperature. Hence its effectiveness depends on the acidity, pH, temperature and substances present in fruit juice.
- According to Indian Fruit Product Order, the maximum amount of sulphur dioxide allowed in fruit juice is **700 ppm**, in squash, crush and cordial **350 ppm** and in RTS and nectar **100ppm.**

II. Benzoic acid

- It is only partially soluble in water hence its salt, sodium benzoate, is used. One part of sodium benzoate is soluble in 1.8 parts of water at ordinary temperature, whereas only 0.34 part of benzoic acid is soluble in 100 parts of water.
- Pure **sodium benzoate** is tasteless and odourless. The antibacterial action of benzoic acid is increased in the presence of carbon dioxide and acid, e.g., *Bacillus subtilis* cannot survive in benzoic acid solution in the presence of carbon dioxide.
- **Benzoic acid** is more effective against yeasts than against moulds. It does not stop lactic acid and acetic acid fermentation. The quantity of benzoic acid required depends on the nature of the product to be preserved, particularly its acidity.

- In case of juices having a pH of 3.5 to 4.0, which is the range of a majority of fruit juices, addition of **0.06 to 0.10%** of sodium benzoate has been found to be sufficient. In case of less acidic juices such as grape juice at least 0.3% is necessary.
- The action of benzoic acid is reduced considerably at pH 5.0. Sodium benzoate in excess of 0.1% may produce a disagreeable burning taste.
- According to F.P.O. its permitted level in RTS and nectar is **100 ppm** and in squash, crush and cordial **600 ppm**.

5. Preservation by Drying

Microorganisms need moisture to grow so when the concentration of water in the food is brought down below a certain level, they are unable to grow. Moisture can be removed by the application of heat as in sun-drying or by mechanical drying (dehydration). Sun- drying is the most popular and oldest method of preservation. In these days, mechanical drying has replaced sun-drying. This is a more rapid process as artificial heat under controlled conditions of temperature, humidity and air flow is provided and fruits and vegetables, e.g., green peas, cauliflower, mango, mahua, etc., are dried to such an extent that the microorganisms present in them fail to survive. In this method, juices are preserved in the form of powder. The juice is sprayed as a very fine mist into an evaporating chamber through which hot air is passed. The temperature of the chamber and the-flow of air are so regulated that dried juice falls to the floor of the chamber in the form of a dry powder. The powder is collected and packed in dry containers which are then closed airtight. The powder when dissolved in water makes a fruit drink almost similar to the original fresh juice. Fruit juice powders are highly hygroscopic and require special care in packing. All juices cannot, however, be dried readily without special treatment. Mango juice powder is prepared by this technique but the method is very expensive and not popular in India.

6. Preservation by Filtration

In this method, the juices are clarified by settling or by using ordinary filters, and then passed through special filters which are capable of retaining yeasts and bacteria. Various types of germ-proof filters are used for this purpose. Recently this method has come into use in U.S.A., Germany, etc., for preserving apple and grape.

7. Preservation by Carbonation

- Carbonation is the process of dissolving sufficient carbon dioxide in water or beverages so that the product when served gives off the gas as fine bubbles and has a characteristic taste.
- Carbonation adds to the life of a beverage and contributes in some measure to its tang. Fruit juice beverages are generally bottled with carbon dioxide content varying from 1 to 8 g per litre.
- Though this concentration is much lower than that required for complete inhibition of microbial activity (14.6 g/litre), it is sufficient for supplementing the effect of acidity on pathogenic bacteria.

- Another advantage of carbonation is the removal of air thus creating 'an anaerobic condition, which reduces the oxidation of ascorbic acid and prevents browning.
- Moulds and yeasts require oxygen for their growth and become inactive in the presence of carbon dioxide. In ordinary carbonated drinks, the oxygen which is normally present in solution in water in sufficient amount to bring about fermentation, is displaced by carbon dioxide. Although carbonated beverages contain sugar much below 66%, the absence of air and the presence of carbon dioxide in them help to prevent the growth of moulds and yeasts.
- High carbonation should, however, be avoided as it usually destroys the flavour of the juice.
- The keeping quality of carbonated fruit beverages is enhanced by adding about 0.005% sodium benzoate. The level of carbonation required varies according to the type of fruit juice and type of flavour.

8. Preservation with Sugar

- Syrups containing **66%** or more of sugar do not ferment.
- Sugar absorbs most of the available water with the result that there is very little water for the growth of microorganisms hence their multiplication is inhibited, and even those already present die out gradually.
- Dry sugar does not ferment. Thus sugar acts as a preservative by osmosis and not as a true poison for microorganisms. Fruit syrup, jam, jelly, marmalade, preserve, candy, crystallized fruit and glazed fruit are preserved with sugar.

9. Preservation by Fermentation

- Decomposition of carbohydrates by microorganisms or enzymes is called 'fermentation'. This is one of the oldest methods of preservation. By this method, foods are preserved by the alcohol or organic acid formed by microbial action.
- The keeping quality of alcoholic beverages, vinegars and fermented pickles depends upon the presence of alcohol, acetic acid and lactic acid, respectively. Care should be taken to seal the fermented products from air to avoid further unwanted or secondary fermentation. Wines, beers, vinegar, fermented drinks, fermented pickles, etc., are prepared by these processes.
- Fourteen per cent alcohol acts as a preservative in wines because yeasts, etc., cannot grow at that concentration. About 2 per cent acetic acid prevents spoilage in many products.

10. Preservation by Salt

- Salt at a concentration of **15 to 25%** is sufficient to preserve most products. It inhibits enzymatic browning and discoloration and also acts as an antioxidant. Salt in the form of brine is used for canning and pickling of vegetables which contain very little sugar and hence sufficient lactic acid cannot b formed by fermentation to act as preservative. It exerts its preservative action by (i) causing high osmotic pressure resulting in the plasmolysis of microbial cell! (ii) Dehydrating food as well as microorganisms by drawing out and tying up the moisture by ion hydration,

(iii) ionizing to yield the chloride ion which is harmful to microorganisms, (iv) reducing the solubility of oxygen in water, sensitizing the cells against carbon dioxide, and interfering with the action of proteolytic enzymes.

11. Preservation by Acids

- Low acid foods are spoilt rapidly. A highly acidic environment inhibits the growth of food spoilage organisms. Lowering the protein of certain foods by anaerobic fermentation, action on carbohydrates producing lactic acid is one the methods of food preservation. The same spoilage inhibitory effects can by produced by acidic additives such as vinegar or citric acid. Nutrient losses through fermentation are small. In fact, in certain cases, the nutrient levels are increase, particularly through microbial vitamin and protein synthesis. Acids are added to or allowed to form in foods to preserve them. Acetic (vinegar), citric (lime juice) and lactic acids are commonly used for preservation. About 2% acetic acid prevents spoilage of many products. Onions are bottled in vinegar with a little salt. Vinegar is also added to pickles, chutneys, sauces and ketchups. Citric acid is added to many fruit squashes, jams and jellies to increase the acidity and prevent mould growth.

12. Preservation by Oil and Spices

- A layer of oil on the surface of any food produces anaerobic conditions which prevent the growth of moulds and yeasts. Thus pickles in which enough oil is added to form a layer at the top can be preserved for long periods. Spices like turmeric, pepper, and asafoetida have little bacteriostatic effect and their ability to prevent growth of other microorganism is questionable. Their primary function is to impart their characteristic flavour to the food.

13. Preservation by Antibiotics

- Certain metabolic products of microorganisms have been found to have germicidal effect and are termed as antibiotics. Their use in medicine for controlling certain disease-producing organisms in the body is well known. Some antibiotics are also used to preserve fruits, vegetables and their products.
- Nisin is an antibiotic produced by *Streptococcus lactis,* an organism commonly found in milk, curd, cheese and other fermented milk products. It is non- toxic and has no adverse effect on the sensory qualities of food.
- It is widely used in the food industry especially for preservation of acid foods in which it is more stable. It is commonly used in canning of mushrooms, tomatoes and milk products. 'Nisin suppresses the growth of spoilage organisms, mainly the gas-producing) spore-forming bacteria and toxin-producing *Clostridium botulinum.* Subtilin, an antibiotic obtained from certain strains of *Bacillus subtilis,* is used in preservation of asparagus, corn and peas.
- It is most effective against gram-positive bacteria and spore-forming organisms. Canned peas and tomatoes containing 10 and 20 ppm of subtilin respectively were found to be free of micro-organisms.
- Subtilin and nisin effectively reduce the thermal process requirements necessary to control the spoilage of several food products.

- Pimaricin, an antifungal antibiotic, can be used for treating fruits and fruit juices.
- At present the above three antibiotics are permitted only in such foods as are cooked prior to use and in the process of cooking the residual antibiotic is expected to be destroyed.

14. Preservation by Irradiation

- Sterilization of food by ionizing radiations is a recently developed method of preservation which has not yet gained general acceptance. The unacceptable flavour of some irradiated foods and the fear that radioactivity might be induced in such food has come in the way of its greater use. The harmful effects on the human body of radiation from nuclear explosions have given rise to such apprehension in the minds of many people.

Freezing process types of freezing changes during cold storage-thawing; Canning of fruits and vegetables-process-unit operations

- **Freezing** as a preservation method probably was observed by prehistoric people during cold weather; and, until frozen storage cabinets were developed in the late 1800s, naturally occurring snow and ice were used to freeze foods.
- **Clarence Birdseye** was one of the first to experiment with quick freezing as a way to retain fresh taste and texture. In the 1930s, his products were introduced to U.S. consumers.

There are various methods of freezing:

1. Sharp Freezing (Slow Freezing)

- This technique, first used in 1861, involves freezing by circulation of air, either naturally or with the aid of fans.
- The temperature may vary from -15 to -29°C and freezing may take from 3 to 72 hours.
- The ice crystals formed are large and rupture the cells. The thawed tissue cannot regain its original water content. The first products to be sharp frozen were meat and butter.
- Nowadays freezer rooms are maintained at - 23 to -29°C or even lower, in contrast to the earlier temperature of -18°C.

2. Quick Freezing

- In this process the food attains the temperature of maximum ice crystal formation (0 to -4°C) in 30 minutes or less.
- Such a speed results in formation of very small ice crystals and hence minimum disturbance' of cell structure. Most foods are quick frozen by one of the following three methods:

A. By direct immersion

- Since liquids are good heat conductors, food can be frozen rapidly by direct immersion in a liquid such as-brine 'or sugar solution at low temperature, Berries in sugar solution" packed fruit juices and concentrates are frozen in this manner.
- The refrigeration medium must be edible and capable of remaining unfrozen at -18°C and slightly below. Direct immersion equipments such as Ottesen Brine Freezer, Zarotschenzeff "Fog" Freezer, T.v.A. Freezer, Bartlett Freezer, etc., of commercial importance earlier, are not used today.

B. By indirect contact with refrigerant

- Indirect freezing may be defined as freezing by contact of the product with a metal surface which is itself cooled by freezing brine or other refrigerating media.
- This is an old method of freezing in which the food or package is kept in contact with the passage through which the refrigerant at -18 to -46°C flows.

C. By air blast (Air blast freezing)

- This refers to vigorous circulation of cold air in order to freeze the product. Freezing is done by placing the foodstuffs on trays or on a belt which are then passed slowly through an insulated tunnel containing air in it.
- Here the air temperature is approximately -18 to - 34°C or even lower. This process is economical and a variety of sizes and shapes can be accommodated.

(i) Fluidized bed freeing

- This is a modification of air blast freezing.
- The foodstuff is fluidized to form a bed of particles, and then frozen.

(ii) Plate freezing

- In this method, food products are placed in contact with a cold surface. enzyme action.

3. Cryogenic Freezing

- Although most foods retain their quality when quick frozen by the above methods, a few (mushrooms, sliced tomatoes, whole strawberries and raspberries) require ultrafast freezing.
- Such materials are subjected to cryogenic freezing which is defined as freezing at very low temperature **(below -60°C).**
- The refrigerants used at present in cryogenic- freezing are liquid nitrogen and liquid carbon dioxide. In the former case, freezing may be achieved by (i) immersion in the liquid, (ii) spraying of liquid, or (iii) circulation of its vapour over the product to be frozen.

4. Dehydro-freezing

- This is a process where freezing is preceded by partial dehydration.
- In case of some fruits and vegetables about 50 % of the moisture is removed by dehydration prior to freezing.

- This has been found to improve the quality of the food. Dehydration does not cause deterioration and dehydrofrozen foods are relatively more stable.

5. Freeze-drying

- In this process food is first frozen at -18°C on trays in the lower chamber of a freeze drier and the frozen material dried (initially at 30°C for 24 hours and then at 20°C) under high vacuum (0.1 mm Hg) in the upper chamber.
- Direct sublimation of the ice takes place without passing through the intermediate liquid stage. The product is highly hygroscopic, excellent in taste and flavour and can be reconstituted readily.
- Mango pulp, orange juice concentrate, passion fruit juice and guava pulp are dehydrated by this method.

Canning

- The process of sealing foodstuffs hermetically in containers and sterilizing them by heat for long storage is known as canning.
- **In 1804, Appert in France** invented a process of sealing foods hermetically in containers and sterilizing them by heat.
- Appert is known as the **'Father of Canning'**. This work formed the foundation for modern canning procedure. In honour of the inventor, canning is also known as appetizing.
- **Saddington in England** was the first to describe a method of canning of foods in 1807.
- In 1810, Peter Durand, another Englishman, obtained the first British patent on canning of foods in tin containers. In 1817, William Underwood introduced canning of fruits on a commercial scale in the USA.
- Fruits and vegetables are canned in the season when the raw material is available in plenty. The canned products are sold in the off-season and give better returns to the grower.

Principle and Process of Canning

1. Selection of fruits and vegetables

(i) Fruits and vegetables should be absolutely fresh.

(ii) Fruits should be ripe, but firm, and uniformly mature. Over-ripe fruits should be rejected because they are infected with microorganisms and give a poor quality product. Unripe fruits should be rejected because they generally shrivel and toughen on canning.

(iii) All vegetables except tomatoes should be tender.

(iv) Tomatoes should be firm, fully ripe and of deep red colour.

(v) Fruits and vegetables should be free from dirt.

(vi) They should be free from blemishes, insect damage or mechanical injury.

2. Grading

The selected fruits and vegetables are graded according to size and colour to obtain uniform quality. This is done by hand or by machines such as screw grader and roller grader. Fruits like berries, plums and cherries are graded whole, while peaches, pears, apricots, mangoes, pineapples, etc., are generally graded after cutting into pieces or slices.

3. Washing

It is important to remove pesticide spray residue and dust from fruits and vegetables. One gram of soil contains 10^{12} spores of microorganisms. Therefore, removal of microorganisms by washing with water is essential. Fruits and vegetables can be washed in different ways. Root crops that loosen in soil are washed by soaking in water containing 25 to 50 ppm chlorine (as detergent). Other methods of washing are spray washing, steam washing, etc.

4. Peeling

The objective of peeling is to remove the outer layer. Peeling may be done in various ways.

(i) **Hand peeling:** It is done mostly in case of fruits of irregular shape, e.g., mango and papaya, where mechanical peeling is not possible.

(ii) **Steam peeling:** Free-stone and clingstone peaches are steam peeled in different ways. The former are cut and steam washed. Potatoes and tomatoes are peeled by steam or boiling water.

(iii) **Mechanical peeling:** This is done in case of apples, peaches, pineapples and cherries and also for root vegetables like carrots, turnips and potatoes.

(iv) **Lye peeling:** Fruits like peaches, apricots, sweet oranges, mandarin oranges and vegetables like carrots and sweet potatoes are peeled by dipping them in 1 to 2 per cent boiling caustic soda solution (lye) for 30 seconds to 2 minutes depending on their nature and maturity. Hot lye loosens the skin from the flesh by dissolving the pectin. The peel is then removed easily by hand. Any trace of alkali is removed by washing the fruit or vegetable thoroughly in running cold water or dipping it for a few seconds in 0.5 per cent citric acid solution. This is a quick method where by cost and wastage in peeling is reduced.

(v) **Flame peeling:** It is used only for garlic and onion which have a papery outer covering. This is just burnt off. Vegetables like peas are shelled, carrots are scraped, and beans are snipped or trimmed.

5. Cutting

Pieces of the size required for canning are cut. Seed, stone and core are removed. Some fruits like plum from which the seeds cannot be taken out easily are canned whole.

6. Blanching

- It is also known as scalding, parboiling or precooking.
- Fruits are generally not blanched leaving the oxidizing enzyme system active.

- Sometimes fruit is plunged for a given time-from half to, say, five minutes, according to variety into water at from 180°F to 200°F, and then immediately cooled by immersion in cold water.
- The object is to soften the texture and so enable a greater weight to be pressed into the container without damage to the individual fruit.
- Blanching is Usually done in case of vegetables by exposing them to boiling water or steam for 2 to 5 minutes, followed by cooling.

7. Cooling

After blanching, the vegetables are dipped in cold water for better handling and keeping them in good condition.

8. Filling

Before filling, cans are washed with hot water and sterilized but in developing countries these are subjected to a jet of steam to remove dust and foreign material. Automatic, large can-filling machines are used in advanced countries but choice grades of fruits are normally filled by hand to prevent bruising in India. Hand filling is the common practice. After filling, covering with syrup or brine is done and this process is called syruping or brining.

9. Exhausting

The process of removal of air from cans is known as exhausting. After filling and lidding or clinching, exhausting is essential. The major advantages of exhausting are as under:

(i) Corrosion of the tinplate and pin holing during storage is avoided.
(ii) Minimizes discolouration by preventing oxidation.
(iii) Helps in better retention of vitamins particularly vitamin C.
(iv) Prevents building of cans when stored in hot climate or at high altitude.
(v) Reduces chemical reaction between the container and the contents.
(vi) Prevents development of excessive pressure and strain during sterilization.

Containers are exhausted either by heating or mechanically. The heat treatment method is generally used. The cans are passed through a tank of hot water at 82 to 87°C or move on a belt through a covered steam box. In the water exhaust box, the cans are placed in such a manner that the level of water is 4-5 cm below their tops. The exhaust box is heated till the temperature of water reaches 82 to 100°C and the centre of the can shows a temperature of about 79°C. The time of exhausting varies from 6 to 1 a minutes, depending on the nature of the product. In the case of glass jars or bottles, vacuum closing machines are generally used. The bottles or jars are placed in a closed chamber in which a high vacuum is maintained.

It is preferable to exhaust the cans at a lower temperature for a longer period to ensure uniform heating of the contents without softening them into pulp. Exhausting at high temperature should be avoided because. The higher the temperature, the more is the volume of water vapour formed, and consequently the greater the vacuum produced in the can.

10. Sealing

Immediately after exhausting the cans are sealed airtight by means of a can sealer. In case of glass jars a rubber ring should be placed between the mouth of the jar and the lid, so that it can be sealed airtight. During sealing the temperature should not fall below 74°C.

11. Processing

- Heating of foods for preserving is known as processing, however, in canning technology processing means heating or cooling of canned foods to inactivate bacteria.
- Many bacterial spores can be killed by either high or very low temperatures. Such drastic treatment, however, affects the quality of food.
- Processing time and temperature should be adequate to eliminate all bacterial growth. Moreover, over-cooking should be avoided as it spoils the flavour as well as the appearance of the product.
- Almost all fruits and vegetables can be processed satisfactorily at a temperature of 100°C, i.e., in boiling water.
- The presence of acid retards the growth of bacteria and their spores. Further, they do not thrive in heavy sugar syrup which is normally used for canning of fruits.
- Vegetables (except the more acidic ones like tomato and rhubarb) which are non-acidic in nature, have a hard texture, and proximity to soil which may infect them with spore-bearing organisms are processed at higher temperatures of 115 to 121°C.

Table 17.8: *Fruits and vegetables can be classified into the following four groups according to their pH value*

Class	pH	Product
Low acid (called non-acid)	above 5.0	Vegetables such as peas, lima bean, asparagus, cauliflower, potato, spinach, beet, corn, french bean'
Medium acid	4.5-5.0	Turnip, carrot, okra, cabbage, pumpkin, beet, green bean, etc., and products like soups and sauces
Acid	3.7-4.5	Tomato, pear, banana, mango, jackfruit, pineapple, sweet cherry, peach, apple
High acid	below 3.7	Juice, rhubarb, prune, sauerkraut, pickle, chutney

- Bacterial spores can be more easily destroyed at pH 3.0 (fruits) than at pH 5.0 to 6.0 (vegetables, except tomato and rhubarb). Bacterial spores do not grow or germinate below pH 4.5.
- Thus, a canned product having a pH less than 4.5 can be processed in boiling water but a product with pH above 4.5 requires processing at 115 at 121°C under a pressure of 0.70 to 1.05 kg/cm^2 (10 to 15 lb/sq inch). It is essential that the centre of the can should attain these high temperatures.

- The sealed cans are placed in the cookers, keeping the level of water 2.5 to 5.0 cm above the top of the cans.

12. Cooling

- After processing. the cans are cooled rapidly to about **39°C** to stop the cooking process and to prevent stack-burning.
- Generally the first method, i.e., dipping the cans in cold water, is used.
- If canned products are not cooled immediately after processing, peaches and pears become dark in colour, tomatoes turn brownish and bitter in taste, peas become pulpy with cooked taste and many vegetables develop flat sour (become sour).

13. Storage

- After labelling the cans, they should be packed in strong wooden cases or corrugated cardboard cartons and stored in a cool and dry place.

Containers for packing of canned products

- Both tin and glass containers are used in the canning industry, but tin containers are preferred.

1. Tin containers

Tin cans are made of thin steel plate of low carbon content, lightly coated on both sides with tin metal. It is difficult to coat the steel plate uniformly and during the process of manufacture small microscopic spots are always left uncoated, although the coating may appear perfect to the eye. The contents of the can may react with these uncoated spots resulting in discolouration of the product or corrosion of the tin plate. When the corrosion is severe, black stains of iron sulphide are produced. It is necessary, therefore, to coat the inside of the can with some material (lacquer) which prevents discolouration but does not affect the flavour or wholesomeness of the contents. This process is known as "lacquering".

Two types of lacquers are used:

(i) **Acid-resistant:** Acid-resistant lacquer is a golden coloured enamel and cans coated with it are called R-enamel or A.R cans. These cans are used for packing acid fruits which are of two kinds : (a) those whose colouring matter is insoluble in water, e.g., peach, pineapple, apricot, grapefruit, and (b) those in which it is water-soluble, e.g., raspberry, strawberry, red plum and coloured grapes. Fruits of group (a) are packed in plain cans and those of group (b) in lacquered cans.

(ii) **Sulphur-resistant:** This lacquer is also of a golden colour and cans coated with it are called C-enamel or S.R. cans. They are meant for non-acid foods only and should not be used for any highly acid product as acid eats into the lacquer. These cans are used for pea, corn, lima bean, red kidney bean, etc.

Table 17.9: *The sizes of cans in general use*

Trade name of can	Size (mm)
Al	68 × 102
1-lb Jam	78 × 90
AJ-T	78 × 119
A2	87 × 114
1-lb Butter	103 × 70
A 2-lb Jam	103 × 102
A 21/2	103 × 119
7-lb Jam	157 × 148
A 10	157 × 178

Points to be Remembered

S. No.	Particulars	Remarks
1.	FPO (Food Product Order) was passed by the Govt. of India in 1955	
2.	The FPO license is given by Central Government	
3.	FPO permits the use of only two chemicals as preservative potassium-meta bisulphite and sodium benzoate	
4.	Fruit products preserved by potassium-meta-bi sulphite	
5.	Coloured products preserved by sodium benzoate	
6.	How much quantity of colour is permitted in preservative in preservation of food	200 ppm
7.	According to FPO, maximum amount of KMS allowed in RTS	100 ppm
8.	According to FPO, maximum amount of KMS allowed in fruit juice	700 ppm
9.	What is the permissible limit of SO_2 in raisin grapes	750 ppm
10.	Sulphur containing food is stored in	SR cans
11.	Which is a class one preservative	Sugar
12.	Sugar used as a preservative have concentration of	70%
13.	The principle followed in freezing is	Crystallization

S. No.	Particulars	Remarks
14.	Most suitable radiation for disinfection of food is	Gamma rays
15.	Lye peeling is done by	1% hot alkali solution
16.	Salt concentration of pickles	Above 15%
17.	Vinegar concentration in preservation should not be more than	2%
18.	Waxing of fruit is done mainly to reduce	Transpiration and respiration
19.	Pre cooling is meant to	Remove field heat
20.	Physiological process responsible for rapid deterioration of freshly harvested fruits and vegetables	Respiration
21.	Tomato sauce must have TSS	30%
22.	For jelly making fruit type is selected	Firm ripe
23.	The major chemicals compounds responsible for the specific aromas of most ripening fruits are	Esters of aliphatic alcohol's and short chain fatty acids
24.	Which enzyme is responsible for fruit softening	Pectinase
25.	Grapes varieties which are suitable for raisin making have TSS	Above 19° brix
26.	Ethylene gas is commonly used to initiate ripening and to degreen many fruits except	Grapes
27.	Loss in firmness in mango fruit during ripening process is due to	Hydrolysis of pectin
28.	Metallic colours are	Banded in preservation
29.	Yeast is used in making	Cider
30.	Which amino acids is the percussor of ethylene production	Methionine
31.	Father of canning	Nicolas Appert (France)
32.	Jelmeter test is used for determination of	Pectin content
33.	Determination of TSS by	Hand Refractometer
34.	Determination of salt concentration in any substances by	Salometer
35.	In washing solution cl concentration	50 ppm
36.	Ripening of mango commercially done by	Ethrel
37.	Gas emitted during fruit ripening is	Ethylene
38.	Vapour Heat Treatment done in	Mango

S. No.	Particulars	Remarks
39.	Hydrocooling at 12-15°C with 500 ppm Bavistin increases shelf-life of	Mango
40.	About 36% of vegetables decay by	Soft rot bacteria
41.	About 30% of fruits decay by	Penicillium spp.
42.	Alcohol content in cider	4-6%
43.	Potato is stored at temperature	2-4° C
44.	The term appetizing is used for	Canning
45.	Bio degradable plastic is	Polyhydroxy butyrate
46.	Cryo- preservation is associated with	Liquid nitrogen
47.	Flower preservative is	KNO_3 (Potassium Nitrate)
48.	Pectin is measured by	Jell meter
49.	In lye peeling which chemical is used	NaOH (1%)
50.	Removal of outer covering of fruits is known as	Peeling
51.	Mango pieces are dipped before pickling in the solution of	2 % salt
52.	Product made by yeast fermentation is	Cider
53.	Bacteria are involved in production of	Vinegar
54.	Lactic bacteria is used in	Pickle
55.	Pineapple variety suitable for canning is	Kew
56.	ICMR recommended how many grams of vegetables in our daily diet	300 g
57.	Lye peeling is done at a temperature of	93° C
58.	In pickle, white cottony surface is due to	Rhizopus
59.	In storage, soft rot is caused by	*Erwinia*
60.	Ethylene concentration for de-greening is	20 ppm

CHAPTER 18

Recent Advances in Crop Improvement

R.K. Jaiswal
Associate professor (Horticulture)
RVSKVV, R.A.K. College of Agriculture, Sehore (M.P.)

Crop-improvement programme focuses on development of new crop varieties and hybrids with wider adaptability and higher yield along with tolerance/ resistance to various insect-pests and diseases and for overcoming adverse impact of abiotic stresses. To realize envisioned goal, greater emphasis is laid on the development of genomic resources for targeted traits and crops, besides pre-breeding for identification of desirable genes in related and wild species to broaden genetic-base by transferring desirable alleles from alien sources. Suitable biotechnological tools such as marker-assisted selections, transgenic crop variety development and many such novel tools and techniques have also been deployed for faster and accurate identification of promising crop genotypes. High-yielding vegetable crops varieties and hybrids with tolerance to pests /diseases and to various farming constraints such as soil and water salinity, soil acidity, drought, flood and such other factors have been developed. For different designated agro-climatic regions/zones of the country, improved varieties of vegetable crops have been notified for cultivation.

Recent Advances

Arka Samrat tomato was recommended as bacterial wilt resistant tomato hybrid for Karnataka, Tamil Nadu, Andhra Pradesh and Kerala. It has 100 tonnes/ha yield potential.

The plants of **Arka Avinash brinjal** are tall and spreading with dark green foliage and long green fruits, resistant to bacterial wilt, green tender fruits with good keeping and cooking quality. It yields 36 tonnes/ha. The plants of **Arka Harshitha brinjal** are tall and spreading with dark green foliage and fruits, resistant to bacterial wilt, tender fruits with good keeping and cooking quality. Yield 40 tonnes/ha.

The plants of **Arka Unnathi brinjal** are tall and erect with dark green foliage and fruits with green calyx, fruits borne in clusters, resistant to bacterial wilt, tender fruits with good keeping quality and good cooking quality. It is suitable for both kharif and rabi.

The yields potential is 34 tonnes/ha. **Chilli**, Arka Khyati, a CGMS based high-yielding F_1 hybrid having light green, medium pungent and smooth fruits which turn deep red on maturity, yielding 40-45 tonnes/ha (fresh) and 5-5.5 tonnes/ha (dry) in 180 days, has been recommended for cultivation in Karnataka.

Pea 'Arka Priya' was recommended as powdery mildew resistant variety for zones I, IV and VIII. Arka Nirmal, Arka Harini, Arka Mayur, Arka Tapas, Arka Uttam, and Arka Chaitra have been released.

Sem or Indian bean, Arka Pradhan, Arka Krishna, Arka Adarsh, Arka Prasidhi, Arka Bhavani, Arka Vistar have been released.

French bean, Arka Sharath, is round fruited, string less, smooth, crispy, fleshy with no parchment has been recommended for release by SVEC for Karnataka. It is suitable for steamed beans, plants are bushy and photo insensitive and suitable for both kharif and rabi seasons. It gives maximum number of pods per plant (44.5) compared to checks. It has high pod yield potential of 18.5 tonnes/ha in 70 days.

Yard long bean, Arka Mangala, is with tall plants (3-4 m), long (80 cm) light green, stingless, round and tender with crisp textured pods. It comes to first harvest in 60 days. Pod yield is 25 t/ha in 100 days. It is suitable both for kharif and rabi.

Pumpkin, Thar Kavi, has been developed by hybridization followed by selection from lines CM16 × CM19. Fruits are small (900g) with 7-8 kg yield/ plant. It is moderately resistant to fruit fly, powdery mildew and pumpkin mosaic virus under field conditions.

Muskmelon, Sel-3, is a selection, yield of 31.60 tonnes/ha with elongated global fruits, golden yellow smooth rind, creamy-white flesh and crisp and juicy texture.

An early and high-yielding variety of **long melon** is recommended for Bihar, Uttar Pradesh, Jharkhand and Punjab. The average fruit length and fruit weight is 30 cm and 50-60g, respectively. The fruit is crispy, light green with smooth prominent ridges. It has an average yield of 175-200 q/ha.

Two antioxidant rich vegetable **amaranth** varieties, viz. Arka Samraksha and Arka Varna, were developed. Arka Samraksha is uprooting type, high-yielding variety with green leaves and stem, high antioxidant activity of 499 mg (AEAC units) and minimum nitrate (27.3 mg) and oxalates (1.34 g/100 g fresh weight of leaves) content, yields 10.9 t/ha in 30-35 days. Arka Varna is uprooting type, high-yielding amaranth variety with green leaves and pink stem, high antioxidant activity of 417mg (AEAC units), nitrate (37.6 mg/100 g fresh leaf) and oxalates (1.42g/100 g fresh weight of leaves), yields 10.6 t/ha in 30-35 days.

Arka Bheem, a tri-parental synthetic variety of onion has been recommended for cultivation in Karnataka. It has red to pinkish red, elongated globe shaped bulbs, with an average bulb weight of 120 g. It yields 42 tonnes/ha in 130 days.

Potato

Kufri Mohan has been recommended for release. This variety produces attractive white-cream, ovoid, uniform tubers and is suitable for northern and eastern plains of the country. Medium maturing (90 days) variety, suitable for fresh consumption.

With mealy texture, pleasant flavour and free from discolouration after cooking, it produces high tuber yield (35-40 tonnes/ ha). It is resistant to late blight (*Phytophthora infestans*) and recommended for cultivation in Indo-Gangetic plains. Further, based on consistently good performance over the years, seven advanced breeding lines (hybrids) namely J/8-85, J/8-91, J/8-119, PS/09-16, PS/09-9 (table purposes); MP/9-28 (chips) and HT/07-1329 (heat stress) were identified for multi-location trials.

Tuber crops

Greater yam clone, Dah-9/196, produced highest yield (30.2 tonnes/ha), followed by Da-293 under nontrailing conditions. Among white yam hybrids, Drh- 1150, produced highest tuber yield (61.70 tonnes/ha), followed by Drh-1125 (59.70 t/ha). The dwarf white yam hybrids, viz. Drd-1038, Drd-1110, Drd-1835, Drd- 920, Drd-1089 and Drd-1078 had excellent cooking quality.

Source: DARE/ICAR Annual Report 2016-17

Of the 10 determinate F1 hybrids of **tomato** evaluated, Hybrid 369 with Arka Vikas fruit type possessed triple disease resistance (ToLCV+BW+EB), whereas Hybrid 371 possessed combined resistance to ToLCV and early blight and Hybrid 373 recorded the highest yield (24 tonnes/ha). Similarly of the 25 indeterminate F1 hybrids evaluated, maximum yield was recorded in IIHR 2042 × IIHR 2834 (76 tonnes/ha), followed by IIHR 2042 × IIHR 2856 (76 tonnes/ha) and PH 4225 (73 tonnes/ha). Fruit firmness was highest in IIHR 2867 × IIHR 2834 (8.7 kg/ cm2). These hybrids possessed triple disease resistance (ToLCV+BW+EB).

Six promising populations of **chilli** with combined resistance to thrips, CMV and ChiVMV were identified and advanced. Four advanced breeding lines of brinjal, viz. IIHR 37-36-4-4, (38 tonnes/ha), IIHR 37-36-4-20, (35 tonnes/ha), IIHR 37-36-13- 7 (32 tonnes/ha) and IIHR 37-36- 3-4 (30 tonnes/ha) were promising for high yield and resistance to bacterial wilt.

A total of 116 advanced lines of **okra** were evaluated. Of them, IIHR 11-1-50 and IIHR 299-1 were identified for high yield (23.54 and 27 kg/3.4 m2 plot, respectively) along with combined resistance to YVMV + powdery mildew (no incidence) during summer. **Kashi Vardaan (VRO-25)** is an early, medium tall (120-125 cm) hybrid, taking 42-44 days for first flowering. Its fruits are harvested 47-100 days after sowing. It has yield potential of 150-155 q/ha, resistant to yellow vein mosaic virus (YVMV) and okra enation leaf curl virus (OELCV). It is recommended for cultivation in Uttar Pradesh, Bihar, Jharkhand and Punjab.

Of the 20 selective crosses of **onion** (four male sterile, disease resistant lines and seven male fertile disease resistant lines), two F_1 progenies, viz. PBRMS 319 × PBRC 339 (26.50 tonnes/ ha and PDI 10.50%) and

PBRMS 318 × PBRC 338 (24.80 tonnes/ha and PDI 12.20%) were found tolerant to purple blotch disease during rainy season Bhima Safed (NRCWO-3), a white onion line with 22-25 tonnes/ha bulb yield was recommended for cultivation during *kharif* in Chhattisgarh, Gujarat, Karnataka, Madhya Pradesh, Maharashtra, Odhisa, Rajasthan and Tamil Nadu. Onion, Bhima Super, Bhima Dark Red and Bhima Shweta, with 500-

600 sets/m2 production were recommended for set production in *kharif* season in areas prone to heavy rainfall. These varieties have bulb yield potential of 22-25 tonnes/ha in *kharif*.

Four advanced breeding lines of **onion**, viz. DOGR Hy-5, a F1 hybrid with uniform, dark red and global bulbs and thin neck, 20.11% more bulb yield (387.25 q/ha) than the best check Bhima Dark Red (322.4 q/ ha), matures 101 days after transplanting, free from double bulbs and bolters, suitable for *kharif* cultivation and; RGP-1, an open-pollinated line with red and flat globe bulbs having thin neck and free of doubles and bolters, 17.5% more yield (328.6 q/ha) than the best check Bhima Shakti (279.65 q/ ha), matures 118 days after transplanting and suitable for *rabi* cultivation; RGP-2, an open-pollinated line having uniform, global, dark red bulbs with thin neck and free of doubles and bolters, suitable for *rabi* season, with 32% more bulb yield (369.15 q/ha) than best check Bhima Shakti (279.65 q/ha) and matures 116 days after transplanting, were recommended for multi locational testing.

Potato

A potato variety, **Kufri Lalit,** with red peel and light yellow pulp was released for commercial cultivation in Bihar, West Bengal, Asom, Odhisa and Jharkhand. It is field resistant to late blight.

SOURCE: DARE/ICAR ANNUAL REPORT 2015-16

A number of varieties/hybrids of vegetable crops were identified for release (*see* table). Two F_1 **tomato** hybrids, viz. **Hybrid-369** and **Hybrid-371**, with triple resistance to ToLCV+BW+EB diseases, high yield (80 tonnes/ha) and firm fruits were developed.

Brinjal variety, CARl Brinjal 1, with green oblong fruits, 25-30 tonnes/ha yield and resistant to bacterial wilt was developed and released for cultivation for Andaman and Nicobar region.

Arka Neelachal Prabha (CHCL 92), a mid-season, (65-75 days) and high-yielding **chilli** variety (493 g/ plant equivalent to 109 q/ha) with dark green and medium long (5-6 cm) fruits was identified for cultivation for east coast.

Arka Arjun, a **French bean** variety resistant to Mungbean yellow mosaic virus (MYMV) with a yield potential of 17.4 tonnes/ha was identified for summer cultivation (up to 35°C) in plateau. The pods are green, round, smooth and stringless. Arka Sharath, another French bean variety (round, fleshy and stringless pods with 20 tonnes/ha green pod yield) was recommended for release in Kamataka.

Two varieties of **amaranth**, **Arka Samraksha** (10-12 tonnes/ha yield) with high antioxidant activity (-500 mg/l00 g fresh weigh), low nitrates (27.3 mg/l00 g fresh weight) and low oxalates (1.34g/l00g fresh weight) content, and **Arka Varna** (10-12 tonnes/ha yield) with high antioxidant activity (417 mg/l00 g fresh weight), low nitrates (38.2 mg/l00 g fresh weight) and low oxalates (1.42 g/l00 g fresh weight) content were identified for release (as against normal antioxidant activity, nitrates and oxalates content ranging from 150--200 mg, 75-80 mg and 2.5 to 3.5 g/l00 g fresh weight, respectively in existing amaranth cultivars). These varieties are suitable for round the year cultivation in Kamataka.

A fertile inter-specific **spine gourd** hybrid *(Momordica subangulata)* and **teasel gourd** *(Momordica dioica)* with 30-45 g fruit weight, amenable for easy propagation through stem and root cuttings, improved fruit setting (90% due to natural pollination) extended harvesting period (April-November) and higher yield (5-7 kg/plant) was identified.

Vegetable	Variety	Salient features
Pea *(Pisum sativum)*	Kashi Ageti	Early-maturing (35-37 days for 50% flowering) and 12-12.5 tonnes/ha yield
Tomato *(Lycopersicon esculentum)*	Kashi Abhay	Semi-determinate hybrid, uniform maturity and moderately tolerant to ToLCV, 60-70 tonnes/ha yield
Chilli *(Capsicum annum)*	Kashi Tez	A CMS-based early-maturing, dual purpose hybrid, tolerant to anthracnose and thrips and 14-15 tonnes/ ha green fruit yield
Brinjal *(Solanum melongena)*	Kashi Uttam	Round-fruited, high-yielding (50-55 tonnes/ ha), tolerant to fruit- and shoot-borer and resistant to lodging
Pointed Gourd *(Trichosanthes dioica* L.)	Kashi Suphal	Fruits contain few seeds, fleshy, attractive with 18- 21.5 tonnes/ha yield
Pepo *(Cucurbita pepo)*	Kashi Shubhangi	Early-maturing, fruits are dark green, elongated, borne in 8-10 numbers/plant and yield 60-67.5 tonnes/ha
Satputia *(Luffa acutangula)*	Kashi Khushi	High-yielding (5.24-6.27 kg fruits/plant in 8-10 pickings), bearing -140 fruits/plant in clusters (5-6 fruits/cluster)
Ivy gourd *(Coccinia indica)*	Kashi Bharpoor	High-yielding (25-28 kg/plant) and bears oval, light green fruits (3,200-3,500 fruits/plant)
Kartoli *(Momordica cochinchinensis)*	Kashi Haritika	Early-maturing and high-yielding (30-35 tonnes/ha) variety
Kakrol *(Momordica dioica)*	Kashi Gautam	Early-maturity (75-85 DAS) and bears 45-63 fruits/ plant (2.5-4.5 kg/plant)
Indian Bean *(Dolichos lablab)*	Kashi Khushhaal	Semi-pole type, early-maturing. Pods 13.5 cm long, dark green, shiny, slightly curved with 4-5 coffee coloured seeds. First picking starts 107 days after sowing and yields 30-38 tonnes/ha
French Bean *(Phaseolus vulgaris)*	Kashi Sampann	Late-maturing, plants bushy, high-yielding (45-50 round, light green pods). It is tolerant to GYM V and high temperature (32 38°C) and yields 27-27.5 ton nest ha
Watermelon *(Citrullus lanatus)*	Kashi Pitamber	Early-maturing (76-81 DAS), fruits small (2.50-3.42 kg) with yellow rind from ovary stage and yields 8.3- 12.5 kg/plant

CARI Poi Red, a multi-cut poi (*Basella alba* L.) variety with higher yield (50--52 tonnes/ha), glossy red leaves, better adaptability to tropical island conditions and rich in anthocyanin was identified at Port Blair. It yields 15-22 tonnes/ha in first harvest.

Potato

Kufri Lalit, a potato hybrid, resistant to late blight with red-peel, yellow flesh, round, medium deep-eyed tubers and 25.5 tonnes/ha yield (15% increase over Kufri Lalima) was identified for cultivation in Indo-Gangetic plains and recommended for release.

Tuber crops

Two **triploid cassava hybrids**, viz. Sree Athulya (high yielding 39 tonnes/ha) with a high starch content (34.8%) and Sree Apoorva with higher yield (37.6 tonnes/ha) and 33.3% starch content were recommended for release and cultivation in Tamil Nadu and Andhra Pradesh and Tamil Nadu and Kerala, respectively.

Similarly, two **sweet potato** varieties, Swarna and Aparna, with 20-25 tonnes/ha tuber yield in 110-120 days duration were released by ICAR-CIARI, Port Blair.

SOURCE: DARE/ICAR ANNUAL REPORT 2014-15

In **tomato** (determinate), Kashi Aman, showed high level of resistance to both monopartite and bipartite viruses both under artificial and open field screenings. It yielded 50-60 tonnes/ha and was recommended for cultivation in Punjab, Uttar Pradesh, Bihar and Jharkhand. Six F_1 hybrids pyramided with ToLCV resistant genes (Ty1+Ty2+Ty3) were evaluated for triple disease resistance to ToLCV+BW+EB. H-329 (56tonnes/ha), H-367 (53tonnes/ha) and H-363 (48tonnes /ha) were high-yielding with triple disease resistance.

In **brinjal,** for the first time, bacterial wilt resistant lines of were selected and advanced to F_3 and F_4 generations. Progenies of two cross combinations, IIHR- 3 × IIHR-108-37-36-4-1 and IIHR-3 × IIHR-108-37- 36-1-3, were promising for yield. Fertility of sterile F_1 interspecific hybrid in a reciprocal cross of *Solanum macrocarpon* × *Solanum melongena* was restored by backcrossing with *Solanum macrocarpon*.

In **okra**, four lines (IIHR-296-22-10-11-598, IIHR- 291-14-11-585, IIHR-294-1-10-1-595 and IIHR-285-6- 10-11-138) resistant to Yellow Vein Mosaic Virus were identified under naturally hot spot condition at Attur, Tamil Nadu. Among inter-specific crosses, 100 % fruit setting was observed in *A. tuberculatus* × *A. esculentus* followed by *A. tetraphyllus* × *A. esculentus* (97.3 %).

In **carrot**, of the ten male sterile lines evaluated, MS 32-3-3, MS 32-2-2, MS 50-7 and MS 40-2, were 100% sterile and stable with good root yield and quality characters. Two maintainer lines, MF 43-7-7 (root length 13.67 cm and root weight 86g) and MF 38-1 (root length 11.00 cm and root weight 70g), were stable for fertility. Carrot lines, SH-C-11, SH-C-52, SH-C-51 and SH-C-141-1, were identified with yield potential of 396.66, 385.55, 382.22 and 380.00 q/ha, respectively and were superior to Nantes and Shalimar Carrot-1 in yield and beta carotene content.

In **garden pea**, Arka Apoorva, a dual purpose (whole pod and as salad) variety with crisp and sweet pods, 12 tonnes green pod /ha yield in 90 days, combined resistance to powdery mildew and rust was identified for release. Of six advanced breeding lines for high temperature tolerance, Arka Ajit × Arka Sampoorna- IPS-3BK recorded maximum pod yield (5.8tonnes/ ha), followed by 7-6 × KTP-4-IPS-12BK (5.6tonnes/ ha). In dwarf types, IIHR 18 × Oregon 1-2 gave maximum pod yield (5.6 tonnes/ha) in 90 days, whereas Arka Ajit and IIHR 544 recorded 4.4 and 2.35tonnes/ ha, respectively. In *kharif*, Arka Ajit × Arka Sampoorna-IPS-3BK gave maximum pod yield (7.5 tonnes/ha) followed by IIHR 7-6 × KTP 4-IPS12BK (7.0 tonnes/ha). In dwarf types, IIHR 18 × Oregon 1-2 recorded maximum pod yield (8 tonnes/ha) in 90 days, whereas in Arka Ajit and IIHR 544 it was 6.43 and 2.65 tonnes/ha, respectively.

In **French bean**, for the first time, an advanced breeding line (IC525260 × IC525283-07-1-6-5) resistant to Mung Bean Yellow Mosaic Virus (MYMV) was identified and successfully field demonstrated. It has yield potential of 17.5tonnes/ha and 42 pods /plant. French bean, IIHR PB-1, IIHR PB-2, IIHRPB-7 (pole types) and rust resistant (IIHR 31 and Arka Anoop) were crossed and evaluation of F1 progenies showed rust resistance to be a dominant trait.

In **watermelon**, line 42-174 recorded no incidence of water melon bud necrosis virus (WBNV) and produced round striped fruits with red pulp and 8%TSS. Arka Manik, Arka Muthu, Sugar Baby and IIHR-14 (yellow fleshed) varieties were treated with 0.2, 0.3 or 0.4% colchicines with or without PEG at cotyledon stage for six consecutive days which gave rise to 804 tetraploids. These were selfed and advanced to next generation.

In ***Momordica* species**, a seedless interspecific hybrid (*M. dioica* × *M. cochinchinensis*) was developed combining the desirable attributes of spine gourd and sweet gourd in addition to production of seedless fruits of bigger size (>20g) compared to normal spine gourd (<15g).

In **Onion**, two F1 hybrids (DOGR Hy-1 and DOGR Hy-2) suitable for *rabi* cultivation were developed. The bulbs of DOGR Hy-1 are light red and flat-globular, with early maturity and yield of 41.30 tonnes/ha, while bulbs of DOGR Hy-2 are dark red and globular, early in maturity and yield of 34.96tonnes/ha compared to Bhima Kiran (28.91 tonnes/ha).

Indian spinach (Basella), CARI Poi Selection, is a new variety of Poi or Indian spinach (*Basella alba* L.), was developed through mass selection from local germplasm. It has broad and glossy leaves, short internodes and high yield (55-60 tonnes/ha) compared to local types (35-38 tonnes/ha).

Tuber crops

Potato, Kufri Gaurav and Kufri Garima were notified for commercial cultivation. Nine advanced hybrids for early and medium maturity were introduced.

SOURCE: DARE/ICAR ANNUAL REPORT 2013-14

Improved **chilli** varieties (Arka Kyati), cauliflower (Arka Vimal and Arka Spoorthi), pole type dolichos (Arka Swagath), yard long bean (Arka Mangala) and garden peas (Arka Priya and Arka Pramodh) were identified for release. Yellow vein mosaic virus (YVMV) disease resistant lines in okra (IIHR 285-54-11-1, IIHR 296-22-4 and IIHR 299-52-5), powdery mildew resistant lines in ridge gourd (RGDMR 1 and RGDMR 2) and mungbean yellow mosaic virus resistant lines in French bean (IC 525283 × IC 525260)–09-1-5- 2) were further evaluated. Two varieties of chilli (Kashi Sinduri and Kashi Gaurav) and one hybrid of tomato (Kashi Abhimaan) were notified by the central varietal release committee.

Three **onion** varieties (Bhima Shubra, Bhima Shweta and Bhima Shakti) and one garlic (Bhima Purple) variety were released. Further, haploid induction in onion was tried through gynogenesis. Root tip squash from a few gynogenic regenerants were evaluated. In garlic, five collections of long day type having higher yield and quality were made. The clone, CITH-G 5, recorded highest plant height (109.75 cm) and number of leaves/ plant (8.25), while highest yield was recorded in CITH- G 2 (212.25 q/ha), closely followed by CITH-G 1 (202.99 q/ha).

Marker-assisted selection was carried out to identify pyramided lines in F6 and F7 families, segregating for three major genes against tomato leaf curl virus (TLCV). The marker assays included the CAPS marker for *Ty-1*and SCAR markers for *Ty-2* and *Ty-3* genes. A total of six lines carrying all three major genes in homozygous state were identified. Evaluation of these lines for agronomic and morphological traits was completed in two seasons. Pyramided line, D6-1-4, was observed superior to other lines in resisting TLCV infection.

To develop moisture-stress tolerant **tomato**, '*BcZAT12*' gene, (Accession no. DQ166621), driven by drought- specific promoter '*lea1*' cloned in *Agrobacterium tumefaciens* strain GV 3101, was used for transformation. A total 16 lines from 10 independent events were obtained and rooted on 100 mg/litre rooting medium. These lines were tested for gene integration by PCR and southern blot using P32 labelled *BcZAT12* probes. Physiological and biochemical data of these events revealed significant moisture tolerance.

SOURCE: DARE/ICAR ANNUAL REPORT 2012–13

In **tomato**, two triple resistant F1 hybrids (Arka Samrat and Arka Rakshak), bacterial blight, ToLCV, early blight with a yield potential of 101-119 tonnes/ ha were identified for release.

In **onion**, a tri-parental derived synthetic variety, Arka Bheem, with red to pinkish red elongated globe-shaped bulbs of 120 g weight and yield potential of 47 tonnes/ha with 130 days maturity has been identified. A white onion variety, Arka Swadista, suitable for fermented preservation with uniform white oval globe bulbs (weight 35-40 g) and TSS of 18-20 % and yield potential of 16-18 tonnes/ha in 105 days maturity have been identified. Bhima Shakti and Sel. 126 (IARI) were identified for release in Zone III (Delhi, Uttar Pradesh, Haryana, Bihar and Punjab), Zone IV (Rajasthan and Gujarat), Zone V (Madhya Pradesh, Chhattisgarh and Odisha) and Zone VI (Maharashtra, Karnataka and Andhra Pradesh) and Bhima Shweta for release in Zone III, V and VI. In garlic, temperate garlic VGP 5 (VPKAS, Almora) is identified for release in Zone I (Jammu and Kashmir, Himachal Pradesh and Uttarakhand), Bhima Purple (DOGR) was identified for release in Zone III, and VI and G 189 (NHRDF) was identified for release in Zone III, IV and VI.

Arka Sharath, a bushy **French bean** variety with smooth pods suitable for steamed beans with a pod yield of 18.5 tonnes/ha in 70 days was identified. A **cowpea** variety, Kashi Unnati, was identified and recommended for agroclimatic Zone IV. This variety is dwarf and bush type, photoperiod-insensitive, early- maturing and suitable for sowing in both spring-summer and rainy seasons. It flowers in 40-45 days and pods become ready for harvesting in 50-55 days. The pods are green, cylindrical, pulpy and free from parchment. The variety is resistant to golden mosaic virus and has a yield potential of 150 q/ha. **Okra** Kashi Kranti, has been identified and recommended for Zone IV. It takes 40-42days for 50% flowering; fruits 8-10 cm long, 10-12 g in weight, dark green with five ridges. The variety has a yield potential of 90 q/ha. Kashi Gaurav **chilli** variety has been identified and recommended for Zone II. The plants are bushy, tolerant to thrips and mites, dark green foliage, good combiner, 50% flowering at 35-40 days after transplanting, dark green fruits when young and dark red when ripe, 9- 11 cm long, 1.1-1.2 cm thick, pendant and pungent with a yield (red ripe) potential of 110 q/ha.

In *lablab* bean, a bush vegetable poded variety, Arka Soumya, with a pod yield potential of 19 tonnes/ ha in 90 days has been identified. A new triploid seedless **watermelon** variety, Arka Madhura, having a yield potential of 60 tonnes/ha with TSS of 14% and suitable for protected cultivation has been identified. An early **cauliflower** variety, Arka Spoorthi, with good quality compact curds, weighing 332 g ready for harvesting in54-56 days with a yield potential of 166 q/ha and moderately resistance to alternaria and downy mildew has been identified for release.

Potato and tuber crops

The advanced evaluation trial conducted with cassava mosaic disease (CMD) resistant **cassava** hybrids revealed that seven lines, CMR 1, 70, 73, 106, 109,120 and 129, showed stability in yield (35-40 tonnes/ha) and high starch content (25-30%). In cassava, clone CPT 32 recorded highest (6.1%) true protein content, followed by CPT 13 (5.9%) and CPS 30 (5.2%) on dry-weight basis as compared to 1.9% in M4 and 2.2% in Sree Padmanabha.

The white-fleshed clones of **sweet potato**, IGSP 22 and IGSP 10-6, were identified as the best with respect to tuber yield of 21-30 and 18-27 tonnes/ha respectively. In **yams**, accession Da 11 recorded highest true protein content of 13.3 and 4.0% on dry- and fresh-weight basis respectively. Da 11 and Da 68 also exhibited field tolerance to anthracnose disease. In **aroids**, seven advanced hybrid selections of *Amorphophallus* having stability in yield (38-42.5 tonnes/ha), good cooking quality and free from diseases were identified. In addition, four early harvestable *Amorphophallus* hybrids with excellent cooking quality, namely Am H 1, Am H 1(b), Am H 5 and Am H 102 suitable for 7th month harvest, have also been identified. Thirty hybrids of **taro** including a novel dwarf type with a mean yield of 12.5 tonnes/ha and good cooking quality were isolated from hybrid progeny.

In **potato**, clone YY6/3 C-II has been registered as an elite genotype having PVY extreme resistant gene in triplex state. The transgenic potatoes with durable resistance to late blight and suitable for cold chipping have been developed.

SOURCE: DARE/ICAR ANNUAL REPORT 2011–12

Arka Samrat and Rakshak in **tomato**; Arka Muthu, Arka Akash and Arka Aishwarya in **watermelon**; Arka Sharath in **French bean**; Arka Neelachal Sabuja and Arka Neelachal Kunkhi in **ivy gourd**; Arka Neelachal Kirti in **pointed gourd**; Arka Neelachal Shree in **spine gourd**; Arka Neelachal Gaurav in **teasel gourd** and Arka Dhanvantri and Arka Aswini in **velvet bean** were identified for release. **Chilli** high-yielding F1 hybrid Arka Harita, tolerant to powdery mildew and chilli veinal mottle potyvirus, and high-yielding **carrot** variety, Arka Suraj, were recommended for release in Southern Karnataka. Kashi Harittima, a variety of **pole bean**, was recommended for release. In **Dolicos**, Arka Soumya, Arka Sambhran and Arka Amogh; in **Amaranth**, Arka Samraksha and Arka Varuna were recommended for release. Multiple disease resistant (LCV, bacterial wilt, early blight, root-knot nematode) line of tomato was found promising. In chilli and Capsicum, powdery mildew resistant line and in brinjal bacterial wilt resistant line were found promising.

In **onion**, nine new red and white type improved varieties, Bhima Kiran, Bhima Shakti, Bhima Shweta, Bhima Shubra, Arka Ujjwal, Arka Swadista, Arka Vishwas, Arka Sona and Syn 6 for different seasons with better quality and high yield potential were identified for release. Onion varicty, RO 597, was identified for release in Zones III and VI, while Line 355 was identified for release in Zones III, IV and VI. In **muskmelon**, 60 lines, **bottle gourd** (20 lines) and **chilli** (45 lines) were evaluated for morphological and yield parameters. Efforts are being made to incorporate heat tolerance and virus resistant genes from *Citrullus colocynthis* into *C. lanatus.* Four varieties, Thar Mahi (sword bean), Thar Kartiki and Thar Maghi (Indian bean) and Thar Bhadavai (cluster bean) were released for arid region.

The varieties LS 2 and Iceberg of **lettuce**, SH-K-28,SH-K 30, CITH-K 6 and SH-K 29 of **kale**, CITH-P-4 of **parsley,** Ford Hook of **celery**, CITH-CC 3 of **Chinese cabbage**; Green Head and Palam Samridhi of **broccoli**; CITH Green of **artichoke** and CITH-A 1 of **Asparagus** were found promising for high yield and net returns.

In **tomato**, 16 transgenic events for resistance to tospo virus (PBNV) were generated in four genetic backgrounds, 8 events for resistance to TLCV. Twenty events in four genetic backgrounds for resistance to TLCV and tospo virus were approved for trial by RCGM. In **watermelon**, four events for resistance to bud necrosis were also approved for field trials. Two transgenic events with chitinase gene in tomato, two transgenic events with cry gene in tomato and brinjal and two transgenic events with Dreb 2A were advanced to T3.

Suitable tomato varieties/hybrids having resistance to bacterial wilt, were developed for protected condition by making eight F1 crosses using Swarna Lalima having resistance to bacterial wilt as female parent and eight exotic lines as male parent. One promising F1 hybrid Swarna Lalima EC 596741 (yield 187.79 tonnes/ha, resistant to bacterial wilt, fruit weight 65 g, TSS 5.2 °B, titratable acidity 0.3%) was identified. Harvesting period ranged from 180 to 200 days.

Potato and tuber crops

In **potato**, *Solanum verrucosum* was identified to have high yield and resistance to late blight. The hybrid, JX 576, was released as Kufri Gaurav for north-western plains. The variety possess higher N, P and K efficiency, thereby enabling lesser use of N, P, K fertilizer and restoration of soil health. Similarly, processing hybrid, MP/01 916, (released as Kufri Chipsona 4) was found promising for Karnataka, West Bengal and Madhya Pradesh. Hybrid, MS/4-2261 has 15.41% dry matter with good keeping quality and field resistance to late blight. LBY 15 and LBY 17 having combined resistances to late blight and PVY were selected for multilocational trials. Hybrid, OS/01-516, resistant to cyst nematode and late blight out yielded Kufri Giriraj at Ootacamund. Sequencing of chromosome 2 covered 2.4 Mbp sequence with 877 large contigs. Transgenic lines of potato with durable late blight resistance using RB gene were selected for further evaluation. A cold chipping line developed using RNAi gene was further evaluated and found highly promising. Putative somatic hybrids developed earlier were confirmed using molecular and phenotypic markers.

Sixteen cassava landraces with very good chip making quality were identified, of which, Ci-24 had low cyanogens and was free from symptoms of cassava mosaic disease. Eighty-seven accessions were evaluated for drought tolerance, of which 34 were found tolerant.

In **sweet potato**, 10 orange-fleshed clones showed total carotenoids of 11–14 mg/100g fresh weight and β-carotene of 9–13 mg/100g fresh weight and the dry- matter content was 21–25%. Clones, CIP-440127 and Samrat, were promising for salt tolerance at Erasama in Orissa and Canning in West Bengal, respectively. The Single Marker Analysis (SMA) of DNA isolated from 250 hybrid sweet potato seedlings using 5 ISSR primers resulted in identification of one marker ISSR Y818 3/1 associated with β-carotene. Six accessions of *D. alata,* viz. Da 10, Da 11, Da 60, Da 62, Da 64 and Da 157, were found to be tolerant to anthracnose disease from screening of field gene bank of greater yam. The *Amorphophallus* hybrid, Am H 102, gave significantly higher yield of 43.8 tonnes/ha with resistance to collar rot and in taro, hybrid Co H 4-3 with high yield (22.2 tonnes/ha) have been identified for on-farm trials.

A **cassava** transformation protocol was optimized. Cassava transformation was done with three hairpin constructs of Sri Lankan cassava mosaic virus (SLCMV)/ Indian cassava mosaic virus (ICMV) in cassava cultivar H 226. Transformation efficiency was0.6–1%. Full genome of SPLCV (2.8 kb) was amplified through rolling circle amplification and amplified product (2.8kb) has been cloned in pUC18 vector.

SOURCE: DARE/ICAR ANNUAL REPORT 2010–11

Onion: Bhima Red variety has been developed through bulb to row selection method from a base population of B-780. After transplanting it takes 120-140days for harvesting. The bulbs possess attractive red colour, round shape with higher marketable bulbs. This variety yields around 30 tonnes/ha in *rabi* season. This variety is also suitable for late *kharif* season with a yield potential of 50 tonnes/ha.

Garlic: Bhima Omkar line has been identified for high yield and better quality. It has been developed through clonal selection. Its bulbs are medium in size, compact and white in colour. Average marketable yield is 7.7 tonnes/ha with an average of 18-31 cloves per bulb.

Potato: Two hybrids, Kufri Nilima and Kufri Frysona, were recommended for release as varieties.

Kufri Nilima is a medium-maturing hybrid, resistant to late blight and cyst nematodes. It is suitable for cultivation in Nilgiris hills of Tamil Nadu.

Kufri Frysona produces oblong tubers of attractive shape with shallow eyes and white flesh colour. Its tubers are free from most of the external and internal defects. The average total tuber yield is 39.8 tonnes/ha with an average French fry grade tuber yield of 25.8 tonnes/ha. The tubers contain on an average 22%dry-matter content and have very low reducing sugar content (< 100 mg/100 g fresh tuber weight) and negligible enzymatic browning. The variety is suitable for planting in the main season in Indo-Gangetic plains and matures 100–110 days after planting.

SOURCE: DARE/ICAR ANNUAL REPORT 2009–10

Recent Initiatives by Central Potato Research Institute, Shimla

Potato Genome Sequencing

The complex genome of potato has been deciphered by a consortium of 26 international institutes belonging to 14 countries. CPRI from India was the proud partner of this consortium.

A hybrid approach including Sanger, Illumina, and 454 technologies was adopted for sequencing and assembly of 727 Mb which constitutes about 86% of the complete potato genome. A total of 39,031 protein-coding genes have been predicted in the sequence out of which about 800 genes encode disease-resistance related proteins (Xun et al. 2011). The sequence information will lead to identification of candidate genes and their markers which will be utilised for improving agronomic traits of potato, such as improved quality, yield, heat tolerance and disease resistance. It will also reveal new insights into the potential mechanisms of tuber initiation and development, host-parasite interaction, abiotic stress tolerance mechanism etc.

Potato Germplasm Repository

(a) Added 481 accessions of cultivated potatoes and 271 accessions of 47 wild species in the national active germplasm repository from Europe and America

Accessions were evaluated for agronomic traits, abiotic and biotic stresses (adaptability, late blight, bacterial wilt, stem necrosis, viruses (PVX, PVY, PVS, PVA, PVM, PLRV, PALCV), leaf hopper and mites, potato tuber moth, cyst nematodes, keeping quality, chipping, vitamin C, heat, water and nutrient use efficiency and promising accessions for each trait were identified/documented. Till date, about 50 potato varieties have been released by CPRI using the germplasm lines available in its repository (Annual reports (2006-10); Kumar *et al.*, 2005; Kumar *et al.*, 2008).

(b) Developed and characterized andigena core collection of 78 accessions out of 740 accessions using phenotypic and molecular markers (SSR)

The core collection of Andigena germplasm was developed by recording data on 21 traits of 740 accessions. The data was analyzed by Power Core software and 78 accessions representing total diversity was constituted as core collection. In subsequent year the data were analyzed by SPSS for cluster analysis. Based on three sampling strategies within cluster the optimum core set of 78 accessions was constituted. Power core was found to be the best method for core set analysis (Annual Report 2010-11).

Protoplast isolation, fusion and regeneration of somatic hybrid

Gene prospecting through somatic hybridization

Developed two interspecific somatic hybrids of potato *Solanum tuberosum* dihaploid C-13 (+) *S. etuberosum*, and C-13 (+) *S. pinnatisectum* resistant to Potato Virus Y and late blight through protoplast fusion to overcome the sexual barriers imposed by difference in ploidy and endosperm balance number (EBN).

Phenotypic characters of a somatic hybrid and its parents Many novel genes for late blight resistance, virus resistance, high dry matter content and other useful traits are available in diploid wild potato species like *Solanum pinnatisectum, S. etuberosum, S. cardiphyllum, S. acaule, S. brachistotrichum, S. jamesii, S. polyadenium, S. stoloniferum* etc. that are sexually incompatible with cultivated potato. To overcome this sexual barrier, somatic hybrids of those wild species with cultivated potato have been developed and characterized. The derived tetraploid genotypes can be exploited as pre-breeding material for introgression breeding (Sarkar *et al.* 2011; Tiwari *et al.* 2010).

Marker assisted pre-breeding

A unique parental line having extreme resistance gene to PVY (Ryadg) in triplex state developed through marker assisted pre-breeding which has been registered as INGR 10143 by the Plant Germplasm Registration Committee of ICAR.

Severe inbreeding depression and tetraploid nature of cultivated potato, preclude development of homozygous parental line. Therefore, marker assisted selection (MAS) was used to develop a unique parental line having PVY extreme resistance gene (Ryadg) in triplex state. This clone can be used as a parent in breeding programme for PVY resistance, since almost all the progeny derived from any cross with this parent will have at least one *Ryadg* allele conferring extreme resistance to PVY (Annual report 2010-11, pp. 45).

Phenotypic characters of a somatic hybrid and its parents

Segregation of SCAR marker in the progeny of test cross, QB/A 9-120 x Triplex clone.

Gene Cloning

Fourteen genes/promoters possessing important traits were cloned for use in genetic transformation studies without any IPR obligation. The following genes and genetic elements have been cloned in the institute and are being used for genetic transformation work:

S. No.	Gene/Promoter	Source organism/Variety	Trait
1.	*cry1Ab*/GBSS	*Bacillus thuringiensis* (Synthetic)	Tolerance to potato tuber moth
2.	*cry9Aa2*/CaMV35S	*Bacillus thuringiensis* subsp. *galleriae*	Tolerance to potato tuber moth
3.	*osmotin*/CaMV35S	*Solanum chacoense*	Tolerance to late blight and salinity
4.	*CP*/CaMV35S	Potato leaf roll virus	Resistance to potato leaf roll virus

S. No.	Gene/Promoter	Source organism/Variety	Trait
5.	PTGS construct of *MP*/CaMV35S	Potato stem necrosis virus	Resistance to potato stem necrosis virus
6.	glgCm/CaMV35S	*E. coli*	Increase in starch quantity
7.	*AmA1*/CaMV35S & *AmA1*/GBSS	*Amaranthus hypochondriacus*	Improvement of protein quality and quantity
8.	*inhh*/CaMV35S & *inhh*/GBSS	Tobacco	Reduction of cold-induced sweetening
9.	RNAi construct of vacuolar *invertase*/ CaMV35S	Potato	Reduction of cold-induced sweetening
10.	RNAi construct of *GBSS*/CaMV35S	Potato	Improvement of starch quality
11.	RNAi construct of *GA20oxi*/CaMV35S	Potato	Dwarf potato
12.	RNAi construct of rep gene	Potato apical leaf curl virus	Resistance to potato apical leaf curl virus
13.	Plastid transformation vector (pSKC21)	Tobacco/ *Bacillus thuringiensis* subsp. *galleriae*	Resistance to tuber moth
14.	Gene cassettes with fused *cry1Ab+cry1B* genes	*Bacillus thuringiensis*	Resistance to tuber moth

Transgenics for tailor-made variety development

Developed seven different events of transgenic potatoes with important agronomic traits viz., late blight durable resistance, reduction of cold induced sweetening, high protein content, resistance to Potato Virus Y, Potato Apical Leaf Curl Virus, Potato Tuber Moth, and altered plant architecture. Though the above traits are agronomically desirable, generally they cannot be manipulated by conventional breeding. Genetic engineering technique was, therefore, exploited for developing designer varieties with the following specific traits.

Target trait	Gene/Construct (source)/IPR issues	Status
Late blight resistance	*RB* of *Solanum bulbocastanum*/ MTA with ABSPII, Cornell University	Promising transgenic genotypes expressing the RB gene and conferring field resistance to late blight have been identified for hills and plains. BRL-1 trial with promising lines in progress (Shandil *et al.* 2008).
Reduction of cold induced sweetening	Invertase inhibitor gene from tobacco (*ntinhh*)/Own construct	One promising transgenic line of Kufri Chipsona-1 with excellent cold-chipping attributes and no yield reduction has been identified. BRL-1 trial with promising lines in progress.

Target trait	Gene/Construct (source)/IPR issues	Status
Increase in protein content	*AmA1* gene of *Amaranthus hypochondriacus*/ Patented gene of NIPGR, New Delhi	One transgenic line each of Kufri Chipsona 2 and Kufri Badshah with higher protein content and improved protein quality have been identified.
Dwarf architecture	RNAi for *GA20-oxidase1* gene/Own construct	Positive transgenic lines identified; glass house evaluation showed reduction of plant height in 20 transgenic lines.
Resistance to PVY	CP gene/Own construct	Positive transgenic lines identified by molecular analysis; glass house evaluation showed very good level of PVY in CP sense transgenics .
Resistance to Potato Apical Leaf Curl Virus	Rep gene/Own construct	Positive transgenic lines of Kufri Pukhraj identified by molecular analysis; glass house evaluation showed very high level of PALCV resistance in two transgenic lines (Tomar *et al.* 2011).
Resistance to potato tuber moth	Synthetic *cry1Ab* gene (ICAR property) driven by GBSS promoter	Positive transgenic lines identified; glass house evaluation showed very good level of resistance to potato tuber moth in one transgenic line of Kufri Badshah (Kumar *et al.* 2009).

Development of meiotic tetraploid (MTs) for introgression of late blight resistance

Developed 92 meiotic tetraploids using unilateral sexual polyploidization (USP) method involving three wild species (*S. chacoense, S. microdontum* and *S. verrucosum*). Two of these meiotic tetraploids are in advanced breeding trials (F1C6) besides using them as parental material in late blight breeding programme.

Development of multiple disease resistant potato hybrids

Developed two advanced hybrids (LBY-15 & LBY-17) having combined resistance to *late blight and Potato Virus Y using* marker assisted selection approach. These are being evaluated in multilocational trials under AICRP (Potato).

Two indigenous, highly late blight resistant genotypes, *viz*., Kufri Girdhari and QB/A9-120 and germplasm accessions YY-3, YY-6 and YY-13 having PVY extreme resistance gene in duplex condition were used as parental lines. The advanced clones were selected on the basis of SCAR marker and immunological response besides performance under field conditions (Bhardwaj *et al*., 2007; Kaushik *et al*., 2008; Annual report 2010).

Gene stacking for multiple disease resistance

Identified seven potato genotypes having multiple resistance genes for Late Blight (*R1* & *R3*), Potato Virus Y (*Ryadg*) & Cyst Nematodes (*HC*, *H1* & *Gro1*) using molecular markers.

Potato genotypes possessing late blight resistance genes, R1 and R3a were characterized through molecular markers. Based on the MAS results parental lines possessing these genes were hybridized and genotypes with combined (R1 and R3a) genes in single host background were developed. Further, the potato genetic stocks possessing resistance to late blight (*R1*, *R2*, *R3*, *Cos A-29 & 30* and *R8*), Potato Virus Y (*Ryadg*) and Golden cyst nematode (*HC*, *H1* and *Gro1*) have been identified through molecular markers for developing potato varieties with combined resistance to all the three diseases (Bhardwaj *et al*., 2011).

Development of heat tolerant genotypes

Identified four heat tolerant clones with good agronomic traits. These have been introduced into AICRP for multi-location testing.

Based on superior performance under early planting conditions (heat stress) at Ladol in Gujarat, four CIP advance clones namely CP4054, CP4184, CP4197 and CP4206 were identified and introduced in AICRP during 2011 for multi-location testing (Singh *et al.,*2010).

SOURCE: CENTRAL POTATO RESEARCH INSTITUTE, SHIMLA

Recent Initiatives by Indian Agricultural Research Institute, New Delhi

Cole Crops

Cauliflower

Ninety-five SI hybrids of early group cauliflower were evaluated. Hybrid 351 aa × Hyb 85 recorded highest marketable curd (1140 g) as well as net curd (880 g) weight. Out of 14 CMS hybrids, 8498 × Pusa Deepali recorded highest marketable curd and net curd (1206.7 and 950 g, respectively) weights. In mid- group, out of 36 SI hybrids CC 32 × FC 278 recorded highest marketable curd (1850 g) and net curd (1510 g) weights. Out of 136 CMS hybrids evaluated, Hybrid 8401 × 310- 22 recorded highest marketable curd and net curd (1972 and 1567 g) weights, followed by DB 15 × 340 (1950 and 1660 g), respectively. In snowball group, out of 18 F1 hybrids evaluated, highest marketable yield was recorded in KTH 122 (75.55 t/ha).

Under resistance breeding programme, a total of 263 RILs of various generations (F2:3, F2:4, F2:5, F2:6) were evaluated for downy mildew resistance. Seven and 10 plants with highly resistant category (0 score) were obtained in F2:6 generation of Pusa Himjyoti × BR 2 combination and F2:5 generation of BR-36 (Layna × 522), respectively. A total of 152 NILs of various generations (F2:3, F2:4) were phenotyped for resistance to downy mildew disease. Twenty six and 20 plants with highly resistance category (0 score) were obtained in F 2:4 generation of 309 × BR 2 × 309 and F2:3 of 3-5-1-1 × 466 × 3-5-1-1, respectively. Twenty one inbred lines were assessed for resistance to black rot and downy mildew diseases. Genotypes BR 202-2, BR 161, BR 207, AL 3 and AL 15 were found to be resistant against downy mildew and AL 15, BR 161 and BR 1 against black rot disease.

Cabbage

- **Variety released:** At IARI Regional Station, Katrain one self-incompatibility system based F1 hybrid, KGMR 1 was released as Pusa Cabbage Hybrid 1 by the Central Sub-Committee on Crop Standards, Notification and Release of Varieties for Horticultural Crops for Agroclimatic Zones-I & IV. This is an early maturing (55-60 days) hybrid with round and compact head and yielding 35-40 t/ha. The hybrid has shown tolerance to black rot with better field staying capacity after maturity.
- **Promising genotypes:** Two SI based cabbage hybrids KTCBH 51 and KTCBH 81 were evaluated in the AVT-II under the AICRP (VC) trials. One CMS based cabbage hybrid evaluated in IET gave superior performance with a yield of 65.2 t/ha at IARI Regional Station, Katrain. A flat head hybrid H 208-122 (62.8 t/ ha) was found promising followed by H 645-111 (55.8 t/ha) and H 208-5 (52.7 t/ha). It was observed that hybrid S 645 × S 691 had maximum heterosis for yield (46.7%) followed by S 645 × S 696 (45.5%) and S 624 × S 645 (37.2%). Out of 20 CMS based hybrids, Hybrid 991- 5 was found promising with 69.8 t/ha yield followed by hybrids 854-6 (67.1 t/ha) and 871-111 (65.5 t/ha).In red cabbage, out of 6 hybrids developed by using 3 CMS lines, hybrids 931 × KR and 931 × RCGA were found promising with 29.2 and 22.2 t/ha yields, respectively.

Cucurbitaceous Crops

Bitter Gourd

- **Genotype registered:** One predominately gynoecious line PreGy 1 which showed high female: male ratio (5:1 to 7:1) with a yield of 27.5 t/ha as against 17.8 t/ha in Pusa Do Mausami (check) was registered at NBPGR as INGR 12014.
- **Promising genotypes:** Two entries (DBGS 37 and DBGS 57) were promoted to AVT-II under AICRP-VC trial. One variety DBGS 102 and two hybrids DBGH 12 (gyn × mon) and DBGH 263 (mon × mon) were promoted to AVT I under coordinated trials. One sequential fruited line DBGS 201 (Sel 1) produced 30.6 t/ha fruits as compared to 17.5 t/ha in Pusa Do Mausami (check).Ten gynoecious based and one monoecious based hybrid were evaluated and compared with hybrids of private seed companies. Among these hybrids, DBGy 201 × S 2 (Gy× Mon) gave maximum yield (23.8 t/ha) followed by hybrid S 2 × S 63 (Mon × Mon) (21.3 t/ha).

Four varieties were evaluated under insect proof net house in *kharif* season (July to November), the variety DBGS 54 yielded 474 kg fruits in 100 square meter area. Among the hybrids evaluated under low cost polyhouse during winter season (December-April) the hybrid DBGy 201 × DBGS 54 produced 418.50 kg fruits in 100 square meter area.

Cucumber

- **Variety identified:** First extra early improved variety of cucumber Pusa Barkha was identified for *kharif* season cultivation in North Indian plains. It has field tolerance to high humidity, high temperature and downy mildew disease with an average fruit yield of 18.8 t/ha.

- **Promising genotypes:** Cucumber selections DC 54 and DC 78 and F1 hybrid DCH 6 were advanced to AVT-II of AICRP (VC). During spring summer season selections DC 54 and DC 78 yielded 18.5 and 17.9 t/ha, respectively. Tropical gynoecious lines DGC 3, DGC 10 and DGC 102 showed stable performance at higher temperature under net house. Out of 30 F1 hybrids evaluated, DCH 6 and DCH 9 yielded 21.1 and 20.5 t/ ha, showed an increase of 20.6% and 17.1% higher yield than the check Pant Sankar Khira 1 (17.5 t/ha), respectively. Under *kharif* trial, selections DC 70 (PDI=7.2%) and DC 77 (PDI= 8.6%) showed high resistance to downy mildew disease. The F1 hybrids DCH 16 (21.6 t/ha) and DCH 19 (21.2 t/ha) also showed high yield and resistance to downy mildew disease. The parthenocarpic lines Sel. DPaC 6, DPaC 9 and DPaC 10, predominantly gynoecious line DC 128 and gherkin lines DG 5, DG 8 and DG 11 were observed highly promising for yield and resistance to downy mildew under polyhouse during winter.

Luffa

- **Genotype registered:** Luffa (sponge gourd) genotype DSG 6 (IC-0588956; INGR 12013), resistant to *tomato leaf curl New Delhi virus* was registered at NBPGR.
- **Promising genotypes:** Selections DSG 43 (15.4 t/ ha) and DSG 48 (14.7 t/ha) and F1 hybrids DSGH 3 (16.8 t/ha) and DSGH 9 (16.2 t/ha) were found to be highly promising for yield. DSG 48, DSG 104 and F1 hybrids DSGH 3 and DSGH 9 are advanced to AVT-II of AICRP (VC). In ridge gourd, DRG 74 and F1 hybrid DRGH 4 were found promising with an average yield of 17.5 t/ ha and 18.9 t/ ha, respectively. A gynoecious genetic stock was developed which segregates into gynoecious and hermaphrodite (*Satputia*) in the ratio of 1:1 and maintained by utilizing Satputia as pollen parent. Ridge gourd Sel. DRG 7 was advanced to AVT-II of AICRP (VC).

Pumpkin

- Selections DPU 48, DPU 12 and DPU 37 yielded 35.33, 33.14 and 30.60 t/ha and showed an increase of 35.88%, 27.46% and 17.69 % over the check Pusa Vishwas (26.00 t/ha), respectively.

Muskmelon and watermelon

- The muskmelon genotype DM 148 was found to be highest yielder (22.7 t/ha) with a yield advantage of 18% over Pusa Madhuras along with good quality (TSS 11.6). The other superior lines with respect to yield were DM 151(21.6 t/ha) and DM 154 (20.8t/ha) with desirable fruit quality attributes. Among F1 hybrid combination, DMH 3, DMH 5 and DMH 12 were found promising for fruit yield, early maturity and quality traits. The F1s between muskmelon lines (Hara Madhu and Pusa Madhuras) and snap melon lines (DSM 11 and DSM 12) were found to have moderate resistance to *Fusarium* wilt suggesting dominant nature of inheritance. Icebox type watermelon lines, PWMH 25-4-3 (7.8 mg/100 g) and IPWM 34-1-2(7.2 mg/100 g), containing high lycopene were selected.

Minor Cucurbits

- In summer squash, selection DS 8 (15.14 t/ha) was found to be the most promising and the yield advantage was 31.42 % higher than that of the local check (11.52 t/ ha). In long melon, selections DLM 27 and DLM 34 yielded 26.10 and 23.94 t/ha, respectively. In round melon, selections DRM 26 and DRM 44 yielded 6.73 and 6.59 t/ha, respectively.

Solanaceous Crops

Brinjal

- The selections DBR 190 (green round fruited) and DBSR 195 were found promising with a yield of 43.8 and 36.2 t/ha, respectively. Among 72 F1 hybrids evaluated, DBHR 9 (purple round), DBHR 38 (purple round) and DBHL 155 (white oblong) were found promising with a yield of 54.4, 52.8 and 51.7 t/ha, respectively. Lines DBR 569 and BL 1 were found tolerant to phomopsis blight.

Tomato

- DT 7 (determinate type) was promoted to AVT-II of AICRP (VC). During *kharif* season, selections SPS 2, SPS 3, SPS 5, BAR 13, H 88-2, TH 348-2-11, H 88-1-2 and hybrid combinations TH 348 × H 88, H 86 × H-88 and H 86 × H 24 were found resistant to ToLCV. Out of 96 genotypes, F1 crosses PSH × PS, 120 × PS and 3900×PS recorded fruit setting at high day and night temperature and yielded 21, 20, and19.5 t/ha, respectively. Selection Lp 2 and Pusa Sadabahar were found tolerant to low temperature. In cherry tomato three red fruited and one orange fruited lines were found promising.

Chilli

- Out of 74 genotypes DCH 10-05 and DCH 11-08 were found promising with a yield of 18.84 t/ha and 17.71 t/ha, respectively. Two genotypes, DCS-LCV-P1 and KA-LCV-P2, for leaf curl tolerance and one,VNR- 314-F4-P3, for high temperature tolerance were identified.

Capsicum

- At IARI Regional Station, Katrain, thirteen germplasm lines were found promising and genotypes KT 5 (29.1 t/ha), Banglore 42 (28.0 t/ha) and Banglore 38 (25.9 t/ha) were found promising having medium sized green fruits. In paprika, genotype DP 2 ranked first in fruit yield/ha. Out of twenty five F1 cross combinations evaluated, three, namely, Sel-12-2 × Palam Bell (45.0 t/ha), Yolo Wonder × SolanBharpoor (39.3 t/ha) and Yellow Capsicum × Yolo Wonder (37.8t/ ha) were found to surpass the standard checks Nishat and Pusa Deepti for marketable fruit yield and other traits.

Root and Bulbous Crops

Tropical Carrot

- **Variety identified:** Pusa Vasuda is the first public sector tropical carrot hybrid developed using CMS system. Its roots are smooth, attractive, vigorous, self-coloured, red, sweet, juicy, rich in total carotenoids, lycopene, TSS and minerals. It is suitable for *salad*, juice extraction, cooking and carotenoid extraction industry and gives an average yield of 40 t/ha.
- **Promising lines and hybrids:** Under hot and humid condition, 27 genotypes were evaluated by sowing in the first week August. Five genotypes were found promising having normal desirable roots with least forking and an average root weight ranging from 113.33 g (IPC 10) to 180 g (IPC 31 P-1). Nine hybrids were of high quality for root shape, surface, external and internal colour and external appearance, and root weight varied from 86.7 to 240.0 g. Under October sowing conditions, maximum root length was observed in IPC 104 orange (27.87 cm) followed by IPC 100 (27.33 cm), root diameter was highest in IPC 55 Red (4.88 cm) followed by IPC 104 orange (4.79 cm), and root weight was maximum in IPC 104 orange (253.3 g) followed by PM Dark orange (220 g). Among 87 CMS lines, highest root length was noted in IPC 126 P-8 (31 cm) followed by IPC 55-1-2 P-2 (29.6 cm), root diameter was maximum in IPC 126-16-1 P-1 (5.70 cm) followed by IPC 55-1-2 P-2 (5.54 cm), and root weight was highest in IPC 126-16-1 P-1 and IPC 98a BP P-5 (300 g).

Temperate Carrot

- CMS based hybrids, KTCH 1020 (49.5 t/ha), KTCH 1022 (44.1 t/ha) and KTCH 96-113 (43.7 t/ha) superseded the check hybrid Pusa Nayanjyoti (42.5 t/ ha) with 16.5%, 3.8% and 2.8% increase in yield, respectively.

Radish

- **Pusa Jamuni:** It is a purple fleshed nutritionally rich radish variety. It has distinct advantage in root size, shape, yield and consumer preference over the existing varieties. It contains high anthocyanins and ascorbic acid with an average yield 50 t/ha.
- **Pusa Gulabi:** It is a pink fleshed nutritionally rich radish variety having medium root size, cylindrical shape, optimal yield and consumer preference over the existing varieties. It contains high carotenoids, anthocyanins and optimal ascorbic acid and gives an average yield of 60 t/ha.

Onion

- **Pusa Riddhi:** Bulbs of this variety are compact in nature, flat globe in shape, and dark red in colour. The equatorial diameter of bulbs ranges from 4.5 to 6.0 cm, polar diameter ranges from 4.8 to 6.3 cm and single bulb weight ranges from 70 to 100 g. It is pungent and rich in antioxidant (quercetin 107.42 mg/100 g). The variety is suitable for *kharif* and *rabi* seasons. The variety is also suitable for storage and export with an average yield of 31.66 t/ha, an average increase of 28.60% over the check (28.53 t/ha).

- **Pusa Soumya:** This is the first bunching onion variety proposed for commercial cultivation in India. This multi-cut variety is suitable for green onion production round the year. It produces bluish green leaves and clumps can be separated for further multiplication. The variety is least affected by pests and diseases and has an average yield potential of 26.38 t/ ha from single harvest.
- **Promising genotypes:** During rabi season trial, Sel. 121 (Pink) (33.7 t/ha), Sel. 397 (dark red) (31.5 t/ ha) and Sel. 131 (White) (30.2 t/ha) were found promising compared to Pusa Red (20.8 t/ha). Hybrids developed using tropical short day male sterile line 121(A), namely, Sel. 121(A) × EG (39.7t/ha), Sel.121(A) × I-33 (37.5 t/ha) and Sel. 121 (A) × Sel. 131(35.6 t/ha) were found superior for bulb yield to the best commercial hybrid Flare (32.8 t/ha). Sel. 121(A) × Sel. 383 was found superior for anthocyanin, CUPRAC, K and Fe content. In *kharif* season white coloured selection, Sel. 106 (yield: 25.81 t/ha and water productivity 8.5 kg/m3) was found most promising next to Agrifound Dark Red. (28.94 t/ha and 9.8 kg/ m3). Sel. 383 was found promising for FRAP and CUPRAC (1.21ìmol trolox/g and 2.78 moltrolox/g, respectively), Sel. 157 (371.02 mg GAE/ 100g) for total phenol content and Sel. 102-1 for TSS (12.830 Brix).

Leguminous Crops

Garden pea

- Out of nine early maturing varieties, GP 17 was found promising and recorded highest pod yield (3.02- 4.0 t/ha) under high temperature. At IARI, Regional Station, Karnal, GP 17 also recorded highest pod yield (8.38 t/ha) against Arkel (7.14 t/ha) under November sowing. However, GP 473, a medium maturing and highly resistant to powdery mildew variety, recorded 9.85 t/ha pod yield at IARI, Karnal during 2011-12. Among new early maturing lines, GP 901 and GP 902 were found completely resistant to powdery mildew having long curved green pods with 8-9 seeds/pod. Among edible podded peas, GPE 1 (6.54 t/ha) and GPE 4 (7.33 t/ ha) were found highly resistant to powdery mildew disease.

Broad bean

- **Pusa Udit:** Pods of Pusa Udit are extra-long, flattish and light green. Fresh seeds have attractive green colour and good in taste. This is a dual purpose broad bean variety. Both tender pods and dried seeds are edible. The variety is suitable for packaging and transport. It has an average yield potential of 17.63 t/ ha, which is 88.52% higher than that of earlier released variety Pusa Sumeet.

Minor Legumes

- **Dolichos bean** DB 7 (dark green with purple sutures), DB 10 (long white straight pod), DB 14 (pink medium) and DB 15 (pink long), and medium maturing lines DB 3 (long green pod with purple suture), DB 5 (purple-red pod), DB 12 (light green) and DB 9 (long green straight pod) were found promising.

- **Cowpea:** The lines CP 5, CP11, CP 19, CP 21, CP 24, CP 26, CP 55, CP 56 and Pusa Sukomal were found resistant to Cowpea Golden Mosaic Virus.

Malvaceous Crop

Okra

- Out of 164 genotypes of okra evaluated for yield, fruit quality traits and YVMV resistance, DOV 66, DOV 64 and DOV 62 were found resistant to YVMV up to 90 days of sowing. Out of 82 F1 cross combinations, DOH 1 and DOH 2 were earliest (50 % flowering in 43 days), had dark green fruits with heavy fruiting and yielded 21.5 and 22.7 t/ha, respectively.

Leafy Vegetables

Palak

- Among six genotypes of *palak*, Selection GS recorded 28 % higher vitamin C than Pusa Bharati. In spinach, Selection VS type was found high yielding and late in bolting with higher vitamin C and carotenoid content 24.3% and 8.5%, respectively, than Virginia Sivoy.

Fenugreek

- Out of 19 *methi* genotypes, BJF 8 gave maximum yield (14. 4 t/ha) followed by Selection MPB (13.9 t/ ha), and vitamin C content ranged from 7.56 to 51.25 mg/100 g whereas total carotenoid from 15.8 to 60 mg/ 100 g.

Lettuce

- Progambo was observed to have the highest leaf yield of 1.75 kg/ plant followed by Gallega (1.00 kg). Stem Lettuce Angustana had the highest CUPRAC (1.984 μmoltrolox/g) and FRAP content (4.696 μmoltrolox/g), Reina De Mai Pleine Terre had the highest lycopene content (18.72 mg/100g) while Wonder Von Stuttgart had the highest total carotenoid of 45.129 mg/100g. The lowest phenolic content (41.94) was present in Arctic Kwig.

Haploid Development in Vegetable *Brassicas* Using Anther Culture

- Anther culture was used in cabbage and cauliflower to develop haploids. Initial inoculation of anthers at 35°C for two days was beneficial for embryogenesis. The highest per cent embryogenesis was recorded on Nitsch media supplemented with 140 g/l sucrose, 800 mg/l Glutamine, 100 mg/l L-Serine, 30 mg/l Glutathione and 2 mg/L AgNO3. The anther derived embryos were successfully regenerated and multiplied on MS media supplemented 2.5 mg/L thidiazuron and 0.5 mg/l NAA. The anther derived plants were successfully rooted and hardened.

Embryo Rescue for Transfer of Genes from Alien Species to Cauliflower

- For transferring black rot resistance gene from alien *Brassica* species namely, *B. carinata*, *B. nigra* and *B. napus*, BC1 made in cauliflower and is being advanced through embryo culture. For transferring sterile cytoplasm from alien species into *Brassica oleracea*, crosses were attempted with *Diplotaxis catholica, D. sifolia, D. erucoides* and *D. berthaultii* which are being advanced through embryo culture. Embryo cultured plants of cauliflower carrying male sterile cytoplasm from *Trachystoma ballii* and Anand cytoplasm are in rooting stage and BC1 stage, respectively.

Virus Resistance in Tomato

- The *S. habrochaites* genotype LA1777 was identified for high level of resistance to TOLCNDV, and inter specific F1 between commercial susceptible varieties Pusa Rohini and Pusa 120 were made. These F1 plants showed resistance to this virus. SCAR marker TGO 302 which differentiates the parents was used for selection of desirable plants in segregating generation to develop prebreeding lines. PCR based molecular markers were validated for TOLCNDV resistance locus (Ty1, Ty2 and Ty3), and tomato lines containing combination of different resistant locus were selected from segregating population. Screening of these lines for leaf curl virus in the field condition during *kharif* season led to the identification of 42 lines with desirable alleles and high level of resistance. The resistance for GBNV was identified in *S. peruvainum* accessions EC 368644-3 and EC 379243-2. These lines also showed resistance to TOLCNDV.

Transcriptome Analysis of Bitter Gourd

- Global transcriptome sequencing led to the identification 3,414 unigenes from DBGY201_Female plant and 3,045 from DBGY201_Hermaphrodite plant of bitter gourd. A total of 477 unigenes were found to be significantly differentially expressed between flowers of female and hermaphrodite plants, of which 237 unigenes were down regulated, while 59 unigenes were up-regulated in DBGY201_Hermaphrodite plants.

Garden Pea

- Genetic diversity was assessed among 28 garden pea genotypes with sixteen RAPD primers which produced 79 polymorphic markers showing 73.15% polymorphism. The polymorphism information content (PIC) was the highest for the RAPD primer OPN14 (0.74) indicating that it can be used for fingerprinting of genotypes.

Lettuce

- Diversity study of 62 lettuce lines was carried out with RAPD markers. *A dendrogram* was constructed using unweighted pair group method of arithmetic mean (UPGMA). The Genotypes Wo Suen and Revolution were highly divergent as demonstrated by the marker system.

SOURCE: ANNUAL REPORT 2012-13, INDIAN AGRICULTURAL RESEARCH INSTITUTE, NEW DELHI

Recent Initiatives by Indian Institute of Vegetable Research, Varanasi

Genetic Improvement of SolanaceousVegetables

Tomato

- **Generation advancement:** A total of 123 cross combinations *viz*; F7 to F8 :37, F6- to F7:6, F5 to F6:8, F4 to F5 : F3 to F4 : 27 F2 to F3 : 27 and F1(BC1 for heat tolerant-1) were advanced.
- **Sources for early blight (*Alternaria solani*) resistance/ tolerance:** Twenty five genotypes were screened under artificial conditions increased with *Alternaria solani*and 23 genotypes were screened under field conditions. Under artificial conditions, EC-520061 showed highly resistant, WIR-3957 showed resistant, EC-520078 showed moderately resistant reaction and other genotypes have shown moderately susceptible to highly susceptible reaction. Under field conditions, EC-520061 and WIR-3928 have showed moderately resistant disease reaction and others have showed moderately susceptible to highly susceptible reaction.

Brinjal

- **Hybridization**: Thirty-two intra-specific cross-combinations were developed. Under distant hybridization programme, inter-specific crosses were made utilizing nine cultivars and *S. incanum, S. khasianum* and *S. gilo.* In some crosses where premature fruit drop was observed, embryo rescue was attempted.
- **Generations advancement**: A total of 216 segregating populations(F2: 19; F3: 45; F4: 29; F5: 50; F6: 30; F7: 22; F8: 15; F9: 6) were advanced. An advanced material IVBL-22 has been identified for multi-location testing under AICRP (VC) trial for year 2013-14 in IET.
- **Evaluation of RILs**: Morphological characterization of 114 RILs was taken up at two locations (IIVR, Varanasi and SPC, Sargatia) for agri-horticultural traits. One hundred and twelve SSR primers were tested for parental polymorphism survey, of which, 21 were identified polymorphic among the parental lines of the RILs. For identification of more polymorphic primers, samples have been sent for trancriptome analysis.
- **Maintenance breeding**: Seeds of Kashi Sandesh (550 g), Kashi Komal (260 g), Kashi Prakash (400g) and Kashi Taru (600 g) were multiplied for distribution to farmers and for multi-location demonstration/evaluation by public and private organizations. Parental lines of hybrids and PR-5 (400 g) and BR-14 (600 g) were also multiplied.

Chillies

- **F1 hybridsevaluation:** Fifty two F1s were evaluated for yield and its contributing traits. The number of fruits per plant ranged from 6.33 to 248.80. The top five F1s in terms of yield of green fruits were GMS 3 x Kashi Sinduri (319.19 q/ha), California Wonder x EC-519636 (272q/ha), VR-339 x Kashi Gaurav (251q/ha), A-2 x VR-339 (231q/ ha) and A-7 x VR-339 (210q/ha. The cross California Wonder x EC 519636 showed fruits of stuff pickle type whereas fruits of GMS 2 x Kashi Sinduri has appealing dark red colour which may be utilised for paprika oleoresin purpose.

- **Development of pre-breeding lines for resistance, quality and yield attributes in chilli:** Under the pre-breeding lines development activity, 32 interspecific combinations were made.
- **Screening of chilli germplasm for pepper leaf curl virus resistance:** 372 germplasm were evaluated for pepper leaf curl virus resistance under field conditions. It was observed that none of the genotypeswas immune, resistant or moderately resistant. However, the genotypes BS-35, GKC-29 and Bhut Jolokia showed resistant.
- **Screening against thrips and mites infestation:** 131 germplasm were screened for thrips and mite infestation under field conditions. Among them, the genotypes Japani Longi, Co-5677 and LCA353 were tolerant.
- **Estimation of ascorbic acid in green chilli:** Seventeen lines were evaluated for ascorbic acid content in green fruits. The amount of ascorbic acid ranged between 313.01 (AKC89/38) and 70.125 mg/100 FW (PBC-81) with an average of 187.55 mg/100 FW. The other promising lines with respect to ascorbic acid content were VNS-4 (262.65), CCA-4261 (235.88), PBC-904 (228.01), EC
- 519625 (208.25), Kashi Sinduri-535 (187.64), and
- Kashi Gaurav (193.37).
- **Generation advancement:** A total of 148 segregating populations F2 : 86, F3 : 28, F4 : 7, F5 : 6, F6 : 10 and F7 : 16 were advanced.
- **Phenotyping of F_2:3 populations:** The F_3 population consisting of 239 individual derived from a cross between Kashi Sinduri (long, wrinkled, nonpungent fruits with high oleoresin content) and AKC 89/38 (round, smooth and pungent fruits) is being used to generate the genetic map and QTL analysis. A total of 14 important quantitative and qualitative traits *viz.* flower orientation, length, diameter of fruit, pedicel length, fruit surface, fruit shape, fruit shape at blossom end, fresh ripen fruit colour, presence of anthocyanin on the fruit, length and width of leaf, pericarp thickness, number of fruit per plant, fresh weight of fruit were measured in the progeny in order to map gene/ QTL controlling these traits. The frequency distribution was also plotted showing the normal distribution curve for most of the traits which implied that these traits are governed by polygenes.
- **Polymorphism survey:** A total of 350 SSR primers were analyzed in Kashi Sinduri and AKC 89/38. Out of 350 SSR primer pairs, 32 were polymorphic and the percentage of polymorphic bands was about 9.14. Some of the SSR markers showing polymorphism between parental lines have been depicted in Figure.
- **Maintenance breeding:** The cytoplasmic male sterile lines were maintained through sib mating with corresponding maintainer lines whereas one genetic male sterile line was maintained in heterozygous condition. The released variety of chillies and parental lines of hybrids were maintained under cages of nylon net

ooo

CHAPTER 19

Recent Advances in Crop Production

G.L. Sharma
Scientist (Horticulture)
Indira Gandhi Krishi Vishwavidyalaya (IGKVV), Raipur, Chhattisgarh

To maximize yielding ability of improved varieties and hybrids of different vegetable crops, cost-effective and efficient production technologies have been developed for different agro-ecological conditions.

Recent Advances

In **tomato**, drip fertigation with 80 kg K_2O/ha in nine splits registered maximum fruit yield (2.38 kg/ plant; 58.17 t/ha), 70% moreover conventional fertilization. However, maximum potassium use efficiency (98.5 kg yield/ kg K_2O) was with K fertigation @ 60 kg/ha. The maximum dry matter production (215.53 g/ plant) was recorded with K fertigation at100 kg/ha.

Tuber Crops

Compost (24.66 t/ha) prepared from **cassava** starch factory solid waste was an alternative to farmyard manure (26.64t/ha), green manuring in situ with cowpea (27.18t/ha), crop residue incorporation (25.03t/ha), vermincompost (22.15t/ha) and coir pith compost (21.78t/ha), NPK up to75%(26.55t/ha), $MgSO_4$ @ 20kg/ ha (27.94 t/ ha) and $ZnSO_4$@ 12.5 kg/ha (24.44 t/ha).

Methodology for **cassava** acreage estimation by remote sensing and GIS was developed using Landsat 8 OLI satellite imagery along with inverse multiquadratic based Possibilisticc-Meansclassifier. At optimized weighted constant for inverse multiquadratic based Possibilistic c-Means for 4 date combination, the total estimated area was found to be 4234.1 and 2175.63 ha for white Thailand and Mulluvadi, respectively. The total estimated area under cassava in Salem district in 2014 was 8324.2 ha.

SOURCE: DARE/ICAR ANNUAL REPORT 2016-17

In **onion** cv. Arka Kalyan and Arka Nikethan, both ultradried seeds (H"3% moisture) and seeds with 5.2% moisture maintained > 90% germination and vigour even after 50 months irrespective of storage temperatures (ambient and at 15 0C).

Spray of 5000 ppm ethrel 45 and 60 days after planting resulted in reduced (55%) scape length in onion.

SOURCE: DARE/ICAR ANNUAL REPORT 2015-16

Grafting **tomato** hybrid, Arka Rakshak on brinjal rootstock enhanced survival under excess moisture, while bell pepper hybrid, Indra, raised on chilli rootstock produced significantly higher yield in net house.

Integrated nutrient management (INM) in **onion** helped reducing the use of inorganic fertilizers by 25%. A combined application of 110: 40 : 60: 40 kg NPKS along with organic manure equivalent to 15 tonnes FYM and *Azospirillum* and Phosphate Solubilising Bacteria (PSB) @ 5 kg each/ha recorded higher marketable bulb yield.

SOURCE: DARE/ICAR ANNUAL REPORT 2014-15

In an attempt on **organic production of vegetables**, integrated use of poultry manure @ 2.5tonnes/ha + vermicompost @ 3.5tonnes/ha + bio-inoculation with *Rhizobium/Azotobacter* and PSB, recorded significantly higher yield of tomato (335q/ ha), cabbage (354.6q/ ha) and cowpea (106.2q/ha) as compared with inorganic nutrient management (tomato-268.5q/ha, cabbage- 309.5q/ha and cowpea- 80q/ha) grown respectively in *kharif*, *rabi* and *zaid* in a crop rotation sequence.

Performance of four **capsicum** hybrids/varieties under naturally-ventilated net house (white, 40mesh) was better with respect to growth and yield (386.28- 506q/ ha) as compared to open field (98-125q/ha) in *rabi*. The performance of Swarna (yellow) and Indra (red) varieties was better as compared to Cab-1201 and Popti. Better yield in pumpkin (5.5kg/plant) and bitter gourd (1.25kg/plant) were realized due to application of neem compost, followed by green manuring with subabul (*Leucaena leucocephala*) and lowest in control.

In **onion**, N and K uptake was higher at 15-60 days after transplanting (DAT), attaining peak at 25- 30 DAT (active vegctative growth) followed by sharp decrease at 60 DAT. Phosphorus and sulphur uptake was higher at 20-60 DAT with peak at 40-50 DAT (late vegetative growth and bulb initiation stages) and it continued up to 110 DAT.

Vernalization significantly enhanced the number of scapes and reduced the days to first scape emergence in **onion** Bhima Kiran and Bhima Super. On exposure to 10°C for 10 days, the seed yield in Bhima Kiran and Bhima Super was improved by 157 and 123% respectively, compared to untreated control.

Virus-free **garlic** plantlets were produced through meristem culture from virus indexed explants using 0.1 – 0.3 mm meristems. *In vitro* bulbils were induced from virus-free plantlets on medium containing 6% sucrose. Before planting, cold treatment of bulbils increased germination and reduced the period required for complete germination compared with untreated control.

Soil application of **poultry manure** at the rate of 10 tonnes/ha in **onion and garlic** produced relatively higher yields compared to other treatments. However, marketable bulb yield recorded in organic farming was 17-48 % lesser than inorganic farming. The status of available nutrients in soil after harvesting showed more available nutrients with application of inorganic fertilizers over organic sources.

Tuber Crops

In **coleus**, growing sunhemp for 45 days and incorporating into soil, application of FYM @ 80 kg/ ha, raising nursery with *Trichoderma* (1kg/1000 kg of FYM) enriched FYM and planting 40 days old rooted cuttings at a spacing of 60cm × 40cm in a raised bed recorded a higher dry tuberous root yield (1851 kg/ha) with a benefit : cost ratio of 2.6.

Integrated application of organic manures (FYM @10 tonnes/ha), secondary elements ($MgSO_4$@25kg/ ha) and micronutrients ($ZnSO_4$@10kg/ha) along with half of the recommended doses of NPK (40:30:40kg/ ha) significantly enhanced the cormel yield (13.9 tonnes/ ha) in taro as compared to control (without nutrient application – 6.39 tonnes/ha).

Drip irrigation equal to cumulative pan evaporation (CPE) up to stolon formation and thereafter, 1.50 times CPE till early tuber bulking and 1.25 times up to maturity produced significantly higher potato yield (36 tonnes/ha).

SOURCE: DARE/ICAR ANNUAL REPORT 2013-14

Significantly higher yields of 11.45 and 34.98 tonnes/ ha were achieved in cowpea and tomato, respectively with zero tillage on permanent ridges and residue retention, as compared to conventional tillage with flat planting and residue removal. In cowpea-tomato cropping system, net benefit obtained was` 163,208/ha/year with zero tillage on permanent ridges and residue retention as against ` 111,429/ha/year under conventional tillage with flat planting and residue removal. Organic carbon, labile organic carbon, non-labile organic carbon, carbon pool index and carbon management index were higher by 1.98 g/kg, 0.06 g/kg, 1.80 g/kg, 0.34 and 17.5, respectively, under zero tillage with residue retention compared to conventional tillage with flat planting and residue removal.

Seed pelleting techniques for onion, tomato and carrot were standardized. Pelleted onion, carrot and tomato seeds could be stored for 3 months under ambient conditions without reduction in seed germination and vigour. A carrier-based microbial product which contains N fixing, P and Zn solubilizing and plant growth promoting microbes was developed, evaluated and commercialized. The specialty of this technology is that farmers don't need to separately apply N fixing, phosphorous solubilizing and growth promoting bacteria individually. Farmers can apply this product conveniently, either through seed, soil, water or nursery media like coco peat.

Tuber Crops

Drip irrigation at 100% CPE was found best for obtaining higher tuber yield in cassava. Further, organic farming of cassava produced 9% higher tuber yield (29.4 tonnes/ha) over conventional practice (26.9 tonnes/ha). Decision support systems Plausible Potato Growing Seasons Estimator (PPGSE) and Yield Estimator were developed for spatial and temporal diversification of potato cultivation. These give the growing seasons and their durations, climatic features of seasons and estimated yield potential for important locations in India. Further, winter potato acreage and production were estimated through remote sensing, GIS and crop modelling. The acreage forecast under winter potato during 2011–12 in Punjab, Uttar Pradesh, Bihar and West Bengal was 90.6, 516, 322.5 and 345 thousand hectares, respectively, while the total production was predicted as 2,071.1, 11,439.7, 5,262.4 and 8,266.2 thousand tonnes, respectively.

SOURCE: DARE/ICAR ANNUAL REPORT 2012–13

For enhancing productivity and input-use efficiency in tomato, seedlings were raised in bed + foliar spraying of water-soluble fertilizers (WSF 0.5%) + seedlings treated with biofertilizers (*Azospirillum* and phosphorus- solubilizing bacteria (PSB) @ 2 kg/ha + foliar application of micronutrient mixture (0.1%) + plastic mulching of beds resulted in maximum fruit yield of tomato (520 q/ha) and nitrogen-use efficiency of 1.989 q/kg N. The minimum fruit yield (281.3 q/ha) was recorded under the farmers' practices. Under IPNM module for *kharif* **cowpea**, cultivar Kashi Kanchan registered maximum yield (142.5 q/ha) with the application of FYM @ 10 tonnes/ha + NPK (30:30:30 kg/ha) + biofertilizer (PSB) @ 2 kg/ha (seed treatment). The minimum yield of 120.7 q/ha was recorded in control, i.e. with 60:60:60 NPK kg/ha. Similarly, application of vermicompost @ 2.5 tonnes/ha + NPK (60:30:30 kg/ ha) + biofertilizers (*Azospirillum*) @ 2 kg/ha (seedling root treatment) recorded maximum head weight (1.37 kg) and marketable yield (389.7q/ ha) in variety Pusa Snow Ball K 1 as against 1.12 kg head weight and 352.5 q/ha yield in the control (NPK 120:60:60 kg/ha).

Alternate furrow irrigation and polythene mulch in **tomato** cv. Kashi Vishesh revealed that maximum yield (43.52 tonnes/ha) was obtained in well-watered plots combined with black polythene mulch; however maximum water-use efficiency of 9.42 q/ ha-cm was achieved when irrigation was given in alternate furrows and mulched with black polythene. With use of black and transparent polythene, 34 and 20% water was saved respectively, whereas about 40% water saving was noticed under alternate furrow irrigation over normal flood irrigation. In studies on off-season production of muskmelon, cv. Kanchan sown in October-end gave highest yield of 56 tonnes/ha. Precision farming studies in *Capsicum* involving raised bed method of cultivation, mulching, drip irrigation, fertigation and foliar nutrition of micronutrients resulted in highest yield of 46.5 tonnes/ha, compared to 27.5 tonnes/ha in furrow irrigated, non-mulched, and non-fertigated plot.

The integrated nutrient management (INM) for sustainable **onion** production revealed that combined application of 75% recommended dose of fertilizer (RDF) + 7.5 tonnes farmyard manure + 3.75 tonnes poultry manure/ha recorded significantly higher marketable yield (46.9 tonnes/ha) over inorganic fertilizer applied treatments (44.6 tonnes/ha). Application of 100% RDF along with 20 tonnes/ha of FYM registered 45.5 tonnes/ha marketable bulb yields. The highest pyruvic acid content (5.50 μmoles/g fresh weight) was recorded in inorganic fertilizers applied treatment. Bulbs harvested from INM treatments recorded lower pyruvic acid level (less than 4 micromoles/g fresh weight) than inorganic fertilizer alone applied treatment. Storage losses were significantly less in only inorganic fertilizers and biofertilizers (15.60%) applied treatments followed by 75% RDF and 15 tonnes/ha FYM (16.46%) over other treatments.

Among various organic manures applied, farmyard manure package recorded relatively higher yield (21.5 tonnes/ha in onion and 3.33 tonnes/ha in garlic) than other organic manures evaluated. However, almost 21- 40% lesser marketable yield was recorded in organic system in comparison to inorganic production system in both the crops. The influence of organic sources of nutrients on soil biological properties revealed that soil bacteria, fungi and actinomycetes population was higher in soil nourished with organic manures than inorganic fertilizers applied plots. The drip irrigation at 100% PE daily recorded the highest onion seed yield with higher cost : benefit ratio of 1:3.81. There was48-60% saving of water in drip irrigation over surface irrigation. For weed control during *rabi* season, soil application of oxyflurofen 23.5% EC @1.5 ml/ litre before planting and one hand-weeding 55 days after transplanting recorded higher marketable bulb yield of 43.5 tonnes/ha.

Potato and Tuber Crops

For mass multiplication of mini-tubers of potato, an aeroponic system with locally-available materials has been developed. On an average of about 60 mini tubers per plant were produced under the areoponic system. The germination of aeroponically produced mini-tubers varied from 92.3 to 100% and the yield from their progeny ranged from 93.4 to 173.2 g/plant.

The protocol for estimation of potato acreage and production through remote sensing, GIS and crop modeling in Punjab, Uttar Pradesh, Bihar and West Bengal has been standardized. Water-use efficiency of potato has been worked out as 177, 170 and 93 kg tubers per ha-mm water with drip, sprinkler and furrow irrigation methods respectively.

A nutrient decision support system website, cassava site-specific nutrient management (CASSNUM) containing all details about SSNM of cassava including N, P and K management enabling the farmers/extension officials to determine the amount of fertilizers to be applied for individual field for a particular yield level, considering the yield potential of locality, native soil fertility and cultivar used, has been developed. Application of 20 and 30 kg/ha of ZnSO4 and MgSO4, respectively, along with recommended doses of NPK and FYM was found essential to produce higher and sustainable crop yields of sweet potato. In **yams** and **taro**, organic farming was on a par with conventional practice.

SOURCE: DARE/ICAR ANNUAL REPORT 2011–12

Organic production system by use of crop rotation, crop residues, organic manures and biofertilizers improved the organic carbon in soil and soil micro flora, beside quality of vegetables. Raised bed planting coupled with paddy straw mulching improved yield and water-use efficiency in **tomato**. In **cucumber**, single-stem training in greenhouse cultivation irrespective of varieties recorded highest yield (628.87 q/ ha), followed by double stem. In capsicum, three levels of pruning recorded highest number of fruits/ plant (7.22) and fruit yield (660.56 q/ha), followed by double stem. The elevated CO_2 (550 ppm) influenced overall growth, water-use efficiency and total biomass in onion Arka Kalyan. The total biomass was maximum at 90 DAP (18.6 g/plant) compared to the control (13.4 g/plant). Its increment at elevated CO_2 levels varied from 52 to 81% at different growth stages. Two cultivars of **okra**, viz. Arka Anamika and US 7109, were grown in naturally-ventilated polyhouse during winter to overcome seasonal barrier in productivity. Okra fruit yield was significantly higher (28.1 tonnes/ha) in mid-October planted crops compared to mid-November planted crops (21.5 tonnes/ha). Out of 172 plant growth-promoting rhizobacterial isolates, *Azospirillum* (AzosH10), *Azotobacter* (Azotbel2), Phosphobacteria (Psbel 6) and *Pseudomonas* (Mpf2) were found to be superior in rhizosphere competence, growth promotion and plant nutrient uptake as compared to others.

The pre-emergence application of Oxyflurophen @ 0.15 kg a.i./ha, 10 days after transplanting recorded the highest marketable yield combined with higher cost : benefit ratio of 2.23 and maximum weed control efficiency in onion. The seed pelleting with DAP 30 g + $ZnSO_4$ 0.10 g + Bavistin 3 g/kg of seeds recorded significantly higher yield combined with higher cost : benefit ratio of 4.3 and 4 on raised and flat beds, respectively. Application of Pendimethalin 30 EC, a pre-emergence herbicide @ 2 ml/ litre at the time of sowing effectively controlled weed population in onion nursery compared to other herbicides.

Potato and Tuber Crops

Thematic maps were prepared for the areas climatically suitable for growing *kharif* potato in India and growing seed crop in the north-eastern states. **Potato** acreage and production were estimated through remote sensing GIS and crop modeling in Punjab, Uttar Pradesh, Bihar and West Bengal. In long-term manurial/ fertilizer experiment, productivity of potato was higher in maize-potato-onion (54.5 tonnes/ha), followed by rice-potato-wheat. New prototype of aeroponics for faster multiplication of potato mini tuber production was designed, developed and tested successfully. A decision support system (DSS) for precise application of nutrients (N) was developed for facilitating location- specific nutrient management. A database of potential productivity of popular potato cultivars in different agroclimatic zones was created. For monitoring health of seeds and plants a dipstick kit was developed which can be used by even farmers to check the presence of virus.

The site-specific fertilizer recommendations were developed for **cassava** using the yield-uptake relationships. A website on site-specific nutrient management and online fertilizer calculation software was developed for uploading in CTCRI website. The demonstration and popularization of INM practices involving secondary and micronutrients in cassava, soil test-based application of Zn, NPK and FYM was observed to be the most economical (benefit : cost ratio 1.75), followed by application of Mg, NPK and FYM (benefit : cost ratio 1.62) compared to farmers' practice (benefit : cost ratio1.09). Organic farming technology for elephant-foot yam validated under on- farm trials proved significantly superior with higher corm yield (34.60 tonnes/ha) and additional returns (` 43,651/ha) over conventional practices.

SOURCE: DARE/ICAR ANNUAL REPORT 2010–11

Potato: At Deesa, 75% recommended dose of NPK (full dose of P and K and half of N as basal and remaining half of N applied 30, 37, 44 and 51 days after planting) through drip fertigation is recommended as it saved 25% NPK fertilizers, increasing potato productivity. The drip system should be operated for 45 minutes during November - January and 65 minutes during February - March at alternate days with the discharge rate of 4 litres/ hour. Irrigation at 25- 30 mm CPE and mulching with paddy straw or any other locally available organic mulch material @ 5 tonnes/ha is recommended for higher yield of potato.

Cassava: Organic manure sources, *viz.* vermicompost @ 3.91 tonnes/ha (tuber yield 31.42 tonnes/ha) and coir pith compost @ 4.6 tonnes/ ha (27.04 tonnes/ha) were equally efficient as a substitute to FYM 12.5 tonnes/ha (30.97 tonnes/ ha) in cassava.

Sweet potato: The INM was worked out for sweet potato at Kalyani, Rajendranagar, Udaipur and Dharwad. The INM has been developed for taro cultivation. Significantly highest tuber yield of sweet potato (10.72 tonnes/ha) was recorded with the application of lime + FYM + NPK + ZnSO4 with a yield response of 15 % over that of FYM + NPK, closely followed by lime + FYM + NPK + MgSO4(10.65 tonnes/ha). Application of 50, 100 and 150 % of NPK showed an increase of 63, 138 and 136 % tuber yield of sweet potato over the control.

Elephant-foot yam: Organic farming continued its superiority for the fifth consecutive year, producing significantly higher corm yield (34.81 tonnes/ha). The crude protein content of organic corms was significantly higher. The INM schedule including biofertilizers gave higher corm yield in different centres.

Intercropping Systems

- Sweet potato + red gram intercropping system (1.8 m strips each accommodating 3 rows) recorded significantly higher tuber equivalent yield (13.01 tonnes/ha) compared to other cropping system, followed by sweet potato as a sole crop (12.96 tonnes/ha). Maximum net return and benefit: cost ratio were recorded with sweet potato + red gram intercropping system.
- Elephant-foot yam intercropped in orchards gave higher net returns with sapota/ mango orchard with recommended dose of NPK at Coimbatore, Dholi and Ranchi.
- For Dharwad and adjoining areas of Karnataka, kharif potato-rabi sorghum crop sequence was recommended for higher productivity and profitability with 20 tonnes/ha FYM and 100% of recommended doses of NPK to potato crop.
- For Kalyani and adjoining areas (West Bengal) in potato-okra-paddy cropping sequence, application of FYM @ 20 tonnes/ha and 100% recommended doses of NPK to potato and 50% of recommended doses of NPK to okra and 75% recommended dose of NPK to paddy was recommended for higher productivity and net return.
- Potato + garlic intercropping system in a 1:1 ratio with 100% recommended doses of NPK fertilizers to both the crops was recommended for Chhindwara and adjoining areas of Madhya Pradesh.
- For Kota and adjoining areas of Rajashtan, potato + methi (every fourth row of potato replaced by two rows of methi at 30 cm spacing) was recommended for higher productivity and net return.
- At Faizabad, intercropping of potato with garlic in a 3:1 ratio with recommended doses of NPK for both the crops was recommended.

- In Nilgiris, potato-French bean combination has been found to reduce the build-up of potato cyst nematodes, enhancing soil organic carbon content due to nitrogen-fixing capacity of beans.

SOURCE: DARE/ICAR ANNUAL REPORT 2009–10

Recent Initiatives by Central Potato Research Institute, Shimla

Agro-techniques for processing potatoes

Technologies for production of processing potatoes standardized.

Agro-techniques *viz.* optimum date of planting (2nd week of October in North-western plains and 3rd week of October in West-central plains), crop geometry (67.5 x 20 cm), nutrient requirement (270 kg N, 80 kg P205 and 150 kg K20/ha; with 50% N and full PK to be applied at planting and rest at earthing up) and water management (irrigation at 20 mm cumulative pan evaporation) practices were standardised (Kumar *et al.*, 2007 and Kumar *et al.*, 2008).

Micro-irrigation and fertigation in potato

Micro-irrigation and fertigation technology standardized to make potato production feasible in the future era of water shortage.

In order to economise on water and fertilizers, micro irrigation and fertigation techniques have been developed and standardized for potato, which economize on water by 40-50% and fertilizer NPK by 25-30% with 25-30% higher yields, over furrow irrigation method (Singh and Singh, 2007, Singh *et al.*, 2010).

Management of seed dormancy

Prolonged seed dormancy by using 1,4 DMN and improved seed quality under non-refrigerated hill situations.

In absence of refrigerated stores in hilly areas, potato seed is stored under diffused light conditions, which results in development of long and lanky sprouts by planting time. 1,4 DMN (1,4-dimethylnaphthalene), a naturally occurring substance present in potato tuber skin, when sprayed on seed potatoes stored under diffused light in hills, delayed dormancy release up to 120 days and resulted in healthy and vigorous sprouts by planting time (CPRI Annual Report 2010-11). The technology is being perfected for transferring to the end users.

Variety-specific tissue culture protocols developed

The standard MS medium was modified for micro propagation of recalcitrant potato cultivar Kufri Jyoti. The modified medium contains increased level of ammonium nitrate from 20. 63 mM to 25.79 mM supplemented with GA3 (0.58 μM) and NAA (0.1 μM) was found to be the best for *in vitro* multiplication of Kufri Jyoti. (Venkatasalam *et al* 2011).

SOURCE: CENTRAL POTATO RESEARCH INSTITUTE, SHIMLA

Recent Initiatives by Indian Institute of Vegetable Research, Varanasi

- Region specific suitable package of practices including optimum sowing time, nursery raising, transplanting, fertilizer dose, seed rate, spacing, transplanting methods, intercultural practices and weed management for realizing optimum potential of improved varieties of vegetable crops have been standardized.

- An integrated nutrient management module on tomato was developed with application of NPK @ (120:60:60 kg/ha) + Pressmud @ 5 t/ha + Azotobacter root dipping before transplanting + foliar spraying of ferrous ammonium sulphate @ 20ppm at three critical growth stages i.e. 30, 45, and 75 days after transplanting gave an yield of 160 t/ha as compared to 60 t/ha in control.
- Soil application of pressmud @ 5 t/ha plus *Azospirillium* and phosphate solubilizing microorganism @ 10 kg/ha coupled with foliar spray of Zn and B @ 50 ppm at 30, 45 and 75 days after transplanting realized 39% additional yield in brinjal as compared to control (N: P:K::120:60:60).
- In case of indeterminate tomato, three-tier staking as support system gave 30% increase in yield.
- In cabbage, residue-free production technologies with an average yield of 650 q/ha and head maturity within 45-50 days has been developed by adopting organic farming techniques.
- The use of low tunnel plastic, soil solarization and net houses were found to be useful for raising healthy and disease-free seedlings of vegetable crops. The techniques also reduced the incidence of TLCV of tomato and LCV in chilli and capsicum.
- Organic manures (FYM or Digested sludge @ 10 t/ha) and biofertilizers (PSM or VAM) individually or in combination significantly enhanced the vitamin C and total carotenoids in broccoli, Chinese cabbage, pea and lettuce as compared to recommended fertilizers.
- Maximum yield of 114.93 t/ha in tomato cv. Tolstoi was recorded when plants were spaced at 60 x 40 cm and drip irrigation was scheduled at alternate day. The maximum WUE (51.19 to 52.17 q/ha cm) was recorded under 60 x 60 cm spacing.
- In a fertigation study on tomato, the highest fruit yield of 67.73 and 69.38 tonnes/ha was obtained when drip irrigation was scheduled at 100% ETo, and N (urea) was fertigated in 8 or 10 split doses, respectively.
- To manage the weed infestation in kharif crop of okra, application of metalochlor @ 1.0 kg/ha has been found most effective and economical and gave maximum yield and C: B ratio (1:2.86).
- Trials on use of micronutrients in major vegetable crops indicated that two foliar spray of Zn (100-200 ppm), Boron (5 ppm) and molybdenum (5-10 ppm) has significant effect in French bean, cow pea, okra and cauliflower as for as yield and quality were concerned.
- Variability of some important antioxidants and minerals were analyzed in between and within the crucifer accessions. Total carbohydrate content ranged between 2.54-4.03g/100g, protein from 0.41-3.57 % of fresh weight, fiber content ranged from 0.60 to 3.62 g/100g, Vitamin C content ranged from 22.16-82.14 mg /100g and b-carotene ranged from 1.56 to 9.09 mg/100g fresh weight. In general kale, broccoli and Brussels sprouts ranked high in their antioxidant capacity containing significant amounts of vitamin C and b-carotene.

- The soil application of Azotobacter @ 15 kg/ha coupled with NPK @ 150:60:80 kg/ha influenced the pericarp thickness, shelf life and yield in tomato hybrids. The shelf life of tomato was enhanced by 3-5 days at ambient temperature. Seedling inoculation with Azotobacter before transplanting along with soil application significantly prolonged the shelf life by at least 2 days depending upon variety as compared to lone soil application.
- Okra, cowpea and tomato can be successfully dried after blanching and drying at 50 – 60°C for 8-10 hrs. Dried vegetables had good sensory quality after rehydration and remained acceptable for 8-10 months at room temperature.
- Osmotic diffusion treatment of bitter gourd, cauliflower, broccoli, cabbage, pointed gourd and ivy gourd after blanching in sodium chloride (1-4%) solution at 40-60°C for 2-3 hrs followed by drying in hot air oven at 50-60°C for 10-12 hrs maintained good sensory quality after rehydration and the product remained acceptable for 8-10 months at room temperature.
- The processing techniques for preparation of ready-to-eat bitter gourd, pointed gourd and ivy gourd chips have been standardized.

SOURCE: INDIAN INSTITUTE OF VEGETABLE RESEARCH, VARANASI

Recent Initiatives by Indian Agricultural Research Institute, New Delhi

1. Development of Seed Production Technology for Hybrid Bitter Gourd

Seed production technology was standardized for bitter gourd hybrids, Pusa Hybrid 1 and Pusa Hybrid 2. Stigma remained receptive up to one day both under open field and net house conditions, but fruit and seed setting was higher when pollination was done between 7 am and 12 noon. GA3 (50ppm) sprayed at three-leaf stage, tendril initiation and bud initiation was found to be the most effective for induction of more female flowers, higher fruit and seed setting, number of fruits/ vine and seed yield for hybrid seed production of Pusa Hybrid-1. Hand pollination done once per flower was found to be the best for higher fruit setting as well as seed yield. Storability studies indicated spring summer-produced seeds had higher germination, vigour and storability as compared to *kharif* season.

2. To Study the Effect of Different Growing Conditions on Stage of Harvest in Hybrid Seed Production in Cucumber

The effect of stage of fruit harvest, i.e., 25, 30, 35, 40, 45 days after anthesis in cucumber cv. Pant Shankar Khira-1on fruit and seed quality was studied under naturally ventilated poly house (NVPH), insect proof net house (IPNH) and open field conditions (OFC). These three environments significantly influenced the fruit and seed development process. As a result under IPNH the fruit weight in summer (580.33g) and *kharif* (684.00g); number of filled seeds in summer (154.46) and *kharif* (169.33) were higher compared to NVPH and OFC.

3. Effect of Berry Pruning and Trusses Retention on Seed Yield & Quality Attributes of Cherry Tomato under Different Environmental Conditions

In the experiment conducted under different environmental conditions at the Center for Protected Cultivation Technology during winter summer season 2011-12 on cherry tomato Pusa selection-1, it was observed that 10 fruits/5 trusses recorded the highest berry weight followed by 10 fruits/7 trusses and 15 fruits/5 trusses under environmentally controlled poly house (ECPH) followed by NVPH and IPNH. The findings indicated that pruning of berries/truss as well as retaining of trusses/plant positively influenced the weight, width and length of berries, and enhanced seed germination and seed weight under ECPH followed by NVPH and IPNH conditions, when compared to un-pruned trusses.

4. Effect of Pollination Time on Fruit Set and Seed Yield under Different Conditions in Hybrid Seed Production of Cucumber

The hand pollination was carried out at 7.00, 9.00 and 11.00 AM under three sets of conditions viz., ECPH, NVPH and IPNH. During summer season, pollination at 7.00 am gave higher fruit set (3.76) and number of fruit developed to maturity (2.30). Additionally, the number of filled seeds per fruit, 100-seed weight, seed yield per fruit was also significantly higher at 7.00 AM pollination.

5. Effect of crop geometry and fertigation frequency on the yield of *kharif* Onion

An experiment was conducted to study the effect of crop geometries (S1 = 15 cm x 10 cm; S2= 12 cm x 10 cm; S3 = 10 cm x 10 cm; S4 = 7.5 cm x 10 cm) and frequency of fertigation (F1 =every alternate day, i.e. 3 times in a week; F2= two times in a week; F3= once in a week) on the yield, irrigation water use efficiency and cost of the drip system in two onion cultivars (AFDR and N 53). Lateral to lateral distance was 1.20 m in all treatments, to maintain the cost of the system uniform. Highest yield was observed in S4F1 (30 t/ha) in AFDR. There was 9.8, 16.1 and 23.3 % decrease in the onion yield in S1, S2 and S3, respectively, in comparison to S4. Irrigation water productivity of 9.9 kg/m3 of water was observed in S4F1. Increase in fertigation frequency from once weekly to every alternate day increased the yield by about 15 %. Water requirement for onion was estimated to be 309 mm. However, because of adequate rainfall, only 132 mm irrigation water was applied through drip.

6. Evaluation of different protected structures for off-season cucumber production

Naturally ventilated polyhouse, fan-pad poly house and insect-proof net house for production of parthenocarpic cucumber (var. Isatis and Kian) were evaluated during 2012-13. Cucumber seedlings (25 days old) were transplanted on 15th Sept. under different protected structures on raised beds along with drip fertigation system. The variety Isatis gave higher average fruits yield (15.50 kg/m2) as compared to Kian (14.15kg/m2) under naturally ventilated polyhouse condition. It is concluded that parthenocarpic cucumbers can be grown successfully under naturally ventilated polyhouse during offseason, which is economically feasible for the growers.

7. Studies on summer squash varieties under plastic low-tunnels during winter season

Four Summer squash varieties viz., Australian Green (Green coloured long fruited), Zucchini (Yellow coloured long fruited), Chand and DS 8 (Tinda type round fruited) were evaluated under plastic low tunnels for their off-season cultivation during winter 2012. Among the long fruited yellow coloured varieties Zucchini hybrid produced maximum fruit yield (56.2 t/ha) as compared to variety Australian Green (53.6 t/ ha).

However, among the round shaped varieties, hybrid Chand produced the highest fruit yield (41.5 t/ ha) as compared to DS 8 (35.5 t/ha).

8. Studies on parthenocarpic cucumber with and without plastic mulch under naturally ventilated poly house

Three parthenocarpic cucumber varieties, viz., Kian, Satis, and Hilton were evaluated under naturally ventilated greenhouse equipped with low pressure drip irrigation cum fertigation system during August to December 2012. Out of three varieties, Kian required minimum period for first harvesting (23 days) after transplanting and produced the highest fruit yield (2.40 kg/plant and 7.78 t/1000 m2) with the cost benefit ratio of 1:2.20. Silver/black plastic mulch of 25-30 micron on raised bed at 30 cm x 30 cm spacing along with drip enhanced the fruit yield by 40-50% compared to without mulching under similar conditions.

9. Performance of sweet pepper with conventional and dry liquid drip fertigation under protected cultivation

Eleven N and K fertilizer combinations, including nine combinations of urea and MOP, one ready mix dry liquid fertilizer (DLF) and a control, were considered in sweet pepper and fertilizer was applied weekly in 32 equal split doses through low head low cost drip system while irrigation was scheduled twice a week. The highest average sweet pepper fruit yield was recorded (62.2 t/ha) in case of the conventional fertilizer treatment. The use of urea as a nitrogen source retained more nitrate in surface layer than that of DLF. The net returns increased by 21 per cent (`30,000) by enhancing 25 per cent (`200) nitrogen as compared to the recommended dose of nitrogen (320 kg/ ha). The above fertilizer rate with urea and MOP was found to be the best as far as yield, benefit cost ratio, net income and payback period were concerned. The study will be useful in operation and management of drip fertigation for enhanced production of sweet pepper under protected cultivation.

10. Development of pole type F1 varieties of summer squash suitable for greenhouse and open field conditions

Summer squash is important vegetable of cucurbits, mainly all cucurbits are wine type in nature but summer squash is having bushy type nature. Nowadays promotion of vertical horticulture is required for greenhouse but presently, varieties available are bushy types. Therefore, Breeding program for the development of wine type summer squash was initiated. A vine type parental line was collected and crossed with Australian Green (Bushy type) under green house conditions. F1 between vine type female and bushy type male (Australian Green) was found suitable for protected conditions.

11. Design of partial root zone deficit (PRD) and regulated deficit irrigation (RDI) techniques for greenhouse tomato

The irrigation design was improved by making an arrangement of drip lateral with control valves for 16 mm diameter lateral fixed with dripper discharge of 2 liter per hour. Low pressure drip fertigation scheduling was controlled by two tensiometers installed at 30 cm and 60 cm depth. Good quality tomato with a yield of more than 10 kg per plant was achieved with 50% less water by using water saving techniques of PRD and RDI under naturally ventilated greenhouse with low pressure drip fertigation.

12. Design and installation of solar energy operated greenhouse

A 200 square meter solar energy operated greenhouse equipped with fan-pad cooling system and low pressure drip fertigation system was designed and installed at CPCT farm. Solar photovoltaic power plant of 2.6 kWp was installed to energize the greenhouse of 200 square meter area. The greenhouse has two exhaust fans each of 500 watts and 0.5 hp pump and motor for cooling pad. Parthenocarpic cucumber was grown successfully in soil-less cultivation system inside the solar energy operated greenhouse.

13. Development of Two Bed and Eight- Row Carrot planter

The two bed and eight-row carrot planter was designed and developed for precise planting of carrot seeds on raised beds with a ridge height of 20 cm. The top and bottom width of bed were 350 and 700 mm, respectively. The planter plants the carrot in four rows on each bed at a row to row spacing of 7.5 cm. An inclined plate metering device setting was used which consisted of circular outer casing, seed plate with cells on periphery, and a cut-off brush. The trapezoidal shape seed hopper of size 165x190 mm (top) and 130x105 mm (bottom) with the side wall slope of 40° to the horizontal was fabricated. The planting geometry adopted on the bed was 75x50 mm for carrot. The average depth of seed placement was 2.25 cm. The field capacity of the machines is 0.5 ha/h.

14. Development of Two-Bed Carrot Harvester

A prototype of a two bed harvester was designed and developed for harvesting carrot grown on raised bed. The width of coverage of the carrot harvester is 2000 mm and it is adjustable as per harvesting width requirement. The optimum design specifications of the carrot harvester arc: rake angle of 25°, length of soil separator 600 mm and angle of soil separator 20°. The field capacity of carrot harvester is 0.5 ha/h with a field by a 35 hp tractor.

15. Development of Tractor Operated Garlic Planter

A nine row tractor operated garlic planter was developed. The planter consists of main frame, vertical plate metering system, reversible tyne furrow openers and seed tubes. The metering system is driven by the ground wheel of 380 mm diameter and power transmission is by a set of chains and sprockets. The vertical metering plates are of 150 mm diameter and each having eight hemispherical cups of diameter 25 mm. The developed planter has two seed boxes. Row to row spacing is 150 mm and desired plant spacing is 75 mm. The field capacity of the planter is 0.2 ha/h at a working speed of 2 kmph. The vertical plate metering system was tested over sticky belt and the clove spacing recorded. The quality of feed index was computed as 88%. The miss index and multiple index were 2% and 10%, respectively. No visible clove damage was recorded in the laboratory testing. Performance evaluation of the prototype is being carried out in the experimental farm of the Institute.

16. Design and Development of Pneumatic Precision Planter for Vegetable Crops

The effect of depth of seed placement for germination and plant stand was studied for design of the planter components. Performance parameters were recorded for radish, spinach and cauliflower for three different depths (10, 15 and 20 mm). No significant difference was observed for plant population in case of radish and cauliflower, whereas significant difference was recorded for spinach with 15 mm depth of planting with higher plant population. Performance parameters were also recorded for okra at two different depths (20 and 40 mm). Dry (8.16% m.c.) as well as soaked (66.27% m.c.) okra seed were used for the experiment. It was observed that almost 50 per cent germination occurred on 5th day for soaked okra seed, whereas the germination for dry seed was 30 per cent on 5th day. However, on 21st day there was no significant difference in germination and dry matter for both the treatments. Planter performance was better in case of dry okra seed. No significant difference was observed in yield for both the treatments except the numerical higher values for dry okra seed planting. On the basis of the observations, the design of the experimental metering system was carried out with all the sub-systems.

17. Development of Prototype Garlic Harvester

In order to reduce the high labour cost, design parameters of a garlic mechanical harvester were optimised through evaluation of relevant biometric and engineering properties. The influence of soil-machine parameters on mechanical harvesting of garlic was determined through field experiments of harvesting system set-up at three levels of soil moisture content (15, 12 and 9%), rake angle (10,15 and 200) and speed of operation (1.5, 3.0 and 4.5 km/h), and optimum design values determined. A 4-row tractor operated garlic harvester was designed accordingly and field evaluated in sandy-loam soil. Crop harvesting and bulb damage percentages were 96.12 and 5.94%, respectively. The machine had soil separation index of 0.26 with power requirement of 4.54 kW, which is adequate for small/ medium size tractors commonly used. The machine could cover 1.9 ha per day of 8 h of operation. Breakeven use of the prototype garlic harvester was estimated to be 218.12 h/year, with a payback period of 3.63 years. Estimated operational cost of the machine is 55 % lower than that of manual harvesting.

CHAPTER 20

Recent Advances in Crop Health Management

B.K. Patidar
Assistant Professor (Plant Pathology)
K.N.K. College of Horticulture Mandsaur (M.P.)

Recent Advances

In **cabbage**, a module comprising foliar application of azadhiractin 0.03% @ 5ml/l, rynaxpyr 18.5 SC @ 0.15 ml/l, novaluron 10 EC @ 1.5ml/l, emamectin benzoate 5 SG @ 0.35g/l at 10-15 days interval was highly effective with 88.43% reduction in DBM larvae and 69.64 % increase in marketable yield with cost : benefit ratio 1:8.68. Carbofuran@1kg, a.i./ha + copper oxychloride@0.25% + TRB4+CRB2 @1% was highly effective reducing galls by 64.5%, soil population 28.0% as well as disease severity scale to 1, gall index 3 and reproductive factor 0.2 over inoculated control where corresponding values were as 0.0, 0.0, and 3.5 and 7 and 1.3, respectively. These treatments also increased the plant height (41.3%), root length (47.7cm) and yield (77.5%) compared to inoculated control. Flonicamid 50 WG @ 0.4g/l was effective with 93.26 and 85.91% reduction in leafhoppers (*Amrasca biguttula biguttula*) and whitefly (*Bemisia tabaci*), respectively. Its residue dissipation followed first order kinetics with half-lives 3 - 3.5 days. The pre-harvest interval (PHI) was 16 and 20 days for recommended and double the recommended doses, respectively. The residues dissipated below MRL of 0.01 mg/kg on the same day. The dietary exposure of measured residues was lower than the MPI of 0.576 mg/person/day on all the sampling days for a single and double dose.

A strain of ***Trichoderma asperellum*** inhibited the mycelial growth of *Pythium aphanidermatum, P. debaryanum* (causing damping off disease), *Sclerotium rolfsii* Sr1 and *S .rolfsii* Sr3 (causing collar rot disease) by 43.57, 38.16, 42.56 and 54.87% respectively after 6 days of its inoculation. It also inhibited the mycelia growth of *Fusarium oxysporum* f.sp.*lycopersici* (causing wilt disease) and *Alernaria solani* (early blight disease) by 26.27 and 24.17%. Its identity was worked out and the sequence deposited in GenBank (NCBI) with an Accession No. KT824429. It showed 100% compatibility with mancozeb, azoxystrobin, cymoxanil + mancozeb, metalxyl + mancozeb at 100, 200 or 300 ppm.

SOURCE: DARE/ICAR ANNUAL REPORT 2016-17

Detection of Seed borne Tobamoviruses in Vegetable Crops

Seeds collected from infected plants (cucumber, capsicum and tomato) were tested for different tobamo viruses by ELISA and RT-PCR. There was externally seed borne infection of CGMMV (8%), PMMV (12%) and ToMV(14%) and none of the viruses were detected in embryo and cotyledon indicating absence of internal seed transmission of CGMMV, PMMV and ToMV in cucumber, capsicum and tomato.

Sequences of S-RNA and M-RNA of Indian isolate of Iris Yellow Spot Virus (IYSV), a tospovirus affecting seed onion crop were determined that showed > 95% identity with other known isolates and occurrence of intra- and inter- species recombination at intergenic region.

SOURCE: DARE/ICAR ANNUAL REPORT 2015-16

Capsicum seeds, substrate (used for producing nursery seedlings) and seed beds under shade net were treated with *Bacillus subtilis* I % w.p and 1 % AS., *P. putida* 1 % w.p and 1 % AS and *Bacillus pumilis* 1 % W.P and 1 % AS. All these bioagent formulations proved to be effective in the management of nematodes, viz. *M. incognita* and *R. reniformis* on Capsicum.

For molecular characterization of **begomoviruses infecting chilli**, complete nucleotide sequences of DNAA were determined. Cloning, sequencing and sequence analysis indicated that each sequence contained the same arrangement of ORFs which is typical for whitefly transmitted begomoviruses in general, with 2 ORFs (AV1 and AV2) in virion sense DNA and four (AC1, replicase, AC2, AC3 and AC4) in complimentary sense DNA. The AV1 (coat protein), AV2 (Precoat protein), AC2 and AC3 genes were of the same size in each sequence and only small differences were noted in the AC1 (Rep) and AC4 genes. Comparison of full-length sequences indicated that chilli leaf curl virus (ChLCV), chilli leaf curl multan virus (ChLCMV), chilli leaf curl Pakistan virus (ChLCPV), pepper leaf curl Bangladesh virus (PepLCBV), tomato leaf curl Sri Lanka virus (ToLCSLV) and tomato leaf curl New Delhi virus were associated with chilli leaf curl disease, based on DNA-A sequence nucleotide identity of 90.5-97.7% with the above six viruses.

Integrated module, comprising seedling dip with Imidacloprid I ml/litre of water, spraying of Buprofezin 1ml/litre at 25 DAT, Fipronil 2 g/litre at 35 DAT, *Verticillium lecanii* 5 g/litre at 45 DAT, Chlorfenapyr 1 ml/litre at 55 DAT, Neem oil 1 % at 65 DAT and subsequent rotation of the same was most effective for mite reduction in chilli over the control.

An entomopathogenic fungus, ***Isaria jarinosa,*** causing natural epizootics on whitefly grown under polyhouse was isolated and characterized. The intergenic spacer region (IGS) of the fungus was amplified by using primers *26S311* F/CNS 1. Post- sequencing amplified PCR product was subjected to BLAST (NCB!) analysis and the nucleotide sequence data were deposited in Gen Bank (Acc. No. KJ013353). The incidences of begomoviral diseases are rapidly increasing and are being reported from different parts of India. Two hundred leaf samples were collected from farmers' fields around Varanasi region and were analysed through PCR using begomoviruses specific primers. The study indicated that all samples were associated with new strains of tomato leaf curl J oydebpur virus (ToLCJ O V) and Tomato leaf curl Bangladesh betasatellite (ToLCBDB).

A severe outbreak of anthracnose was recorded in Kamataka in which symptoms of anthracnose appeared on leaves, neck and bulbs of onion. Dieback was observed on severely infected leaves, which led to the collapse of plants. Twisting and curling of leaves were associated symptoms of onion anthracnose. Thirty-seven isolates of *Colletotrichum* recovered from diseased leaves and neck were characterized based on morphological characteristics and partial sequences of ACT, TUB2, CHS-l, CAL, CYLH3, GAPDH, GS and ITS genes. The combined multilocus sequence analysis coupled with a critical examination of the phenotypic characters revealed the presence of two species, *C. siamense* (27 isolates) and *C. truncatum* (10 isolates). Pathogenicity of two species were confirmed on onion seedlings cv. Arka Kalyan. This is claimed to be the first report of C. *siamense* and C. *truncatum* causing anthracnose disease in onion.

Pumpkin yellow vein mosaic disease (PYVMD) is caused by three different bipartite begomoviruses, namely Tomato leaf curl New Delhi virus (ToLCNDV), Squash leaf curl china virus (SLCCNV) and Tomato leaf curl Palampur virus (ToLCPa V) in combination with four different beta satellites namely Pepper leaf curl betasatellite (PepLCB), Papaya leaf curl betasatellite (PaLCB), Tomato leaf curl betasatellite (ToLCB) and Ludwigia leaf distortion betasatellite (LuLDB). Molecular studies indicated that all these viruses have typical features of Old-World bipartite begomoviruses with the presence of stem-loop structure within an intergenic region (IR), and seven open reading frames (ORFs) [AVI and AV2 in the virion sense and ACl, AC2, AC3, AC4 and AC5 in complementary sense].

Formulations of *Chaetomium* sp. were developed for the biological management of early blight of tomato caused by *Alternaria solani* through induced systemic resistance. Molecular characterization of begamoviruses and tobamoviruses infecting chilli and tomato respectively, was completed.

Tuber Crops

JHULSACAST a potato late blight forecasting model for plains of West Bengal and web-based Decision Support System for management of late blight in western Uttar Pradesh were developed. Indo-Blightcast, a web based forecasting model, developed is applicable pan India. Development of triplex lateral flow immune assay (LFIA) was done for simultaneous detection of viruses, PYX, PYA and PVM, duplex RT-PCR for CMV and TMV and multiplex RT-PCR for simultaneous detection of PAMV, PVS, PVM, PLRV and PYX in potato.

SOURCE: DARE/ICAR ANNUAL REPORT 2014-15

Molecular characterization of begomoviruses in French bean

Characterization of a new strain of tobacco curly shoot virus (TbCSV) infecting common bean revealed it to be a recombinant between tobacco curly shoot virus, mungbean yellow mosaic virus, tomato leaf curl Jodhpur virus, tobacco leaf curl Yunnan virus and *Ageratum* enation virus like ancestors, transmitted both by whitefly and inoculation of infected sap. An analysis of the whole genome of this begomovirus revealed close association with TbCSV (89.1-94.5% sequence similarity) infecting Solanaceous and other weed crops. Absence of DNA-B and association virus isolate with betasatellite confirmed this as monopartite begomovirus. The identified betasatellite shared a highest sequence identity (53.9-93.9%) with tomato leaf curl betasatellite.

Summer squash showing severe yellow mosaic disease symptoms was confirmed as infected by begomovirus through ELISA, dot blot, IC-PCR and PCR amplification. The comparison of coat protein gene sequence revealed that virus isolate shared highest nucleotide identity of 99% with squash leaf curl china virus (SLCCNV) infecting pumpkin. The minimum acquisition access period (AAP) and inoculation access period (IAP) required by *Bemisia tabaci* transmitting SLCCNV were 15 and 10 minutes, respectively with females being more efficient than males in transmitting the virus.

In **cucurbits,** the incidence of Phytoplasma in bitter gourd, bottle gourd, cucumber, ridge gourd and watermelon ranged from 0.5 to 13.7%. SEM of infected samples revealed the presence of pleomorphic bodies in the cell which are Phytoplasma. For further diagnosis, PCR based diagnostics were developed using Phytoplasma-specific 16SrRNA primers. In addition, amplified 1.8kb and 0.7kb PCR fragment from infected plants was identified using phytoplasma specific 16SrRNA and ISR primers.

Promising Bioagents Identified

A promising **parasitoid (*Apanteles paludicole*)** with 45% parasitization during August-September was recorded from plume moth (*Sphenerches caffer*) infesting bottle gourd. Similarly, nymphal parasitoids *Aenasius bombawalei* and *Promuscidea unfasciati* were identified from mealybug (*Phenacoccus solenopsis*) infesting brinjal, tomato, okra, pointed gourd and chilli with maximum cumulative parasitization in tomato (36%). Another parasitoid (*Diaeretiella rapae*) showed differential response for parasitization to different aphid species, maximum being on *Myzus persicae* (59%).

The RT-PCR based detection protocol was standardized for detection of GCLV by using primers flanking its coat protein gene. Similarly, N gene of IYSV was characterized for recombinant antigen production for the preparation of indigenous ELISA kits in onion.

Tuber Crops

Serological and nucleic acid-based diagnostic techniques were standardized for sweet potato leaf curl virus (SPLCV) and dasheen mosaic virus (DsMV) in *Amorphophallus* and yam mild mosaic virus (YMMV) infecting *Dioscorea* sp. Early and accurate detection technique for *Phytophthora colocasiae* was developed through PCR using species specific-primers.

Potato

Late blight forecasting model, JHULSACAST, was developed for *tarai* region of Uttarakhand. Protocols for detection of PLRV and PVY in aphids (RT-PCR),

ToLCNDV-potato in whiteflies (PC-PCR), duplex RTPCR for PVA and PVS, PVY and PVS, and real time PCR for PVX and phylotype detection in *Ralstonia solanacearum* were standardized.

SOURCE: DARE/ICAR ANNUAL REPORT 2013-14

Crop protection research has brought out tangible and effective methods to suppress various pests in major agro-ecosystems. Indigenous knowledge in this regard has been tested and validated for its suitability.

The PCR-based diagnosis was developed for quick detection and identification of Phytoplasma infecting Cucurbitaceous crops (ash gourd, bitter gourd, bottle gourd and cucumber). *Candidatus phytoplasmaasteries* specific primers based on rDNA conserved region were designed and synthesized. This technique is highly suitable as phytoplasma can be detected before symptom appearance, and even in insect vectors which transmits phytoplasma. In okra, use of petroleum 2T oil @ 2 ml/ litre + 5 g surfactant gave superior control of mites (52. 8%) and jassids (52.1%) over the control. The lowest bacterial wilt incidence (20%) was reported in tomato grown on cocopith: vermicompost: lime (1 : 1 : 0.01), while it was 56.7% in the control plots.

Invasive mealy bug, *Phenacoccus solenopsis*, is emerging as one of the major sucking pests in vegetable crops. The major biological control agents identified for this invasive pest were nymphal endoparasitoids (*Aenasius bombawalei*) Hayat and *Promuscidea unfasciati* Girault, and their cumulative parasitization was 22.35%. These endoparasitoids attack the second or third instar nymphs of mealy bug and complete theirlife-cycle during fifth instar of their host. Different insecticide molecules, viz. Imidaclorpid 70 WG @ 0.1 g/litre, 17.8 SL @ 0.5 ml/litre and Chlorpyriphos 20 SC @ 2 ml/litre were highly effective against mealy bugs.

A severe incidence of minor and sporadic lepidopteran leaf eating caterpillar *Diaphania indica* on cucumber and bitter gourd was observed from August to October. The infestation was 60–70%. *Cotesia glomerata* was identified as an important larval endoparasitoid of *D. indica*. The microbial pesticides, *Bacillus thuringiensis* var. *kurstaki* @ 2 g/litre and *Metarrhizium anisopliae* @ 5 g/litre and insecticide molecules rynaxpyr 18.5 SC @ 0.3 ml/litre, flubendamide 40 SC @ 0.5 ml/ litre, emmamectin benzoate 5 SG @ 0.5 g/litre, spinosad 45 SC@ 0.5 ml/litre, fipronil 5 SC @ 0.25 ml/litre and thiacloprid 21.7 SC @ 0.6 ml/litre were found to be most effective against leaf-eating caterpillar (*D. indica*).

The vegetable samples collected from farmers' fields and local markets showed pesticide residues (23%) in which 14% contained residue levels exceeding the prescribed MRL. Cabbage and cauliflower grown in the dry season, accounted for 43% of the samples found to contain residues followed by brinjal and bhendi. Among the organochlorine (OC) compounds, -endosulfan, - endosulfan and endosulfan sulfate were detected in 7% of the samples and synthetic pyrethroid (SP) compound residues, such as -cypermethrin and -cyhalothrin were detected in 12% of the samples tested. Organophosphorus (OP) compound residues, such as profenophos, chlorpyriphos, monocrotophos and triazophos were found in 12% of the samples found in brinjal, cauliflower, bhendi, green chilli and French bean.

Tuber Crops

The new population of ***Phytophthora infestans*** (mt DNA haplotype Ia), which was introduced in 2002, has almost replaced the old population (Ib haplotype) in all the locations. There is no host-specificity among *P. infestans* isolates of tomato and potato. Further, effect of elevated temperature on efficacy of R genes revealed that resistance of R1 and its combinations may be eroded if *P. infestans* adapts to higher temperature. Validated molecular markers for late blight, potato virus Y and cyst nematode resistance in 165 genotypes and identified 18 genotypes with PVY resistance (*RYadg*), 84 genotypes with late blight resistance genes (*R1* and *R3a*) and 79 genotypes with cyst nematode resistance genes (*HC,H1* and *Gro*1-4). Besides, 16 genotypes were identified possessing multiple resistant genes for late blight, PVY and cyst nematodes.

Potato hybrids (LBY 15, LBY 17 and SM/92-338) possessing combined resistance to late blight and potato virus Y, and hybrid, OS/01-497, having combined resistance to late blight and cyst nematode were introduced for multilocational trials. The application of Pencycuron (0.25%) on potato tubers at planting was most economical and effective against black scurf of potato, while that of Thiacloprid (0.4%) alone or in combination with summer oil against sucking pests.

Tuber information café (TIC) in Malayalam, web- based early-warning system for mealy bug incidence in cassava (http://www.ctcri.org/mbug.php), yield prediction model-based on fuzzy logic for elephant-foot yam and a learning system for the same crop using Artificial Neural Networks were developed.

SOURCE: DARE/ICAR ANNUAL REPORT 2012–13

Potato and Tuber Crops

A PCR-based protocol was validated to detect latent infection of *Phytophthora infestans* in seed **potato** tubers. This could detect the infection up to 20 mm away from the disease lesion. The PCR and RT-PCR protocols for PALCV, PVX, PLRV, PVY, PVS and PSTVd and real time PCR protocols for PALCV, PLRV and PVY have been standardized.

Soil solarization together with incorporation of stable bleaching powder @ 3 q/ha was found effective for control of russet scab disease of potato. Application of Thiacloprid (0.3 and 0.4%) along with summer oil (0.06%) gave longer protection against aphids, while application of Thiacloprid alone was effective against whitefly. Monocropping of potato hybrid OS/93 D 204 in potato cyst nematode infested plots in Nilgiris resulted in the lowest PCN population, followed by the recommended practice of growing potato + French bean intercropping in the autumn season.

SOURCE: DARE/ICAR ANNUAL REPORT 2011–12

In **onion**, re-infestation of thrips is mainly through external migration of adult thrips and eggs that were laid inside the leaves. A predatory thrip, *Aeolothrips mongolicus* was found effectively feeding on thrips in onion plants. Fipronil @60 g a.i./ha was identified as an effective insecticide against thrips. To control thrips in onion during *rabi* season, three sprays of Fipronil @1.5 ml/litre at 15 days interval from 30 days after transplanting were effective.

Soil application of neem cake, followed by spraying of neem soap and neem seed powder were effective for control of leafhopper in okra during *kharif*, but only moderately effective during summer when the pest incidence was very high. Sprinkler irrigation was found to significantly reduce the number of chilli thrips without insecticide spray, while weekly spraying of *M. anisoplaie* @ 1 × 109 spores/ml with sunflower oil @ 0.01%, followed by weekly release of *Blaptostethus pallescens* @ 1/ plant recorded significant reduction in thrips population in chilli and *Capsicum*. The CP gene primer was designed for leaf curl disease of chilli. This gene was cloned and sequenced and submitted to National Centre for Biotechnology Information Database. Similarly, planting of maize and sunhemp in four rows as barrier crop was effective in reducing yellow vein mosaic virus (YVMV) incidence in okra.

Potato and Tuber Crops

In **potato**, dipstick assay was standardized for detection of five viruses, viz. PVX, PVA, PVS, PVM and PVY. Kufri Badshah and Kufri Pukhraj were transformed with replicase gene of potato apical leaf curl virus. Kufri Pushkar was found highly resistant to russet scab and pitted type common scab. Soil application of stable bleaching powder @ 3 q/ha was most effective in managing russet scab. Thiacloprid 240 SC was found effective for managing *Myzus persicae*, *Aphis gossipii* and white flies. Spraying of neem and *ratna jayot*extracts in cow urine @ 10% caused 88 and 83% mortality of glasshouse whitefly. Integrated package of practices for managing late blight and bacterial wilt diseases were recommended specifically for the Hassan area of Karnataka.

SOURCE: DARE/ICAR ANNUAL REPORT 2010–11

Agricultural Acarology

The extent of loss of dry chilli due to yellow mite infestation was 11.39% in Guntur variety and 18.56% in *Byadgi Kaddi* in Karnataka. Byadgi Kaddi variety recorded more number of mites as well as more mite damage compared to Guntur. Screening of chilli lines for reaction against yellow mite, *Polyphagotarsonemus latus*, revealed that PBC-61, Udaipur-2, BVC-47 and BVC-53 were free from yellow mite infestation.

Investigations on biological control of spider mite, *T. urticae*, using phytoseiid predator, *Neoseilus longispinosus* clearly indicated that French bean plants provided optimum food for spider mites up to 50 days and hence the prey mites can be mass produced on French bean plants up to 50 days. When 20 days old French bean plants were infested with 5 spider mites per leaflet and with prey : predator ratio of 100:1, maximum number of predators (up to 125 per leaflet) could be harvested when French bean plants were 65 days old.

Propargite 57 EC at 850 g ai/ha significantly reduced yellow mite on chilli after three rounds of spraying with increased yield of 24.01 tonnes of green chilli fruits/ ha. Spiromesifen 96 g ai/ha, Diafenthiuron 450 g ai/ha, Milbemectin, 4 g ai/ ha, Chlorfenapyr 75g ai/ha and Fenazaquin (125g ai/ha) were significantly more effective in reducing yellow mite population in chilli up to 14 days after each application. Effectiveness of mineral oil at 1 % was more evident with second application when the yellow mite population became low and it was next to synthetic insecticides in effectiveness.

Chilli lines tolerant to mite in Karnataka were: PBC 631, Udaipur-2, BVC-47, CA-14, 7A, 8A X CA14, CMS 2B, PJ X 8O, CMS 7A, CMS8A, CMS6B, CMS 6A and CMS7B. Chilli lines Suryamukhi 4, Canning Suryamukhi 6 and 7, Suryamukhi Bullet 5 and 6, Canning Bullet 3,BC-CH-SL-4, Black Cluster, CH-1, K-1, CO-3 and CO-4 showed moderate resistance to *Polyphagatarsonemus latus* lower yellow mite infestation in CA-71, CH3, CO3, CH1, Punjab Surkh andBC-CH-SL-4 lines of chilli Okra varieties, AOL-04-U2 and JOL-05-07, were found tolerant or less susceptible to spider mite, *T. macfarlanei* due to low leaf hair density in Gujarat.

The efficacy of horticultural mineral oil at 0.75- 1%, was modest with 62% reduction in spider mite population on okra. The effectiveness of mineral oil (causing 80% reduction) on chilli against yellow mite was comparable to that of newer chemical acaricides.

At Kalyani, *T. urticae* tolerant varieties have been identified in pointed gourd (BCPG-5), cowpea (Shwetha and Lola) and okra (P-7 and DSU-1), recording less number of mites on leaves.

Tuber Crops

Marigold and yam bean were found to be effective barrier crops for sweet potato weevil. Bio-intensive management of taro leaf blight was found to be effective in suppression of disease, increasing cormel yield at Dapoli, Dholi, Kalyani, Raipur and Rajendranagar. IDM package for elephant-foot yam was found to be effective in suppressing the diseases and enhancing corm yield. Yam bean seed extract was found to be effective as a biopesticide against aphids and leaf-eating caterpillars. Sweet potato weevil was found in all sweet-potato growing areas. Sweet potato viruses were also noticed in certain areas. Cassava mosaic occurring in serious proportions in Tamil Nadu, Kerala and Andhra Pradesh, was also observed in Faizabad. Taro leaf blight and elephant-foot yam diseases were found common in all taro-growing areas.

SOURCE: DARE/ICAR ANNUAL REPORT 2009–10

Recent Initiatives by Central Potato Research Institute, Shimla

Development of Molecular Diagnostics of Potato Pathogens

Developed PCR based protocols for detection of potato viruses (Gawande *et al.*, 2007; Kaushal *et al.*, 2007 & 2010), bacterial wilt (Grover *et al.*, 2009) and late blight pathogens (Sharma *et al.*, 2010), in seed stocks.

Developed portable dipstick kits for detection of **five major potato viruses at field level using gold nanoparticles**

These kits are portable and easy to use by any stakeholder including farmers (Chakrabarti *et al.*, 2010). This will help in improving the health of potato crop.

PGPR formulation Bio-B5

Identified and patented a biofertilizer-cum-bio pesticide formulation (Bio-B5) for eco-friendly management of soil- and tuber-borne diseases as well as for yield enhancement

It consists of a strain of *Bacillus subtilis* (IMI No. 360643) that releases three antibiotic metabolites namely, iturin, viomycin, and chalcone. The formulation has a killing effect against major pathogens. It also inhibits growth of *Ralstonia solanacearum*, the cause of bacterial wilt of potato against which no chemical or resistant variety is effective till date (Sunaina and Ajay 2005, 2007).

Molecular profiling of potato pathogens

Revealed major population shift of *Phytophthora infestans* and *Ralstonia solanacearum* using molecular markers

A major shift in the distribution pattern of *Phytophthora infestans* and *Ralstonia solanacearum* was detected through molecular characterization of samples collected from different agro-climatic zones of India. Results revealed that the 1a mt haplotype, which was detected in 2002, is gradually replacing the earlier Ib population (Chimote *et al*., 2010). Similarly, molecular phylotyping of *Ralstonia solanacearum* revealed that the potato race (race 3/biovar 2), which was virtually absent in sub-tropical regions, is now establishing in Madhya Pradesh, West Bengal, and Karnataka (Grover *et al*., 2006; Sagar *et al*., 2010).

Potato tuber moth (PTM) management in non-refrigerated storage systems

Developed schedule for management of potato tuber moth under non-refrigerated storage condition

In the absence of refrigerated storages as in case of mid hills of Himachal Pradesh, farmers use rustic stores for avoiding distress sale at harvest and such conditions favour heavy infestation of potato tuber moth (PTM) during storage. Spraying of potatoes with CIPC (chlorpropham) prior to storage under these rustic stores drastically reduced the infestation of PTM (no infestation compared to 30% in untreated potatoes) in five villages of Kangra district and simultaneously resulted in 95% sprout inhibition, thereby enhancing the keeping quality also (Chandel *et al*. 2008, Chandla *et al*. 2008).

Recent Initiatives by Indian Institute of Vegetable Research, Varanasi

Brinjal

- Brinjal shoot and fruit borer (BSFB) can be managed by selection of oblong/small fruited cluster bearing varieties and hybrids, installing sex pheromone baited plastic funnel trap at 10 m distance from 20 DAT, the pheromone septa should be changed at 45 days interval, along with clean cultivation through clipping of infested shoot with larvae inside at weekly interval from 15 DAT until the shoot infestation is lost and removal of freshly damaged fruits at each harvest. In case of severe infestation need based foliar spray of neem seed kernel extract (4%) may be given.

- Intercropping of brinjal (2 rows) with coriander (one row) or fennel (1 row) minimizes the incidence of brinjal shoot and fruit borer.
- Seed treatment with imidacloprid @ 2.5g/kg reduces the leafhopper and little leaf incidence.
- Spraying carbendazim @ 0.1% gave good control of Phomopsis blight.
- Seed treatment with captan @ 0.25% and soil treatment with *Trichoderma* powder @ 5kg/ha gave good control against damping off.
- Soil drenching twice with *Trichoderma* @ 1% near collar region provided protection against collar rot and wilt diseases.

Tomato

- Planting of two rows of African marigold as trap crop with every 14 rows of tomato along with, installation of *H. armigera* pheromone traps @ 4/ha to monitor the initial attack of pest, innundative release of *Trichogramma brassiliensis* @ 2, 50,000 parasitized eggs/ha at 10 days interval at the beginning of flowering and fruiting and spray of *Ha* NPV @ 350 LE/ha twice after flowering can manage the fruit borer. Use of synthetic pyrethroids and endosulfan alternatively with NSKE (4%) is effective against fruit borer.
- Marigold was also identified as suitable intercrop/border crop for leafminers as well as fruit borer management apart from harboring activity of major parasitoid of leafminer *Neochrysocharis Formosa,* the fruit borer parasitoids *Microplitis* sp, *Cotesia* sp., *Campoletis chlorideae* and a polyembryonic parasitoid, *Copidosoma* sp.
- Seed treatment with imidacloprid, netting of nursery, spray 4% NSKE minimize whitefly population, TLCV and leaf miner incidence. Pre-transplanting application of imidacloprid @ 0.3 ml/lit in nursery protects the newly planted crop from whitefly infestation.
- Soil application of neem cake @ 100g +*Trichoderma*@10g/ sq. m in nursery bed controlled post-emergence damping off completely.
- Summer ploughing, green manuring with sunhemp/dhaincha, application of *Trichoderma* powder @ 5 kg/ha and soil drenching twice with *T. viride* @ 1% controlled wilt disease.
- Single spray of copper oxychloride@ 0.3% in kharif and mancozeb@ 0.25% in winter gave good control against early blight.
- *Alternaria* blight appears in 11th standard meteorological week (SMW) at diurnal temperature 14.3-32.1°C and progressed maximum in 14th SMW at an avg. temperature of 24.7°C.

Chilli

- The mite infestation in early crop growth phase reduces the yield to a greater extent.
- At cellular level, the difference in damage symptoms caused by mites and thrips were established.
- Actinomycetes, a bacterium (Gram negative) a fungus A. phytoseid predator *Amblyseius* sp. and anthocorid bug (*Orius* Sp) was found frequently associated with *P. latus*. Actinomycetes and bacteria caused mortality to the extent 20%.

- The desirable varieties/lines with higher tolerance index were identified as Japanese longi, NEC, VR 338, G-4, PBC-535 and VNS-4.
- Chlorfenapyr @ 1.5 ml/l, abamectin @ 1.5 ml/l and spiromesifen @ 0.75 ml/l were found highly effective against yellow mite and thrips in chilli.
- Soil application of neem cake @ 100g and *Trichoderma*@10 g/m^2 in nursery followed by seed treatment with carbendazim @ 2.5 g/kg of seed controlled damping off.
- Two sprays of flusilazole @ 0.1% gave a good control of anthracnose.
- Seedling dip in imidacloprid @ 0.25% solution for 30 minute followed by seedling dip in *Trichoderma* @ 1% for 10 minutes gave protection against soil-borne pathogen and whitefly.
- Die-back and anthracnose: *Colletotrichum capsici* infection initiated from last week of September and progressed to 3rd week of October. 22-32°C, >90% RH with rainfall are the predisposing factors.

Okra

- June first week sowing is most ideal to avoid the damage due to shoot and fruit borer during rainy season.
- Seed treatment with imidacloprid @ 3 g/kg protects the crop from jassid damage upto 50-60 days of sowing.
- Foliar spraying of dicofol @ 2.5ml/l and wettable sulphur (0.2%) alternatively at 10 days interval effectively controls spider mite.
- Rove beetle, *Paederus varicornis* is identified as an important predator of jassids
- 2-3 foliar spray of difenconazole or hexaconazole @ 0.05% gave appreciable protection against leaf spot and powdery mildew diseases.
- Tolerant varieties to YVMV like VRO-5 and VRO-6 should be used.
- Seed treatment with carbendazim @ 0.3% should be done for good establishment of the seedlings.

Cole Crops

- Chinese cabbage was identified as an ideal trap crop against DBM in cabbage and cauliflower crop.
- Diamondback moth population in cabbage can be managed by growing paired mustard rows with every 25 cabbage rows. The first row of mustard should be sown 15 days before and second 25 days after cabbage planting. Dichlorovos 0.1% is sprayed at weekly intervals on the mustard to kill the trapped DBM larvae. If required NSKE (5%) may be sprayed to control the larval population in main crop.
- Sowing of coriander as border crop and foliar spray of entomopathogenic fungal (*Verticillium lecani*) formulation @ 500 g/ha at 10 days interval is most effective in the management of cabbage aphid.
- Spray of *Bt* formulations (Delfin®/Dipel®) @ 500 g/ha at 35 and 50 days after planting were found effective against DBM in cabbage.
- The *Spodoptera litura* in early cauliflower and cabbage can be managed by hand removal and destruction of egg masses followed by three foliar sprays of SlNPV

@ 300 LE/ha mixed with sticker (0.05%) and jaggery (2%) at an interval of 10 days. NPV spraying should be done at early morning and late evening to reduce the adverse effect of UV irradiation on NPV solution.

- Nursery application of neem cake @100g/m^2 and seed soaking in streptocycline @100 ppm protected the damping off and bacterial rot pathogens.
- Removal of lower leaves in the morning, burning them and a subsequent spray of mancozeb @ 0.25% along with a sticker @ 0.1% gave a good control against *Alternaria* leaf spot.

Cucurbits

- Fruit fly can be managed through mass trapping of adult males using mineral water bottle traps (25-30 traps/ha) baited with cuelure (1 part), ethyl alcohol (6 parts) and carbaryl (2 parts) saturated wood blocks at 15 mt distance from trap to trap coinciding with initiation of flowering along with bait spray containing 10% molasses and 0.2% carbaryl applied in one plant per 100 m^2 area.
- Food bait consisting of rotten banana /pumpkin pulp (1000g) + yeast (10 g) + malathion (10 ml) + citric acid (3 g) may be used in flat earthen pots to attract and kill the adult fruit flies.
- Bait spray containing 20 ml malathion (50 EC) + 20 liters water +500 g molasses attract adult flies and control the population.
- Spraying of carbaryl 50 WP @ 2 g/lit at cotyledon stage control red pumpkin beetles.

Pea

- Two foliar sprays of penconazole@0.05% or flusilazole @0.1% gave a good control against powdery mildew disease.

Recent Initiatives by Indian Agricultural Research Institute, New Delhi

Management of Vegetable Diseases

Out of 29 cauliflower and 19 cabbage genotypes screened for black rot resistance by artificial inoculation, none showed complete resistance, while cauliflower genotypes, DC 5, KT 15, 1385, Grandess, KT 2 MR, Pusa Meghna and cabbage genotypes KGMR 1, 83-6, AC 204, Pusa Ageti, Kinner Red, Red Cabbage, Red Rock Mammoth, S 645, S 696 and KTCBH 81 exhibited partial resistance. The disease management module comprising the use of yellow coloured seeds, nursery root dipping with Bavistin (0.5%) followed by *Trichoderma viride* enriched compost application 15 days after transplanting and need based fungicide/ insecticide application was found to be most promising for seed crop of capsicum var. California Wonder. Seed treatment with metalaxyl, one foliar spray of propiconazole (0.1%) and removal of 4 lower leaves at 60 DAS were found to be most effective for the management of mustard diseases, and only 11.6% *Alternaria* blight disease was recorded in comparison to 53.3% in control plots.

Seed health testing

Vegetable crop seeds when tested for their health, showed a total of 31 fungi and the fungal infection varied from 0.2% in tomato seed cv. Pusa Ruby to 6.7% in bitter gourd cv. PDM. Maximum fungal infection on the seed samples was due to *Alternaria alternata* (23.5%) followed by *Aspergillus flavus* (22.5%), *Rhizopus* sp. (13.04%) and *Aspergillus niger* (10.6%). Twenty six fungi were recorded on the seed of paddy varieties and maximum fungal infection was of *Aspergillus flavus* (26.7%) followed by *Alternaria padwickii* (18.6%), *Curvularia lunata* (8.9%) and *Fusarium moniliforme* (6.9%). Seed germination of the paddy samples varied from 6.0 to 91.8% and seed vigour index varied from 60 to 1682.

Incidence of viral disease on vegetables

In cucurbits (bottle gourd) occurrence of CMV (3%), PRSV-W (8%) and ZYMV (1%) was observed. In tomato fields, maximum incidence was of GBNV (15%). Thrips (*Thrips tabaci*) on tomato and white fly (*Bemisia tabaci*) on cucumber were observed.

Management of viral diseases

Mulching of plastic showed comparatively less aphid and thrips population in capsicum fields. Bottle gourd plants sown in June and sprayed fortnightly with imidacloprid (0.01%) showed significantly lowest incidence of virus diseases (2.71%) and aphids (3.28/pl.) with highest yield (35.69 t/ha)). Management of virus diseases in tomato revealed that CMV, GBNV and ToLCV could be controlled by spray of Agroneem (2ml/l) /Dimethoate (0.05%). In capsicum, Mulching + Neem oil (2%) + Imidacloprid (0.01%) was most effective in reducing the incidence of PVY, PMMoV and CMV. The combined incidence of virus diseases was 8.33 % in treated plot as compared to control (23.80%).

Viral and Phytoplasmal Diagnosis

Garlic common latent virus (GarCLV)

The coat protein (CP) gene (960 bp) of five Indian *Garlic common latent virus* (GarCLV) isolates was sequenced. Sequence comparisons revealed that the CP of GarCLV isolates from India shared 98-100% amino acid sequence identity and were grouped with isolates from Australia, Brazil, Japan and South Korea.

Bean common mosaic virus (BCMV)

Association of BCMV with French bean samples collected from Sikkim was established through DAS-ELISA and RTPCR. Coat protein gene of BCMV from cv. Allopetri was cloned and sequenced (NCBI Genebank accession no. JQ753313). *In silico* recombination analysis of 76 sequences of BCMV revealed evidence for 11 putative recombinants. However, Jammu and Kashmir as well as Sikkim isolates of BCMV exhibited no recombination indicating their independent evolution/stability.

Lettuce Mosaic Virus (LMV)

Mild mosaic and stunted growth symptoms were observed in 19 out of 38 plants of lettuce (*Lactuca sativa* L, Accession No. EC687345, variety NVRS-10:001818) at IARI experimental farm, New Delhi. Leaf-dip electron microscopy revealed flexuous virus particles measuring 750 × 13 nm suggesting the association of a potyvirus. The coat protein gene (834bp) (CP) of the virus infecting lettuce shared 96 to 100% identity at amino acid level with the corresponding regions of LMV isolates, suggesting natural infection of LMV infecting *Lactuca sativa.*

Dual antigen construct for simultaneous diagnosis of Potato virus Y (PVY) and Potato virus X (PVX)

The conserved coat protein (CP) gene regions of PVY (246 bp) and PVX (243 bp) were sub-cloned in frame in pET-28b (+) vector to create a dual expression construct. Cocktail polyclonal antibodies (PAbs) generated from fusion protein (~20 kDa) efficiently detected mixed infection of PVY and PVX in infected potato sample in DAC-ELISA. The Cocktail PAb was further validated with 70 field potato samples, where it simultaneously detected both PVY and PVX in 66 samples with the O.D. range 0.10-2.15. This is the first demonstration of engineered fusion antigen to PVX and PVY, the most important viruses affecting potato all over the world.

Insect Pest Management

Three foliar sprays of profenophos+cypermethrin mixture @ 440 g a.i./ha at fortnightly interval recorded lowest fruit damage in bitter gourd due to fruit fly, *Bactrocera cucurbitae* (10.71% on number and 8.23% weight basis). Okra border-cropped with maize had significantly reduced leafhopper population.

Brinjal border-cropped with maize followed by sprays of flubendiamide @ 25 g a.i./ha, chlorantraniliprole @ 10 g a.i./ha and novaluron @ 75 g a.i./ha in sequence at fortnightly interval was most successful in minimizing shoot and fruit borer damage in brinjal fruits.

Nematology

Metham Sodium for the Management of Root Knot Nematode in Polyhouse

Root-knot nematodes were effectively reduced in tomato cultivation under polyhouse conditions, on applying Metham sodium @ 300 l/ha with a dilution 1:7 (30ml/m2) deep into furrow (25cm depth) one month before transplanting of tomato seedlings.

Bioefficacy of Indazoles in Tomato Infected with *Meloidogyne incognita* Race 1

L16 {6-(4-fluorophenyl)-4,5-dihydro-4-Phenyl- 2H-indazol-3-ol} @ 125 ppm onwards showed increase in plant growth of tomato and maximum reduction (84.5%) in number of galls compared to the control.

Management through Transgenic Approach

Several transgenic brinjal lines have been developed to express the double stranded RNA of four esophageal gland genes and two FMRFamide like peptide genes of *Meloidogyne incognita* independently to bring about their silencing in the nematodes. Evaluation on the effect of gene silencing showed 60 to 80% reduction in multiplication factor of the nematode indicating their potential in reducing the population pressure in the soil. QPCR analysis indicated good expression of different genes in various primary events and also in the plants of T1 generation.

CHAPTER 21

Glossary of Common Horticultural Terms

S.L. Chawala

Associate Professor (Horticulture)
ASPEE College of Horticulture and Forestry,
Navsari Agricultural University, Navsari, Gujarat

- **Abscission layer:** Specialized cells, usually at the base of a leaf stalk or fruit stem, that trigger both the separation of the leaf or fruit and the development of scar tissue to protect the plant.
- **Abscission:** The dropping of leaves, flowers or fruit by a plant. Can result from natural growth processes (e.g., fruit ripening) or from external factors such as temperature or chemicals.
- **Absorption:** The intake of water and other materials through root or leaf cells.
- **Accelerator:** Any substance used to assist the decomposition of organic matter in the preparation of garden compost.
- **Accumulated heat units:** Number of hours in a growing season. Usually calculated at temperatures above 50°F, but can be calculated at other temperatures, depending on the crop.
- **Acid soil:** Soil with pH below 7 on a pH scale of 1 to 14. The lower the pH, the more acidic the soil.
- **Active ingredient:** The chemical in a pesticide formulation that actually kills the target pest.
- **Additive:** A substance that, when added to a pesticide, reduces the surface tension between two unlike materials (e.g., spray droplets and a plant surface), thus improving adherence. Also called an adjuvant or surfactant.
- **Adventitious bud:** A bud in an unusual place on a plant, often on internodes. May be the result of injury. Suckers and water sprouts usually grow from adventitious buds.
- **Adventitious root:** A root in an unusual place, often where a branch contacts soil or damp material. A plant cannot be reproduced from cuttings or layering unless adventitious roots develop.

- **Adventitious:** Growth not ordinarily expected, usually the result of stress or injury. A plant's normal growth comes from meristematic tissue, but adventitious growth comes from non-meristematic tissue.
- **Aeration:** Introduction of air to compacted soil by mechanically removing plugs of topsoil. Aeration helps oxygen, water, fertilizer and organic matter to reach roots. Also, incorporating air into compost by turning the pile. Incorporating air into compost helps the plant material decompose more quickly.
- **Aeration:** Mechanically loosening or puncturing soil to increase permeability to water and air.
- **Aerial Phytophthora:** A highly transmissible fungal disease which will devastate Madagascar Periwinkle (vinca) if they are planted before mid-May, receive less than full sun and are given overhead watering.
- **Aerobic:** Active in the presence of free oxygen.
- **Aerobic:** Requiring available oxygen to live. For example, composting is performed by aerobic organisms; but rotting occurs under anaerobic conditions.
- **After-ripening:** The seed maturation process that must be completed before germination can occur.
- **Aggregation:** The process by which individual particles of sand, silt and clay cluster and bind together to form soil peds.
- **Agronomy:** Applied agricultural science dealing with rural economy and husbandry. In recent years it has concentrated mainly in the theory and practice of crop production and soil management.
- **Alkaline Soil:** In such soils, elements like iron and zinc are not readily available to plants. The characteristic reaction of soils derived from limestone.
- **Alkaline soil:** Soil with pH above 7 on a pH scale of 1 to 14. The higher the reading, the more alkaline the soil.
- **Allelopathy:** The excretion by some plants of compounds from their leaves and/or roots that inhibit the growth of other plants.
- **Alternaria:** A genus of fungi containing plant pathogens. Causes the following: Citrus fruit black rot, early blight of potato or tomato, gray leaf spot on cabbage, poinsettia blight, tomato stem canker, etc.
- **Ammonium (NH_4+):** A plant available form of nitrogen contained in many fertilizers and generated in the soil by the breakdown of organic matter.
- **Anaerobic:** Active in the absence of free oxygen.
- **Anaerobic:** Requiring the absence of oxygen to live.
- **Anion:** A negatively charged ion. Plant nutrient examples include nitrate (NO_3), phosphate ($H_2PO_4^-$) and sulphate (SO_4^-).
- **Annual:** A plant that germinates, grows, flowers, produces seed and dies in the course of a single growing season.
- **Anther:** The pollen-bearing part of a flower's male sexual organ. The filament supports the anther; together they are referred to as the stamen.

- **Anthracnose:** A disease caused by the fungus *Colletotrichum*, not *Gloeosporium* as stated by some. it is characterized by sunken, dark lesions or spots in leaves. May be observed in: Ash, Elm, Mulberry, Red Oak, etc.
- **Anvil pruner:** A pruning tool that cuts a branch between one sharpened blade and a flat, anvil-shaped piece of metal. Has a tendency to crush rather than make a smooth cut.
- **Apex:** The tip of a stem or root.
- **Apical dominance:** The inhibition of lateral bud growth by the presence of the hormone auxin in a plant's terminal bud. Removing the growing tip removes auxin and promotes lateral bud break and subsequent branching, usually directly below the cut.
- **Arboretum:** An area devoted to specimen plantings of trees and shrubs.
- **Aspect:** Direction of exposure to sunlight.
- **Assimilation:** Building of cell matter from inorganic and organic materials (carbohydrates and sugars).
- **Attractant:** A material that lures pests.
- **Auxin:** One of the best known and most important plant hormones. Most abundantly produced in a plant's actively growing tips. Generally stimulates growth by cell division in the tip region and by cell elongation lower down the shoot. Growth of lateral buds is strongly inhibited by the normal concentration of auxin in the growing tip.
- **Available water supply:** Soil water that is available for plant uptake. Excludes water bound tightly to soil particles.
- **Axil:** The upper angle formed by a leaf's stalk (petiole) and the internodes above it on a stem.
- **Axillary bud:** A bud that forms in an axil.
- ***Bacillus thuringiensis* (Bt):** A bacterium used as a biological control agent for many insect pests.
- **Bacterium:** Usually an unicellular, prokaryotic organism without a nucleus. Some cause diseases of plants.
- **Bacterium:** A single-celled, microscopic organism having a cell wall but no chlorophyll. Reproduces by cell division.
- **Balled and burlapped (B&B):** A plant dug with soil. The root ball is enclosed with burlap or a synthetic material.
- **Band:** To apply a pesticide or fertilizer in a strip over or along each crop row.
- **Bare-root (BR):** A plant with little or no soil around its roots; a common method of selling deciduous plants and small evergreens.
- **Basal break:** New growth that develops at the base of a branch or near a plant's crown.
- **Basal:** (1) At or near the base of a branch or trunk. (2) At or near a plant's crown.
- **Bed:** An area of raw soil that is usually planted with flowers or vegetables. Also known as flower beds.

- **Beneficial insect:** An insect that helps gardening efforts. May pollinate flowers, eat harmful insects or parasitize them, or break down plant material in the soil, thereby releasing its nutrients. Some insects are both harmful and beneficial. For example, butterflies can be pollinators in their adult form but destructive in their larval (caterpillar) form.
- **Berry:** The fleshy fruit of cane fruits, bush fruits and strawberries.
- **Biennial bearing:** Producing fruit in alternate years.
- **Biennial:** A plant that germinates and produces foliage and roots during its first growing season, then produces flowers and seeds and usually dies during its second growing season.
- **Biosolids:** A by-product of wastewater treatment sometimes used as a fertilizer.
- **Blanch:** To exclude light from plants or parts of plants to render them white or tender. Often done to cauliflower, endive, celery and leeks.
- **Blight:** A fast-spreading fungal or bacterial disease which causes dead patches on stems or leaves and can kill the whole plant.
- **Blight:** Rapid, extensive discoloration, wilting and death of plant tissue.
- **Blossom end rot:** In tomatoes, not a rot, but a dying back of the distal (farthest from stem) end of the fruit. Occurs when there is insufficient soluble calcium in the soil and inconsistent watering.
- **Blotch:** A blot or spot (usually superficial and irregular in shape) on leaves, shoots, or fruit.
- **Bolting:** Producing seed or flowering pre-maturely, usually due to heat. For example, cool-weather crops such as lettuce bolt during summer. Leaf crops are discouraged from bolting by removal of flower heads.
- **Bonsai:** One of the fine arts of gardening; growing carefully trained, dwarfed plants in containers selected to harmonize with the plants. Branches are pruned and roots trimmed to create the desired effect.
- **Border:** Usually features decorative plants or garden accessories such as bricks and gravel which are located alongside another landscape feature such as a fence or patio.
- **Botanical insecticide:** An insecticide, such as rotenone or pyrethrum, derived from a plant. Most botanicals biodegrade quickly. Most, but not all, have low toxicity to mammals.
- **Botrytis:** A genus of fungi containing plant pathogens.
- **Botrytis:** A fungal disease promoted by cool, moist weather. Also known as gray mold or fruit rot.
- **Bract:** A modified leaf, usually small, but sometimes large and brightly coloured, growing at the base of a flower or on its stalk. Clearly seen on dogwoods and poinsettias.
- **Bramble:** A spiny cane bush with berry fruits (e.g., raspberries and blackberries).
- **Branch:** A subsidiary stem arising from a plant's main stem or from another branch.
- **Break:** (1) Any new growth coming from a bud. (2) See Bud break.

- **Broadcast:** (1) To sow seed by scattering it over the soil surface. (2) To apply a pesticide or fertilizer uniformly to an entire, specific area by scattering or spraying it.
- **Broadleaf evergreen:** A non-needled evergreen.
- **BTU (British thermal unit):** Amount of heat required to raise the temperature of 1 pound of water 1°F.
- **Bud break:** The resumption of growth by resting buds.
- **Bud head:** A swollen or enlarged area where a bud was grafted to a stock.
- **Bud scale:** A modified leaf that forms a protective covering for a bud.
- **Bud union:** The suture line where a bud or scion was grafted to a stock. Sometimes called a *graft union.*
- **Bud:** A small protuberance on a stem or branch, sometimes enclosed in protective scales, containing an undeveloped shoot, leaf, or flower.
- **Budding:** The grafting of a bud onto stock of a different plant. The bud is the scion.
- **Budstick:** A shoot or twig used as a source of buds for budding.
- **Bulb:** A true bulb is thought to be a complete or nearly complete miniature of a plant that is enclosed in fleshy leaves called scales, which contain a provision of reserve food. If you were to cut a hyacinth bulb in half (bud to root) it is most interesting to see that the flower is housed there, waiting to emerge into our world. Wrapped around a bulb is a thin outside covering called a tunic. These are like flaky leaves. The basil plate which is located at the base of the bulb holds the food storing scales together, and new roots will sprout from the outside edge of the basil plate once the bulb ends its dormant period and begins its growth cycle. Daffodils, hyacinth, lily, tulips and onions are what are called true bulbs.
- **Bulb:** An underground storage organ consisting of a thin, flattened stem surrounded by layers of fleshy, dried leaf bases. Roots are attached to the bottom.
- **Bulbil:** A small bulb like organ that sometimes forms in place of flowers.
- **Bulblet:** (1) An underground bulbil. (2) A tiny bulb produced at the base of a mother bulb.
- **C:N ratio:** The ratio of carbon to nitrogen in organic materials. Materials with a high C:N ratio (high in carbon) are good bulking agents in compost piles, while those with a low C:N ratio (high in nitrogen) are good energy sources.
- **Calcium carbonate ($CaCO_3$):** A compound found in limestone, ashes, bones and shells; the primary component of lime.
- **Callus:** Tissue that develops when cambium or other meristematic tissue is wounded.
- **Calorie:** Amount of heat required to raise the temperature of 1 cubic centimetre of water 1°C.
- **Calyx:** The collective term for the sepals (the cup, usually green, between a flower and its stem).
- **Cambium:** The living, growing layer of cells between the xylem and phloem. In woody plants, it is located just beneath the bark.

- **Candelabrum:** A strong, dominant rose cane with accelerated growth that originates from a bud union and explodes with many blooms.
- **Candle:** On a pine tree, new terminal growth from which needles emerge.
- **Cane:** The externally woody, internally pithy stem of a bramble or vine.
- **Canker:** A sunken lesion caused by the breakdown of tissues from the phloem and outward.
- **Canker:** A localized lesion on a limb or trunk, usually due to disease or injury. Part of the bark or wood appears to be eaten away or is sunken.
- **Canopy:** (1) The top branches and foliage of a plant. (2) The shape producing structure of a tree or shrub.
- **Capillary Force:** The action by which water molecules bind to the surfaces of soil particles and to each other, thus holding water in fine pores against the force of gravity.
- **Capitulum:** (1) A dense, short, compact cluster of sessile flowers, as in composite plants or clover. (2) A very dense grouping of flower buds, as in broccoli.
- **Catfacing:** Disfigurement or malformation of a fruit. Fruits typically affected include tomatoes and strawberries. Al- though not fully understood, cat-facing is thought to be caused by insects or adverse weather during fruit development.
- **Cation exchange capacity (CEC):** A soil's capacity to hold cations as a storehouse of reserve nutrients.
- **Cation:** A positively charged ion. Plant nutrient examples include calcium (Ca++) and potassium (K+). (See also Anion.)
- **Cell:** The smallest structure in a plant.
- **Centralleader:** (1) A trunk or stem extending up through the axis of a tree or shrub and clearly emerging at the top.
- **Cercus (pl., cerci):** A thread-like or sometimes forceps-like tail near the tip of an insect's abdomen (usually a pair).
- **Chelate:** A complex organic substance that holds micronutrients, usually iron, in a form available for absorption by plants.
- **Chilling:** The chilling hours a particular clone needs before over-wintering buds will open.
- **Chlorophyll:** The green matter in plants and the essential molecule for photosynthesis.
- **Chlorophyll:** The green pigment in plants. Responsible for trapping light energy for photosynthesis.
- **Chloroplast:** A specialized component of certain cells. Contains chlorophyll and is responsible for photosynthesis.
- **Chlorosis:** A foliar symptom of a mineral deficiency, stress or disease; the absence of chlorophyll. Inter-veinal chlorosis (the veins remain darker than the remainder of the leaf blade) usually indicates a lack of iron.
- **Chlorosis:** Yellowing of a leaf.

- **Clay:** The smallest type of soil particle (less than 0.002 mm in diameter).
- **Climber:** A plant that climbs on its own by twining or using gripping pads, tendrils, or some other method to attach itself to a structure or another plant. Plants that must be trained to a support are properly called trailing plants, not climbers.
- **Cloche:** A plastic, glass, or plexiglass plant cover used to warm the growing environment and protect plants from frost.
- **Clone:** A plant group whose members have all been derived from a single individual through constant propagation by vegetative (asexual) means, e.g., by buds, bulbs, grafts, cuttings or laboratory tissue culture.
- **Cold frame:** A plastic-, glass-, or plexiglass-covered frame that relies on sunlight as a source of heat to warm the growing environment for tender plants.
- **Cole crops:** A group of vegetables belonging to the cabbage family; plants of the genus *Brassica*, including cauliflower, broccoli, cabbage, turnips and brussels sprouts.
- **Coleoptera:** An insect family made up of species having horny front wings that fit over their hindwings. Includes beetles and weevils.
- **Collar:** A swollen area at the base of a branch where it connects to a trunk. Contains special tissue that prevents decay from moving downward from the branch into the trunk.
- **Compaction:** Pressure that squeezes soil into layers that resist root penetration and water movement. Often the result of foot or machine traffic.
- **Companion planting:** The practice of growing two or more types of plants together in the hope that the combination will discourage disease and insect pests.
- **Compatible:** Different varieties or species that set fruit when cross-pollinated or make a successful graft union when intergrafted.
- **Complete metamorphosis:** A type of insect development in which the insect passes through the stages of egg, larva, pupa and adult. The larva usually is different in form from the adult.
- **Compost:** Humus made by decomposing vegetative matter in a compost bin or pile.
- **Compost:** The product created by the breakdown of organic waste under conditions manipulated by humans. Used to improve both the texture and fertility of garden soil.
- **Compound bud:** More than one bud on the same side of a node. Usually, unless growth is extremely vigorous, only one of the buds develops, and its branch may have a very sharp angle of attachment. If it is removed, a wider angled shoot usually is formed from the second (accessory) bud. Ashes and walnuts are examples of plants that typically have compound buds.
- **Conifer:** A cone-bearing tree or shrub, usually evergreen. Pine, spruce, fir, cedar, yew and juniper are examples.
- **Conk:** A fungal fruiting structure (e.g., shelf or bracket fungi) formed on rotting woody plants.

- **Contact pesticide:** A pesticide that kills on contact.
- **Contract:** It is a signed document that outlines the work to be performed by the installer and a schedule of payments to be made by the customer, and the circumstances under which the work will be done and payments will be made.
- **Cordon:**(1) A method of espaliering fruit trees, vines, etc. to horizontal, vertical, or angled wire or wooden supports so maximum branch surface is exposed to the sun, resulting in maximum fruit production. (2) A branch attached to such a support.
- **Corm:** An underground storage organ consisting of the swollen base of a stem with roots attached to the underside. Crocus and gladiolus are examples of plants that form corms.
- **Cormel:** A small, under developed corm, usually attached to a larger corm.
- **Cornicle:** A short, blunt horn or tube (sometimes button-like) on the top and near the end of an aphid's abdomen. Emits a waxy liquid that helps protect against enemies.
- **Corolla:** Collectively, all of a flower's petals.
- **Cortex:** Cells that make up the primary tissue of roots or stems.
- **Corymb:** A usually flat-topped flower cluster in which the individual flower stalks grow upward from various points on the main stem to approximately the same level.
- **Cotyledon:** A seed leaf; the first leaf from a sprouting seed. Monocots have one cotyledon; dicots have two.
- **Cotyledon:** Usually the first seedling leaf to appear when a seed germinates. Some flowers, such as lilies, will only produce one leaf, while most others will produce two.
- **Cover crop:** A crop dug into the soil to return valuable organic matter and nitrogen to the soil. Legumes such as clover, cowpeas and vetch are common cover crops.
- **Cross-pollination:** The transfer of pollen from one plant variety to another.
- **Cross-pollination:** The fertilization of an ovary on one plant with pollen from an- other plant, producing an offspring with a genetic makeup distinct from that of either parent.
- **Crotch angle:** The angle formed between a trunk and a main scaffold limb. The strongest angle is 45° to 60°.
- **Crown Gall:** A grotesquely-enlarged, abnormal body on the stem of a plant near the soil surface. Usually caused by a bacterial infection.
- **Crown:** (1) Collectively, the branches and foliage of a tree or shrub. (2) The thickened base of a plant's stem or trunk to which the roots are attached.
- **Crown:** The part of a plant where the roots and stem meet, usually at soil level.
- **Cultivar:** A specially cultivated variety of a plant that most often is reproduced vegetatively. For example, Transparent'is a cultivar of apple.
- **Cultivars:** A cultivated variety of plant, often bred for a desired trait, such as pest or disease resistance, flower colour, fruit colour and/or persistence, habit/size, foliage colour/texture, etc.
- **Cuticle:** (1) A relatively impermeable surface layer on the epidermis of leaves and fruits. (2) The outer layer of an insect's body.

- **Cutin:** (1) A waxy substance on plant surfaces that tends to make the surface waterproof and can protect leaves from dehydration and disease. (2) A waxy substance on an insect's cuticle that protects the insect from dehydration.
- **Cutting:** A plant propagation method wherein a part of a plant is cut and dipped in a rooting hormone to eventually grow into a new plant.
- **Cutting:** A piece of leaf, stem or root removed from a plant and prompted to develop into a new plant that is genetically identical to the parent plant.
- **Cyme:** A flower stalk on which the florets start blooming from the top of the stem and progress toward the bottom.
- **Cyst:** The swollen, egg-containing female body of certain nematodes. Can be seen on the outside of infected roots.
- **Damping off:** Death of a seedling by rot occurring when there is inadequate light and ventilation, as well as too much water and pathogen.
- **Damping-off:** A disease caused by many different organisms. In the most conspicuous cases, a seedling's stem collapses at or near the soil surface, and the seedling topples. Another type rots seedlings before they emerge from the soil or causes seeds to decay before germinating.
- **Day-neutral plant:** A species capable of flowering without regard to day length. (See also Short-day plant, Long-day plant.)
- **Deadhead:** To remove individual, spent flowers from a plant for the purpose of preventing senescence and prolonging blooming. For effective results, the ovary behind the flower must be removed as well.
- **Deciduous:** A plant that sheds all of its leaves annually.
- **Decomposition:** The breakdown of organic materials by microorganisms.
- **Defoliation:** The un-natural loss of a plant's leaves, generally to the detriment of its health. Can be caused by high wind, excess heat, drought, frost, chemicals, insects, or disease.
- **Dehorning:** A drastic method of pruning a neglected tree or shrub. Entails the removal of large branches, especially high in the crown, a few at a time over several seasons.
- **Dermaptera:** An in-sect family made up of species having chewing mouthparts and a pair of large, forceps-like appendages near the tail. Wingless or with one or two pairs of inconspicuous wings. Earwigs are an example.
- **Desiccation:** Drying out of tissue.
- **Determinate:** A plant growth habit in which stems stop growing at a certain height and produce a flower cluster at the tip. Determinate tomatoes, for example, are short, early-fruiting, have concentrated fruit set and do not require staking.
- **Dethatch:** To remove thatch (a tightly intermingled layer of stems, leaves and roots, living and dead, that forms between the soil surface and green vegetation of grass).
- **Diatomaceous earth:** The fossilized remains of diatoms (a type of tiny algae) used to kill insect pests, snails and slugs.
- **Dicotyledon, dicot:** A plant having two cotyledons.

- **Dieback:** Progressive death of shoots, branches, or roots, generally starting at the tips.
- **Differentiation:** A change in composition, structure and function of cells and tissues during growth.
- **Dioecious:** A plant species having male and female flowers on separate plants. An example is holly. For female holly to produce berries, a male holly must be in the vicinity to provide pollen.
- **Disbud:** The selective removal of some flower buds so that the remaining buds receive more of the plant's energy and produce larger, showier flowers. Roses, chrysanthemums and camellias are often disbudded.
- **Division:** The breaking or cutting apart of a plant's crown for the purpose of producing additional plants, all genetically identical to the parent plant.
- **Dormancy:** The annual period when a plant's growth processes greatly slow down.
- **Dormant bud:** A bud formed during a growing season that remains at rest during the following winter or dry season. If it does not expand during the following growing season, it is termed latent.
- **Dormant oil:** Horticultural oil applied during the dormant season to control insect pests and diseases.
- **Dormant:** Resting or not growing. A deciduous tree is dormant in the winter.
- **Double worked:** Grafted twice, i.e., grafted to an intermediate stock.
- **Double, semidouble:** A flower with more than the normal number of petals, sepals, bracts, or florets. May be designated botanically by the terms flore pleno, plena or pleniflora.
- **Downy mildew:** A fungal disease of plants characterized by visible, off-white spore masses on the leaf underside.
- **Drainage:** The movement of water through the soil. With good drainage, water disappears from a planting hole in less than a few hours. If water remains standing overnight, drainage is poor.
- **Drainage:** The ability of soil to transmit water through the surface and subsoil.
- **Drip irrigation:** This is the practice of applying water slowly through various types of pipes, tubes or specialized hoses. The simplest form would be the use of soaker hoses commonly found in garden centers. More sophisticated systems can be installed by homeowners or irrigation contractors that are preset to go on by a computer.
- **Drip zone:** The area from the trunk of a tree or shrub to the edge of its canopy. Most, but not all, of a plant's feeder roots are located within this area.
- **Dripline:** An imaginary line on the ground directly beneath the outermost tips of a plant's foliage. Rain tends to drip from leaves onto this line.
- **Dwarfed:** Restricted plant size without loss of health and vigour.
- **Early blight:** The progressive yellowing and dying of the leaves from the ground up. In tomato, caused by *Alternaria solani*.
- **Ecology:** The science of relationships between organisms and their environment.

- **Economic threshold:** The level at which pest damage justifies the cost of control. In home gardening, the threshold may be aesthetic rather than economic.
- **Edging:** A shallow trench or physical barrier of metal, wood, brick or synthetic material used to define the border between lawn turf and another area, such as paving or a flower bed.
- **Emasculate:** To remove a flower's anthers.
- **Embryo:** The dormant, immature plant within a seed; the "germ" referred to in wheat germ.
- **Enation:** Epidermal out growths on leaves or stems.
- **Endophytes:** Fungi that live in some grasses (called endophytic) and make them harmful or deadly to a variety of above ground grass-eating insects.
- **Endosperm:** The nutritive tissue within the seed of a flowering plant. Surrounds and is absorbed by the embryo.
- **Enzyme:** A biological catalyst that aids in conversion of food from one form to another.
- **Epidermis:** The outermost layer of cells covering a plant's leaves, roots and young parts.
- **Epinasty:** An abnormal downward curving growth or movement of a leaf, leaf part or stem.
- **Erwinia:** The bacterium which is the causal agent of Fireblight. Historically, it was the first plant disease identified as being caused by a bacterium.
- **Espalier:** The training of a tree or shrub to grow flat on a trellis or wall. Espalier patterns may be very precise and formal or more natural and informal.
- **Etiolation:** Development of yellow, long, spindly growth on a plant as a result of insufficient light.
- **Evergreen:** A plant that never loses all of its foliage at the same time.
- **Excise:** To remove or extract, as an embryo from a seed or ovule.
- **Excurrent:** A tree form in which the main trunk remains dominant with small, more or less horizontal branches. Fir and sweet gum are examples.
- **Exfoliating:** Peeling off in shreds or thin layers, as in bark from a tree.
- **Exoskeleton:** The outer support structure of an insect.
- **Exotic:** Non-native.
- **Exposure:** The intensity, duration and variation in sun, wind and temperature that characterize any particular lawn or planting site.
- **Fallow:** To keep part of a garden unplanted or in a cover crop during the growing season.
- **Family:** A broad group of plants with common characteristics.
- **Family:** One group used in classifying organisms. Families consist of a number of similar genera (plural for genus).
- **Fasciation:** Distortion of a plant that results in thin, flattened and sometimes curved shoots.

- **Feeder roots:** Fine roots and root branches with a large absorbing area (root hairs). Responsible for taking up the majority of a plant's water and nutrients from the soil.
- **Fertility (soil):** The presence of minerals necessary for plant life.
- **Fertilization:**(1) The fusion of male and female germ cells following pollination. (2) The addition of plant nutrients to the environment around a plant.
- **Fertilizer analysis:** The amount of nitrogen, phosphorus (as P_2O_5) and potassium (as K_2O) in a fertilizer expressed as a percent of total fertilizer weight. Nitrogen (N) is always listed first, phosphorus (P) second and potassium (K) third.
- **Fertilizer:** A product containing one, two or all three plant macronutrients (nitrogen, phosphorus and/or potassium).
- **Fertilizer:** A natural or synthetic product added to the soil or sprayed on plants to supply plant nutrients.
- **Fibrous root:** A root system that branches in all directions, often directly from the plant's crown, rather than branching in a hierarchical fashion from a central root. (See also Taproot.)
- **Filament:** The stalk supporting a flower's anthers.
- **Fireblight:** The Rose Family disease caused by *Erwinia amylovora*, marked by the progressive blackening and dying of shoot tips and leaves, which can kill the whole plant.
- **Flagging:** Loss of turgor and drooping of plant parts, usually as a result of water stress.
- **Floricane:** Second-year growth of cane- berries. Produces fruits on laterals.
- **Foliar fertilization, foliar feeding:** Fertilization of a plant by applying a diluted soluble fertilizer, such as fish emulsion or kelp, directly to the leaves.
- **Force:** To bring a plant into early growth, generally by raising the temperature or transplanting it to a warmer situation. Tulips and paper whites are examples of plants that are often forced.
- **Form:** (1) A naturally occurring characteristic different from other plants in the same population. (2) The growth habit (shape) of a plant.
- **Formal:** (1) A garden that is laid out in precise symmetrical patterns. (2) A flower, such as some camellias, that consists of layers of regularly overlapping petals.
- **Frond:** Specifically, the foliage of ferns, but often applied to any foliage that looks fern-like, such as palm leaves.
- **Fruit:** The edible portion of a plant that is closely associated with a flower. Botanically, a fruit is a ripened, mature ovary.
- **Fruiting habit:** The location and manner in which fruit is borne on woody plants.
- **Full shade:** A site that receives no direct sun during the growing season.
- **Full sun:** A site that receives at least eight hours of direct sun each day during the growing season.
- **Fungi:** Plural of fungus.

- **Fungicide:** Any material capable of killing fungi. Sulphur and copper sulphate are two common mineral fungicides.
- **Fungicide:** Something which kills fungi. Most of these products, however, do not kill the spores.
- **Fungus:** A kingdom of eukaryotic (see prokaryote), non-photosynthetic, organisms. They are not a part of the plant kingdom. Some cause diseases of plants.
- **Fungus:** A plant organism that lacks chlorophyll, reproduces via spores and usually has filamentous growth. Examples are molds, yeasts and mushrooms.
- **Fusarium:** A genus of plant pathogenic fungi; some are saprophytes. *Fusarium oxysporum*, which causes wilts, has numerous forms and affects many economically important plants.
- **Gall:** A growth on plant stems or leaves caused by abnormal cell growth stimulated by the sucking of some insects (e.g., aphids) or by viral, fungal or bacterial infection.
- **Genus:** A group of related species, each of which is distinct and unlikely to cross with any other. A group of genera forms a family, and a group of families forms an order.
- **Geotropism:** The turning or curving of a plant's parts in response to gravity. A root growing downward is an example. Geotropism is controlled largely by the hormone auxin.
- **Germinate:** The beginning of growth in a seed; to sprout.
- **Germination:** The initial sprouting stage of a seed.
- **Girdling:** The cutting, removing, or clamping of bark all the way around a trunk or branch. Sometimes, girdling is done deliberately to kill an unwanted tree, but often it results from feeding by insects or rodents. Wires and ties used to support a tree can cause girdling, as can string trimmers.
- **Glabrous:** Hairless, but not necessarily smooth.
- **Glaucous:** Covered with a grayish, bluish, or whitish waxy coating that is easily rubbed off. Blue spruce needles are an example of glaucous leaves.
- **Grade:** The degree and direction of slope on an area of ground.
- **Graft incompatibility:** The eventual failure of the graft union due to pathogens or physiologic reasons usually causing the death of the scion (part being propagated).
- **Grafting:** The act of inserting a shoot or bud of one plant into the trunk, branch, or roots of another, where it grows and becomes a permanent part of the plant.
- **Gravitational water:** Water in excess of a soil's capacity which drains downward to groundwater.
- **Green cone:** An enclosed composting unit often used for composting food waste.
- **Ground cover:** A plant, such as ivy, vinca or juniper, used to cover the soil and form a continuous low mass of foliage. Often used as a substitute for turf grass, especially in shade, embankments, or other areas where turf grasses don't perform well or are difficult to maintain. (Can also mean material used to cover the ground and prevent weeds from reaching the surface).

- **Groundcover:** Plants used in lieu of grass for holding soil and providing leaf texture.
- **Growing season:** The period between the beginning of growth in the spring and the cessation of growth in the fall.
- **Growth regulator:** A compound applied to a plant to alter its growth in a specific way. May be a natural or synthetic substance.
- **Guard cell:** Leaf epidermal cells that open and close to let water, oxygen and carbon dioxide pass through the stomata.
- **Haltere:** A small, knob-like organ (sometimes shaped like a baseball bat or bowling pin) located on the thorax of insects of the order Diptera. Takes the place of hindwings and helps balance the insect in flight.
- **Harden/Hardening out:** A method used to aclimatize seedlings and plants that have been started indoors to be transplanted outdoors. Place the plants in a sheltered, sunny place outside for a few hours at a time and return them to the house, greenhouse or garage each night. Extending the outside time each day until the risk of frost has passed and the plant can be moved to a permanent outside location. Use of a cold frame can help to speed up the process if the growing season is short.
- **Hardening off:** The process of gradually exposing seedlings started indoors to outdoor conditions before transplanting.
- **Hardpan:** An impervious layer of soil or rock that prevents root growth and downward drainage of water.
- **Hardscaping:** The installation of non-plant features in the landscape. These include features such as walls, walks, driveways, kerbs, patios, etc.
- **Hardy/Hardiness:** A plant's ability to survive the winter without protection from the cold. The temperature range in which a plant will grow and thrive. Cold is usually the main consideration; however, heat and humidity can also cause plants to fail.
- **Hardy:** Frost or freeze-tolerant. In horticulture, this term does not mean tough or resistant to insect pests or disease.
- **Haustorium:** A modified hyphal branch of a parasitic plant. Grows into a host plant's cell to absorb food and water.
- **Head:** (1) To cut off part of a shoot or limb rather than removing it completely at a branch point. (2) The part of a tree from which the main scaffold limbs originate.
- **Heartwood:** The center cylinder of xylem tissue in a woody stem.
- **Heeling in:** The temporary burying of a newly dug plant's roots to prevent their drying until a new planting site is prepared. Nurseries heel in bare root berries, trees and shrubs.
- **Hemiptera:** An insect family made up of species generally having sucking mouthparts and four wings. Wings are thick at the base and membranous at the free end. Includes true bugs, stink bugs, assassin bugs and back swimmers.
- **Herbaceous perennial:** A plant that dies back in the winter and re-grows from the crown in spring.

- **Herbaceous:** A plant with soft upper growth rather than woody growth that is found in shrubs. Annuals, biennials and perennials may be herbaceous.
- **Herbicide:** A plant killer. Herbicide injury on non-targeted plants can mimic certain diseases.
- **Herbicide:** A chemical used to kill undesirable plants.
- **Heterozygous :** Having mixed hereditary factors, not a pure line.
- **Homoptera:** An insect family made up of species having sucking mouthparts and usually two pairs of wings. Includes aphids, scales, leaf hoppers and cicadas.
- **Homozygous:** Having purity of type, a pure line.
- **Honeydew:** A sticky substance excreted by aphids and some other insects.
- **Hormone:** A naturally occurring compound that alters plant growth in a specific manner. (See also Growth regulator.)
- **Horticultural oil:** Oil made from petroleum products, vegetable oil, or fish oil, used to control insect pests and diseases. Oils work by smothering insects and their eggs and by protectively coating buds against pathogen entry.
- **Horticulture:** The branch of the science of agriculture that relates to cultivating gardens or orchards, including the growing of vegetables, fruits, flowers and ornamental shrubs and trees.
- **Horticulturist:** Someone who has extensive knowledge of plants, their care and their requirements for survival.
- **Host:** A plant on which an insect or disease completes all or part of its life-cycle.
- **Hot composting:** A fast composting process that produces finished compost in 6 to 8 weeks. High temperatures are maintained by mixing balanced volumes of energy materials and bulking agents, by keeping the pile moist, and by turning it frequently to keep it aerated.
- **Hotbed:** An enclosed bed for propagating or protecting plants. Has a source of heat to supplement solar energy.
- **Humus:** The end product of decomposed animal or vegetable matter. (See also Compost.)
- **Humus:** Thoroughly decayed organic matter. Added to lawns gardens or beds, it will increase a soil's water-holding capacity, improve aeration and support beneficial microbial life in the soil.
- **Hybrid:** A crossbred cultivar that is not likely to breed true from seeds. Some methods of propagations such as division or cutting should produce identical copies of a hybrid parent.
- **Hybrid:** A cross between two varieties or species, whether of the same genus or two genera.
- **Hydroponics:** A method of growing plants without soil. Plants usually are suspended in water or polymers.
- **Hymenoptera:** An insect family made up of species having four membranous wings, of which the front pair are larger. Includes bees, wasps, sawflies and ants.
- **Hypha:** A single filament of a fungus.

- **Hypocotyl:**The first leaf- like structure that appears on a germinating seed. Grows upward in response to light.
- **Immobilization:** The process by which soil microorganisms use available nitrogen as they break down materials with a high C:N ratio, thus reducing the amount of nitrogen available to plants.
- **Immune:** A plant that does not become diseased by a specific pathogen.
- **Immune:** Absolutely not susceptible to a disease.
- **Incompatible:** Kinds or varieties of a species that do not successfully cross-pollinate or intergraft.
- **Incubation:** A period of development during which a pathogen changes to a form that can penetrate or infect a new host plant.
- **Indeterminate:** A plant growth habit in which stems keep growing in length indefinitely. For example, indeterminate tomatoes are tall, late-fruiting and require staking for improved yield.
- **Infection:** The condition reached when a pathogen has invaded plant tissue and established a parasitic relationship between itself and its host.
- **Infiltration:** The movement of water into soil.
- **Inflorescence collective:** A group of individual flowers. The grouping can take many forms, such as a spike (flowers closely packed along a vertical stem, e.g., snapdragons), an umbel or corymb (flowers forming a flattened dome, e.g., yarrow), a panicle (a complex hierarchical arrangement of flowers, e.g., hydrangeas) or a capitulum (tightly packed disc flowers, e.g., the center of a daisy).
- **Inoculation:** The introduction of a pathogen to a host plant's tissue.
- **Inoculum:** The parts of a pathogen that infect plants.
- **Insect:** The largest group of arthropods. They can damage or kill plants directly or indirectly, as disease vectors.
- **Insectary plant:** A plant that attracts beneficial insects.
- **Insecticidal soap:** A specially formulated soap that is only minimally damaging to plants, but kills insects. Usually works by causing an insect's outer shell to crack, resulting in drying out of its interior organs.
- **Insecticide:** Any material that kills insects. Includes numerous botanical, mineral and chemical products, both organic and synthetic.
- **Insecticide:** Something which kills insects.
- **Instar:** The stage of an insect's life between molts.
- **Integrated pest management:** A method of managing pests that combines cultural, biological, mechanical and chemical controls, while taking into account the impact of control methods on the environment.
- **Intensive gardening:** The practice of maximizing use of garden space, for example by using trellises, intercropping, succession planting and raised beds.
- **Intercropping, interplanting:** The practice of mixing plants to break up pure stands of a single crop.
- **Internode, interstem:** The portion of a stem between two nodes.

- **Interstem, interstock:** The middle piece of a graft combination made up of more than two parts, i.e., the piece between the scion and the rootstock. Often has a dwarfing effect.
- **Invasive:** Growing vigorously and out-competing other plants in the same area; difficult to control.
- **Ion:** An atom or molecule with either positive or negative charges.
- **Irrigation:** This is the practice of applying supplemental water to plant and lawn areas.
- **Isoptera:** An insect family made up of species having soft bodies, strong mandibles and well developed laws. Workers and soldiers are wing- less and sterile. Termites are an example.
- **Joint:** A node; the place on a stem where a bud, leaf, or branch forms.
- **Juvenile stage:** (1) The early or vegetative phase of plant growth characterized by carbohydrate utilization. (2) The first stage of an insect's life cycle, either a larva or a nymph.
- **Key:** A tool for plant or animal classification and identification. Consists of a series of paired statements that move from general to specific descriptions.
- **Knot garden:** A formal garden in which two or more kinds of plants with different coloured foliage, often herbs, are planted and pruned so they interweave and form a knot pattern.
- **Landscape Architect/Designer:** A person who has education and/or training in the design of residential and commercial properties. Jobs can include plantings, walkways, walls, water features, landscape lighting, fences, and other similar features. LDs are often employed by a design-build firm that does most the work themselves. They may contract some tasks to other companies. They may have a Landscape Architecture degree from an accredited college.
- **Landscape contractor:** A firm that implements landscape plans prepared by Landscape Architects or Landscape Designers. They may bid on a job or work with the designer throughout the design process. Depending on their in-house expertise, they may hire sub-contractors to perform specialized tasks such as masonry, paving, carpentry, etc.
- **Landscape fabric:** A synthetic fabric that is usually water-permeable, it is spread under paths or mulch to serve as a weed barrier.
- **Landscape:** This is 'the lay of the land' which includes plants and other physical features such as lawn or meadow areas, waterways, pavement, utilities, structures, etc.
- **Larva:** The immature form of an insect that undergoes complete metamorphosis. Different from the adult in form. Also called caterpillar.
- **Latent bud:** A bud that does not break during the season after it is formed.
- **Lateral bud:** A bud on the side, rather than the tip of a stem.
- **Lateral:** A branch attached to and subordinate to another branch or trunk.
- **Lawn restoration:** Improving a lawn by a combination of fertilization, aeration, turf plug planting, and/or seeding without killing or removing all existing turf.

- **Layering:** A form of propagation in which shoots that are still attached to the mother plant are layer on top or just under the soil surface so that rooting occurs. Once rooted, the layered section can be cut from the mother plant and moved elsewhere. Strawberry plants are good candidates for layering.
- **Layering:** A method of stimulating adventitious roots to form on a stem. There are two primary methods of layering. In ground layering, a low-growing branch is bent to the ground and covered by soil. In air layering, moist rooting medium is wrapped around a node on an aboveground stem.
- **Leaching:** Movement of water and soluble nutrients down through the soil profile.
- **Leader:** A developing stem or trunk that is longer and more vigorous than laterals.
- **Leaf scar:** A visible, thickened crescent or line on a stem where a leaf was attached.
- **Leaflet:** A single division of a compound leaf.
- **Lenticel:** A small opening on the surface of fruits, stems and roots that allows exchange of gases between internal tissues and the atmosphere.
- **Lepidoptera:** An insect family made up of species having four wings covered with minute scales. Members undergo complete metamorphosis through the egg, larva, pupa and adult stages. Includes butterflies and moths.
- **Lesion:** Generally, a localized site of damaged or diseased plant tissue. Can be caused by bacteria, fungi, viruses or be environmental in origin (freeze injury, mechanical abrasion, sun scald, etc.)
- **Lime/Limestone:** A white or grayish mineral compound used to reduce soil acidity and to supply calcium for plant growth.
- **Lime:** A rock powder consisting primarily of calcium carbonate. Used to raise soil pH (decrease acidity).
- **Loam:** A soil with roughly equal influence from sand, silt and clay particles.
- **Loam:** An ideal soil type for growing, loam contains an equal balance of sand, silt and clay. It has benefits of both clay and sand in that it will hold moisture and organic compounds, yet it is porous and will drain well.
- **Lodge:** To fall over, usually due to rain or wind. Corn and tall grasses are examples of plants susceptible to lodging.
- **Long-day plant:** A plant requiring more than 12 hours of continuous daylight to stimulate a change in growth, e.g., a shift from the vegetative to reproductive phase.
- **Macronutrient:** Collectively, primary and secondary nutrients.
- **Macropore:** A large soil pore. Macro- pores include earthworm and root channels and control a soil's permeability and aeration.
- **Mandible:** The first pair of jaws on insects: stout and tooth-like which helps in chewing insects, needle- or sword-shaped which helps in sucking insects; lateral (left and right) upper jaws of biting insects.
- **Maturity:** Ripeness, usually the state of development that results in maximum quality.
- **Meristem:** Plant tissue in the process of formation; vegetative cells in a state of active division and growth, e.g., those at the apex of growing stems and roots.

- **Mesophyll:** A leaf's inner tissue, located between the upper and lower epidermis, where raw materials (carbon dioxide and water vapour) are held for use in photosynthesis.
- **Metamorphosis:** The process by which an insect develops.
- **Microclimate:** The climate of a small area within a larger climate area. For example, a backyard can have a climate different from that of the surrounding neighbourhood, or there may be climate differences within a backyard. Microclimates can influence plant growth and should be considered in plant selection and care.
- **Micronutrient:** A nutrient used by plants in small amounts, less than 1 part per million (boron, chlorine, copper, iron, manganese, molybdenum and zinc). Also called a trace element.
- **Micropore:** A fine soil pore, typically a fraction of a millimetre in diameter. Microspores are responsible for a soil's ability to hold water.
- **Mildew:** A visible fungal growth on the exterior of a plant.
- **Mite:** A small arachnid; usually found in large numbers sucking sap on leaf undersides. Mite infestations are sometimes confused with diseases.
- **Miticide:** Something which kills mites.
- **Modified central leader, modified leader:** A system of pruning used primarily on fruit trees. The central leader is encouraged for the first few years, then suppressed. This system allows for well placed scaffolds and strong crotches, but keeps the tree's crown relatively close to the ground for easy harvesting.
- **Molt:** The shedding of skin during insect growth. The form assumed between molts is called an instar.
- **Monocotyledon, monocot:** A plant having one cotyledon.
- **Monoecious:** A species having both male and female flowers on the same plant. Pecans, avocados and squash are examples of monoecious plants.
- **Morphology:** The study of the form of plants or plant parts.
- **Mosaic:** Nonuniform foliage coloration with a more or less distinct intermingling of normal green and light green or yellowish patches.
- **Mosaic:** Well-defined, generally angular, discoloured zones on leaves. One symptom of a viral infection.
- **Mottle:** An irregular pattern of light and dark areas.
- **Mulch:** A layer of bark, peat moss, compost, shredded leaves, hay or straw, lawn clippings, gravel, paper, plastic or other material spread over the soil around the base of plants. During the growing season, mulch can help retard evaporation, inhibit weeds and moderate soil temperature.
- **Mulch:** Any material placed on the soil surface to conserve soil moisture, moderate soil temperature and/or control weeds. Wood chips, bark chips and shredded leaves are examples; inorganic materials such as rocks or newspapers also are used.
- **Mutation:** A genetic change within an organism or its parts that changes its characteristics. Also called a bud sport or sport.

- **Mycelia:** Masses of fungal threads (hyphae) that make up the vegetative body of a fungus.
- **Mycology:** The study of fungi.
- **Mycorrhizae:** Beneficial fungi that infect plant roots and increase their ability to take up nutrients from the soil.
- **Native plant:** A plant indigenous to a specific habitat or area.
- **Naturalize:** (1) To design a garden with the aim of creating a natural scene. Planting generally is done randomly, and space is left for plants to spread at will. (2) The process whereby plants spread and fill in naturally.
- **Necrosis:** Tissue death.
- **Nematicide:** Something which kills or protects against nematodes.
- **Nematode:** A microscopic roundworm, usually living in the soil. Many feed on plant roots and can be disease pathogens or vectors. Others are beneficial parasites of insect pests.
- **Nematode:** Tiny, wormlike animals which can penetrate roots and cause galls (abnormal swelling of plant tissue) and other malformations Nutrient – One of the elements required by a plant for its growth.
- **Nitrate (NO_3^-):** A plant-available form of nitrogen contained in many fertilizers and generated in the soil by the breakdown of organic matter. Excess nitrate in soil can leach to groundwater.
- **Nitrifier:** A microbe that converts ammonium to nitrate.
- **Nitrogen (N):** A primary plant nutrient, especially important for foliage and stem growth.
- **Nitrogen cycle:** The sequence of biochemical changes undergone by nitrogen as it moves from living organisms, to decomposing organic matter, to inorganic forms and back to living organisms.
- **Nitrogen fixation:** The conversion of atmospheric nitrogen into plant available forms by *Rhizobia* bacteria living on the roots of legumes.
- **Node:** A joint in plants from which leaves emerge.
- **Node:** The point on a plant where a branch, bud, or leaf develops. On younger branches, it usually is marked by a slight swelling. The space on the stem between nodes is an internode.
- **Nonpoint source:** A relatively small, nonspecific source of pollutants that, when added to other sources, may pose a significant threat to the environment. (See also Point source.)
- **Non-selective pesticide:** A pesticide that kills most plants or animals.
- **N-P-K:** Acronym for the three major plant nutrients contained in manure, compost and fertilizers. N stands for nitrogen, P for phosphorus and K for potassium.
- **Nucleus:** The organelle within a cell that contains chromosomes and thus controls various cellular processes, including division into new cells.
- **Nutrient:** Any substance, especially in the soil, that is essential for and promotes plant growth.

- **Nutrients:** Nitrogen, phosphorus, potassium, calcium, magnesium, sulphur, iron and other elements needed by growing plants and supplied by minerals and organic matter in soil and by fertilizers.
- **Nymph:** The immature stage of an insect that undergoes simple metamorphosis. Usually similar in form to the adult.
- **Offset:** A new shoot that forms at the base of a plant or in a leaf axil.
- **Open-pollinated seeds:** Seeds produced from natural, random pollination so that the resulting plants are varied.
- **Organelle:** A structure within a cell, such as a chloroplast, that performs a specific function.
- **Organic matter:** Plant and animal residues, such as leaves, trimmings and manure, in various stages of decomposition.
- **Organic fertilizer:** A natural fertilizer material that has undergone little or no processing. Can include plant, animal and/or mineral materials
- **Organic matter:** Any material originating from a living organism (peat moss, plant residue, compost, ground bark, manure, etc.).
- **Organic:** In horticulture or agriculture, systems in which no chemicals or pesticides or fertilizers are used. Acceptable organic growth methods include natural plant derived or mineral substances along with the acceptance that certain pest and diseases be allowed in the garden.
- **Ornamental plant:** A plant grown for beautification, screening, accent, specimen, colour or other aesthetic reasons.
- **Ornamental:** Plants grown for aesthetics and not for consumption or economic use.
- **Orthoptera:** An insect family made up of species having a tough, leathery shell, membranous hindwings and straight forewings. Includes locusts, crickets, grasshoppers and katydids.
- **Osmosis:** Passage of materials through a membrane from an area of high concentration to an area of lower concentration.
- **Ovary:** The part of a flower containing ovules that will develop into seeds upon fertilization. Along with the style and stigma, it makes up the pistil (female sexual organ).
- **Over seeding:** Spreading seed over established turf that has been prepared for restoration.
- **Ovule:** Within the ovary, a body that will develop into seed after fertilization.
- **Oxidation:** The chemical process by which sugars and starches are converted to energy. In plants, known as respiration.
- **Palisade mesophyll:** The cells just beneath a leaf's upper epidermis that contain most of the leaf's chlorophyll and are responsible for most photosynthesis.
- **Palmate:** (1) A leaf whose veins radiate outward from a single point somewhat like the fingers of a hand. (2) A form of espalier training.

- **Parasite:** An organism on or in another one, deriving some or all of its nourishment from its host.
- **Parasite:** Any animal or plant that lives in or on another animal or plant and withdraws nutrients from its host.
- **Parasitic seed plant:** A plant that lives parasitically on other seed plants. An example is mistletoe.
- **Parterre:** A formal garden in which shrubs, flowers and paths form a geometric pattern of matched pairs.
- **Parthenocarpic:** Seedless.
- **Pathogen:** A disease-causing organism.
- **Pathogen:** Any organism that causes disease. Generally applied to bacteria, viruses, fungi, nematodes and parasitic plants.
- **Pathology:** The study of diseases.
- **Ped:** A cluster of individual soil particles.
- **Pedicel:** The stem of an individual flower.
- **Peduncle:** The main stem supporting a cluster of flowers (as opposed to a pedicel, which is the stem of an individual flower).
- **Pendulous:** More or less hanging or declined.
- **Perennial:** A plant that lives more than two years or three seasons and normally flowers annually. Many die down during the winter but the roots are unaffected by frost and new growth appears as the weather improves and the temperature rises.
- **Perianth:** Collectively, all external flower parts.
- **Permeability:** The rate at which water moves through a soil.
- **Persistent:** (1) Adhering to a position instead of falling, whether dead or alive, e.g., flowers or leaves. (2) A pesticide that retains its chemical properties in the soil for a long time.
- **Pesticide:** Synthetic substances used to kill and assortment of pests that might include weeds and insects. Herbicides, fungicides and insecticides are all types of pesticides.
- **Petals:** The usually showy structures around a flower's reproductive organs.
- **Petiole:** The stalk of a leaf.
- **pH :**A measure of acidity or alkalinity. The scale is logarithmic; a change in 1 pH unit is a 10-fold change. Values from 0 to 7 indicate acidity, and values from 7 to 14 indicate alkalinity. For example, 7 is neutral, 6 is acid (10 times the hydrogen ion concentration of pH 7), and 5 is very strongly acid (100 times the hydrogen ion concentration of pH 7).
- **pH:** The "potency of Hydrogen". A number from 0 to 14, the negative logarithm of the hydrogen ion concentration of a solution. A pH of 7 is neutral. A reading below 7 is acidic, one higher than 7 is alkaline.
- **Phenological stag:** Crop development stage.
- **Pheromone:** A vapour or liquid emitted by an insect that causes a specific response from a receiving in-sect. Some pheromones are used to find a mate. Synthetic pheromones are used as attractants in insect traps.

- **Phloem:** Photosynthate conducting tissue.
- **Phosphate:** The form of phosphorus listed in most fertilizer analyses (P_2O_5).
- **Phosphorus (P):** A primary plant nutrient, especially important for flower production. In fertilizer, usually expressed as phosphate (P_2O_5).
- **Photoperiod:** The amount of time a plant is exposed to light.
- **Photosynthate:** A food product (sugar or starch) created through photosynthesis.
- **Photosynthesis:** The light-mediated reaction in green plants where water and carbon dioxide become carbohydrates and oxygen.
- **Phototropism:** A growth response to light. Growth of a plant toward a light source is the most common example.
- **Phytophthora:** A very large and quite variable fungal genus containing plant pathogens.
- **Phytoplasma:** A microscopic, bacteria-like organism that lacks a cell wall. Previously called mycoplasma.
- **Phytotoxic:** Toxic to a plant (phyto = plant).
- **Picotee:** A pattern of flower petal coloration in which the edges of the petal are a contrasting colour to the body.
- **Pinch:** To remove a growing tip from a stem, thus causing axillary shoots or buds to develop.
- **Pistil:** The female sexual organ of a flowering plant, made up of the stigma, style and ovary.
- **Plant nutrition:** A plant's need for and use of basic chemical elements.
- **Pleach:** To intertwine branches of trees, vines, or shrubs to form an arbour or hedge.
- **Pleniflora, pleno, plena:** A term used in botanical names to indicate a double-flowered variety.
- **Plugs:** A plug plant is any developing plant whose growth cycle is initiated well in advance of its actual planting. Plug plants are designed to be introduced to your garden or landscape with substantial growth already established. Once assimilated the plant can begin to produce flowers, fruit or vegetables much sooner than traditional seed planting.
- **Point source:** A single, identifiable source of pollutants such as a factory or municipal sewage system.
- **Pollard:** A method of tree pruning that involves heading back severely to main branches each year so as to produce a thick, close growth of young branches.
- **Pollen:** A plant's male sex cells, which are held on the anther for transfer to a stigma by insects, wind or some other mechanism.
- **Pollenizer:** A plant whose pollen sets fruit on another plant.
- **Pollination:** The transfer of pollen from the stamen (male part of the flower) to the pistil (female part of the flower), which results in the formation of a seed.
- **Pollination:** The transfer of pollen from a male anther to a female stigma, enabling fruits to set and develop.

- **Pollinator:** An agent such as an insect that transfers pollen from a male anther to a female stigma.
- **Pome fruit:** A fruit having a core, such as an apple, pear, or quince.
- **Pomology:** The science of fruits and the art of fruit culture, especially tree fruits.
- poses. This process may take months or longer. Cold composting does not kill weed seeds or pathogens.
- **Post emergent:** A product applied after crops or weeds emerge from the soil.
- **Potash:** The form of potassium listed in most fertilizer analyses (K_20).
- **Potassium (K):** A primary plant nutrient, especially important for developing strong roots and stems. In fertilizers, usually expressed as potash (K20).
- **Pre-emergent:** A product applied before crops or weeds emerge from the soil.
- **Pre-harvest interval:** The period of time that must pass from the time a pesticide is applied to a crop until the crop is safe to pick and use.
- **Predator:** An animal that eats another animals.
- **Pre-plant:** A product applied before a crop is planted.
- **Prickle:** A rigid, straight, or hooked out-growth of bark or stems. Often called a thorn, but technically different. Roses are examples of plants with prickles.
- **Primary nutrient:** A nutrient required by plants in a relatively large amount (nitrogen, phosphorus and potassium).
- **Primocane:** First-year growth, usually vegetative, on caneberries. Only fall bearing raspberries produce fruit on primocanes in late summer.
- **Processed fertilizer:** A fertilizer that is manufactured or is refined from natural ingredients to be more concentrated and more available to plants.
- **Propagate:** To start new plants by seeding, budding, grafting, dividing, etc.
- **Prune:** To remove plant parts to improve a plant's health, appearance or productivity.
- **Pruning:** Using pruning shears to remove or cut back the branches of woody plants.
- **Pseudobulb:** A thickened, aboveground, modified stem that serves as a storage organ. Found in some orchids.
- **Pseudomonas:** The most important member of the group, Pseudomonas syringae It causes bacterial stem canker in stone fruits and dozens of other diseases.
- **Pubescen:** Hairy.
- **Pupa:** The stage between larva and adult in insects that go through complete metamorphosis.
- **Pythium:** A genus of fungi, closely related to Phytophthora, which contains plant pathogens. Causes blights, rots and damping off.
- **Quarantine:** A regulation forbidding sale or shipment of plants or plant parts, usually to prevent disease, insect, nematode or weed invasion of an area.
- **Quick-release fertilizer:** A fertilizer that contains nutrients in plant-available forms such as ammonium and nitrate.
- **Quote:** The proposed price to perform the work outlined on an accompanying plan or document. Also referred to as a bid.

- **Raceme:** A flower stalk on which the florets start blooming from the bottom of the stem and progress toward the top.
- **Radial spacing:** The horizontal spacing of branches around a trunk.
- **Radicle:** The first part of a seedling to emerge from the seed. Grows down- ward and develops into the primary root.
- **Raised bed:** An elevated garden bed offering better drainage, aeration and warmer soil than a conventional bed.
- **Relative humidity:** The ratio of water vapour in the air to the amount of water the air could hold at the current temperature and pressure.
- **Resistant:** A plant having qualities that make it retard the activities of a pathogen or insect pest.
- **Respiration:** The process within plants of converting sugars and starches into energy.
- **Retaining wall:** A wall built to stabilize a slope and keep soil from sliding or eroding downhill.
- **Reversion growth:** A stem that originates from and has the characteristics of the plant's rootstock.
- **Rhizobia bacteria:** Bacteria that live in association with roots of legumes and convert atmospheric nitrogen to plant-available forms, a process known as nitrogen fixation.
- **Rhizoctonia:** A fungal group containing plant pathogens. *Rhizoctonia solani* is the causal agent of Brown Patch in 'Raleigh' St. Augustine grass.
- **Rhizome:** A thickened underground stem that grows horizontally with bud eyes on top and roots below. Bearded iris is an example of a plant that produces rhizomes.
- **Rhizosphere:** The thin layer of soil immediately surrounding plant roots.
- **Root cutting:** A section of root prepared for the purpose of vegetative propagation.
- **Root hair:** A delicate, elongated epidermal cell that occurs just behind a root's growing tip. Root hairs increase the root's surface area and absorptive capacity.
- **Root pruning:** The cutting or removal of some of a plant's roots.
- **Root-bound:** A condition in which a plant's roots have completely filled its container. Typically, the roots begin to encircle the pot's outer edge. Further growth is prevented until the plant is removed from the container.
- **Rootstock:** The root or stem onto which a scion or interstock is grafted.
- **Rosette disease:** A zinc deficiency of pecans growing in alkaline soils. Shoot tips do not elongate normally and the leaves are crowded together, almost like a rosette.
- **Rosette:** A small cluster of leaves radially arranged in an overlapping pattern.
- **Rot:** Decomposition and destruction of tissue.
- **Rotation:** The practice of growing different plants in different locations each year to prevent the buildup of soil borne diseases and insect pests.
- **Row cover:** A sheet of synthetic material used to cover plants in order to retain heat and exclude insect pests.
- **Rugose:** Wrinkled.

- **Russet:** Yellowish-brown or reddish- brown scar tissue on a fruit's surface.
- **Rust:** A plant disease caused by a fungus and named because of the orange, orange-red to rusty colour of the spores.
- **Sand:** The coarsest type of soil particle (0.05 to 2 mm in diameter).
- **Sanitation:** The process of removing sources of plant pathogens from a growing area, for example, by cleaning up plant debris and sterilizing the tools and growing media.
- **Saprophyte:** A plant or fungus which draws nutrients from dead organic matter.
- **Scab:** (1) A crust-like disease lesion. (2) A specific disease that causes scab lesions.
- **Scab:** A lesion characterized by a thickening and malformation of the outermost layers of plant tissue.
- **Scaffold, scaffold branches:** The principal branches of a tree or shrub arising from the trunk or another main branch to form the plant's framework.
- **Scale:** (1) A modified leaf that protects a bud. (2) A type of insect pest.
- **Scarification:** Nicking, sanding, or otherwise compromising the hard outer coating of a seed to increase its water intake and thus promote germination.
- **Scion:** A cutting or bud that is grafted to the stock of another plant.
- **Secondary nutrient:** A nutrient needed by plants in a moderate amount: calcium, magnesium and sulphur.
- **Seed coat:** A hard outer covering that protects a seed from disease and insects. Also prevents water from entering the seed and initiating germination before the proper time.
- **Seedling:** A young plant grown from seed.
- **Selective pesticide:** A pesticide that kills only certain kinds of plants or animals; for example, 2,4-D kills broadleaf lawn weeds but leaves grass largely un- harmed.
- **Self-fruitful:** A plant that bears fruit through self-pollination.
- **Self-pollinating:** Plants that do not require pollen from another plant to produce fruit.
- **Self-unfruitful:** A plant that requires another variety for pollination.
- **Senescence:** The aging process. Also used to describe a plant that is in the process of going dormant for the season, although technically only the parts that are dying (the leaves) are becoming senescent.
- **Sepal:** An appendage at a flower's base, typically green or greenish and more or less leafy in appearance.
- **Separation:** The process of removing new bulbs or corms from their parent for purposes of propagation.
- **Sessile:** Stalkless and attached directly at the base, as in sessile leaves.
- **Shear:** To cut back a plant (as opposed to selective pruning or deadheading). Often used to regenerate plants with many small stems, where deadheading would be too time consuming.
- **Shoot:** One season's branch growth. The bud scale scars (ring of small ridges) on a branch mark the start of a season's growth.

- **Short-day plant:** A plant requiring more than 12 hours of continuous darkness to stimulate a change in growth, e.g., a shift from the vegetative to reproductive phase.
- **Shoulder ring:** One of the ridges around the base of a branch where it attaches to a trunk or to another branch.
- **Shrub:** A woody plant that grows to a height of 3 to 12 feet. May have one or several stems with foliage extending nearly to the ground.
- **Side-dress:** To apply fertilizer to the soil around a growing plant.
- **Sign:** Direct evidence of a damaging factor (for example, a pest or pathogen itself, secretions, insect webbing or chemical residue).
- **Signal word:** An indication of toxicity on pesticide labels. Pesticides labelled "Caution" are the least toxic, those labeled "Warning" are more so and those labelled "Danger" are the most toxic.
- **Silt:** A type of soil particle that is intermediate in size between sand and clay (0.002 to 0.05 mm in diameter).
- **Simple metamorphosis:** A type of insect development in which the insect passes through the stages of egg, nymph and adult. The nymph usually resembles the adult.
- **Slow-release fertilizer:** A fertilizer material that must be converted into a plant-available form by soil micro-organisms.
- **Smut:** A plant disease caused by a fungus, visible as a body of very dark spores.
- **Soft pinch:** To remove only the succulent tip of a shoot, usually with the finger-tips.
- **Soil salinity:** A measure of the total soluble salts in a soil.
- **Soil solution:** The solution of water and dissolved minerals found in soil pores.
- **Soil structure:** The arrangement of aggregates (peds) in a soil.
- **Soil texture:** How coarse or fine a soil is. Texture is determined by the proportions of sand, silt and clay in the soil.
- **Soil:** A natural, biologically active mixture of weathered rock fragments and organic matter at the earth's surface.
- **Soilless mix:** A sterile potting medium consisting of ingredients such as sphagnum peat moss and vermiculite.
- **Soluble salt:** A compound often remaining in the soil from irrigation water, fertilizer, compost or manure applications.
- **Sonic repeller:** A sonic-wave emitting unit said to disrupt the activities of small mammals. Not proven to be effective.
- **Species:** The basic unit of plant or animal classification. Plants within a species have several characteristics in common. Most important, they can cross with one another, but normally not with members of other species. Classification of species is quite fluid, with periodic revision by botanists.
- **Specific epithet:** The second word in a plant name. (The word "species" refers to plants, not the word in the name.)

- **Specimen:** An individual plant with outstanding characteristics (leaves, flowers or bark), generally used as a focal point in a landscape.
- **Spore:** (1) The reproductive body of a fungus or other lower plant, containing one or more cells. (2) A bacterial cell modified to survive in an adverse environment.
- **Spore:** A small, hard reproductive or resting phase of a fungus, bacterium, fern, etc.
- **Spot treatment:** To apply a pesticide to a small section or area of a crop.
- **Spur:** On fruit trees, a short, compact twig with little or no internodal development on which flowers and fruit are borne.
- **Staking:** The practice of driving a support into the ground next to a plant to support it in its growth.
- **Stamen:** The male reproductive organ of a flower, consisting of a filament and a pollen containing anther.
- **Stamen:** The male, pollen-producing part of a flower consisting of the anther and its supporting filament.
- **Standard:** A plant pruned so that it consists of a single bare, vertical stem, atop which a shaped mass of foliage, usually globular, is maintained.
- **Stem cutting:** A section of a stem prepared for vegetative propagation.
- **Sterile:** (1) Material that is free of disease organisms (pathogens), as in potting medium. (2) A plant that is unable to produce viable seeds.
- **Stigma:** The part of a female sex organ that receives pollen. Supported by the style, through which it is connected to the ovary. Often sticky when receptive.
- **Stipules:** A pair of appendages found on many leaves where the petiole meets the stem.
- **Stolon:** A horizontal stem running along, but above, the soil surface and producing roots and leaves where its nodes contact the soil. Straw- berries are an example of a plant with stolons. Also called a runner.
- **Stoma:** Tiny openings in a leaf's epidermis that allow water, oxygen and carbon dioxide to pass into and out of the plant.
- **Stone fruit:** A fleshy fruit, such as a peach, plum, or cherry, usually having a single hard stone that encloses a seed. Also called a drupe.
- **Strain:** A variation within a cultivar or variety
- **Stratification:** The exposure of seeds to moisture and low temperature to overcome dormancy.
- **Style:** The part of a plant's female sex organ that supports the stigma and connects it to the ovary.
- **Stylet:** A nematode's lance-like or needle- like mouthpart. Used to puncture and feed from plant cells.
- **Subspecies:** A major division of a species, more general in classification than a cultivar or variety.
- **Succession planting:** The practice of planting new crops in areas vacated by harvested crops.

- **Succession:** The progression of a plant community to a stable mixture of plants.
- **Sucker:** A shoot or stem that originates underground from a plant's roots or trunk, or from a rootstock below the graft union. (See also Reversion growth.)
- **Summer oil:** A light, refined horticultural oil used during the growing season to control insect pests and diseases.
- **Sunscald:** Winter or summer injury to the trunk of a woody plant caused by hot sun and fluctuating temperatures. Typically, sunscalded bark splits and separates from the trunk.
- **Sustainable gardening:** Gardening practices that allow plants to thrive with minimal inputs of labour, water, fertilizer and pesticides.
- **Symbiotic:** Mutually beneficial.
- **Symptom:** A change in a plant's growth or appearance in response to living or nonliving damaging factors.
- **Systemic pesticide:** A pesticide that moves throughout a target organism's system to cause death.
- **Taproot:** A thick central root attached directly to a plant's crown. Taproots branch little if at all.
- **Taxonomy:** Classification or naming of plants or animals.
- **Temporary branch:** (1) A small shoot or branch left on a young tree's trunk for protection and nourishment. (2) A low lateral allowed to remain until a tree is tall enough to have scaffolds at the desired height.
- **Tender:** Not tolerant of frost and cold temperatures. In horticulture, tender does not mean weak or susceptible to insect pests or diseases.
- **Tendril:** A slender projection used for clinging, usually a modified leaf. Easily seen on vines such as grapes and clematis.
- **Terminal bud:** The bud at the tip of a stem, trunk, or branch. Its development extends the plant's growth.
- **Terminal:** The tip (apex), usually of a branch or shoot.
- **Thatch:** A layer of undecomposed stems and roots near the soil surface which can harbour pathogens and block the flow of water into the root zone.
- **Thatch:** A tightly intermingled layer of stems, leaves and roots, living and dead, that forms between the soil surface and green vegetation of grass.
- **Thermoperiod:** The change in temperature from day to night.
- **Thermophili:** High-temperature, as in microorganisms that break down organic matter in a hot compost pile.
- **Thin:**(1) To remove an entire shoot or limb where it originates. (2) To selectively remove plants or fruits to allow remaining plants or fruits to develop.
- **Thorn:** A hard, sharp-pointed, leafless branch. Hawthorn is an example of a plant that produces thorns.
- **Tiller:** A shoot that arises from a plant's crown. Generally associated with grass species.

- **Tissue culture:** The process of generating new plants by placing small pieces of plant material onto a sterile medium. Also called *embryo culture*.
- **Tolerant:** A plant that will produce a normal yield even if infested by a disease or insect pest.
- **Tolerant:** Able to live in the presence of a disease or pest and not be affected.
- **Topiary:** A tree or shrub shaped and sheared into an ornamental, unnatural form, usually a geometric shape or the shape of an animal.
- **Totipotency:** The ability of any cell to develop into a complete plant.
- **Trace elements:** Elements required by plants for their mineral nutrition in minute quantities than either the macronutrients or micronutrients.
- **Transpiration:** The process of losing water in the form of vapour through the stomata.
- **Transplant:** To remove plants from one place and replant them in another (or from a container into the ground.
- **Tree:** A woody plant that typically grows more than 12 feet tall and has only one main stem or trunk.
- **Trellis:** Latticework used to support climbing plants .
- **Trickle irrigation:** This is the practice of applying water slowly through various types of pipes, tubes or specialized hoses. The simplest form would be the use of soaker hoses commonly found in garden centers. More sophisticated systems can be installed by homeowners or irrigation contractors that are preset to go on by a computer.
- **Tropism:** The tendency of a plant part to turn in response to an external stimulus, either by attraction or repulsion, as a leaf turns toward light.
- **Trunk:** The main stem of a tree. Also called a bole.
- **Truss:** A flower cluster, usually growing at the terminal of a stem or branch.
- **Tuber:** An underground storage organ made up of stem tissue. Contains buds on the surface, from which shoots may arise. Potatoes are an example.
- **Tuberous root:** An underground storage organ made up of root tissue. Sprouts only from the point at which it was attached to the stem of the parent plant. Dahlias are an example.
- **Tubers:** Are rough skinned and produce roots from many parts of its surface. New plants come to life from eyes or growth buds from this short, fat, bumpy exterior. Anemone, Aladium, Gloxinia and Tuberous Begonia all belong to the tuber family. To reproduce, Begonias grow larger and produce growth buds while Caladium grow new tubers from the sides of the original. Ranunculus and Dahlia are just a few that are part of the tuberous root family and in fact are truly real roots. Their food supply is kept in the root tissue and like tubers, they produce buds from which new plants grow. The growing area is often called the crown and the buds are restricted to the neck of the root where they grow on the base of the old stem.
- **Turgor:** Cellular water pressure responsible for keeping cells firm.
- **Twig:** A young stem (1 year old or less) that is in the dormant winter stage (has no leaves).

- **Umbel:** A group of flowers growing from a common point on a stem. Usually found on the lower portion of a shoot, it does not expand under normal growth stimuli. It will be stimulated into breaking if the growth above it is damaged or pruned away.
- **Vaporization:** The evaporation of the active ingredient in a pesticide during or after application.
- **Variety:** A strain of a plant having distinctive features that persist over successive generations in the absence of human intervention. Generally, variety applies to naturally occurring strains, while cultivar applies to horticulturally developed strains.
- **Vascular tissue:** Water-, nutrient- and photosynthate conducting tissue.
- **Vector:** A transmitter or carrier of disease.
- **Vector:** An organism which transmits an infectious agent.
- **Vegetative propagation:** The increase of plants by asexual means using vegetative parts. Normally results in a population of identical individuals. Can occur by either natural means (e.g., bulblets, cormels, offsets, plantlets or runners) or artificial means (e.g., cuttings, division, budding, grafting or layering).
- **Veinal necrosis:** A symptom of Oak Wilt in live oaks. The interior portions of the leaf are discoloured, yellow to reddish-brown, while the margins remain green.
- **Venation:** The arrangement of veins in a leaf.
- **Vermiculite:** A lightweight, flaky mineral called "mica" that has been heated to the point of expansion. The sponge-like granules are then capable of holding both water and air. This amendment is added to potting mixtures and container gardens to improve root growth due to aeration and moisture retention.
- **Vernation:** The arrangement of new leaves within an older leaf sheath (e.g., on a grass plant).
- **Vertical spacing:** The vertical space between branches on a tree.
- **Verticillium:** A genus of fungi containing plant pathogens. They generally invade vascular tissues and cause wilt diseases.
- **Viability:** A seed's ability to germinate.
- **Virus:** An infectious agent too small to see with a compound microscope. Multiplies only in living cells.
- **Virus:** Nucleic acid surrounded by a protein envelope. It can reproduce only within a host, usually with disastrous consequences.
- **Water sprout:** A vigorous shoot originating above ground on a plant's trunk, older wood, or bud union. Usually breaks from a latent bud. Often the result of heavy pruning.
- **Water-holding capacity (WHC):** The ability of a soil's micropores to hold water for plant use.
- **Waterlogged:** Soil that is saturated with water.
- **Water-soaking:** Lesions that appear wet and dark and usually are sunken and/or translucent. Often a symptom of bacterial disease.

- **Weed:** Any undesirable plant or grass species; any plant growing where you don't want it.
- **Weed-and-feed:** A combination fertilizer and herbicide sometimes used on lawns.
- **Wilt disease:** A loss of turgor (water pressure) in shoot tips or leaves caused by a pathogen interfering with vascular tissues.
- **Wilt:** (1) Lack of freshness and turgor and drooping of leaves from lack of water. (2) A vascular disease that interrupts a plant's normal uptake and distribution of water.
- **Wilting point (WP):** The point at which water content within plant cells is low enough that cellular turgor is lost and the plant wilts.
- **Witch's broom:** A crowded mass of abnormal branching and rebranching; usually pathological, sometimes propagated as a novelty.
- **Witches' broom:** Abnormal brush like development of many weak shoots.
- **Woody perennial:** A plant that goes dormant in the winter and begins growth in spring from aboveground stems.
- **Xeric:** A plant or landscape that conserves water. Most xeric plants need minimal supplemental water after an establishment period (18 to 24 months after planting) unless there is extreme drought.
- **Xylem:** Water- and nutrient conducting tissue.

Objective Type Questions

Akhilesh Tiwari[1] and R. K. Saraf[2]
[1]*Senior Scientist (Horticulture)*, [2]*Principal Scientist (Horticulture)*
JNKVV – Dryland Horticulture Research & Training Center,
Garhakota, Distt. – Sagar (M.P.)

A. Multiple Choice Questions:

1. Indian Institute of Vegetable Research is situated in:
   A. Varanasi B. Lucknow C. Bhopal D. Jaipur
2. Use of radiation technology as a tool in the development of high yielding varieties is called:
   A. Nuclear Agriculture B. Biogas Agriculture
   C. Bioinformatics D. Solarization
3. The Queen of Spice is considered to be:
   A. Black Pepper B. Cardamom C. Clove D. Cumin
4. TRIP stands for:
   A. Total Respiration Index Potentials
   B. Trade Related Intellectual Property Rights
   C. Technology for Rice Intensification Process
   D. Time Response Index Properties
5. Orobanchae is a:
   A. Parasitic fungi B. Parasitic bacteria
   C. Parasitic weed D. Parasitic insect
6. G.J. Mendel studied on:
   A. Drosophila B. Beans C. Cowpeas D. Peas
7. Light traps are used for the collection of:
   A. Positively phototropic insects
   B. Negatively phototropic insects
   C. Geotropic insects
   D. Hydrotropic insects

8. In self-incompatibility system:
   A. Self-pollination takes place
   B. Pollens not formed
   C. Pollens do not germinate on self-pollination
   D. Pollens are sterile
9. The base composition in RNA is:
   A. Cytosine, Thymine, Adenine, Guanine
   B. Cytosine, Uracil, Adenine, Guanine
   C. Cytosine, Thymine, Adenine, Histidin
   D. Cytosine, Uracil, Adenine, Thymine
10. Dicliny means:
    A. Hermaphrodite flowers
    B. Cleistogamous flowers
    C. Flowers are either staminate or pistillate
    D. Male sterile flowers
11. Taking a genotype or group of genotypes of a plant into a new environment where they are not being grown before, is called:
    A. Plant introduction
    B. Genotype interaction
    C. Selection
    D. Hybridization of plants
12. A system of breeding in which repeated crosses are made with donor parents to transfer of specific characters to a well adapted variety is referred to as:
    A. Pedigree method
    B. Bulk method
    C. Backcross method
    D. Recurrent selection method
13. Colchicine is mainly used for induction of:
    A. Hybrid
    B. Pureline
    C. Inbreeding depression
    D. Polyploidy
14. DUS test stands for:
    A. Distinctness, Uniformity, Stability
    B. Development, Utility, Security
    C. Diffusion, Ultra-permeability, Solubility
    D. Development, Uniformity, Security
15. Distant Hybridization is applied for:
    A. Interspecific hybridization
    B. Intergeneric hybridization
    C. Heterosis
    D. Both A & B
16. Somatic hybridization refers to:
    A. Fusion of egg cell
    B. Fusion of protoplast
    C. Crossing over
    D. Mitosis
17. Artificial seed means:
    A. Encapsulation of somatic embryos
    B. Encapsulation of cytoplasm
    C. Encapsulation of nucleus
    D. Encapsulation of mature seed
18. Storage of any cells or tissues at -196°C is known as:
    A. Ultra Preservation
    B. Deep preservation
    C. Cryo-Preservation
    D. Neo Preservation

19. Lysimeter measures:
    A. Amount of direct solar radiation B. Soil moisture tension
    B. Evapotranspiration D. Wind direction
20. The one of the following is not under Hi-Tech Horticulture:
    A. Protected Cultivation B. High Density Planting System
    C. Micro-irrigation system D. Alley Cropping system
21. Growing of two or more crops simultaneously on the same piece of land with a definite row pattern is called:
    A. Multiple cropping B. Intercropping
    C. Sequence cropping D. Relay cropping
22. Allelopathy refers to:
    A. Harmful effect of one allele to another allele
    B. Harmful effect of one crop to another crop
    C. Harmful effect of terminator gene
    D. Harmful effect of monocropping system
23. The capacity of the soil to produce crops of economic value is called:
    A. Soil fertility B. Soil productivity
    C. Soil consistency D. Soil competency
24. Relative proportion of sand, silt and clay refers to:
    A. Soil texture B. Soil structure
    C. Soil plasticity D. Soil colloids
25. Acidic soil can be converted into neutral soil by addition of:
    A. CaCO B. $ZnSO_4$
    C. CuSO D. $FeSO_4$
26. The most abundant among the microbes present in the soil are:
    A. Fungi B. Actinomycetes
    C. Algae D. Bacteria
27. The nodules of legumes contain bacteroids to produce a reddish protein called:
    A. Myoglobin B. Haemoglobin
    C. Leghaemoglobin D. Aspargine
28. Deficiency of nitrogen mainly occurs in:
    A. Older leaves B. Middle leaves
    C. New leaves D. Growing points
29. The water that is used by plants for absorption is:
    A. Hygroscopic water B. Gravitational water
    C. Capillary water D. Water retention
30. The salt tolerant crop is:
    A. Pea B. Lemon C. Sugarbeet D. Cowpea

31. Muriate of potash contains:
A. 16% K B. 25% K C. 40% K D. 60% K

32. Organic manures are applied in the field:
A. 15-30 days before sowing
B. alongwith sowing
C. 15-20 days after sowing
D. the time of flowering

33. Bulky organic manure is:
A. FYM B. Oil cakes C. Boan meal D. Fish meal

34. Application of fertilizers along with irrigation water is called:
A. Microjet irrigation
B. Cablegation
C. Fertigation
D. Corrugation

35. Dryland farming refers to:
A. Cultivation of crops as rainfed conditions
B. Cultivation of crops with assured irrigation system
C. Cultivation of crops in more than 1200 mm rainfall
D. Cultivation of crops in no rain condition

36. Pendimethalin can be used for weed control in soybean as:
A. Pre emergence
B. Pre plant incorporation
C. Post emergence
D. Through herbigation

37. Powdery mildew of pea is controlled by apraying:
A. Sulphur fungicide
B. Mercury fungicides
C. Copper fungicides
D. Miscellaneous fungicides

38. The correctly matched pairs of crop and disease is:
A. Cabbage – Early Blight
B. Cucumber – Club Root
C. Potato – Black Wart
D. Tomato – White Rust

39. Severity of disease club root in cabbage can be reduced by application of:
A. Lime B. Gypsum C. FYM D. Urea

40. Death of tissue due to virus infection is described as:
A. Mottling B. Necrosis C. Mosaic D. Curling

41. Economic Threshold Level is always:
A. Equal to Economic Injury Level
B. Lower than Economic Injury Level
C. More than Economic Injury Level
D. At maximum infestation level

42. Integrated Pest Management refers to:
A. Use of natural enemies to control insect-pests
B. Use of resistant varieties
C. Keeping pest populations below injurious levels
D. Control of insects by organic means

43. Leaf Curl Virus is transmitted by:
   A. White Fly B. Jassid C. Aphid D. House fly
44. Chenopodiaceae is the family of:
   A. Carrot B. Sweet potato C. Spinach D. Leek
45. Browning of cauliflower is due to deficiency of:
   A. Calcium B. Phosphorus
   C. Boron D. Manganese
46. In grafting the upper part of the stem is called:
   A. Bud B. Apex C. Scion D. Stock
47. Brix refers to:
   A. Percentage of total soluble solids in juice
   B. Percentage of reducing sugars in juice
   C. Percentage of non-reducing sugars in juice
   D. Percentage of total sugars in juice
48. Vegetables are rich sources of:
   A. Carbohydrates B. Fat
   C. Protein D. Vitamins and Minerals
49. Which are called the protective food?
   A. Cereal crops B. Oilseed crops
   C. Vegetable crops D. All of the above
50. According to the recommendation by ICMR, an average man should consume vegetables:
   A. 200 g/day B. 250 g/day C. 300 g/day D. 400 g/day
51. The flavour of cole crops is due to:
   A. Volatile sulphur compound B. Volatile aldehyde
   C. Volatile ketones D. None of the above
52. Trypsin inhibitors are toxic substances found in:
   A. Cole crops B. Beans C. Bulb crops D. Root crops
53. Which mineral acts as antioxidants:
   A. Selenium B. Sodium C. Potassium D. Fluorine
54. The substances which neutralize free radicles in the body are called as:
   A. Antioxidants B. Antitranspirants
   C. Phytoalexins D. Antibodies
55. Bioflavonoids works as by:
   A. Increase the efficiency of Vit 'C'
   B. Decrease the efficiency of Vit 'C'
   C. Neutralize the vitamin 'C'
   D. None of the above

56. Cheratin is found in:
A. Bottle gourd
B. Bitter gourd
C. Ridge gourd
D. Pointed gourd

57. Diosgenin is found in:
A. Bitter gourd
B. Celery
C. Dioscoria yam
D. Carrot

58. Celery contains compounds effective against hypertension is:
A. Dithiolthiones
B. 3-n-butyl pthalide
C. Propyl disulphide
D. Allicin

59. India ranks .... in world vegetable production.
A. 1st
B. 2nd
C. 3rd
D. 4th

60. Vivipary means
A. Seed germinates inside the fruit while it is still on the parent tree
B. Seed germinates as usual
C. Seed dormancy is very long
D. None of the above

61. Which of the following is draught tolerant vegetable crop?
A. Bitter gourd
B. Sweet potato
C. Okra
D. Brinjal

62. Seed Plot Technique in potato was proposed by:
A. Dr. M. S. Swaminathan
B. Dr. Hargovind Khurana
C. Dr. Pushkarnath
D. Dr. B. P. Pal

63. Vegetable forcing is the growing of vegetables under:
A. Laboratory
B. Open field
C. Glass house
D. All of the above

64. The tomato variety which sets fruit in low temperature at 8°C:
A. Pusa Ruby
B. Pusa Sheetal
C. Pant T – 3
D. Punjab Chhuhara

65. Most common sex expression in cucurbits is:
A. Monoecious
B. Dioecious
C. Gynoecious
D. Hermaphrodite

66. In watermelon the most important judging indication for maturity is:
A. Days after pollination
B. Dull sound when fruit is thumped
C. Metallic sound when fruit is thumped
D. None of the above

67. Pride of India is a variety of:
A. Apple
B. Rose
C. Cabbage
D. Chrysanthemum

68. Cauliflower, cabbage, broccoli, kohlrabi and brussels sprouts have been derived from wild cabbage that still grows in coastal regions of:
    A. Latin America B. Mesopotamia
    C. Europe and Northern Africa D. South Asia and South Africa
69. .... deteriorates drastically upon inbreeding:
    A. Carrot B. Okra C. Brinjal D. Pea
70. The staminate .... Plants out yield pistillate plants by about 25%:
    A. Asparagus B. Lettuce C. Spinach D. Beet root
71. A tuber represents:
    A. Enlarged underground root B. Enlarged underground stem
    C. Enlarged underground nodules D. None of the above
72. In which crop has the transgenic plant been developed by incorporating Bt gene through genetic engineering technology:
    A. Pea B. Bottle gourd C. Brinjal D. Radish
73. California Wonder is a variety of:
    A. Tomato B. Brinjal C. Sweet Pepper D. Chilli
74. Sex expression in pointed gourd plant is:
    A. Monoecious B. Dioecious
    C. Andromonoecious D. none of the above
75. Gynoecious lines are available in:
    A. Cucumber B. Watermelon
    C. Bitter gourd D. none of the above
76. Okra species which has been extensively used to breed cultivars resistant to yellow vein mosaic virus is:
    A. *Abelmoschus tuberculatus* B. *Abelmoschus ficulneus*
    C. *Abelmoschus manihot* ssp. *manihot* D. None of the above
77. Most of the sweet potato varieties have been developed by:
    A. Mutation breeding B. Clonal selection
    C. Hybridization D. None of the above
78. Potato stored at –1 to 0°C suffers from internal breakdown known as:
    A. Black heart B. Potato scab
    C. Potato wart D. None of the above
79. Which type of self incompatibility is found in cabbage:
    A. Heteromorphic B. Homomorphic
    C. Gametophytic D. Sporophytic
80. Yellow colour of onion is due to:
    A Cuminol B. Curcumin
    C. Quercetin D. Capsanthin

81. In onion, smell is occur due to presence of:
 A. Sulphur B. Phosphorus C. Ferrus D. Calcium
82. Seed plot technique in potato is used to produce:
 A. Insect free seeds B. Virus free seeds
 C. Big size tubers D. Tubers for consumption
83. Vitamin A is abundantly found in:
 A. Potato B. Cowpea
 C. Leafy vegetables D. Onion
84. The ascorbic acid content in radish is significantly increased by the application of:
 A. Nitrogen B. Phosphorus C. Potash D. Sulphur
85. The pungency in radish is due to:
 A. Isothiocynate B. HCN
 C. Allyl Propyl Disulphide D. Allicin
86. Pink skinned radish is due to the presence of:
 A. Lycopene B. Xanthophyll C. Anthocyanin D. Ascorbin
87. The concentration of isothiocyanate in radish is higher at:
 A. Near root tip and peel B. Toot top and flesh
 C. Root top and near root tip D. Peel and flesh
88. The rat-tailed radish is mainly grown for its:
 A. Fleshy roots B. Pods C. Leaves D. Flowers
89. The origin of radish is:
 A. Mediterranean region B. Middle East
 C. China D. India
90. The edible portion of radish develops from:
 A. Primary roots and hypocotyle B. Secondary roots
 C. Auxillary roots D. All of the above
91. The inflorescence of radish is called:
 A. Cyames B. Racemes C. Siliqua D. Umbel
92. The pods of rat-tailed radish varies from:
 A. 2-3 cm B. 5-10 cm C. 10-15 cm D. 20-100 cm
93. Jamunapari Giant is a variety of:
 A. Carrot B. Beet root C. Radish D. Turnip
94. After Independence, the first vegetable breeding station for temperate crops of ICAR was established at:
 A. Katrain, Himachal Pradesh B. Ladakh, Jammu & Kashmir
 C. Srinagar, Jammu & Kashmir D. Darjeeling, West Bengal
95. All India Coordinated Research Project for Improvement of Vegetable Crops was started in:
 A. 1970-71 B. 1974-75 C. 1984-85 D. 1998-99

96. Asian Vegetable Research and Development Center is situated in:
A. Thailand B. India C. Taiwan D. China

97. International Potato Centre (CIP) is situated in:
A. Mexico B. India C. West Indies D. Peru

98. Nitrates and Oxalates are the toxic compounds naturally occuring in:
A. Brinjal B. Spinach
C. Cucumber D. Pointed gourd

99. The most ancient type of vegetable garden is:
A. Kitchen garden B. Truck garden
C. Market garden D. Vegetable forcing

100. .... type of vegetable garden is followed in the Dal Lake of the Kashmir valley:
A. Hydroponics B. Kitchen garden
C. Floating garden D. None of the above

101. The seed stalks formation in spinach starts at:
A. 45 days B. 55 days C. 65 days D. 75 days

102. *Artiplex hortensis* L. or mountain spinach is being grown on small scale especially in:
A. Maharashtra and Karnataka
B. Gujrat and MP
C. Tamilnadu and Kerala
D. Himachal Pradesh and Jammu & Kashmir

103. The brinjal seed is germinated at:
A. 5ºC B. 10ºC C. 15ºC D. 20ºC

104. The optimum temperature at which pumpkin seeds are sown:
A. 15.6ºC B. 20.6ºC C. 24.6ºC D. 26.6ºC

105. 100 g of squash seeds may contain:
A. 100 seeds B. 200 seeds C. 250 seeds D. 350 seeds

106. The seed rate of long melon is:
A. 1-2 kg/ha B. 2-3 kg/ha C. 3-4 kg/ha D. 4-5 kg/ha

107. Flowering in pumpkin starts at:
A. 30-40 days after sowing B. 50-60 days after sowing
C. 60-80 days after sowing D. 85-95 days after sowing

108. Pusa Meghdoot is a variety of:
A. Bottle gourd B. Bitter gourd
C. Snake gourd D. Ridge gourd

109. The number of seeds of bitter gourd per 100 g are about:
A. 350 B. 450 C. 550 D. 650

110. Carrot seeds will not germinate below:
A. 7.9ºC B. 6.9ºC C. 5.9ºC D. 3.9ºC

111. The length of sweet potato cuttings should normally be:
A. 10 cm B. 20 cm C. 30 cm D. 40 cm

112. The vegetative growth of potato is best at:
A. 10ºC B. 15ºC C. 20ºC D. 24ºC

113. Potato tuber production is the maximum at:
A. 15ºC B. 20ºC C. 25ºC D. 30ºC

114. The weight of elephant foot yam corms as planting material should be:
A. 50-70 g B. 100-200 g C. 250-300 g D. 500 g

115. The 'corm' of elephant foot yam as planting material should be taken from:
A. 1st year crop B. 2nd year crop
C. 3rd year crop D. 4th year crop

116. The best storage parameters of garlic is:
A. 0 – 2.2ºC Temp and 60% RH B. 5ºC Temp and 80% RH
C. 11ºC Temp and 70% RH D. 11ºC temp and 60% RH

117. Hardy vegetable crops is:
A. Tomato B. Okra C. Knol-Khol D. Brinjal

118. All Green, Jobner Green and Pusa Jyoti are the improved varieties of :
A. Fenugreek B. Palak
C. Coriander D. None of the above

119. Chekurmanis is commonly known as:
A. Proteinous vegetable
B. Carbohydrate rich vegetables
C. Multivitamin and multimineral packed vegetables
D. None of the above

120. Chekurmanis is propagated by:
A. Seed B. Cuttings C. Bulbs D. Rhizomes

121. The viability of cucumber seeds retain for:
A. 1 year B. 2-3 years C. 4 years D. 5 years

122. Sweet gourd is a:
A. Monoecious B. Dioecious
C. Hermaphrodite D. None of the above

123. Ratooning in the pointed gourd gives flower:
A. Earlier than the planted crop B. Later than the planted crop
C. At the same time D. None of the above

124. Nantes is the .... Variety of carrot.
A. European B. Asiatic
C. Both D. None of the above

125. Sweet potato is a native of:
A. India B. China
C. Tropical America D. Middle East

126. Premature producing of seed stalk in onion is known as:
A. Blanching B. Bolting C. Bracting D. Training
127. Leaf fall in potato occurs due to deficiency of:
A. Calcium B. Boron C. Manganese D. Sulphur
128. Heart rot in sugarbeet is due to deficiency of:
A. Calcium B. Magnesium C. Boron D. Zinc
129. Seed rate of celery per hectare is:
A. 100 g B. 125 g C. 200 g D. 400g
130. Potato stored at -1 to 0ºC suffers from internal breakdown known as:
A. Wart B. Cavity
C. Collapsing D. Black heart
131. Heavy irrigation causes cabbage head:
A. Rotting B. Splitting C. Bolting D. Loosening
132. Pusa Early Bunching and Kasuri selection are the varieties of:
A. Fenugreek B. Coriander C. Palak D. Turnip
133. Paraheliortrophy is shown by:
A. Cowpea B. Tomato C. Cabbage D. Potato
134. Changes of leaf orientation with the response of high temperature and drought is termed as:
A. Thermoperiodism B. Photosensitivity
C. Paraheliotrophy D. Photorespiration
135. Muskmelon crosses with:
A. Cucumber B. Watermelon
C. Long melon D. None of the above
136. For tomato ketchup the tomato should have:
A. High TSS B. Green colour
C. Yellow colour D. None of the above
137. Powdery mildew is a major problem of:
A. Tomato B. Brinjal C. Cowpea D. Pea
138. Which of the following is a draught tolerant vegetable crop?
A. Bitter gourd B. Sweet potato C. Okra D. Brinjal
139. The plant growth regulator used for root formation is:
A. Gibberellin B. IBA C. Cytokinin D. Ethylene
140. The most serious disease of colocasia is:
A. Blight B. Cercospora leaf spot
C. Fusarium wilt D. Rust
141. Leaves of colocasia are an excellent source of:
A. Vitamin C B. Vitamin A
C. Vitamin B D. None of the above

142. Rhubarb is harvested for the full harvest period in the:
A. 1st year B. 2nd year C. 3rd year D. 4th year

143. Seed of the asparagus is sown in the nursery in the month of:
A. Feb-March B. April-May
C. June-July D. None of the above

144. Optimum temperature for germination of asparagus seed is:
A. 30ºC B. 20ºC C. 15ºC D. 10ºC

145. Curly dwarf disease in globe artichoke is caused by:
A. Mycoplasma B. Bacteria C. Fungus D. Virus

146. Horse radish is propagated by:
A. Root cuttings B. Seeds C. Stem cuttings D. Rhizomes

147. Jerusalem artichoke is cultivated for its:
A. Succulent leaves B. Flower buds
C. Tubers D. Non of the above

148. Globe artichoke is propagated by means of:
A. Suckers B. Root cuttings C. Seeds D. Rhizomes

149. Edible portion of globe artichoke is:
A. Root B. Fruits C. Flower buds D. Leaves

150. Rhubarb is grown for its:
A. Thick leaf stalk B. Tubers C. Tender stem D. Fruits

151. Asparagus starts giving sizeable crop after a period of:
A. 1 year B. 2 years C. 3 years D. 4 years

152. Edible part of asparagus is the:
A. Spear B. Leaf
C. Tender root D. Inflorescence

153. Perennial cucurbitaceous crop is:
A. Bottle gourd B. Ash gourd C. Chayote D. Pumpkin

154. Sex expression in yams is:
A. Dioecious B. Andromonoecious
C. Gynomonoecious D. Monoecious

155. Tapioca is propagated by:
A. Seed B. Tuber
C. Stem cutting D. None of the above

156. Tapioca belongs to the family:
A. Euphorbiaceae B. Dioscoreaceae
C. Araceae D. None of the above

157. Spinach beet belongs to the botanical variety:
A. *B. vulgaris* var. *bengalensis* B. *B. vulgaris* var. *cicla*
C. *B. vulgaris* var. *maritima* D. None of the above

158. Spinach beet is of out-breeding nature which is largely due to:
A. Entomophily B. Anemophily
C. Hydrophily D. None of the above

159. The plants of the type with reddish petioles, stems and leaves belong to the species:
A. *Basella alba* B. *B. rubra*
C. *B. cordifolia* D. None of the above

160. In curry leaf, the plant parts mostly eaten are:
A. Leaves B. Tender stem
C. Flower and fruits D. None of the above

161. Poi or Indian spinach belongs to the family:
A. Chenopodiaceae B. Leguminoseae
C. Basellaceae D. None of the above

162. Badichaulai belongs to the species:
A. Blitum B. Tricolor
C. Caudatus D. None of the above

163. The most common leafy vegetable grown during the summer in India is:
A. Amaranth B. Portulaca
C. Indian spinach D. None of the above

164. Cross pollination in palak is due to:
A. Insects B. Water
C. Wind D. None of the above

165. Premature seedling of lettuce is generally influenced by:
A. Long days B. High temperature
C. Low temperature D. None of the above

166. Cracked stem in celery is caused by the deficiency:
A. Long days B. High temperature
C. Low temperature D. None of the above

167. Aster yellows disease of celery is caused by:
A. Fungus B. Bacteria C. Virus D. Algae

168. Celery belongs to the family:
A. Compositae B. Umbelliferae
C. Cruciferae D. Chenopodiaceae

169. Plant parts of celery consumed as salads are:
A. Stem B. Leaf blade C. Petiole D. Root

170. Lettuce seed does not germinate properly when the soil temperature is above:
A. 20ºC B. 25ºC C. 30ºC D. 35ºC

171. The most populai salad crop is:
A. Lettuce B. Celery C. Parsley D. Parsnip

172. Ripe seeds of okra are sometimes used as a substitute for:
A. Coffee B. Cocoa C. Coriander D. Cacti

173. Okra variety Punjab-8 has been evolved by:
A. Interspecific hybridization B. Irradiation induced mutation
C. Chemical induced mutation D. Pureline selection

174. Okra species which has been extensively used to breed cultivars resistant to yellow vein mosaic virus is:
A. *Abelmoschus tuberculatus* B. *Abelmoschus ficulneus*
C. *Abelmoschus manihot* ssp. *manihot* D. All of the above

175. Most of the sweet potato varieties have been developed by:
A. Mutation breeding B. Clonal selection
C. Hybridization D. Bulk method

176. Flowering of sweet potato is stimulated by:
A. Short days and long nights B. Long days and short nights
C. Equal day and night lengths D. None of the above

177. Sweet potato is considered to be:
A. Diploid B. Triploid C. Tetraploid D. Hexaploid

178. Quality of sweet potato is mostly dependent on its:
A. Sugar content B. Protein content
C. Vitamin content D. Fat content

179. Curing of harvested sweet potato tubers takes place if the tubers are kept for 10 days at:
A. 20°C and 90% R.H. B. 40°C and 70% R.H.
C. 60°C and 40% R.H. D. 80°C and 30% R.H.

180. The sweet potato crop should be earthed up when the crop is about:
A. 4 months old B. 3 months old
C. 2.5 months old D. 1 month old

181. More tuberous root formation in sweet potato occurs if cuttings of vines are prepared from:
A. Apex portion B. Middle portion
C. Basal portion D. All of the above

182. The colour of Pusa Sunehri variety of sweet potato is:
A. White B. Yellow C. Light orange D. Green

183. For normal growth, cucurbits require optimum average monthly temperature from:
A. 25-30°C B. 30-35°C
C. 15-20°C D. None of the above

184. Yellow fleshed watermelon variety is:
A. Durgapura Meetha B. Sugar Babby
C. Durgapura Kesar D. None of the above

185. Male sterility has been commercially exploited for the hybrid seed production in:
A. Cucumber B. Watermelon
C. Muskmelon D. Bottle gourd

186. The main pollinating agent in cucurbits is:
A. Ants B. House fly
C. Honey bees D. All of the above

187. Sex ratio in watermelon is lowered by the spray of:
A. Maleic hydrazide B. Ethephone
C. Cycocel D. None of the above

188. Gynoecious lines of cucumber are maintained by induction of male flowers through spray of:
A. Silver thiosulphate B. Maleic hydrazide
C. Cycocel D. NAA

189. Gynoecious lines are available in:
A. Cucumber B. Watermelon
C. Muskmelon D. Bottle gourd

190. Varietty of sponge gourd recommended by IARI is:
A. Pusa Meghdoot B. Pusa Nasdar
C. Pusa Shubhra D. Pusa Chikni

191. For getting good fruit yield in Parwal planting ratio of female to male plants should be:
A. 15-20: 1 B. 30-40: 1 C. 1 : 1 D. 5 : 1

192. Sex expression in Parwal plant is:
A. Monoecious B. Dioecious
C. Andromonoecious D. Gynoecious

193. Fruits of wild forms of bottle gourd are bitter in taste due to:
A. Solanin B. Tomatine
C. Cucurbitacin D. Momordicin

194. Pumpkin originated in:
A. India B. Africa C. America D. Europe

195. The common pumpkin grown throughout India is:
A. *Cucurbita maxima* B. *C. moschata*
C. *C. pepo* D. None of the above

196. For getting higher number of female flower in bottle gourd, spray the plants at 2-4 leaf stage with boron at a concentration of:
A. 20 ppm B. 10 ppm
C. 3 ppm D. None of the above

197. Bottle gourd originated in:
A. Africa B. America C. India D. China

198. For induction of greater number of female flowers in watermelon, spray seedlings at 2-4 leaf stage with 2,4,5-Triiodobenzoic acid at:

A. 10-15 ppm B. 25-50 ppm
C. 50-100 ppm D. None of the above

199. Watermelon is one of the few vegetables that grow well in acidic soil with pH of:

A. 4.0 B. 5.0
C. 6.0 D. None of the above

200. Watermelon seeds do not germinate satisfactorily below:

A. 15°C B. 21°C
C. 35°C D. None of the above

201. Triploid watermelon variety is:

A. Pusa Bedana B. Asahi Yamato
C. Sugar Baby D. Durgapura Meetha

202. The primary centre of origin of watermelon is:

A. India B. Africa C. America D. China

203. For distant market, melon fruits are picked at:

A. Full slip stage B. Half slip stage
C. Green mature stage D. None of the above

204. In the USA, netted varieties of muskmelon are known as:

A. Cantaloupes B. Honey dew
C. Sugar Baby D. None of the above

205. Muskmelon originated in:

A. America B. Mediterranean region
C. China D. None of the above

206. The F1 variety of cucumber is:

A. Poinsett B. Straight Eight
C. Pusa Sanyog D. None of the above

207. Most common sex expression in cucurbits is:

A. Monoecious B. Dioecious
C. Gynoecious D. Hermaphrodite

208. For seed germination of French bean the optimum temperature is:

A. 10-15°C B. 25-30°C
C. 35-40°C D. None of the above

209. French bean variety highly resistant to angular leaf spot is:

A. SVM-1 B. Kentucky Wonder
C. Contender D. Giant Stringless Green Pod

210. In the sweet type of pea parchment layer in pod wall is:

A. Semi-developed B. Present
C. Absent D. None of the above

211. Pea is very rich in:
A. Carbohydrate B. Fat C. Protein D. Fiber

212. Quality of pea is ascertained by:
A. Refractometer B. Tenderometer
C. Potentiometer D. None of the above

213. The optimum temperature for germination of pea is:
A. 5°C B. 15°C C. 22°C D. 25°C

214. An edible podded variety of garden pea is:
A. Matar Ageta – 6 B. Arkel
C. Pant Sabjimatar – 3 D. Sylvia

215. Garden pea belongs to the subspecies:
A. Arvense B. hortense
C. Elatius D. None of the above

216. The bean of which pollen causes allergy to some people is
A. Broad bean B. Cluster bean
C. Sword bean D. None of the above

217. The bean used for extraction of gum is:
A. Broad bean B. Cluster bean
C. Sword bean D. French Bean

218. Arkel variety of pea is most suitable for:
A. Early season B. Mid season
C. Late season D. None of the above

219. Cowpea is probably a native of:
A. India B. China
C. America D. Mediterranean region

220. The bean which can withstand sufficiently cold temperature is:
A. Broad bean B. Cluster bean
C. Sword bean D. French bean

221. Leek is grown for its:
A. Bulb B. Clove
C. Branched Stem and Leaves D. Roots

222. Seed rate requires for raising seedlings of onion for one hectare is:
A. 2-3 kg/ha B. 4-5 kg/ha
C. 8-10 kg/ha D. 12-15 kg/ha

223. Onion belongs to the family:
A. Liliaceae B. Araceae
C. Alliaceae D. Umbelliferae

224. Roughness of roots in carrot results from:
A. Low temperature and regular water supply
B. High temperature and irregular water supply

C. High temperature and excessive irrigation
D. None of the above

225. Good quality of carrots have a:
A. maximum of cortex/phloem and minimum of core/xylem
B. maximum of core/xylem and minimum of cortex/phloem
C. equal part of cortex/phloem and core/xylem
D. none of the above

226. Wind pollination plays an important role in seed production of:
A. Beetroot B. Turnip C. Radish D. Carrot

227. The system of pollination in radish is:
A. Self-pollination B. Cross-pollination
C. Often cross-pollination D. None of the above

228. Carrot belongs to the family:
A. Liliaceae B. Araceae
C. Alliaceae D. Umbelliferae

229. Self-incompatibility in cole crops is governed by:
A. Single dominant gene B. Cytoplasmic factor s
C. Multiallelic single locus S D. None of the above

230. The fleshy edible part of knolkhol is an enlargement of is:
A. Root B. Stem C. Leaf D. Flower

231. In India, seed of cabbage can be produced in:
A. Hills B. Plain C. Plateau D. Riverbeds

232. Black leg disease of cabbage is caused by:
A. Fungus B. Bacteria
C. Virus D. Mycoplasma

233. Cabbage belongs to the botanical variety:
A. Botrytis B. Capitata
C. Gongylodes D. None of the above

234. Cauliflower belongs to the botanical variety:
A. Botrytis B. Capitata
C. Gongylodes D. None of the above

235. Black rot in cauliflower is caused by:
A. Fungal B. Bacterial
C. Viral D. None of the above

236. Whiptail in cauliflower is caused due to the deficiency of:
A. Boron B. Molybdenum
C. Potassium D. Zinc

237. All cole crops belong to the family:
A. Cruciferae B. Umbelliferae
C. Solanaceae D. Cucurbitaceae

238. Blanching operation is mainly done in:
A. Cauliflower B. Cabbage
C. Knol-khol D. Broccoli

239. Cole crops have originated from the wild species:
A. *Brassica oleracea* var. *botrytis* B. *Brassicaoleracea* var. *rapa*
C. *Brassica oleracea*var. *gongylodes* D. *Brassica oleracea* var. *sylvestris*

240. Cauliflower originated in:
A. Mediterranean region B. Central Asia
C. India D. China

241. Tomato leaf curl virus is transmitted by:
A. White fly B. Aphid C. Thrips D. Mites

242. Frogeye leaf spot is a disease of:
A. Pea B. Cucumber C. Okra D. Chilli

243. The maximum pungency in chilli occurs in the:
A. Peduncle B. Seed C. Pericarp D. Placenta

244. Pungency in chilli is due to:
A. Capsaicin B. Lycopine
C. Curcumin D. Cucurbitacin

245. California Wonder is a variety of:
A. Tomato B. Brinjal C. Sweet pepper D. Chilli

246. Sweet pepper belongs to the species:
A. *Capsicum annuum* var. *annuum* B. *Capsicum annuum* var. *grossum*
C. *Capsicum annuum* var. *frutescens* D. *Capsicum pubescens*

247. Which species of chilli shows purple coloured flowers:
A. *Capsicum frutescens* B. *Capsicum chinense*
C. *Capsicum annuum* D. *Capsicum pubescens*

248. The botanical name of commercial grown chilli is:
A. *Capsicum annuum* var. *annuum* B. *Capsicum annuum* var. *grossum*
C. *Capsicum annuum* var. *frutescens* D. *Capsicum pubescens*

249. The vector causing little leaf of brinjal is:
A. Ants B. Whitefly C. Jassid D. Aphid

250. Little leaf of brinjal is caused by:
A. Fungus B. Bacteria
C. Virus D. Mycoplasma

251. Shoot and fruit borer of brinjal is due to:
A. *Leucinodes orbonalis* B. *Helicoverpa armigera*
C. *Spodoptera litura* D. None of the above

252. The daily mean temperature for successful cultivation of brinjal is:
A. 13-21ºC B. 21-27ºC
C. 25-30ºC D. None of the above

253. The long and slender type of brinjal comes under species:
   A. *Solanum melongena* var. *depressum*
   B. *Solanum melongena* var *serpentinum*
   C. *Solanum melongena* var *esculenta*
   D. *Solanum melongena* var *incanum*

254. Brinjal is a:
   A. long day crop B. short day crop
   C. day neutral crop D. none of the above

255. The mode of pollination of brinjal is:
   A. Self-pollination B. Cross-pollination
   C. Often cross-pollination D. None of the above

256. Which of the following crop shows heterostyly:
   A. Pea B. Cucumber C. Okra D. Brinjal

257. Brinjal is originated in:
   A. India B. China C. Ethiopia D. Australia

258. Aubergine is also known as:
   A. Tomato B. Brinjal C. Chilli D. Potato

259. For distant transportation tomato fruits are picked at the:
   A. Immature green stage B. Mature green stage
   C. Turning stage D. Red ripe stage

260. Pollination and fertilization processes in tomato are impaired due to night temperature below:
   A. 13 °C B. 5 °C C. 10 °C D. 20 °C

261. Pollination is generally done in tomato during:
   A. Morning B. Noon C. Evening D. Night

262. Emasculation is generally done in tomato during:
   A. Morning B. Noon C. Evening D. Night

263. Seed yield of tomato is:
   A. 100-150 kg/ha B. 150-250 kg/ha
   C. 250-400 kg/ha D. 400-600 kg/ha

264. Planting ratio of male:female for hybrid seed production of tomato:
   A. 1: 5 B. 1: 8 C. 1: 10 D. 1: 15

265. The isolation distance of certified seeds of tomato is:
   A. 15 m B. 20 m C. 25 m D. 30 m

266. The method of commercial hybrid seed production of tomato is:
   A. Emasculation and pollination B. Male sterility
   C. Self-incompatibility D. All of the above

267. Which species can be used as donor parent for salt resistance?
   A. *Lycopersicon esculentum* B. *L. pimpinellifolium*
   C. *L. cheesmanii* D. *L. peruvianum*

268. Blossom end rot is the physiological disorder due to deficiency of:
A. Boron B. Calcium
C. Sulphur D. Magnesium

269. A net area of about ....may be required to raise seedlings of tomato for one hectare land:
A. 100 sqm B. 150 sqm C. 200 sqm D. 225 sqm

270. Tomato originated in:
A. India B. China
C. Peru-Bolivia-Equadore D. Ethiopia

271. .... vegetable is the exception of summer season crop:
A. Pea B. Cucumber C. Okra D. Brinjal

272. In ....type of tomato, inflorescence occurs more frequently in almost every internode until terminal ones are formed and elongation ceases at this point:
A. Determinate B. Indeterminate
C. Vine D. None of the above

273. In .... type of tomato, inflorescence cluster occurs at every third internode and the main axis continuous growing indefinitely:
A. Determinate B. Indeterminate
C. Bushy D. None of the above

274. The optimum temperature for growth and lycopene development ranges from:
A. 13-17°C B. 20-24°C C. 26-30°C D. 30-34°C

275. .... vegetable is the exception of winter season crop:
A. Sweet potato B. Potato C. Cabbage D. Carrot

276. The vegetables which can endure frost condition are classified as:
A. Hardy B. Semi hardy
C. Tender D. None of the above

277. In floating garden, floating base is first made from the roots of ....
A. Napier B. Tef C. Doob D. Typha

278. Central Potato Research Institute is situated in:
A. Dehradun B. Jammu C. Nainital D. Shimla

279. Directorate of Onion and Garlic is in:
A. Pune B. Jammu C. Nasik D. Shimla

280. Among the various forcing structures, temperature can be controlled more easily in:
A. Greenhouse B. Cold frames
C. Lath house D. None of the above

281. The classification of vegetables which even though not in all but in majority of cases fulfil the basic requirement, is:
A. Botanical classification B. Based on plant parts used
C. Based on method of culture D. Based on hardiness

282. Sioux is a variety of:
A. Potato B. Tomato C. Chilli D. Brinjal

283. Pant Rituraj is a variety of:
A. Long fruited brinjal
B. Round fruited brinjal
C. Small fruited brinjal
D. None of the above

284. Pusa Sadabahar is …………………type of chilli variety.
A. Annual
B. Biennial
C. Perennial
D. None of the above

285. Kufri Sinduri is the ………………… type of potato variety.
A. Early
B. Medium
C. Late
D. None of the above

286. Which among is a non-exotic varicty of onion?
A. Early Grano
B. Brown Spanish
C. Bermuda Yellow
D. Arka Pitambar

287. Yamuna Safed is a variety of:
A. Onion B. Garlic C. Leek D. Pea

288. Bonneville and Lincoln are the exotic varieties of:
A. Onion B. French bean C. Pea D. Cowpea

289. Detroit Dark Red and Crimson Globe are the exotic varieties of:
A. Radish B. Beet root C. Carrot D. Turnip

290. The botanical name of chhotichaulai is:
A. *Amaranthus tricolor* B. *A. blitum* C. *A. dubius* D. *A. tristis*

291. The botanical name of summer squash is:
A. *Cucurbitamoschata* B. *C. pepo* C. *C. maxima* D. *C. ficifolia*

292. The botanical name of Kakrol is:
A. *Momordica charatia*
B. *M. cochinchinensis*
C. *M. balsamina*
D. None of the above

293. Ash gourd is also known as:
A. Pointed gourd B. Snake gourd C. Wax gourd D. Ivy gourd

294. An example of cucurbitaceous fruit having single seed:
A. Ivy gourd
B. Ash gourd
C. Chow-Chow
D. Pointed gourd

295. The botanical name of small fruited cherry tomato is:
A. *Lycopersiconesculentum form cerasiformae*
B. *Lycopersicon esculentum form pyriformae*
C. *Lycopersicon esculentum form commune*
D. *Lycopersicon esculentum form grandifolium*

296. The chromosome number of tomato is:
A. 20 B. 22 C. 24 D. 26

297. The chromosome number of broad bean is:
A. 12 B. 22 C. 24 D. 26
298. Which spice crop is known as the queen of Spice:
A. Turmeric B. Cumin
C. Black pepper D. Cardamom
299. Golden Revolution is related to:
A. Horticulture B. Fishery
C. Poultry D. None of the above
300. Pride of India is a variety of:
A. Cauliflower B. Cabbage C. Knol-Khol D. Broccoli
301. Which vegetable belongs to warm climate:
A. Cabbage B. Pea C. Tomato D. Potato
302. Primary food materials are synthesized at:
A. Roots B. Stems C. Leaves D. Flowers
303. At what concentration of 2,4-D increases the yield of tomato:
A. 100-150 ppm B. 25-35 ppm
C. 1-5 ppm D. None of the above
304. Which among these is not a growth promoting hormone:
A. Auxins B. Cytokinin
C. Abscisic Acid D. None of the above
305. Seed rate of tomato is:
A. 400-500 g/ha B. 800-900 g/ha C. 1000-1100 g/ha D. 4-5 kg/ha
306. Pusa Himani is a variety of:
A. Onion B. Radish C. Pea D. Potato
307. Sugar Baby is a variety of:
A. Water melon B. Muskmelon
C. Cucumber D. Bottle gourd
308. Potato variety Kufri Chandramukhi matures in:
A. 60 days B. 80-90 days C. 120 days D. 200 days
309. Okra starts fruiting:
A. 30 days after sowing B. 40-45 days after sowing
C. 60 days after sowing D. 100 days after sowing
310. The yield of okra is more in:
A. Zaid season B. Kharif season
C. Rabi season D. All of the above
311. Which among these is a variety of bitter gourd:
A. Arka Suryamukhi B. Arka Harit
C. Pusa Alankar D. None of the above

312. Which of the among is an early variety of cauliflower:
A. Pusa Deepali B. Pusa Kartiki
C. Snowball – 16 D. None of the above

313. The seed rate of cauliflower is:
A. 400-450 g/ha B. 700-750 g/ha C. 1000 g/ha D. 1500 g/ha

314. Which among these is a variety of carrot:
A. Pusa Kesar B. Pusa Himani
C. Pusa Chetki D. Pusa Shubhra

315. Indian Institute of Horticulture Research is situated in:
A. New Delhi B. Varanasi C. Bengaluru D. Lucknow

316. Central Food Technology Research Institute is situated in:
A. Mysore B. Dehradun C. New Delhi D. Lucknow

317. Which among these is not a cole crop:
A. Broccoli B. Cauliflower C. Cabbage D. Onion

318. Brinjal hybrid variety Pusa Anmol is a cross of:
A. Pusa Purple Long x Hyderpur B. PusaKranti x Pusa Uttam
C. Pusa Purple Round x Hyderpur D. None of the above

319. Pusa Sawani is a variety of:
A. Okra B. Tomato C. Brinjal D. Radish

320. Symbiotic bacteria is:
A. Azotobactor B. Rhizobium
C. Clostridium D. None of the above

321. Orobanchae is the root parasite of:
A. Brinjal B. Onion C. Carrot D. Coriander

322. Which crop is susceptible to frost:
A. Tomato B. Onion C. Cabbage D. Spinach

323. The edible portion of cauliflower is:
A. Root B. Stem
C. Leaves D. Inflorescence

324. Round revolution is related to:
A. Tomato B. Potato C. Cabbage D. Brinjal

325. Critical stage of irrigation in potato is:
A. During earthing up B. At the time of germination
C. After the formation of potato tuber D. At the time of tuber formation

326. Tomato can be grown all round the year due to:
A. Short day plant B. Long day plant
C. Day neutral plant D. Photoplastic plant

327. Which is the non bulb forming crop:
A. Onion B. Garlic C. Shallot D. Leek

328. The pungency in onion is due to:
A. Allyl Propyl Disulphide B. Diallyl Disulphide
C. Isothiocynate D. None of the above

329. Which among these is a perennial crop:
A. Tomato B. Drumstick C. Chilli D. Onion

330. The cultivation of vegetables is known as:
A. Agriculture B. Olericulture C. Pisciculture D. Viticulture

331. Hardening off is the:
A. Process of endurance against adverse climate
B. Process of endurance against diseases
C. Process of endurance against insect-pests
D. All of the above

332. The acridity in colocasia is due to the presence of:
A. Quercetin B. Allicin
C. Solanin D. Calcium oxalate

333. The red colour of tomato is due to:
A. Quercetin B. Allicin C. Lycopene D. Capscicin

334. Due to exposure in sunlight potato tubers become green due to:
A. Calcium oxalate B. Solanin C. Allicin D. Quercetin

335. Covering of cauliflower curd by tying of their leaves to retain milky white curd is called:
A. Bolting B. Blanching C. Browning D. Binding

B. Fill in the Blanks

336. Botanical name of tree tomato is __________.
337. Botanical name of sword bean is __________.
338. Globe artichoke is a __________ vegetable crop.
339. Parsnip can germinate at minimum soil temperature __________.
340. French bean cannot develop well below .... base temperature.
341. Very early cauliflower varieties can produce curd at __________ temperature.
342. Late varieties of cauliflower can produce curd at __________ temperature.
343. The plants which can not withstand prolonged cool weather is called __________.
344. Warm season vegetable crops are also known as __________.
345. Kharif onion is a __________ for bulb production.
346. Garlic requires __________ temperature for production of large bulb size.
347. High temperature causes carrot roots __________.
348. Best quality root production of beet root is at __________.
349. Sweet potato is a __________ vegetable crop.

350. The ideal temperature for cassava crop ranges between ________.
351. Female flowers are developed more under ________ in cucurbits.
352. In melons, ________ increases sugar content in the fruit.
353. In tomato, fruits fail to set at ________ temperature.
354. At low temperature below ________ brinjal fruits become deformed.
355. High night temperature of ________ may cause flower drop in chilli.
356. Low night temperature of ________ leads to development of parthenocarpic fruits in chilli.
357. ________ induces flowering in potato.
358. Exposure of temperate vegetables to low temperature of 4-7°C for 6-8 weeks to induce flowering is called as ________.
359. Brussels sprout may require ________ weeks of vernalization.
360. Cabbage and cauliflower requires ________ weeks of vernalization.
361. In snake gourd and pointed gourd, anthesis occurs at ________.
362. Temperature above ________ leads to bolting in palak.
363. Lycopene pigment is completely destroyed at________.
364. ________ is responsible for higher capscisin content in chilli.
365. ________ causes riceyness in cauliflower.
366. ________ temperature during the growth period, beet root show alternate white and coloured rings.
367. ________ condition makes the root tough and pungent before reaching the edible size in temperate radish cultivar.
368. Potato stored at 4-5°C temperature, become undesirably ________.
369. Late blight of tomato and potato is due to ________.
370. Early blight of tomato and potato is due to ________.
371. More than ________ wavelength of light is harmful to plant.
372. Onion requires ________ conditions for induction of flowering.
373. Sweet potato and watermelon are tolerant to ________ condition.
374. Fennel, shallot, rhubarb are tolerant to ________.
375. Palak and Asparagus are tolerant to ________.
376. Beetroot, kale, turnip, bitter gourd, ash gourd and lettuce are tolerant to ________.
377. Water requirements of leafy vegetables are ________.
378. Watermelon, muskmelon, pumpkin and wax gourd require ________.
379. Critical stages of irrigation in cabbage are ________.
380. Critical stages of irrigation in cabbage are ________.
381. Tomato, watermelon and sweet potato are ________.
382. Onion and lettuce are ________ rooted crops.
383. Removal of nutrients is ________ in cabbage and cauliflower.

384. Removal of nutrients is ________________ in okra and cucumber.
385. In general, the vegetable crops whose above ground vegetative parts are consumed require more ________________.
386. The vegetable crops whose underground vegetative parts are consumed require more________________.
387. The vegetable crops whose fruits are consumed require more ________________
388. Cabbage, cauliflower, radish, onion, turnip and garlic require comparatively more ________________.
389. Premature yellowing of lower leaves and new leaves are light green are the symptoms of ________________ deficiency.
390. Purple colouration on leaves is the symptom of ________________ deficiency.
391. Development of grey areas near the leaf margin and eventually a scorch appears around the entire leaf margin are the symptom of ________________ deficiency.
392. ________________ deficiency occurs on acidic and highly leached sandy soil.
393. Loss of green colour at the tips of lower leaves between the veins is due to deficiency of ________________.
394. ________________ deficiency mostly appears from 2nd or 3rd fully mature leaves.
395. Potato variety Kufari Chandramukhi has ________________ of dormancy period.
396. Potato variety Kufari Jyoti has ________________ of dormancy period.
397. Potato dormancy can be broken by the application of ________________
398. Short spell of ________________ RH and ________________ temperature leads to epiphytotic state of late blight.
399. Concentric rings are developed in ________________ disease of potato.
400. ________________ shows the symptoms of dark chocolate coloured hard bodies deposited on the skin, while the flesh remains unaffected in potato.
401. ________________ of potato shows blackening to a depth of 2-5 mm just below the skin.
402. Potato wart is caused by ________________.
403. Bacterial wilt of potato is caused by ________________.
404. Potato scab is caused by ________________.
405. ________________ are the vector of several viruses like potato leaf roll virus, potato virus Y and potato virus A.
406. ________________ is a serious pest of potato at storage and field condition.
407. ________________ in tomato may occur due to irrigation or rainfall after long dry spell and boron deficiency.
408. Outer wall of the grown up fruits continues to develop normally but growth of the remaining internal tissues is retarded is known as ________________ in tomato.
409. Application of ________________ reduces puffiness and improve fruit shape in tomato.

410. Distortion of the blossom end of the fruit gives rise to various ridges, furrows and indentations in tomato fruit is called ______________
411. In tomato fruit, on surface around the calyx and fruit shoulder, tiny yellow spots often appear are called ______________.
412. Serious nursery disease is ______________.
413. Tomato leaf curl virus is transmitted by ______________.
414. Fruit borer is a serious pest of ______________.
415. Phomopsis blight is caused by ______________.
416. Hot water treatment of seeds at ______________ effective against phomopsis blight of brinjal.
417. ______________ is the serious insect-pest of brinjal.
418. Epilachna beetle is the major pest of ______________.
419. ______________ transmit the little leaf of brinjal disease.
420. Ripe fruit rot of chilli is caused by ______________
421. ______________ disease of chilli is transmitted by whitefly and mites.
422. ______________ in cauliflower is caused by deficiency of boron.
423. Deficiency of boron in knol-khol causes ______________.
424. Application of ______________ reduces the incidence of browning in cole crops.
425. Application of ______________ reduces the problem of whip tail of cauliflower.
426. Whip tail is more prevalent in ______________ soil.
427. Buttoning in cauliflower is due to deficiency of ______________.
428. Damage of terminal buds may cause ______________ in cauliflower.
429. Velvety or granular appearance on the curd surface in cauliflower is known as ______________.
430. ______________ is the main reason of riceyness in cauliflower
431. The main cause of leafiness in cauliflower is prevalence of ______________ temperature.
432. ______________ is caused in cauliflower due to boron deficiency and excess supply of nitrogen.
433. Exposure of curds to high light intensity causes ______________ in cauliflower.
434. Tip burning in chinese cabbage is due to ______________ deficiency.
435. Damping off is caused by ______________.
436. Downey mildew of cole crops is caused by ______________.
437. Black leg is a seed borne disease caused by ______________.
438. Black rot is caused by ______________.
439. Diamond back moth mainly attacks on ______________ crops.
440. ______________ is the most damaging stage of cabbage butterfly.
441. Pillow is the physiological disorder of ______________ and due to low calcium level.

442. Leaf silvering in summer squash occurs due to ______________.
443. Arka Manik is the resistant variety of watermelon against ______________
444. ______________ should not be used against powdery mildew in cucurbits.
445. Cucumber mosaic virus is transmitted by ______________.
446. ______________ is a serious pest of cucurbits.
447. Red pumpkin beetles prefer ______________ of cucurbits to feed.
448. Powdery mildew of pea is caused by ______________
449. Anthracnose of pea is caused by ______________
450. Ashy stem blight of beans is caused by ______________
451. Delay in sowing of pea evades infestation of ______________
452. ______________ is a serious insect-pest of peas and beans.
453. Okra seeds do not germinate at or below ______________
454. Powdery mildew of okra is caused by ______________
455. YVMV is a disease of ______________
456. YVMV is transmitted by ______________
457. Shoot &fruit borer is a serious problem of ______________
458. Onion seed has only ______________ year of viability.
459. The favourable condition for purple blotch disease of onion is ______________
460. Small yellow to pale yellow streaks on one side of onion leaves are the symptoms of ______________
461. Onion smut is ______________ borne disease.
462. Onion smut is caused by ______________
463. Soft rot of onion is caused by ______________
464. Thrips is the major problem of ______________
465. Very low respiratory activity is found in ______________
466. Avoidance of application of nitrogen and at the same time essential supply of potassic fertilizers improve the ______________ of vegetables.
467. Rapid removal of field heat from freshly harvested produce is termed as ______________
468. Tubers and bulbs are exposed to relatively high temperature and high relative humidity at farm level after the harvesting is termed as ______________
469. ______________ is the good ethylene absorbent.
470. Zero Energy Cool Chamber has been developed by ______________
471. Zero Energy Cool Chamber works on the principle of ______________
472. Under Zero Energy Cool Chamber, the shelf life of vegetables can be improved by ______________
473. Under Zero Energy Cool Chamber, the temperature can be brought down by ______________
474. Under controlled atmospheric storage, the concentration of O_2 is ______________ and CO_2 is ______________.

475. Storage of vegetables under low atmospheric pressure is referred to as ________________.

476. Export quality of okra fruits should be harvested at green, tender and ________________ size.

477. Onions are packaged in ________________

478. ________________ boxes are more commonly used packaging materials for vegetables.

479. The capacity of the seed to germinate and produce normal seedlings is termed as ________________

480. ________________ is referred to the span upto which seed viability can be retained after harvest.

481. Seeds of chow-chow can be stored along with intact ________________.

482. ________________ seeds can be dried to a low moisture level and would show loss of viability with rise in moisture content.

483. ________________ seeds require relatively high seed moisture content for longevity.

484. Seeds of pointed gourd is ________________ seeds.

485. For each 1% increase in seed moisture between 5-14%, the longevity of the seed is ________________.

486. ________________ refers to hydrate seeds under controlled conditions but preventing the germination.

487. ________________ refers to the application of solid particles that act as a filler with a binder to form a more or less spherical seeds.

488. Isolation distance for production of certified seeds of potato should be ________________.

489. Isolation distance for production of certified seeds of okra should be ________________

490. Isolation distance for production of certified seeds of cabbage should be ________________

491. Isolation distance for production of certified seeds of carrot should be ________________

492. Isolation distance for production of certified seeds of cucumber should be ________________

493. Rhubarb is propagated through ________________

494. Cassava has ________________ sex form.

495. Yams are ________________ as for sex expression.

496. The inflorescence of taro is known as ________________

497. In muskmelon, ________________ sex expression is found.

498. Onion has ________________ type of male sterility.

499. Onion has ________________ flowers where stamens mature before pistil.

500. Elephant foot yam has ________________ flowers where pistil get mature before stamen.

501. _______________ and _______________has high inbreeding depression.
502. _______________ Show very low inbreeding depression.
503. Sioux and Roma are the _______________Of varieties from USA.
504. Pusa Ruby is the cross of _______________
505. _______________ is the method for onion variety development.
506. MDU-1 is the _______________ variety of chilli.
507. Arka Jeet and Arka Rajhans are the varieties of _______________
508. Arka Navneet is the hybrid variety of _______________
509. Punjab Hybrid and Pusa Rasraj are the hybrid varieties of _______________
510. Functional male sterility is found in _______________
511. _______________ can be used for maintenance of female parents on hybridization.
512. Hybrid seed yield of cole crops ranges from _______________
513. The resistance characters available in wild species _______________against TMV in chilli.
514. Cauliflower is sensitive to _______________
515. Heavy irrigation causes _______________ in cabbage head.
516. Pusa Early Bunching and Kasuri Selection are the varieties of _______________
517. Sioux and Marglobe are the varieties of _______________
518. The botanical name of Chinese cabbage is _______________
519. Pusa Chandrima is a variety of _______________
520. Detroit Dark Red is the variety of _______________
521. The botanical name of leek is _______________
522. Garlic is propagated by _______________
523. Seed rate of garlic clove is _______________
524. The botanical name of French bean is _______________
525. Japanese Long green is the variety of _______________
526. Seedless variety of watermelon is _______________
527. Arka Jyoti is the hybrid variety of _______________
528. Seed rate of cucumber, watermelon and muskmelon is _______________
529. Hara Madhu is a variety of _______________
530. Pusa Safed a variety of _______________
531. Lettuce, Celery and Parsley are the _______________ crops.
532. The lowest chromosome number is found in _______________ vegetable crop.
533. The red pigment in beetroot is _______________
534. Beet root is originated from wild species _______________
535. The colour of beetroot is due to presence of red-violet pigment_______________
536. Beetroot is sensitive to _______________ soil.
537. Vegetable pea seeds have _______________ surface.

538. Kidney bean is also known as ____________
539. ____________ is the progenitor of French bean.
540. Most of the cultivars of French bean is ____________
541. Transverse cotyledon cracking is the major problem of ____________
542. Hypocotyl necrosis in French bean is due to low ____________ content.
543. French bean is ready to harvest from ____________ days onwards after germination.
544. ____________ coloured seeds are more susceptible to transverse cotyledon cracking.
545. The botanical name of cowpea is ____________
546. Cowpea is a ____________ crop with respect to photosensitivity.
547. Cowpea is sensitive to ____________
548. Cowpea seeds maintain viability for ____________ under normal storage condition.
549. The root knot nematode is caused by ____________
550. Application of ____________ is good to control nematodes organically.
551. Self-incompatibility can be effectively utilized in ____________ Crops.

Answer Key

1. A	2. A	3. B	4. B	5. C
6. D	7. A	8. C	9. B	10. C
11. A	12. C	13. D	14. A	15.D
16. B	17. A	18. C	19. A	20. D
21.B	22. B	23. B	24. A	25. A
26. B	27. C	28. A	29. C	30. C
31. D	32. A	33. A	34. C	35. A
36. A	37. A	38. C	39. A	40. C
41. B	42. C	43. A	44. C	45. C
46. C	47. A	48. D	49. C	50. C
51. A	52. B	53. A	54. A	55. A
56. B	57. C	58. B	59. B	60. B
61. B	62. C	63. C	64. B	65. A
66. B	67. C	68. C	69. A	70. A
71. B	72. C	73. C	74. B	75. A
76. C	77. B	78. A	79. D	80. C
81. A	82. B	83. C	84. B	85. A
86. C	87. A	88. B	89. A	90. A
91. B	92. D	93. C	94. A	95. A
96. C	97. D	98. B	99. A	100. C
101. C	102. A	103. B	104. D	105. D

106. B	107. C	108. A	109. D	110. D
111. C	112. D	113. B	114. B	115. B
116. A	117. C	118. B	119. C	120. B
121. D	122. B	123. A	124. A	125. C
126. B	127. B	128. C	129. B	130. D
131. B	132. A	133. A	134. C	135. C
136. A	137. D	138. B	139. B	140. A
141. A	142.C	143. A	144. A	145. D
146. A	147. C	148. A	149. C	150. A
151. C	152. A	153. C	154. A	155. C
156. A	157. A	158. B	159. B	160. C
161. C	162. B	163. A	164. A	165. B
166. A	167. C	168. B	169. C	170. C
171. A	172. A	173. C	174. C	175. B
176. A	177. D	178. A	179. D	180. C
181. A	182. C	183. A	184. C	185. C
186. C	187. B	188. A	189. A	190. D
191. A	192. B	193. C	194. C	195. B
196. C	197. C	198. B	199. B	200. B
201. A	202. B	203. C	204. A	205. D
206. C	207. A	208. B	209. A	210. C
211. C	212. B	213. C	214. D	215. B
216. A	217. B	218. A	219. A	220. A
221. C	222. C	223. C	224. B	225. A
226. A	227. B	228. D	229. C	230. B
231. A	232. A	233. B	234. A	235. B
236. B	237. A	238. A	239. D	240. A
241. A	242. D	243. D	244. A	245. C
246. B	247. D	248. A	249. C	250. D
251. A	252. A	253. B	254. C	255. C
256. D	257. A	258. B	259. B	260. A
261. A	262. C	263. A	264. A	265. C
266. A	267. C	268. B	269. D	270. C
271. A	272. A	273. B	274. B	275. A
276. A	277. D	278. D	279. B	280. A
281. C	282. B	283.B	284. C	285. C
286. D	287. B	288. C	289. B	290. B
291. B	292. B	293. C	294. C	295. A
296. C	297. A	298. D	299. A	300. B
301. C	302. C	303. C	304. C	305. A

306. B	307. A	308. B	309.B	310. B
311. B	312. A	313. A	314. A	315. C
316. A	317. D	318. A	319. A	320. B
321. A	322. A	323. D	324. B	325. D
326. C	327. D	328. A	329. B	330. B
331. A	332. D	333. C	334. B	335. B

B. Fill in the blanks

336. *Cyphomandra betacea*
337. *Canavalia gladiata*
338. Temperate perennial
339. 1.6°c
340. 10°c
341. 22 – 27 °c
342. 10 -12 °c
343. Warm season vegetable crop
344. Tender vegetable crop
345. Short day
346. 10 -15 °c
347. Shorter, thicker and less colour
348. 15 -21 °c
349. Tender
350. 25 32 °c
351. Short day condition
352. Lower night temperature
353. 13°c or below
354. 10°c
355. 24°c and above
356. 10-15°c
357. Long day length and cool temperature
358. Vernalization
359. 12 – 16 weeks
360. 10-12 weeks
361. Night
362. 25 °c coupled with long day
363. 40°c 364. High night temperature
365. Fluctuation in temperature
366. Above 25°c
367. Hot weather
368. Sweeter
369. *Phytophthora infestans*
370. *Alternaria solani*
371. 700 nm
372. Long day
373. High acidity
374. High acidity
375. High soil salinity
376. High soil salinity
377. High
378. Less water
379. Stolen formation, tuberization and tuber enlargement
380. Head formation and enlargement
381. Very deep rooted crops
382. Very shallow
383. Very high
384. Low
385. Nitrogen
386. Potash
387. Phosphorus
388. Sulphur
389. Nitrogen
390. Phosphorus
391. Potassium
392. Calcium
393. Magnesium
394. Zinc
395. 4-5 weeks
396. 6-8 weeks
397. 1% thiourea

398. High humidity (more than 80% RH) and cool weather (10-20 °c)
399. Early blight
400. Black scurf
401. Charcoal rot
402. *Synchytriumendobioticum*
403. *Pseudomonas solanacearum*
404. *Streptomyces scabis*
405. Aphids
406. Potato tuber moth
407. Fruit cracking
408. Puffiness
409. Boron
410. Catface
411. Gold flacks
412. Damping off
413. White fly
414. Tomato
415. *Phomopsisvexans*
416. 50°c for 30 min
417. Shoot and fruit borer
418. Brinjal
419. Jassids
420. *Colletotrichumcapscisi*
421. Leaf curl virus
422. Browning
423. Splitting of tuber
424. 10-15 kg/ha borex
425. 1.5 kg sodium or ammonium molybdenum
426. Acidic soil
427.Nitrogen
428. Blindness
429. Riceyness
430. Temperature fluctuation
431. High
432. Hollow stem
433. Pinking
434. Calcium
435. Phythium, Phytophthora and Rhizoctonia
436. Peronosporaparasitica
437. Phomalingum
438. Xanthomonascampestrispv. Campestris
439. Cruciferous crops
440. Caterpillar
441.Cucumber
442. Moisture scarcity
443. Powdery mildew
444. Sulfex
445. Aphids
446. Fruit fly
447. Young seedlings
448. *Erysiphepolygoni*
449. *Colletotrichuml indemuthianum*
450. *Macrophomina phaseolina*
451. Stem fly
452. Pod borer
453. 20°c
454. *Erysiphe cichoracearum*
455. Okra
456. White fly
457. Okra
458. 1 year
459. 28-30°c and 70-90% RH
460. Stemphyllium blight
461. Soil
462. *Urocystis cepulae*
463. *Erwinia carotovora*
464. Onion
465. Potato and onion
466. Keeping quality
467. Precooling
468. Curing
469. $Kmno_4$
470. IARI, New Delhi
471. Evaporative cooling
472. 5-7 days
473. 8-10°c

474. Lowered, elevated
475. Hypobaric storage
476. 6 -8 cm
477. Plastic netted bags
478. CFB
479. Viability
480. Seed longevity
481. Ripe fruits
482. Orthodox
483. Recalcitrant
484. Recalcitrant
485. Halved
486. Priming
487. Pelleting
488. 5 m
489. 200 m
490. 1000 m
491. 800 m
492. 400 m
493. Division of crowns
494. Monoecious
495. Dioecious
496. Spadix
497. Andromonoecious
498. CGMS
499. Protandrous
500. Protogynous
501. Carrot and onion
502. Cucurbits
503. Introduction
504. Sioux x Improved Meeruti
505. Selfing and Massing
506. Mutant
507. Muskmelon
508. Brinjal
509. Muskmelon
510. Tomato
511. Gametocides
512. 500-800 kg/ha
513. *Capsicum chacoense*
514. High acidity
515. Splitting
516. Fenugreek
517. Tomato
518. *Brassica chinensis*
519. Turnip
520. Beet root
521. *Allium porrum*
522. Cloves
523. 500 kg/ha
524. *Phaseolus vulgaris*
525. Cucumber
526. Pusa Bedana
527. Watermelon
528. 2.5-3 kg/ha
529. Muskmelon
530. Sweet potato
531. Salad
532. Broad bean and spinach
533. Betanin
534. *Beta vulgaris* ssp. *Maritima*
535. -Cyanin
536. Soil acidity
537. Wrinkled
538. French bean
539. *Phaseolus vulgaris* ssp. *Aborigineus*
540. Photoinsensitive
541. French bean
542. Calcium
543. 40 days
544. White
545. *Vigna unguiculata*
546. Short day
547. Waterlogging
548. 2 years
549. *Meloidogyne incognita*
550. Neem cake
551. Cole crops

❍❍❍

Appendix–I

ICAR Institutions, Deemed Universities, National Research Centres, National Bureaux & Directorate/Project Directorates

Deemed Universities - 4	
1.	Indian Agricultural Research Institute, New Delhi
2.	National Dairy Research Institute, Karnal
3.	Indian Veterinary Research Institute, Izatnagar
4.	Central Institute on Fisheries Education, Mumbai
Institutions - 60	
1.	Central Island Agricultural Research Institute , Port Blair
2.	Central Arid Zone Research Institute, Jodhpur
3.	Central Avian Research Institute, Izatnagar
4.	Central Inland Fisheries Research Institute, Barrackpore
5.	Central Institute Brackishwater Aquaculture, Chennai
6.	Central Institute for Research on Buffaloes, Hissar
7.	Central Institute for Research on Goats, Makhdoom
8.	Central Institute of Agricultural Engineering, Bhopal
9.	Central Institute for Arid Horticulture, Bikaner
10.	Central Institute of Cotton Research, Nagpur
11.	Central Institute of Fisheries Technology, Cochin
12.	Central Institute of Freshwater Aquaculture, Bhubneshwar
13.	Central Institute of Research on Cotton Technology, Mumbai
14.	Central Institute of Sub Tropical Horticulture, Lucknow
15.	Central Institute of Temperate Horticulture, Srinagar
16.	Central Institute on Post harvest Engineering and Technology, Ludhiana
17.	Central Marine Fisheries Research Institute, Kochi
18.	Central Plantation Crops Research Institute, Kasargod

19.	Central Potato Research Institute, Shimla
20.	Central Research Institute for Jute and Allied Fibres, Barrackpore
21.	Central Research Institute of Dryland Agriculture, Hyderabad
22.	Central Rice Research Institute, Cuttack
23.	Central Sheep and Wool Research Institute, Avikanagar, Rajasthan
24.	Indian Institute of Soil and Water Conservation, Dehradun
25.	Central Soil Salinity Research Institute, Karnal
26.	Central Tobacco Research Institute, Rajahmundry
27.	Central Tuber Crops Research Institute, Trivandrum
28.	ICAR Research Complex for Eastern Region, Patna
29.	ICAR Research Complex for NEH Region, Barapani
30.	Central Coastal Agricultural Research Institute, Ela, Old Goa, Goa
31.	Indian Agricultural Statistics Research Institute, New Delhi
32.	Indian Grassland and Fodder Research Institute, Jhansi
33.	Indian Institute of Agricultural Biotechnology, Ranchi
34.	Indian Institute of Horticultural Research, Bengaluru
35.	Indian Institute of Natural Resins and Gums, Ranchi
36.	Indian Institute of Pulses Research, Kanpur
37.	Indian Institute of Soil Sciences, Bhopal
38.	Indian Institute of Spices Research, Calicut
39.	Indian Institute of Sugarcane Research, Lucknow
40.	Indian Institute of Vegetable Research, Varanasi
41.	National Academy of Agricultural Research & Management, Hyderabad
42.	National Institute of Biotic Stresses Management, Raipur
43.	National Institue of Abiotic Stress Management, Malegaon, Maharashtra
44.	National Institute of Animal Nutrition and Physiology, Bengaluru
45.	National Institute of Research on Jute & Allied Fibre Technology, Kolkata
46.	National Institute of Veterinary Epidemiology and Disease Informatics, Hebbal, Bengaluru
47.	Sugarcane Breeding Institute, Coimbatore
48.	Vivekananda Parvatiya Krishi Anusandhan Sansthan, Almora
49.	Central Institute for Research on Cattle, Meerut, Uttar Pradesh
50.	National Institute of High Security Animal Diseases, Bhopal
51.	Indian Institute of Maize Research,New Delhi
52.	Central Agroforestry Research Institute , Jhansi
53.	National Institute of Agricultural Economics and Policy Research, New Delhi
54.	Indian Institute of Wheat and Barley Research, Karnal

55.	Indian Institute of Farming Systems Research, Modipuram
56.	Indian Institute of Millets Research, Hyderabad
57.	Indian Institute of Oilseeds Research, Hyderabad
58.	Indian Institute of Oil Palm Research, Pedavegi, West Godawari
59.	Indian Institute of Water Management, Bhubaneshwar
60.	Central Institute for Women in Agriculture, Bhubaneshwar
National Research Centres - 15	
1.	National Research Centre for Banana, Trichi
2.	National Research Centre for Citrus, Nagpur
3.	National Research Centre for Grapes, Pune
4.	National Research Centre for Litchi, Muzaffarpur
5.	National Research Centre for Pomegranate, Solapur
6.	National Research Centre on Camel, Bikaner
7.	National Research Centre on Equines, Hisar
8.	National Research Centre on Meat, Hyderabad
9.	National Research Centre on Mithun, Medziphema, Nagaland
10.	National Research Centre on Orchids, Pakyong, Sikkim
11.	National Research Centre on Pig, Guwahati
12.	National Research Centre on Plant Biotechnology, New Delhi
13.	National Research Centre on Seed Spices, Ajmer
14.	National Research Centre on Yak, West Kemang
15.	National Centre for Integrated Pest Management, New Delhi
National Bureaux - 6	
1.	National Bureau of Plant Genetics Resources, New Delhi
2.	National Bureau of Agriculturally Important Micro-organisms, Mau, Uttar Pradesh
3.	National Bureau of Agricultural Insect Resources, Bengaluru
4.	National Bureau of Soil Survey and Land Use Planning, Nagpur
5.	National Bureau of Animal Genetic Resources, Karnal
6.	National Bureau of Fish Genetic Resources, Lucknow
Directorates/Project Directorates - 15	
1.	Directorate of Rice Research, Hyderabad
2.	Directorate of Seed Research, Mau
3.	Directorate of Groundnut Research, Junagarh
4.	Directorate of Soybean Research, Indore
5.	Directorate of Rapeseed & Mustard Research, Bharatpur
6.	Directorate of Mushroom Research, Solan

7.	Directorate on Onion and Garlic Research, Pune
8.	Directorate of Cashew Research, Puttur
9.	Directorate of Medicinal and Aromatic Plants Research, Anand
10.	Directorate of Floricultural Research, Pune, Maharashtra
11.	Directorate of Weed Science Research, Jabalpur
12.	Project Directorate on Foot & Mouth Disease, Mukteshwar
13.	Directorate of Poultry Research, Hyderabad
14.	Directorate of Knowledge Management in Agriculture (DKMA), New Delhi
15.	Directorate of Cold Water Fisheries Research, Bhimtal, Nainital

All India Coordinated Research Projects

1.	AICRP on Nematodes, New Delhi
2.	AICRP on Maize, New Delhi
3.	AICRP Rice, Hyderabad
4.	AICRP on Chickpea, Kanpur
5.	AICRP on MULLARP, Kanpur
6.	AICRP on Pigeon Pea, Kanpur
7.	AICRP on Arid Legumes, Jodhpur
8.	AICRP on Wheat & Barley Improvement Project, Karnal
9.	AICRP Sorghum, Hyderabad
10.	AICRP on Pearl Millets, Jodhpur
11.	AICRP on Small Millets, Bangalore
12.	AICRP on Sugarcane, Lucknow
13.	AICRP on Cotton, Coimbatore
14.	AICRP on Groundnut, Junagarh
15.	AICRP on Soybean, Indore
16.	AICRP on Rapeseed & Mustard, Bharatpur
17.	AICRP on Sunflower, Safflower, Castor, Hyderabad
18.	AICRP on Linseed, Kanpur
19.	AICRP on Sesame and Niger, Jabalpur
20.	AICRP on IPM and Biocontrol, Bangalore
21.	AICRP on Honey Bee Research & Training, Hisar
22.	AICRP -NSP(Crops), Mau
23.	AICRP on Forage Crops, Jhansi
24.	AICRP on Fruits, Bangaluru
25.	AICRP Arid Zone Fruits, Bikaner
26.	AICRP Mushroom, Solan

27.	AICRP Vegetables including NSP vegetable, Varanasi
28.	AICRP Potato, Shimla
29.	AICRP Tuber Crops, Thiruvananthapuram
30.	AICRP Palms, Kasaragod
31.	AICRP Cashew, Puttur
32.	AICRP Spices, Calicut
33.	AICRP on Medicinal and Aromatic Plants including Betelvine, Anand
34.	AICRP on Floriculture, New Delhi
35.	AICRP in Micro Secondary & Pollutant Elements in Soils and Plants, Bhopal
36.	AICRP on Soil Test with Crop Response, Bhopal
37.	AICRP on Long Term Fertilizer Experiments, Bhopal
38.	AICRP on Salt Affected Soils & Use of Saline Water in Agriculture, Karnal
39.	AICRP on Water Management Research, Bhubaneshwar
40.	AICRP on Ground Water Utilisation, Bhubaneshwar
41.	AICRP Dryland Agriculture, Hyderabad
42.	AICRP on Agrometeorology, Hyderabad including Network on Impact adaptation & Vulnerability of Indian Agri. to Climate Change
43.	AICRP Integrated Farming System Research, Modipuram including Network Organic Farming
44.	AICRP Weed Control, Jabalpur
45	AICRP on Agroforestry, Jhansi
46.	AICRP on Farm Implements & Machinery, Bhopal
47.	All India Coordinated Research Project on Ergonomics and Safety in Agriculture
48.	AICRP on Energy in Agriculture and Agro Based Indus.,Bhopal
49.	AICRP on Utilization of Animal Energy (UAE), Bhopal
50.	AICRP on Plasticulture Engineering and Technologies, Ludhiana
51.	AICRP on PHT, Ludhiana
52.	AICRP on Goat Improvement, Mathura
53.	AICRP- Improvement of Feed Sources & Nutrient Utilisation for raising animal production, Bangalore
54.	AICRP on Cattle Research, Meerut
55.	AICRP on Poultry, Hyderabad
56.	AICRP-Pig, Izzatnagar
57.	AICRP Foot and Mouth Disease, Mukteshwar
58.	AICRP ADMAS, Bangalore
59.	AICRP on Home Science, Bhubaneshwar

Network Projects - 19	
1.	All India Network Project on Pesticides Residues, New Delhi
2.	All India Network Project on Underutilised Crops, New Delhi
3.	All India Network Project on Tobacco, Rajahmundry
4.	Network Project on White Grubs & Other Soil Arthropods, Durgapura
5.	Network on Agricultural Acarology, Bangalore
6.	Network on Economic Ornithology, Hyderabad
7.	All India Network Project on Rodent Control, Jodhpur
8.	All India Network Project on Jute and Allied Fibres, Barrackpore
9.	Network project on Improvement of Onion & Garlic, Pune
10.	Network Bio-fertilizers, Bhopal
11.	Network Project on Harvest & Post Harvest and Value Addition to Natural Resins & Gums, Ranchi
12.	Network project on Animal Genetic Resources, Karnal
13.	Network Project on R&D Support for Process Upgradation of Indigenous Milk products for industrial application Karnal
14.	Network Programme on Sheep Improvement, Avikanagar
15.	Network Project on Buffaloes Improvement, Hisar
16.	Network on Gastro Intestinal Parasitism, Izatnagar
17.	Network on Haemorrhagic Septicaemia, Izatnagar
18.	Network Programme Blue Tongue Disease, Izatnagar
19.	Network Project on Conservation of Lac Insect Genetic Resources, Ranchi
Other Projects - 10	
1.	Technology Mission on Cotton (CICR, Nagpur)
2.	Technology Mission on Jute (CRIJAF, Barrackpore)
3.	Continuation, Strengthening and Establishment of Krishi Vigyan Kendras
4.	Strengthening & Development of Higher Agricultural Education in India, New Delhi
5.	Central Agricultutral University, Imphal
6.	Strengthening and Modernization of ICAR Headquarters
7.	Intellectual Property Management & Transfer/Commercialisation of Agricultural Technology (Upscaling of existing component IPR HQ)
8.	Indo US Knowledge Initiative
9.	National Agricultural Innovative Project, New Delhi
10.	National Fund for Basic and Strategic Research, New Delhi

***Source*:** Website of Indian Council of Agricultural Research (ICAR). Available at http://www.icar.org.in/

Appendix–II

International Agricultural Organizations

Name, year of establishment, location of international agricultural organizations

S. No.	Institute	Full name	Year	Location
1.	CIMMYT	International Maize and Wheat Improvement Center	1943	El Batán, Edo Mex, Mexico
2.	ISHS	International Society for Horticultural Science	1959	Belgium
3.	IRRI	International Rice Research Institute (IRRI)	1960	Los Baños, Laguna, Philippines
4.	CIAT	International Center for Tropical Agriculture	1967	Cali, Colombia.
5.	IITA	International Institute of Tropical Agriculture	1967	Ibadan, Nigeria
6.	Africa Rice Center	Formerly known as the West Africa Rice Development Association (WARDA)	1971	Cotonou, Benin
7.	CIP	International Potato Center	1971	Lima, Peru
8.	ICRISAT	International Crops Research Institute for the Semi-Arid Tropics (ICRISAT)	1972	Hyderabad, India
9.	Bioversity International	Bioversity International was originally established by the CGIAR as the International Board for Plant Genetic Resources (IBPGR) in 1974.	1974	Maccarese, Rome
10.	ICARDA	International Center for Agricultural Research in the Dry Area (ICARDA)	1975	Beirut, Lebanon
11.	IFPRI	International Food Policy Research Institute	1975	Washington, D.C., USA
12.	WFC	World Fish Centre	1975	Penang, Malaysia

S. No.	Institute	Full name	Year	Location
13.	World Agroforestry Centre	known as the International Centre for Research in Agroforestry, ICRAF before 2002	1978	Nairobi, Kenya,
14.	IWMI	International Water Management Institute	1985	Colombo Sri Lanka
15.	INIBAP	International Network for the Improvement of Banana and Plantain.	1985	Montpellier, France
16.	IPGRI	International Plant Genetic Resources Institute	1991	Rome, Italy
17.	CGIAR	Consultative Group on International Agricultural Research	1991	Montpellier, France
18.	CIFOR	Center for International Forestry Research	1993	Bogor, Indonesia
19.	ILRI	International Livestock Research Institute	1994	Nairobi, Kenya
20.	BISA	Borlaug Institute for South Asia	2011	Ludhiana in Punjab, Pusa in Bihar and Jabalpur in Madhya Pradesh

International Organizations

Name, year of establishment, HQ of international organizations

S. No.	Abbreviation	Full name	Establishment year	HQ
1.	ILO	International Labour Organization	1919	Geneva, Switzerland
2.	IBRD	International Bank for Reconstruction and Development	1944	Washington, D.C., United States of America
3.	FAO	Food and Agriculture Organization	16 October 1945	Rome, Italy
4.	UN	United Nations	24 October 1945	New York, USA
5.	UNESCO	United Nations Educational, Scientific and Cultural Organization	16 November 1945	Paris, France
6.	IMF	International Monetary Fund	27th December 1945	Washington, D.C., United States
7.	UNICEF	United Nations International Children's Emergency Fund	December 1946	New York City, USA
8.	ISO	International Organization for Standardization	23 February 1947	Geneva, Switzerland

S. No.	Abbreviation	Full name	Establishment year	HQ
9.	ADB	Asian Development Bank	22 August 1966	Manila, Philippines
10.	ASEAN	Association of Southeast Asian Nations	8 August 1967	Jakarta, Indonesia
11.	SAARC	South Asian Association for Regional Cooperation	8 December 1985	Kathmandu, Nepal
12.	WTO	World Trade Organization	1 January 1995	Geneva, Switzerland

❍❍❍

Appendix–III

List of Agricultural Universities

The Agricultural Universities are major partners in growth & development of Agricultural Research and Education under National Agricultural Research System.

S. No.	Name, email & website	Address
1.	**Acharya NG Ranga Agricultural University** **Website:** http://www.angrau.net **Email:** angrau_vc@yahoo.com,raghuvardhanreddy_s@rediffmail.com	Adminstrative Office, Rajendra Nagar, Hyderabad-500030, Andhra Pradesh
2.	**Agriculture University Jodhpur** **Website:** http://www.au-ju.org **Email:** vcunivag@gmail.com	Mandor, Jodhpur-342304
3.	**Agriculture University Kota** **Website:** http://aukota.org **Email:** vcaukota@gmail.com	Borkhera, Kota-324001
4.	**Anand Agricultural University** **Website:** http://www.aau.in **Email:** vc@aau.in, vc_aau@yahoo.com	Anand 388110, Gujarat
5.	**Assam Agricultural University** **Website:** http://www.aau.ac.in **Email:** vc@aau.ac.in,kmbujarbaruah@rediffmail.com	Jorhat 785013, Assam
6.	**Bidhan Chandra Krishi Viswavidyalaya** **Website:** http://www.bckv.edu.in **Email:**bckvvc@gmail.com,sarojsanyal@yahoo.co.in	Mohanpur, Nadia-741252, West Bengal
7.	**Bihar Agricultural University** **Website:** http://www.bausabour.ac.in **Email:** vcbausabour@gmail.com	Sabour, Bhagalpur 813210, Bihar
8.	**Birsa Agricultural University** **Website:** http://www.baujharkhand.org **Email:** vc_bau@rediffmail.com	Kanke, Ranchi-834006, Jharkhand

S. No.	Name, email & website	Address
9.	**Central Agricultural University** **Website:**http://www.cau.org.in **Email:** snpuri04@yahoo.co.in,snpuri@rediffmail.com	P.O. Box 23, Imphal-795004, Manipur
10.	**Chandra Shekar Azad University of Agriculture & Technology** **Website:**http://www.csauk.ac.in **Email:** vc@csauk.ac.in	Kanpur-208002, Uttar Pradesh
11.	**Chaudhary Charan Singh Haryana Agricultural University** **Website:**http://www.hau.ernet.in **Email:** vc@hau.ernet.in	Hisar-125004, Haryana
12.	**CSK Himachal Pradesh Krishi Vishvavidyalaya** **Website:** http://www.hillagric.ac.in **Email:** vc@hillagric.ac.in	Palampur-176062, Himachal Pradesh
13.	**Chhattisgarh Kamdhenu Vishwavidyalaya** **Website:** http://cgkv.ac.in	Anjora ,Durg, **Chhattisgarh**
14.	**Dr Balasaheb Sawant Konkan Krishi Vidyapeeth** **Website:** www.dbskkv.org **Email:** vcbskkv@yahoo.co.in	Dapoli Distt, Ratnagiri 415 712, Maharashtra
15.	**Dr Panjabrao Deshmukh Krishi Vidyapeeth** **Website:** http://www.pdkv.ac.in **Email:** vc@pdkv.ac.in	Krishinagar,Akola-444104, Maharashtra
16.	**Dr Yashwant Singh Parmar Univ of Horticulture & Forestry** **Website:** http://www.yspuniversity.ac.in **Email:** vc@yspuniversity.ac.in, vcuhf@yahoo.com	Solan, Nauni-173230, Himachal Pradesh
17.	**Dr YSR Horticultural University** **Website:** http://www.drysrhu.edu.in **Email:** vc@drysrhu.edu.in	West Godavari Dist., Tadepalligudem-534101 Andhra Pradesh
18.	**Govind Ballabh Pant University of Agriculture & Technology** **Website:** http://www.gbpuat.ac.in **Email:**vcgbpuat@gmail.com	Pantnagar-263145,Distt Udham Singh , Nagar , Uttaranchal
19.	**Guru Angad Dev Veterinary and Animal Science University** **Website:** http://www.gadvasu.in **Email:**vijay_taneja@hotmail.com,vcgadvasu@gmail.com	Ludhiana - 141004, Punjab
20.	**Indira Gandhi Krishi Vishwavidyalaya** **Website:** www.igau.edu.in **Email:** vcigkv@gmail.com	Krishak Nagar, Raipur-492006, Chhattisgarh

S. No.	Name, email & website	Address
21.	**Jawaharlal Nehru Krishi Viswavidyalaya** **Website:** http://www.jnkvv.nic.in **Email:** vst.vcjnkvv@gmail.com	Krishi Nagar, Adhartal Jabalpur-482004, Madhya Pradesh
22.	**Junagadh Agricultural University** **Website:** http://www.jau.in **Email:** vc@jau.in	Univ. Bhavan, Motibagh Junagadh-362001, Gujarat
23.	**Karnataka Veterinary, Animal and Fisheries Sciences University** **Website:** http://www.kvafsu.kar.nic.in **Email:** vckvafsu@yahoo.co.in, dekvafsub@yahoo.com, sskvafsu@yahoo.co.in	Nandinagar, PB No 6, BIDAR 585401, Karnataka
24.	**Kerala Agricultural University** **Website:** http://www.kau.edu **Email:** vc@kau.in, vicechancellorkau@gmail.com	Vellanikara, Trichur 680656, Kerala
25.	**Kerala University of Fisheries & Ocean Studies** **Website:** http://www.kau.edu **Email:** kurup424@gmail.com	Papangad, Kochi-682506 Kerala
26.	**Kerala Veterinary and Animal Sciences University** **Website:** http://www.kcasu.ac.in **Email:** vc@kvasu.ac.in, vc.vetuny@gmail.com	Thiruvananthapuram, 695004, Kerala
27.	**Lala Lajpat Rai University of Veterinary & Animal Sciences** **Website:** http://www.llruvas.edu.in/ **Email:** vc@llruvas.edu.in	Hisar, Haryana
28.	**Nanaji Deshmukh Veterinary Science University** **Website:** http://www.mppcvv.org **Email:** vcnduvs@gmail.com	South Civil Lines, Jabalpur-482001, Madhya Pradesh
29.	**Maharana Pratap Univ. of Agriculture & Technology** **Website:** http://www.mpuat.ac.in **Email:** vc@mpuat.ac.in	Udaipur, Rajasthan 313001
30.	**Maharashtra Animal Science & Fishery University** **Website:** http://www.mafsu.in **Email:** vcmatsu@gmail.com,cadaba_prasad@yahoo.co.in	Seminary Hills, Nagpur-440006, Maharashtra
31.	**Mahatma Phule Krishi Vidyapeeth** **Website:** http://mpkv.mah.nic.in **Email:** vcmpkv@rediffmail.com	Rahuri-413722, Maharashtra
32.	**Manyavar Shri Kanshiram Ji University of Agriculture and Technology** **Website:** http://www.mskjuat.edu.in/ **Email:** vc.mskjuat@gmail.com	Banda-210001, Uttar Pradesh

S. No.	Name, email & website	Address
33.	**Narendra Deva University of Agriculture & Technology** **Website:**http://www.nduat.ernet.in **Email:**vc_nduat2010@yaho.co.in	Kumarganj, Faizabad -224229, Uttar Pradesh
34.	**Navsari Agricultural University** **Website:** http://www.nau.in **Email:** vc_2004@yahoo.co.in	Navsari-396450 Gujarat
35.	**Orissa Univ. of Agriculture & Technology** **Website:** http://www.ouat.ac.in **Email:** ouat_dproy@yahoo.co.in,bsenapati1942@yahoo.com	Bhubaneshwar-751003, Orissa
36.	**Prof. Jayashankar Telangana State Agricultural University** **Website:**www.pjtsau.ac.in **Email:** vcpjtsau@gmail.com	Rajendranagar, Hyderabad - 500 030
37.	**Punjab Agricultural University** **Website:** http://www.pau.edu **Email:** vcpau@pau.edu	Ludhiana-141004, Punjab
38.	**Rajasthan University of Veterinary and Animal Sciences** **Website:** http://rajuvas.org **Email:** vcrajuvas@gmail.com	Bikaner 334006 Rajasthan
39.	**Rajendra Agricultural University** **Website:** http://www.pusavarsity.org.in **Email:** : vcrau@sify.com	Pusa, Samastipur 848125, Bihar
40.	**Rajmata Vijayraje Sciendia Krishi Vishwa Vidyalaya** **Website:** http://www.rvskvv.nic.in **Email:** vcrvskvv@gmail.com	Race Cource Road, Gwalior 474002 Madhya Pradesh
41.	**Rani Laxmi Bai Central Agricultural University** **Website:** **Email:** : ddgedn@icar.org.in	Jhansi, Uttar Pradesh
42.	**Sardar Vallabhbhai Patel University of Agriculture and Technology** **Website:** http://www.svbpmeerut.ac.in **Email:**vc_agunivmeerut@yahoo.com	Modipuram, Meerut - 250110 Uttar Pradesh
43.	**Sardarkrushinagar-Dantiwada Agricultural University** **Website:** http://www.sdau.edu.in **Email:** vc@sdau.edu.in	Sardar Krushinagar, Distt Banaskantha, Gujarat-385506
44.	**Sher-E-Kashmir Univ of Agricultural Sciences & Technology** **Website:** http://www.skuast.org **Email: vc@skuast.org**	Railway Road, Jammu 18009, J&K

S. No.	Name, email & website	Address
45.	**Sher-E-Kashmir Univ of Agricultural Sciences & Technology of Kashmir** **Website:** http://www.skuastkashmir.ac.in **Email:** vc@skuastkashmir.ac.in,skuastkvc@gmail.com	Shalimar Campus, Shrinagar-191121, Jammu & Kashmir
46.	**Sri Karan Narendra Agriculture University** **Website:**http://sknau.ac.in **Email:**nsrdsr@gmail.com, vc@sknau.ac.in	Jobner-303329
47.	**Sri Venkateswara Veterinary University** **Website:**http://www.svvu.edu.in **Email:**prabhakarvrao@yahoo.com	Admn office, Regional Library Building, Tirupati-517502
48.	**Swami Keshwanand Rajasthan Agricultural University** **Website:** http://www.raubikaner.org **Email:** vcrau@raubikaner.org	Bikaner-334006,Rajasthan
49.	**Tamil Nadu Agricultural University** **Website:** http://www.tnau.ac.in **Email:** vc@tnau.ac.in	Coimbatore-641003, Tamil Nadu
50.	**Tamil Nadu Fisheries University** **Website:** http://www.tnfu.org.in **Email:**	Nagapattinam – 611 003, Tamil Nadu
51.	**Tamil Nadu Veterinary & Animal Science University** **Website:** http://www.tanuvas.ac.in **Email:**vc@tanuvas.org.in, karanmgk@gmail.com	Chennai-600051, Tamilnadu
52.	**University of Agricultural Sciences, Bangalore** **Website:** http://www.uasbangalore.edu.in **Email:**vc@uasbangalore.edu.in,knarayanagowda@yahoo.co.in	GKVK,Bengaluru-560065, Karnataka
53.	**University of Agricultural Sciences, Dharwad** **Website:** http://www.uasd.edu **Email:** vc_uasd@rediffmail.com	Dharwad-580005, Karnataka
54.	**University of Agricultural Sciences, Shimoga** **Website:**http://www.uasbangalore.edu.in/asp/agriShimoga.asp **Email:**	Shimoga, Karnataka
55.	**University of Horticultural Sciences** **Website:** http://uhsbagalkot.edu.in/ **Email:** dandinbnm@gmail.com	Sector No 60 Navanagar Bagalkot 587102 Karnataka
56.	**University of Agricultural Sciences** **Website:** http://www.uasraichur.edu.in **Email:** vcuasraichur10@rediffmail.com	PB 329, Raichur – 584101 Karnataka

S. No.	Name, email & website	Address
57.	**UP Pandit Deen Dayal Upadhaya Pashu Chikitsa Vigyan Vishwa Vidhyalaya evam Go Anusandhan Sansthan** **Website:** http://www.upvetuniv.edu.in **Email:** singhambika1945@gmail.com,duvasuvc@gmail.com	Mathura-281001, Uttar Pradesh
58.	**Uttarakhand University of Horticulture and Forestry** **Website:** http://www.uuhf.ac.in/ **Email:**	Pauri Garhwal, Uttarakhand
59.	**Uttar Banga Krishi Viswavidyalaya** **Website:**http://www.ubkv.ac.in **Email:** vcubkv@gmail.com, vcubkv@rediffmail.com	P.O. Pundibari, Dist. Cooch Bihar-736165,West Bengal
60.	**Vasantrao Naik Marathwada Agricultural University** **Website:** http://www.mkv2.mah.nic.in **Email:** vcmau@rediffmail.com	Parbhani-431402, Maharashtra
61.	**West Bengal University of Animal & Fishery Sciences** **Website:**http://www.wbuafscl.ac.in **Email:** wbuafs@wb.nic.in	68 KB Sarani, Kolkata-700037, West Bengal
62.	**Sri Konda Laxman Telangana State Horticultural University** **Email:** vcskltshu@gmail.com,registrarskltshu@gmail.com	Rajendra Nagar Campus, Hyderabad
63.	**Kamdhenu University** Email: vc.kamdhenu.university@gmail.com	Karmayogi Bhavan, Block-1, B1-wing,4th Floor,Room No.414, Sector-10-A, Gandhinagar-382010

Source: Indian Council of Agricultural Research (ICAR). Available http://www.icar.org.in/en/universities.htm

Appendix–IV

List of Authorities/Agencies/Institutions/Functionarles

1. National Horticulture Board (NHB), Gurgaon
2. Coconut Development Board, Kochi
3. Small Farmer's Agri-Business Consortium (SFAC), New Delhi
4. Directorate of Cashew and Cocoa Development (DCCD), Kochi
5. Directorate of Arecanut and Spices Development (DASD), Calicut
6. National Committee on Plasticulture Applications in Agriculture & Horticulture (NCPAAH), New Delhi
7. National Horticulture Research & Development Foundation, Nashik
8. National Seeds Corporation, New Delhi
9. State Farm Corporation of India, New Delhi
10. National Institute for Agricultural Extension Management (MANAGE), Hyderabad
11. Fresh & Healthy Enterprises Ltd. (FHEL), New Delhi
12. Agricultural and Processed Food Products Export Development Authority (APEDA), New Delhi
13. Directorate of Marketing & Inspection (DMI), New Delhi
14. Ministry of Food Processing Industries (MFPI), New Delhi
15. National Medicinal Plants Board (NMPB), New Delhi
16. Central Institute of Horticulture (CIH), Nagaland
17. National Centre for Cold Chain Development (NCCD)
18. National Institute of Food Technology Entrepreneurship and Management (NIFTEM), Sonipat, Haryana
19. United Planters' Association of Southern India (UPASI) Tea Research Foundation (TRF), Coimbatore District, Tamil Nadu, India.

ooo

Appendix–V

List of Various Boards/ organizations

Name, establishment year and HQ of various boards/ organizations related to agriculture

S. No.	Name of board/ organization	Establishment year	HQ	Remarks
1.	Coffee Board of India	1942	Bangalore, Karnataka, India	Functions under the Ministry of Commerce and Industry, Government of India.
2.	Indian Rubber Board	1947	Kottayam, Kerala, India	Government of India
3.	Tea Board of India	1st April 1954	Kolkata	Under the control of the Union Govt.
4.	Central Insecticide Board and Registration Committee	1968	Faridabad	Ministry of Agriculture, Government of India
5.	Coconut Development Board	1979	Kochi, Kerala, India	Functions under Ministry of Agriculture, Government of India
6.	National Horticulture Board	1984	Gurgaon (Haryana)	Functions under Department of Agriculture & Co-operation, Ministry of Agriculture, Govt. of India.
7.	Spices Board of India	26th February 1986	Cochin, Kerala, India	Functions under Ministry of Commerce, Government of India
8.	National Medicinal Plants Board	2000	New Delhi	Department of AYUSH, Ministry of Health & Family Welfare, Government of India
9.	National Development Council	6th August 1952	New Delhi	Government of India

S. No.	Name of board/ organization	Establishment year	HQ	Remarks
10.	National Agricultural Cooperative Marketing Federation of India Ltd	October 2, 1958	New Delhi	
11.	Agricultural and Processed Food Products Export Development Authority (APEDA)	December, 1985	New Delhi	Established by the Government of India under the Agricultural and Processed Food Products Export Development Authority Act passed by the Parliament in December, 1985. The Act (2 of 1986) came into effect from 13th February, 1986 by a notification issued in the Gazette of India
12.	Food Corporation of India (FCI)	1964	New Delhi, India	
13.	Agricultural Prices Commission	January, 1965		Since March 1985, the Commission has been known as Commission for Agricultural Costs and Prices. (CACP)
14.	Indian Farmers Fertiliser Co-operative Limited (IFFCO)	November 3, 1967	New Delhi	
15.	National Fertilizers Limited (NFL)	23rd August, 1974	Noida, India	
16.	National Commission on Agriculture	1976		Government of India
17.	National Environment Tribunal	1995		Function under Ministry of Environment and Forest, Government of India
18.	National Environment Appellate Authority (NEAA)	1997		Function under Ministry of Environment and Forest, Government of India
19.	National Biodiversity Authority	2003	Chennai, Tamil Nadu	The National Biodiversity Authority (NBA) was established in 2003 to implement India's Biological Diversity Act (2002)
20.	National Green Tribunal (NGT)	October 18, 2010	New Delhi	Function under Ministry of Environment and Forest, Government of India
21.	National Commission on Farmers (NCF)	November 18, 2004		

S. No.	Name of board/ organization	Establishment year	HQ	Remarks
22.	Food Safety and Standards Authority of India (FSSAI)	2006	New Delhi	Ministry of Health & Family Welfare, Government of India
23.	International Flower Auction Centre	2007	Bangalore, Karnataka	Government of India
24.	Directorate of Plant Protection, Quarantine and Storage (DPPQS)		Faridabad	Functions under Ministry of Agriculture
25.	Genetic Engineering Approval Committee (GEAC)			Function under Mnistry of Environment and Forest, Government of India

Appendix–VI

List of Various Acts and Rules

S. No.	Acts/rules	Year	Remarks
1.	Land Acquisition Act	1894	The Land Acquisition (Amendment) Bill, 2007, was Passed in the Lok Sabha in 2009
2.	Livestock Importation Act	1898	Amended in 2001
3.	Destructive Insects and Pests (DIP) Act	1914	
4.	Agricultural Produce Act (Grading and Packing)	1937	(Amended in 1986), Act is popularly known as 'Agmark' standards.
5.	Central Committee for Food Standards	1941	
6.	The Rubber Act	1947	
7.	Vegetable Oil Products Order	1947	
8.	Vegetable Oil Products Order	1998	
9.	Minimum Wages Act	1948	
10.	Industries (Development and Regulation) Act (IDRA)	1951	
11.	Prevention of Food Adulteration (PFA) Act	1954	
12.	Food Product Order	1955	
13.	Sugar Control Order	1956	
14.	Solvent Extracted Oil, De-oiled Meal and Edible Flour (Control) Order	1967	
15.	Meat Food Products Order	1973	

S. No.	Acts/rules	Year	Remarks
16.	Milk and Milk Product Order (MMPO)	1992	
17.	Milk & Milk Product Amendment Regulations	2009	
18.	Edible Oils Packaging (Regulation) Order	1998	
19.	Essential Commodities Act	1955	
20.	Fertilizer (Control) Order	1957/1985	
21.	Fertilizer (Movement Control) Order	1960/1973/2001	
22.	Prevention of Cruelty to Animals Act	1960	
23.	National Co-operative Development Corporation (NCDC) Act	1962	
24.	Seeds Act	1966	
25.	Seeds Rules	1968	
26.	Seeds (Control) Order	1983	
27.	National Seed Policy	2002	
28.	Seeds Bill	2004	
29.	Insecticide Act	1968	
30.	Insecticides Rules	1971	
31.	Wildlife Protection Act	1972	The Act was amended in January 2003
32.	Water (Prevention and Control of Pollution) Act	1974	The Act was last amended in 2003
33.	Standards for Weight and Measures Act	1976	
34.	Forest Conservation Act	1980	
35.	The Consumer Protection Act	1986	
36.	Environment Protection Act	1986	The Act was last amended in 1991
37.	Bureau of Indian Standards (BIS) Act	1986	
38.	The Bureau of Indian Standards Rules	1987	
39.	The Bureau of Indian Standards (Certification) Regulations	1988	
40.	Recombinant DNA Safety Guidelines	1990	

S. No.	Acts/rules	Year	Remarks
41.	Revised Recombinant DNA Safety Guidelines	1994	
42.	Revised Guidelines for Research in Transgenic Plants & Guidelines for Toxicity and Allergenicity Evaluation of Transgenic Seeds, Plants and Plant Parts	1998	
43.	Protection of Plant Varieties and Farmers' Rights Act (PPVFRA)	2001	
44.	Multi-State Co-operative Societies (MSCS) Act	2002	
45.	Biological Diversity Act	2002	
46.	Plant Quarantine (Regulation of Import into India) Order	2003	Came into force on January 1, 2004
47.	Biological Diversity Rules	2004	
48.	National Biotechnology Strategy	2005	
49.	Environment Policy	2006	
50.	Mahatma Gandhi National Rural Employment Guarantee Act	2005	The law was initially called the **National Rural Employment Guarantee Act** (**NREGA**) and was renamed with the prefix "Mahatma Gandhi" on 2 October 2009, Gandhi's birth anniversary
51.	Foreign Trade Policy	2006	
52.	Food Safety and Standards Act	2006	Amended in 2008
53.	National Food Security Act	2013	
54.	National Policy on Skills Development	2009	

❍❍❍

Appendix–VII

Area, Production and Productivity of Horticulture Crops - All India

Area in '000 Ha
Production in '000 MT
Productivity in MT/ha

Crops	2013-14			2014-15			2015-16		
	(Final)			(Final)			(1st Estimate)		
Fruits	Area	Production	Productivity	Area	Production	Productivity	Area	Production	Productivity
Almond	21	13	0.62	21	10	0.48	21	10	0.48
Aonla/ Gooseberry	104	1225	11.78	95	1173	12.35	89	1165	13.09
Apple	313	2498	7.98	319	2134	6.69	319	2262	7.09
Banana	803	29725	37.02	822	29221	35.55	837	29895	35.72
Ber	48	663	13.81	42	401	9.55	41	410	10.00
Citrus									
(i) Lime/ Lemon	286	2835	9.91	268	2950	11.01	273	2992	10.96
(ii) Mandarin	330	3431	10.40	299	3699	12.37	316	3669	11.61
(iii) Sweet Orange (Mosambi)	335	3886	11.60	275	4229	15.38	275	4248	15.45
(iv) Others	126	994	7.89	111	777	7.00	112	746	6.66
Citrus Total (i to iv)	**1078**	**11147**	10.34	**953**	**11655**	12.23	**976**	**11655**	11.94
Custard apple	22	165	7.50	30	228	7.60	30	232	7.73
Grapes	119	2585	21.72	123	2823	22.95	124	2827	22.80
Guava	268	3668	13.69	246	3994	16.24	255	4102	16.09
Jackfruit	158	1573	9.96	118	2088	17.69	121	2140	17.69
Kiwi	5	8	1.60	5	8	1.60	5	9	1.80
Litchi	84	585	6.96	83	528	6.21	86	552	6.42
Mango	2516	18431	7.33	2163	18527	8.57	2200	19517	8.87
Papaya	133	5639	42.40	115	4913	42.72	122	5306	43.49

Crops	2013-14			2014-15			2015-16		
	(Final)			(Final)			(1st Estimate)		
Fruits	Area	Production	Productivity	Area	Production	Productivity	Area	Production	Productivity
Passion Fruit	19	124	6.53	19	129	6.79	19	127	6.68
Peach	18	94	5.22	19	97	5.11	19	107	5.63
Pear	42	317	7.55	42	303	7.21	43	320	7.44
Picanut	1	0	0.00	1	0	0.00	1	0	0.00
Pineapple	110	1737	15.79	116	1984	17.10	115	2007	17.45
Plum	23	76	3.30	23	72	3.13	23	84	3.65
Pomegranate	131	1346	10.27	181	1789	9.88	184	1820	9.89
Sapota	177	1744	9.85	106	1339	12.63	108	1372	12.70
Strawberry	0	2	#DIV/0!	1	8	8.00	1	8	8.00
Walnut	122	241	1.98	115	238	2.07	115	239	2.08
Others	901	5372	5.96	349	2938	8.42	335	2851	8.51
Total Fruits	**7216**	**88977**	12.33	**6110**	**86602**	14.17	**6188**	**89018**	14.39
Vegetables			#DIV/0!			#DIV/0!			#DIV/0!
Beans	138	1370	9.93	218	2204	10.11	209	1970	9.43
Bitter gourd	79	807	10.22	76	770	10.13	81	830	10.25
Bottle gourd	103	1819	17.66	108	1826	16.91	131	2083	15.90
Brinjal	711	13558	19.07	673	12589	18.71	670	12780	19.07
Cabbage	400	9039	22.60	386	8585	22.24	380	8529	22.44
Capsicum	30	167	5.57	32	183	5.72	32	179	5.59
Carrot	62	1074	17.32	64	968	15.13	65	950	14.62
Cauliflower	434	8573	19.75	411	7926	19.28	411	7899	19.22
Cucumber	43	678	15.77	43	678	15.77	53	780	14.72
Chillies (Green)	140	1687	12.05	181	1998	11.04	173	1992	11.51
Elephant Foot Yam	5	222	44.40	24	678	28.25	10	383	38.30
Muskmelon	37	761	20.57	42	863	20.55	47	878	18.68
Mushroom		17	#DIV/0!		51	#DIV/0!	0	56	#DIV/0!
Okra/ Ladyfinger	533	6346	11.91	504	5709	11.33	504	5794	11.50
Onion	1204	19402	16.11	1173	18927	16.14	1178	20333	17.26
Parwal/ Pointed gourd	13	169	13.00	18	347	19.28	14	212	15.14
Peas	434	3869	8.91	476	4652	9.77	483	4650	9.63
Potato	1973	41555	21.06	2076	48009	23.13	2085	48096	23.07
Radish	173	2485	14.36	168	2307	13.73	169	2203	13.04
Pumpkin/ Kaddu	20	416	20.80	49	1122	22.90	50	1090	21.80
Sweet Potato	106	1088	10.26	107	1228	11.48	111	1450	13.06

Crops	2013-14			2014-15			2015-16		
	(Final)			(Final)			(1st Estimate)		
Fruits	Area	Production	Productivity	Area	Production	Productivity	Area	Production	Productivity
Tapioca	228	8139	35.70	208	4373	21.02	191	3986	20.87
Tomato	882	18736	21.24	767	16385	21.36	777	18286	23.53
Watermelon	75	1810	24.13	84	2049	24.39	85	1960	23.06
Others	1574	19108	12.14	1654	25053	15.15	1556	21136	13.58
Total Vegetables	**9396**	**162897**	17.34	**9542**	**169478**	17.76	**9465**	**168506**	17.80

Source: http://nhb.gov.in/Default.aspx

Appendix–VIII

Conversion Chart

Conversion chart for length, volume, mass and temperature

From	To	Multiply By
Kilometers (Km)	Miles (Mi)	0.62
Kilometers (Km)	Feet (Ft)	3280.8
Meters (M)	Feet (Ft)	3.28
Centimeters (Cm)	Inches (In)	0.39
Millimeters (Mm)	Inches (In)	0.039
Inches (In)	Meters (M)	0.0254
Inches (In)	Centimeters (Cm)	2.54
Inches (In)	Millimeters (Mm)	25.40
Feet (Ft)	Meters (M)	0.30
Yards (Yd)	Meters (M)	0.91
Yards (Yd)	Kilometers (Km)	0.00091
Miles (Mi)	Kilometers (Km)	1.61

Volume

From	To	Multiply By
Liters (L)	Quarts (Qt)	1.057
Liters (L)	Gallons (Gal)	0.264
Milliliters (Ml)	Cups (C)	0.0042
Milliliters (Ml)	Ounces (Oz)	0.0338
Ounces (Oz)	Milliliters (Ml)	29.57
Cups (C)	Milliliters (Ml)	236.6
Quarts (Qt)	Liters (L)	0.95
Gallons (Gal)	Liters (L)	3.785

Mass

From	To	Multiply By
Kilogram (Kg)	Tons (Ton)	0.0011
Kilogram (Kg)	Pounds (Lb)	2.2046
Grams (G)	Ounces (Oz)	0.035
Grams (G)	Pounds (Lb)	0.002205
Milligrams (Mg)	Ounces (Oz)	0.000035
Ounces (Oz)	Milligrams (Mg)	28350
Ounces	Grams (G)	28.35
Pounds (Lb)	Kilograms (Kg)	0.454
Tons (Ton)	Kilograms (Kg)	907.18

Temperature

From	To	Conversion
Fahrenheit (F)	Celsius (C)	(Temperature (F) – 32)*5/9
Celsius (C)	Fahrenheit (F)	(Temperature (C)*9/5) + 32

Conversion Table

Length

- 1 kilometre (km) = 1000 metres (m)
- 1 km = 0.6214 miles
- 1 m = 1.0936 yards
- 1 m = 3.2808 feet
- 1 mile = 1760 yards
- 1 mile = 1.609 km
- 1 yard = 0.9144 m
- 1 foot = 0.3048 m

Area

- 1 km^2 = 100 hectares (ha)
- 1 km^2 = 0.3861 square mile
- 1 km^2 = 247.105 acres
- 1 m^2 = 10.7639 square feet
- 1 ha = 10,000 m2
- 1 ha = 2.4711 acres
- 1 square mile = 2.59 1 km^2
- 1 acre = 0.4047 ha
- 1 acre = 4046.86 m^2

- 1 acre = 4840 square yard
- 1 square yard = 9 square feet
- 1 square yard = 0.8361 m^2
- 1 square foot = 0.0929 m^2

Weight

- 1 tonne = 1000 kg
- 1 tonne = 1.1023 US ton
- 1 US ton = 0.9072 tonnes
- 1 hg = 100 gram
- 1 kg = 2.2046 pounds (Ib)
- 1 kg = 35.274 ounce (oz)
- 1 lb = 0.4536 kg
- 1 oz = 28.3495 gram

Units

- 1 crore = 10 million
- 1 million = 10 lakh
- 1 lakh = 100000
- 1 billion = 1000 million

References

Abawi, G.S., Providente, R. Crosier, D.C. and Hunter, J.E. 1978.Inheritance of resistance to white mold disease of *Phaseoluscoccineus. J. Hered.* 69**:** 200-202.

Adamson, R.M. 1965. Self- and cross-incompatibility in early round-headed cabbage. *Canadian Journal of Plant Science*, 45:493-97.

Anbinder, I., Reuveni, M., Azari, R., Paran, I., Nahon, S., Shlomo, H., Chen, L., Lapidot, M., and Levin, I. 2009. Molecular dissection of Tomato leaf curl virus resistance in tomato line TY172 from *Solanum peruvianum. Theor. Appl. Genet.* 119**:** 519-530.

Anon. 1971-2012. Annual Reports of All India Coordinated Vegetable Improvement Project, IIVR Varanasi.

Anonymous. 2001. Progress Report, Ministry of Agriculture Govt. of India, NCPAH, New Delhi.

Anonymous.1994. Drip irrigation in India, INCID, New Delhi. Anonymous, 1998. INCID report, New Delhi.

Anu, A., Peter, K.V. 2003. Analysis of seed protein of 29 lines of *Capsicum annuum* L. by polyacrylamide gel electrophoresis. *Genetic Resources and Crop Evolution* 50**:** 239–243.

Arthikala, M.K., Nanjareddy, K., Lara, M. Sreevathsa, R. 2014. Utility of a tissue culture-independent *Agrobacterium*-mediated *in planta* transformation strategy in bell pepper to develop fungal disease resistant plants. *Scientia Horticulturae* 170**:** 61–69.

Arya, P.S. 2001. *Ginger Production Technology*, Kalyani Publishers, New Delhi.

Arya, Prem Singh. 2000. Off-season Vegetable Growing in Hills. APH Publishing Corp. New Delhi, pp. 7-27.

Babu, B.S., Pandravada, S.R., Rao, RDVJP, Anitha, K., Chakrabarty, S.K. and Varaprasad, K.S. 2011. Global sources of pepper genetic resources against arthropods, nematodes and pathogens. *Crop Protection* 30**:** 389–400.

Bains, D.C. 1955. Resistance of cabbage to black rot. *Phytopathology*45**:** 35- 37.Baswana, K.S. 1990. Inheritance of stalk rot resistance in cauliflower. *Euphytica* 57**:** 93-96.

Banerjee, A., Dutta, R., Roy, S. and Ngachan, S.V. 2014. First report of *Chilli veinal mottle virus* in Naga Chilli (*Capsicum chinense*) in Meghalaya, India. *Virus Disease* 25**:** 142–143.

Bannerot, H.O., Boulidard, L., Cauderon, Y. and Tempe, T. 1974. Cytoplasmic male sterility (CMS) transfer from *Raphanus* to *Brassica*. Cruciferae conference, Scottish Horticultural Research Institute, Dundee, Scotland, pp52.

Barnes, W.C. 1966. Development of multiple disease resistant hybrid cucumber. Proc. Am. Soc. Hortic. Sci., 89:390.

Bhardwaj, C. L. and Sud, A. K.1995. A visual Rating Scale for White Rust disease assessment of crucifers crops. *Indian J. Mycol.Pl. Pathol.* 25(3)**:** 181-83

Bohr, G.W. and Whitaker, T.A.1949. A gene for male sterility in muskmelon.

Bose, T. K., Som, M, G, and Kabir, 2002. J. *Vegetable Crops*. Naya Prokash Calcutta.

Bose, T.K., Som, M.G. and Kabir, J. (Ed). 1993. Vegetable Crops. Naya Prokash, Calcutta, 959p.

Bose, T.K. 2002. Vegetable Crops, Volume 1. Naya Prokash, 2002, 668p.

Bosland, P.W. and Baral, J.B. 2007. Bhut Jolokia—the world's hottest known chile pepper is a putative naturally occurring inter specific hybrid. *HortScience* 42: 222–224.

Briard, M., Le Clerc, V. and Peltier, D. 2002. Molecular analysis of the genetic variability within Apiaceae family. *Acta Hort.*521:155-163.

Chadda, K .L. 2004. *Handbook of Horticulture.* Indian Council of Agricultural Research, New Delhi. 998p.

Chadda, K.L. 1993. *Advances in Horticulture* Vol. 9 Plantation and Spice Crops. Malhotra Publishing House.

Chakravarti, B.P., Hegde, S.V. and Gupta, D.K. 1973. Resistance in cabbage to black rot caused by *Xanthomonas campestris* (Pammel) Dowson in India. *Curr. Sci* 42(22): 803-804.

Chand Ramesh. 1996. Agricultural Diversification and Small Farm Development in Western Himalaryan Region, in T. Haque (ed.) *Small Farm Diversification*: *Problems and Prospects,* National Centre for Agricultural Economics and Policy Research, New Delhi,pp.79- 102.

Channappagoudar, S.B. 2007. Studies on *In Vitro* Regeneration and Genetic Transformation in Chilli (*Capsicum annuum* L.). PhD Thesis, University of Agricultural Sciences, Dharwad, India.

Chattopadhyay, B., Singh, A.K., Yadav, T., Fauquet, C.M., Sarin, N.B. and Chakraborty, S. 2008. Infectivity of the cloned components of a begomovirus: DNA beta causing chilli leaf curl disease in India. *Archives of Virology* 153: 533–539.

Chen, J.Y. 2009. Identification of QTLs for phosphorus utilization efficiency in maize (*Zea mays* L.) across P levels. *Euphytica* 167: 245-252.

Chiang, M.S., Chong, C., Landry, B.S. and Crete, R. 1993. Cabbage pp113- 155 In: G. Kalloo and B.O. Bergh (Edited) *Genetic Improvement of Vegetable Crops* Pergamon Press, Oxford, UK.

Cooper, J.L., Till, B.J., Laport, R.G., Darlow, M.C., Kleffner, J.M., Jamai, A., El- Mellouki, T., Liu, S., Ritchie, R., Nielsen, N., Bilyeu, K.D, Meksem, K., Comai, L. and Henikoff, S. 2008. TILLING to detect induced mutations in soybean. *BMC Plant Biology* 8: 9.

Crehu, G. du 1968. Early testing of pollen-stigma compatibility relationships in *Brassica oleracea* by fluorescence. *Proceedings Brassica Meeting of Eucarpia*, Dixon, G.E. (ed.):34-36.

Dash, S.S., Kumar, S. and Singh, J.N. 2001. Cytomorphological characterization of a nuclear male sterile line of chilli pepper (*Capsicum annuum* L.). *Cytologia* 66: 365–71.

Daskalov, H., Russinova, I and Popova, D. 1967. New hybrids and constant varieties of pepper. *Nortic vitic Sci.* Sofa 4: 63-71.

Debbarama, C. Khanna, V.K., Tyagi, W., Rai, M. And Meetei, N.T. 2013. Wide hybridization and embryo rescue for crop improvement in *Capsicum*. *Agrotechnology* doi:10.4172/2168-9881.S11-003.

Desai, H.R., Bandania, K.A., Rai, A.B. and Patel, A.J. 2007. Assessment of yield loss and resistance to yellow mite *Polypagotarsonemus latus* (Bank) in chilli. *Vegetable Science* 34: 46-50.

Deshpande, A.A. and Rawal, R.D. 2007. Resistant sources of chili (*Capsicum annuum* L.) anthracnose fruit rot disease (*Colletotrichum capsici* Syd.) against different isolates collected from commercial chili growing areas of India. *First International Symposium on Chilli Anthracnose*, pp. 43, 17–19 September, Hoam Faculty House, Seoul National University, Seoul, South Korea.

Deshpande, R.B. 1933. Studies in Indian chillies: The inheritance of some characters in *Capsicum annuum* L. *Indian Journal of Agricultural Sciences* 3: 219-300.

Deshpande, R.B. 1935. Studies in Indian chillies. IV. The inheritance of pungency in *Capsicum annuum* L. *Indian Journal of Agricultural Sciences* 5: 613-16.

Dhaliwal, M.S. 2010. Exploitation of male sterility system. In: Kumar R, Rai AB, Rai M and Singh HP, eds., *Advances in Chilli Research*. Studium Press Pvt. Ltd., New Delhi, pp. 133-44.

Dhamayanthi, K.P.M. and Reddy, V.R.K. 2000. Cytogenetic effects of gamma rays and ethyl methane sulphonate in chilli pepper (*Capsicum annuum* L.). *Cytologia* 65: 129–33.

Dickson, M.H. and Wallace, D.H. 1986. Cabbage Breeding pp. 395-432, In: Bassett, M. J. (Edited) 'Breeding Vegetable Crops' AVI Publishing Company Inc. Connecticut (USA).

Dohroo, N.P. 1988. Germplasm reaction of cauliflower to stalk rot (*Sclerotinia sclerotiorum*). *Indian J. Pl. Pathol.* 6(2): 144.

Doullah, M.A.U., Meah, M.B and Okazaki, K. 2006. Development of an effective screening method for partial resistance to *Alternaria brassiciola* (dark leaf spot) in*Brassica rapa*. *European Journal of Plant Pathology*116: 33-43

Dwivedi, N., Kumar, R., Paliwal, R., Kumar, R., Kumar, S., Singh, M. and Singh, R.K. 2013. QTL mapping for important horticultural traits in pepper (*Capsicum annuum* L.). *Journal Plant BiochemistryBiotechnology* doi: 10.1007/s13562-013-0247-1.

El – Shawaf, I.I.S., and Baker, L.R. 1981. Inheritance of parthenocarpic yield in gynoecious pickling cucumber for once – over mechanical harvest by diallel analysis of six genoecious line. *J.Am. Soc. Hort. Sci.* 106:359.

Elias, R., Till, B.J., Mba, C. And Al-Safadi, B. 2009. Optimizing TILLING and EcoTILLING techniques for potato (*Solanum tuberosum* L). *BMC Research Notes* 2:141.

Fageria, M.S., Choudhary, B.R. and Dhaka, R.S. 2003. Vegetable Crops, Production technology Vol II, Kalyani Publishers, New Delhi, 285p.

Fang, Z., Liu, Y., Lou, P. and Liu, G. 2004. Current Trends in Cabbage Breeding pp75-107, In: P.K. Singh, S. K. Dasgupta and S. K. Tripathi (Edited) 'Hybrid Vegetable Development'. The Haworth Press Inc. NewYork, USA.

Fang, Z., Sun, P., Liu, Y., Yang, L., Wang, X., Hou, A. And Bian, C. 1997. A male sterile line with dominant gene (Ms) in cabbage and its utilization for hybrid seed production. *Euphytica*, 97: 265-268.

FAOSTAT.(2012). http://faostat.fao.org/site/339/default.aspx

Fjellstrom, Robert G. and Williams, Paul H.1997. *Fusarium* yellows and Turnip mosaic virus resistance in *Brassica rapa* and *B. Juncea*. *Hortscience* 32(5): 927-930.

Fonseca, G. G., Gandra, E. A. and Sclowitz, L. F. 2008. Oyster mushroom species differentiation through molecular markers RAPD. *Int. J. Pl. Breed. Genet.* 2(1): 13-18.

Fukino, N., Ohara, T., Monforte, A.J., Sugiyama, M., Sakata, Y., Kunihisa, M. and Matsumoto, S. 2008. Identification of QTLs for resistance to powdery mildew and SSR markers diagnostic for powdery mildew resistance genes in melon (*Cucumis melo* L.). *Theoretical Applied Genetics* 118: 165–175.

Gajbhiye, K.S. and Mandal, C.1999. Agro-Ecological Zones, their Soil Resource and Cropping Systems. *National Bureau of Soil Survey and Land Use Planning Nagpur. In: Status of Farm Mechanization in India. pp 1-32.*

Garg, R., Kumar, S., Kumar, R., Loganathan, M., Saha, S., Kumar, S., Rai, A.B. and Roy, B.K. 2013. Novel source of resistance and differential reactions on chilli fruit infected by *Colletotrichum capsici*. *Australasian Plant Pathology* 42: 227–233.

Gill, H. S. and Sharma, S.R. 1996. *"Cole-Crops" in 50 Years of Crop Science Research in India* R. S. Paroda and K. L. Chadha (Technical Editors), ICAR, New Delhi. pp 635-47.

Grout, B.W.W. 1998. Cauliflower: in Biotechnology in Agriculture and Forestry, Volume 6, Crops-II, Y.S.P. Bajaj (Ed), Springer-Verlag, Berlin, pp.211.

Hari Har Ram. 1998. Vegetable Breeding: Principles and Practices. Kalyani Publishers, New Delhi.

Hausche, P.E. and Gabelman, W.H. 1963. Digenic control of male sterility in carrots. (*Daucus carota* L). *Proc. Am. Soc. Hortic. Sci.* **82**, 341-350.

Hayes HK and Jones DF. 1916. First generation crosses in cucumber.

Hegde, D M.1997. Nutrient requirement of solanaceious vegetables: A review: Extension Bulletin, ASPAC, Food and Fertilizer Technology Centre No.441.

Horsfall, J.G. and Barrat, R.W. 1945. An improved grading system for measuring plant diseases. *Phytopathology* 35: 655.

Horst, I., Welham, T., Kelly, S., Kaneko, T., Sato, S., Tabata, S., Parniske, M. and Wang, T.L. 2007. TILLING mutants of *Lotus japonicus* reveal that nitrogen assimilation and fixation can occur in the absence of nodule-enhanced sucrose synthase. *Plant Physiology* 144: 806-20.

Hoser-Krauze, J. 1979. Inheritance of self-incompatibility and the use of it in the production of F1 hybrids of cauliflower. Genetica Polonica, 20: 341- 67.

http://agritech.tnau.ac.in/

http://apeda.gov.in/apedawebsite/index.html

http://jnkvv.org/

http://nhb.gov.in/Default.aspx

http://www.dogr.res.in/index.php/en/

http://www.iari.res.in/

http://www.icar.org.in/

http://www.iihr.ernet.in/

http://www.iivr.org.in/

http://www.rvskvv.net/

http://www.tnau.ac.in/

http://www.yspuniversity.ac.in/

https://www.ncpahindia.com/drip-irrigation

Javier, Ibáñez M. Dolores Vélez, M. and Teresa de Andrés Joaquín Borrego. 2009. Molecular markers for establishing distinctness in vegetatively propagated crops: a case study in grapevine. *Theor. Appl. Genet*. 119:1213–22 .

Jha, T.B., Dafadar, A. and Ghorai, A. 2012. New genetic resource in *Capsicum* L. from Eastern Himalayas. *Plant Genetic Resources: Characterization and Utilization* 10: 141–44.

Ji, Y., Scott, J.W., Hanson, P. Graham, E., and Maxwell, D. 2007. Sources of resistance, inheritance, and location of genetic loci conferring resistance to members of the tomato-infecting begomoviruses. pp. 343–266. In: *Tomato Yellow Leaf Curl Virus Disease*. H. Czosnek, ed. Springer, Dordrecht, the Netherlands.

Kalloo G and Bergh O I. 1993. Root Crops: (In) *Genetic Improvement of Vegetable Crops*, Pergamon Press, p. 62.

Kalloo, G., Srivastava, U., Singh, M. and Kumar, S. 2005. Solanaceous vegetables. In : Dhillon B.S., Tyagi R.K., Saxena S. and Randhawa, eds. *Plant Genetic Resources: Horticultural Crops*, Narosa Publishing House, New Delhi, pp. 19–33.

Kalloo, G. 1988. Vegetable breeding. Panima Educational Book Agency, New Delhi.

Kanwar, H.S., Bhattarai, D.R. and Mehta, D.K. 2010. Seed Technology: Processing, Storage and Marketing, Jain Brothers, New Delhi, 203 p.

Kapoor, K.S., Gill, H.S. and Sharma, S.R. 1985. A technique for artificial inoculation of cauliflower seedlings with *S. sclerotiorum*. *Phytopathology* 112: 191-92.

Kaur, N., Singh, D.J. and Singh, K.D. 2011. Physiological and biochemical traits analysis of *Capsicum annuum* L. germplasm for resistance to *Colletotrichum capsici*. *Journal of Cell and Plant Sciences* 2: 12–21.

Kehie, M., Kumaria, S., Tandon, P. and Ramchiary, N. 2014. Biotechnological advances on *in vitro* capsaicinoids biosynthesis in *capsicum*: A review. *Phytochemistry Reviews* doi: 10.1007/s11101-014-9344-6.

Khare, D. and Bhale, M.S. 2011. Seed Technology. Scientific Publishers, Jodhpur, 260 p.

Khirbhat, S.K., Vajnana, T. and Mehra, R. 2004. Cultural and pathogenic variation among the nine isolates of *Colletotrichum capsici* causing fruit rot of *Capsicum*. *Capsicum Eggplant Newsletter* 24: 131–134.

Kim, *et al.* 2014. Genome sequence of the hot pepper provides insights into the evolution of pungency in *Capsicum* species. *Nature Genetics* doi:10.1038/ng.2877.

Kothari, S.L., Joshi, A., Kachhwaha, S. and Ochoa-Alejo, N. 2010. Chilli peppers — A review on tissue culture and transgenesis. *Biotech. Advances* 28: 35–48.

Kucera, V. 1990. Overcoming self-incompatibility in *Brassica oleracea* with a sodium chloride solution. Sbornik-UVTIZ, Zahradnictvi, 17:13-16.

Kucera, V., Chytilova, V., Vyvadilova, M and Klima, M. 2006. Hybrid breeding of cauliflowers using self-incompatibility and cytoplasmic male sterility. Horticultural Science (Prague) 33:148-152.

Kumar, J.C. and Dhaliwal, M.S..2007. In: *Technique of developing hybrids in vegetable crops.* Agrobios (India) pp. 42-43.

Kumar, A.M., Reddy, K.N., Sreevathsa, R., Ganeshan, G. and Udayakumar, M. 2009. Towards crop improvement in bell pepper (*Capsicum annuum* L.): Transgenics (uid A::hpt II) by a tissue-culture-independent *Agrobacterium*- mediated *in planta* approach. *Scientia Horticulturae* 119: 362–70.

Kumar, O.A., Panda, R.C. and Rao, K.G.R. 1988. Cytogenetics of interspecific hybrids in the genus *Capsicum*. *Euphytica* 39: 47–51.

Kumar, R., Kumar, S., Dwivedi, N., Kumar, S., Rai, A., Singh, M., Yadav, D.S. and Rai, M. (2009). Validation of SCAR markers, diversity analysis of male sterile cytoplasms (S-) and isolation of an alloplasmic S-cytoplasm in *Capsicum*. *Scientia Horticulturae* 120: 167–72.

Kumar, R.V., Sharma, V.K., Chattopadhyay, B. and Chakraborty, S. 2012. An improved plant regeneration and *Agrobacterium*–mediated transformation of red pepper (*Capsicum annuum* L.). *Physiology and Molecular Biology of Plants* doi: 10.1007/s12298-012-0132-8.

Kumar, S., Kumar, R., Kumar, S., Singh, M., Rai, A.B. and Rai, M. 2011. Incidences of leaf curl disease on *Capsicum* germplasm under field conditions. *Indian Journal of Agricultural Science* 81: 187–89.

Kumar, S., Kumar, R. and Singh, J. (2006a). Cayenne/American pepper. In: Peter KV ed., Handbook of Herbs and Spices, Woodhead Publishing, Cambridge, UK, pp. 299–312.

Kumar, S., Kumar, S., Singh, M., Singh, A.K and Rai, M. 2006b. Identification of host plant resistance to pepper leaf curl virus in chilli (*Capsicum* species). *Scientia Horticulturae* 110: 359–61.

Kumar, S., Rai, S.K., Banerjee, M.K. and Kalloo, G. 2001. Cytological mechanisms of male sterility in a nuclear-cytoplasmic line of chilli pepper (*Capsicum annuum* L.). *Capsicum Eggplant Newsletter* 20: 64–67.

Kumar, S., Singh, V., Singh, M., Rai, S., Kumar, S., Rai, S.K. and Rai, M. 2007. Genetics and validation of fertility restoration associated RAPD markers in pepper (*Capsicum annuum* L.). *Scientia Horticulturae* 111: 197–202.

Kunkalikar, S.R., Poojari, S., Arun, B.M., Rajagopalan, P.A., Chen, T.C., Yeh, S.D., Naidu, R.A., Zehr, U.B. and Ravi, K.S. 2011. Importance and genetic diversity of vegetable-infecting Tospoviruses in India. *Phytopathology* 101: 367–76.

Levesque, C. A., Vrain, T. C. and De Boer, S. H. 1994. Development of species specific probe for *Pythium ultimum* using amplified ribosomal DNA. *Phytopathology* 84:474-78.

Lin, S.W., Chou, Y.Y., Shieh, H.C., Ebert, A.W., Kumar, S., Mavlyanova, R., Rouamba, A., Tenkouano, A., Afari-Sefa, V. and Gniffke, P.A. 2013. Pepper (*Capsicum* spp.) germplasm dissemination by AVRDC – The World Vegetable Center: an overview and introspection. *Chronica Horticulturae* 53: 21–27.

Mackay, C.R.1977. A diallel cross method for the recognition of S-allele homozygotes in turnip (*Brassica campestris L.* ssp. *rapifera)*. Heredity, 38:201-208.

Mathur, R., Dangi, R.S., Dass, S.C. and Malhotra, R.C. (2000). The hottest chilli variety in India. *Current Science* 79: 287–88.

McCallum, C.M., Comai, L., Greene, E.A. and Henikoff, S. 2000. Targeting Induced Local Lesions in Genomes (TILLING) for Plant Functional Genomics. *Plant Physiology* 123(2): 439-42.

McDonald, M.B. and Copeland, L.O. 2005. Seed Production: Principles & Practices, Chapman and Hall, New York, 749 pp.

McIntyre, C.L., Mathews, K.L., Rattey, A., Chapman, S.C., Drenth, J., Ghaderi, M., Reynolds, M. and Shorter, R. 2010. Molecular detection of genomic regions associated with grain yield and yield-related components in an elite bread wheat cross evaluated under irrigated and rainfed conditions. *Theoretical Applied Genetics* 120: 527-41.

Meghvanși, M.K., Siddiqui, S., Khan, H., Gupta, V.K., Vairale, M.G., Gogo, H.K. and Singh, L. 2010. Naga Chilli: a potential source of capsaicinoids with broad-spectrum ethnopharmacological applications. *Journal of Ethnopharmacology* 132: 1–14.

Mondal, C.K., Acharyya, O. and Hazra, P. 2013. Biochemical basis of plant defense for leaf curl virus of chilli (*Capsicum annuum*). *Proceeding XV EUCARPIA Meeting on genetics Genetics and Breeding of Capsicum and Eggplant*, 2–4 September, Turin, Italy, pp. 315–22

Mukherjee, S. and Sharma, A.K. 1990. Intraspecific variation of nuclear DNA in *Capsicum annuum* L. *Plant Sciences* 100: 1–6.

Nair, P. K. K. 2002. The Hindu Survey of Indian Agriculture 2002. M/S Kasturi and Sons Ltd., Chennai, India.

Nanda, C. 2011. Dynamics of Anthracnose Disease Causing Pathogen and Inheritance and SSR Marker Assisted Tagging of Resistance to Anthracnose in Chilli (*Capsicum annuum* L.). PhD Thesis, University of Agricultural Sciences, Bengaluru, India.

Narayanamoorthy, A. 1997. Drip Irrigation–A viable option for future irrigation development. *Productivity*, .38 (3).

NHB. 2014. *Handbook of Indian Horticulture Database*, NHB, Gurgaon.

Nieuwhof, M. 1961. Male sterility in Brussels sprouts, cauliflower and cabbage. *Euphytica*, 10:351-56.

Nybe, E.V., Miniraj, N. and Peter, K.V. 2007. *Spices* (Horticultural Science Series Vol.5, New India Publishing Agency, New Delhi.

Odland, M.L. and Noll, C.J. 1950. The utilization of cross-compatibility and self-incompatibility in the production of F1 hybrid cabbage.

Ogura, H. 1968. Studies of the male sterility in Japanese radish with special reference to the utilization of this sterility towards the practical raising of hybrid seeds. *Memoirs of the Faculty of Agriculture,* Kagoshima University, 6:39-78.

Panda, R.C., Kumar, O.A. and Rao, K.G.R. 1984. Cytomorphology of induced octaploid chilli pepper (*Capsicum annuum* L.). *Theor. Appl. Genet.* 68: 567–70.

Pandey, A. K.1998. A note on effect of different culture media on seed germination and growth of tomato seedlings. *Vegetable Science* 25 (2): 181-82.

Pandravada, S.R., Varaprasad, K.S., Reddy, K.J. and Rao, E.S. 2010. Screening and identification of sources of resistance against root-knot nematode (*Meloidogyne javanica*) in chilli (*Capsicum annuum*) germplasm. *Indian Journal of Agricultural Sciences* 80: 92–94.

Parthasarthy, V.A., Prasath, D. and Babu, K.N. 2010. Advances in *Capsicum* biotechnology. In: Kumar R, Rai AB, Rai M and Singh HP (eds.), *Advances in Chilli Research*, Studium Press Pvt. Ltd., New Delhi, pp. 87 117.

Parthasarthy, V.A., Kanoiannam, K. and Srinivasan, V. 2008. Organic Spices. New India Publishing Agency, New Delhi.

Pearson, O.H. 1972. Cytoplasmically inherited male sterility characters and flavour components from the species *Brassica nigra* (L.) Koch x *B. oleracea* L. *Journal of the American Society for Horticultural Science*, 97: 397-402.

Peter, K.V. 2001. *Handbook of Herbs and Spices* Vol. I., Woodhead Publishing Limited, Cambridge England.

Peter, K.V., Nybe, E.V. and Kurien, A. 2005. Yield gaps and constraints in ginger. In: Ravindran P.N. and Nirmal Babu K. (Eds.) *Ginger The genus Zingibel*, pp. 527-32.

Peter, K.V. 2005. Genetics and Breeding of Vegetables. ICAR, New Delhi, pp 375

Peterson, C.E. and Simon, P.W. 1986. Carrot Breeding. In: Basset MJ (ed). *Breeding Vegetable Crops*. AVI Publishing Co. Westport, Connecticut. pp. 321-56.

Porch, T.G., Blair, M.W., Lariguet, Patricia, GCPCE and Broughton, W.J. 2009. Generation of a mutant population for tilling common bean genotype BAT 93. *Journal of the American Society for Horticultur Science* 134(3): 348-55.

Pratap, M., Kumar, B. V., Mohammad, Shaik and Mohammad, S.1997. Effect of herbicides and time of weeding on weed control and fruit yield of tomato. *Crop Research*, 14(1): 113-117.

Prathibha, V.H. 2009. Pathological and Variability Studies on *Colletotrichum* spp. Causing Anthracnose (Fruit Rot) Disease of Chilli (*Capsicum annuum* L.). PhD Thesis. University of Agricultural Sciences, Bengaluru.

Pray, C.E., Ranaswami, B. and Kelley, T. 2001. The impact of economic reform on R & D by the Indian seed industry. *Food Policy* 26: 587–98.

Pruthi, J.S. 1993. *Major Spices of India- Crop Management and Post Harvest Technology.* ICAR, New Delhi.

Purkayastha, J., Alam, S.I., Gogoi, H.K. and Singh, L. 2012b. *Capsicum assamicum* sp. (Solanaceae) from Assam, northeastern India. *Ozean Journal of Applied Science* 5: 55–66.

Purkayastha, J., Alam, S.I., Gogoi, H.K., Singh, L. and Veer, V. 2012a. Molecular characterization of Bhut Jolokia the hottest chili. *Journal of Bioscience* 37: 757–68.

Qin, C. *et al.* 2014. Whole-genome sequencing of cultivated and wild peppers provides insights into *Capsicum* domestication and specialization. *Proceeding of National Academy of Sciences USA.*

Rai, V.P., Kumar, R., Kumar, S., Rai, A., Kumar, A., Singh, M., Singh, S.P., Rai, A.B. and Paliwal, R. 2013. Genetic diversity in *Capsicum* germplasm based on microsatellite and random amplified microsatellite polymorphism markers. *Physiology and Molecular Biology of Plants* 19**:** 575–86.

Rai, V.P., Kumar, R., Singh, S.P., Kumar, S., Kumar, S., Singh, M. and Rai, M. 2014. Monogenic recessive resistance to *Pepper leaf curl virus* in an interspecific cross of *Capsicum. Scientia Horticulturae* 172**:** 34–38.

Rai, V.P., Rai, A.C., Kumar, S., Kumar, R., Kumar, S., Singh, M., Rai, A.B. and Singh, S.P. 2010. Emergence of new variant of chilli leaf curl virus strain in North India. *Vegetable Science* 37**:** 124–28.

Rai, Nagendra and Rai, Mathura. 2006. Cucurbits breeding In: *Heterosis breeding in vegetable crops*. New India Publishing Agency. pp. 321-56.

Ramaiah, K. and Royappa, P.M. 1935. Pungency in chillies - a Mendelian character. *Current Science* 4**:** 236–37.

Ramchiary, N., Kehie, M., Brahma, V., Kumaria, S. and Tandon, P. 2013. Application of genetics and genomics towards *Capsicum* translational research. *Plant Biotechnology Report* doi: 10.1007/s11816-013-0306-z.

Ranganathan, P. and Jagatheeswari, D. 2013. Chromosome studies on garden pepper (*Capsicum frutescens* L.). *International Journal of Research in Botany* 3**:** 1–3.

Rao, A.M., Nanda, C., Pratibha, V.H. and Ramesh, S. 2007. Diversity and distribution of hot pepper *Colletotrichum* spp. isolates in Southern India. *First International Symposium on Chilli Anthracnose*, 17–19 September, Hoam Faculty House, Seoul National University, Seoul, South Korea, pp. 38.

Rao, K.G.R. and Kumar, O.A. 1983. Colchicine induced interchanges in chillies (*Capsicum annuum*L.). *Experientia* 39**:** 76–77.

Rattan, R. S. and Saini, S. S.1979.Euphytica, 28 : 315-17.

Ravindran, P.N., Nirmal, Babu K., Shiva, K.N. and Johny, A.K. 2006. *Advances in Spices Research. Agrobios,* Jodhpur.

Reddy, K.M. and Reddy, M.K. 2010. Breeding for virus resistance, In: Kumar R, Rai A.B., Rai, M. and Singh, H.P. eds., *Advances in Chilli Research*, Studium Press Pvt. Ltd., New Delhi, India, pp. 119–132.

Reddy, M.K., Sadashiva, A.T. and Deshpande, A A. 2002. Cytoplasmic male sterility in chilli (*Capsicum annuum* L.). *Indian Journal of Genetics* 62**:** 363–64.

Riggs, T.J. 1988. Breeding F1 hybrid varieties of vegetables. *Journal of Horticultural Science*, 63:369-82.

Robertson, D., Palmer, J.D., Earle, E.D. and Mutschler, M.A. 1987. Analysis of organelle genomes in somatic hybrids derived from cytoplasmic male sterile (CMS) *Brassica oleracea* and atrazine resistant *Brassica campestris. Theoretical and Applied Genetics*, 74:303.

Rubatzky, B. and Yamaguchi, M. 1999. *World Vegetable*. Chapman and Hall.

Rundfeldt, H. 1960. Untersuchungen zur zuchtung des koptkohls (*B. oleracea* L.var.*capitata*). *Z. Pflanzenz,* 44:30-62.

Sampson, D.R. 1957. The genetics of self- and cross-compatibility in

Sanatombi, K., Sen-Mandi, S. and Sharma, G.J. 2010. DNA profiling of *Capsicum* landraces of Manipur. *Scientia Horticulturae* 124: 405–408.

Selvakumar, R. 2007. Variability among *Colletotrichum capsici* causing chilli anthracnose in north east India. *First International Symposium on Chilli Anthracnose*, 17–19 September, Hoam Faculty House, Seoul National University, Seoul, South Korea, pp. 35.

Senanayake, D.M.J.B., Varma, A. and Mandal, B. 2012. Virus-vector relationships, host range, detection and sequence comparison of chilli leaf curl virus associated with an epidemic of leaf curl disease of chilli in Jodhpur, India. *Journal of Phytopathology* 160: 146–55.

Seshadri, V.S., Sharma, J.C. and Chaudhury, B. 1972. Pusa watermelon varieties assure good yield and quality. *Indian Horticulture*. 16: 21-32.

Shanmugavelu, K,G. Kumar, N. and Peter, K.V. 2002. Production Technology of Spices and Plantation Crops. Agrobios, Jodhpur.

Sharma, T. R. and Tewari, J. P. 1995. Detection of genetic variation in *Alternaria brassicae* by RAPD fingerprints. *J. Plant Biochem. Biotechnol.*4:105-107.

Sharma, A., Kumar, V., Giridhar, P. and Ravishankar, G.A. 2008. Induction of *in vitro* flowering in *Capsicum frutescens* under the influence of silver nitrate and cobalt chloride and pollen transformation. *Electronic Journal of Biotechnology* 11: 1–6.

Sharma, P.N., Kaur, M., Sharma, O.P., Sharma. P. and Pathania, A. 2005. Morphological, pathological and molecular variability in *Colletotrichum capsici*, the cause of fruit rot of chillies in the subtropical region of north-western India. *Journal of Phytopathology* 153: 232–37.

Sharma, S.K. 2000. Breeding for diseases resistance in vegetable crops. CASH (Vegetable), pp. 67-74.

Sharma, T. R. and Tewari, J. P. 1996. Synthesis of *Alternaria brassicae*- specific Dig labeled probe with Polymerase chain reaction. *Phytopathology* 86:566.

Sharma, T. R., Prachi and Singh, B, M. 1999. Applications of polymerase chain reaction in phytopathogenic microbes. *Ind. J. Miicrobiol.*39:79-91.

Sharma, B.R., Swarup, Vishnu and Chatterjee, S.S. 1977. Resistance to black rot disease (*Xanthomonas campestris* (Pam.) Dowson) in cauliflower. *Scientia Hortic*. 7: 1-7.

Sharma, Girish, Kumar, V., Dinesh , Haque, A., Bhat, S.R., Prakash, Shyam and Chopra, V.L .2002. *Brassica* coenospecies: a rich reservoir for genetic resistance to leaf spot caused by *Alternaria brassicae*. *Euphytica* 125: 411-17

Sharma, R.C., Sharma, S.L. and Sharma, Indu. 1982. Varietal reaction of cauliflower against stalk rot caused by *Sclerotia*. *Crop Improv.* 9: 167-68

Shih SL, Tsai WS, Green SK, Singh D (2007). *Tomato leaf curl Joydebpur virus* infecting chilli in India. *Plant Pathology* 56: 341.

Shivegowda ,S.T., Mythili, J.B., Lalitha, A., Saiprasad, G.V.S., Gowda, R. and Gowda, T.K.S. 2002. *In vitro* regeneration and transformation in chilli pepper (*Capsicum annuum* L.). *Journal of Horticultural Science and Biotechnology* 77: 629–34.

Sigareva, M.A. and Earle, E.D. 1997. Direct transfer of a cold tolerant Ogura male sterile cytoplasm into cabbage via protoplast fusion. Theoretical and Applied Genetics, 94: 213-20.

Singh, D., Mehta, R. and Talati, J.G. 2008. Identification of chilli (*Capsicum annuum* L.) genotypes through RAPD. *Vegetable Science* 35: 10–13.

Singh, P., Cheema, D.S., Dhaliwal, M.S. and Garg, N. 2014. Heterosis and combining ability for earliness, plant growth, yield and fruit attributes in hot pepper (*Capsicum annuum* L.) involving genetic and cytoplasmic-genetic male sterile lines. *Scientia Horticulturae* 168: 175–88.

Singh, H. P., Samuel, J. C. and Kumar, A. 2000.Microirrigation in horticultural crops. *Indian Horticulture*, 45(3): 37-43.

Singh, N.P., Bhardwaj, A.K. Kumar, A. and Singh, M. 2011. *Modern Technology on Vegetable Production*. New India Publishing Agency, New Delhi.

Singh, S. and Vidyasagar 2012. Effect of common salt (NaCl) sprays to overcome the self incompatibility in the S- allele lines of *Brassica oleracea* var. *capitata* L. SABRAO Journal of Breeding and Genetics, 44:339-348.

Sivanappan, R. K. 1998. Status, scope and future prospects of microirrigation in India. *Proc. Workshop on microirrigation and sprinkler irrigation system. CBIP New* Delhi, April 28-30, 1998: 1-7

Sivanappan, R.K. 1994. Prospects of Microirrigation in India. *Irrigation and DrainageSystems*. 8: 49-58.

Sobhakumari, V.P. and Lalithakumari, D. 2005. High frequency shoot regeneration and *Agrobacterium* mediated DNA transfer in red chilli (*Capsicum annuum* L.). *Plant Cell Biotechnology Molecular Biology* 6: 9–16.

Spehia, R. S., Thakur, B. C. and Raina, J. N. 2008. Effect of In-situ conservation practices on soil loss, moisture carry over and yield of broccoli under rainfed conditions of mid hill zone of Himachal Pradesh. *Asian J. Soil Sci.* Vol.2 No.2:100-103

Sridhar, R., Shanti, M. L., Singh, U. D., Variar, M., Maiti, D., Reddy, J. N., Sinha, P. K., George, M. L. C., Bernardo, M. A., Bordeos, A., Baraoidan, M. R. and Nelson, R. J. 1996. Molecular characterization of bacterial blight and blast pathogen populations of India and breeding for cultivars with bacterial blight and blast resistant genes. *Proc 5th Annual NRBN meeting*.IARI. NewDelhi.

Srinivasan, M., Kothandaraman, S.V., Vaikuntavasan, P. and Rethinasamy, V 2014. Development of conventional and real-time PCR protocols for specific and sensitive detection of *Colletotrichum capsici* in chilli(*Capsicumannuum* L.). *Phytoparasitica* doi: 10.1007/s12600-013- 0380-3.

Stephenson, P., Baker, D., Girin, T., Perez, A., Amoah, S., King, G.J. and Qstergaard, L. 2010. A rich TILLING resource for studying gene function in *Brassica rapa*. *BMC Plant Biology* 10:62.

Strauss, S. H., Bousquet, J., Hipkins, V. D. and Hong, Y. P. 1992. Biochemical and molecular genetic markers in biosystematic studies of forest trees. *New Forests* 6: 125–158.

Su, Yiwen, Zhao, Shang Yi, and Zhang Yanjum. 1995. Study on morphocytology and microsporogenesis in male sterile line of radish *(Raphanus sativus L.). Acta Horticulturae*. **402:** 173-78

Swarup, V. 2006. *Vegetable Science and Technology in India*. Kalyani Publishers, Ludhiana, 656p.

Thamburaj, S and Singh, N. (Tech. Editors). 2001. *Text Book of Vegetables, Tuber Crops and Spices*. ICAR Publication, New Delhi, 469p.

Thompson, K.F. 1957. Self-incompatibility in marrow stem kale *Brassica oleracea* var. *acephala*. I. Demonstration of sporophytic system. Journal of Genetics, 55:45-60.

Thul, S.T., Darokar, M.P. and Shasany, A.K. 2012. Molecular profiling for genetic variability in *Capsicum* species based on ISSR and RAPD markers. *Molecular Biotechnology* 51: 137–47.

Tilli, S.V and Mirzabekov, A.D. 2001. Advances in the analysis of DNA sequence variations using oligonucleotide microchip technology. *Current Opinion in Biotechnology* 12(1): 53-58.

Timin, I. H. 1981, Tsitplazmatich, Muzhsk, sterl'nos, I Seleksiya Rast. Kiev, Ukrainian.

Tiwari, R.S., and Agarwal, A. 2004. *Production Technology of Spices*. International Book Distributing Co. Lucknow India.

Tiwari, V.P. and Ramanujam, S. 1974. Grow Jwala, a disease resistant high yielding chilli. *Indian Farming* 24: 20.

Tsai, W.S., Shih, S.L., Kenyon, L., Green, S.K., and Jan, F.J. 2011. Temporal distribution and pathogenicity of the predominant tomato-infecting begomoviruses in Taiwan. *Plant Pathology* 60: 787-799.

Varshney, R.K., Mahendar, T., Aruna, R., Nigam, S.N., Neelima, K., Vadez, V. and Hoisington, D.A. 2009. High level of natural variation in a groundnut (*Arachis hypogaea* L.) germplasm collection assayed by selected informative SSR markers. *Plant Breeding* 128:486–494.

Verma, P.K., Rawat, K.K., Das, N. and Pradhan, B. 2013. A botanical enigma of India's hottest chilli Bhoot Jolokia (*Capsicum chinense* Jacq). *New York Science Journal* 6: 49–51.

Verma, S., Dhiman, K. and Srivastava, D.K. 2013. *Agrobacterium*-mediatedgenetic transformation of bell pepper (*Capsicum annuum* L. cv. California Wonder) with *gus* and *nptII* genes. *International Journal of Advanced Biotechnology and Research* 4: 397–403.

Verma, T.S.1990. Advances in breeding of improved varieties in temperate vegetables. All India workshop-cum seminar on vegetable production technology.pp.29-32.

Vidyasagar and Chatterjee S.S. 1984. Early testing of pollen-stigma compatibility in Indian cauliflower by fluorescence microscopy. Vegetable Science 11:113-117.

Wallace, D.H. 1979. Procedures for identifying S-allele genotypes of

Watts, L.E. 1963. Investigations into the breeding system of cauliflower I. Studies on self-incompatibility. *Euphytica,* 12:323-40.

Wehner, T.C. and Miller, C.H. 1985. Effect of gynoecious expression on yield and earliness of fresh market cucumber hybrid. *J. Am. Soc. Nert Sci.,* 110:464.

Weil, C.F. and Monde, R.A. 2007. Getting to the point - Mutations in maize. *Crop Science* 47: S60-S67.

Welch, J.E. and Grimball , E.L. 1947. Male sterility in carrot. Science.**106**: 398.

Werner, S., Diederichsen, E., Frauen, M., Schondelmaier, J. and Jung, C. 2008. Genetic mapping of clubroot resistance genes in oilseed rape. *Theoretical Applied Genetics* 116: 363–72.

Yang, C., Guo, W.Z. and Li, G.Y. 2008. QTL mapping for *Verticillium* wilt resistance at seedling and maturity stages in *Gossypium barbadense* L. *Plant Scienc* 174: 290–98.

Yang, R., Yu, Y.J., Xu, J.B., Chen, G. and Zhang, F.L. 1995. Studies on techniques of spraying NaCl solution on flowers combined with honeybee pollination to overcome self-incompatibility of Chinese cabbage. *Acta Agriculturae Boreali Sinica,* 10:77-81.

Yumnam, J.S., Tyagi, W., Pandey, A., Meetei, N.T. and Rai, M. 2012. Evaluation of genetic diversity of chilli landraces from north eastern India based on morphology, SSR markers and the *Pun1* locus. *Plant Molecular Biology Reporter* doi: 10.1007/s11105-012-0466-y.

Zhang, F.L., Wang, M., Liu, X.C., Zhao, X.Y. and Yang, J.P. 2008. Quantitative trait loci analysis for resistance against turnip mosaic virus based on a doubled-haploid population in Chinese cabbage. *Plant Breeding* 127:82-86.

Zhanguo, Xin Z., Wang, M.L., Barkley, N.A., Burow, G., Franks, C., Pederson, G. and Burke, J. 2008. Applying genotyping (TILLING) and phenotyping analyses to elucidate gene function in a chemically induced sorghum mutant population. *BMC Plant Biology* 8: 103.

❍❍❍

LIST OF CONTRIBUTORS

S. No.	Name of Chapter	Name of Author and address
1.	Importance of Vegetable Crops	S.K. Tyagi Scientist (Horticulture) RVSKVV, Krishi Vigyan Kendra, Khargone (M.P.)
2.	Classification of Vegetable Crops	Dr. S.S. Singh Assistant Professor (Horticulture) MGCGVV, Chitrakoot, Satna (M.P.)
3.	Vegetable Plant Nutrients	Dr. R.C. Jain Associate Professor (Soil Science) RVSKVV, R.A.K. College of Agriculture, Sehore (M.P.)
4.	Manures and Fertilizers	Vikas Chaurasiya Field Officer Indian Farmers Fertiliser Cooperative Limited
5.	Water Management in Vegetable Crops	Dr. V.K. Jain Scientist (Agriculture Engineering) RVSKVV, Krishi Vigyan Kendra, Ashok Nagar (M.P.)
6.	Growth and Development	Dr. I.S. Tomar Senior Scientist and Head RVSKVV, Krishi Vigyan Kendra, Jhabua (M.P.)
7.	Breeding of Vegetable Crops	A.R. Khire Senior Scientist (Plant Breeding & Genetics) RVSKVV, Zonal Agriculture Research Station, Khargone (M.P.)
8.	Basic Information on Vegetable Crops	Dr. Ajit Singh Senior Scientist and Head Krishi Vigyan Kendra, Burhanpur (M.P.)
9.	Climate and Soil Requirements for Vegetable crops	Dr. Vijay Agrawal Deputy Director Horticulture Department of Horticulture & Food Processing Department, Govt. of Madhya Pradesh
10.	Improved Varieties and Hybrids of Vegetable Crops	S.K. Tyagi Scientist (Horticulture) RVSKVV, Krishi Vigyan Kendra, Khargone (M.P.)
11.	Production Technology of Vegetable Crops	S.K. Tyagi Scientist (Horticulture) RVSKVV, Krishi Vigyan Kendra, Khargone (M.P.)
12.	Seed Production of Vegetable Crops	Dr. N.K. Gupta[1] and [2]Dr.Nityamanjari Mishra [1]Professor (Horticulture) College of Agriculture, Indore (M.P.) [2]Jr. Scientist (Horticulture) RRTTS, (OUAT) Keonjhar, Odisha

S. No.	Name of Chapter	Name of Author and address
13.	Pesticides and its Uses	Dr. Ashish Shrivastava Assistant Professor (Plant Pathology) JNKVV, College of Agriculture, Ganj Basoda (M.P.)
14.	Major Diseases of Vegetable Crops	Y.K. Jain Senior Scientist (Plant Pathology) RVSKVV, Zonal Agriculture Research Station, Khargone (M.P.)
15.	Major Insect Pests of Vegetable Crops	Dr. M.L. Sharma Principal Scientist (Entomology) and Associate Director Research RVSKVV, Zonal Agriculture Research Station, Khargone (M.P.)
16.	Physiological and Nutritional Disorders in Vegetable Crops	[1]Dr. Rashmi Vajpayee and [2]Dr. Pragnyashree Mishra [1]Scientist (Horticulture) RVSKVV, College of Agriculture, Gwalior (M.P.) [2]Assistant Professor (Horticulture) OUAT, Odisha
17.	Post Harvest Technology of Vegetable Crops	Dr. S.D. Patil Associate Professor (Horticulture) College of Agriculture, Dhule, Maharashtra
18.	Recent Advances in Crop Improvement	Dr. R.K. Jaiswal Associate professor (Horticulture) RVSKVV, R.A.K. College of Agriculture, Sehore (M.P.)
19.	Recent Advances in Crop Production	Dr. G.L. Sharma Scientist (Horticulture) Indira Gandhi Krishi Vishwavidyalaya (IGKVV), Raipur, Chhattisgarh
20.	Recent Advances in Crop Health Management	B.K. Patidar Assistant Professor (Plant Pathology) K.N.K. College of Horticulture Mandsaur (M.P.)
21.	Glossary of Common Horticultural Terms	Dr. S.L. Chawala Associate Professor (Horticulture) ASPEE College of Horticulture and Forestry, Navsari Agricultural University, Navsari, Gujarat
22.	Objective Type Questions	Dr. R. K. Saraf Principal Scientist (Horticulture) Dryland Horticulture Research & Training Center, (JNKVV), Garhakota, Sagar (M.P.) Dr. Akhilesh Tiwari Senior Scientist (Horticulture) Dryland Horticulture Research & Training Center (JNKVV), Garhakota, Sagar (M.P.)